Tributes
Volume 10

Witnessed Years
Essays in Honour of Petr Hájek

Tributes Series Editor
Dov Gabbay

dov.gabbay@kcl.ac.uk

Witnessed Years
Essays in Honour of Petr Hájek

edited by
Petr Cintula,
Zuzana Haniková
and
Vítězslav Švejdar

ISBN 978-1-904987-63-5

College Publications
Scientific Director: Dov Gabbay
Managing Director: Jane Spurr
Department of Computer Science
King's College London, Strand, London WC2R 2LS, UK

http://www.collegepublications.co.uk

Original cover design by orchid creative   www.orchidcreative.co.uk
Printed by Lightning Source, Milton Keynes, UK

CONTENTS

Preface

THE EDITORS

This book is a gift to Petr Hájek, given by his colleagues, students, family and friends. This means that Petr is an intended reader, but not the only intended reader. Rather, the book bears witness to the imprint he has left in each of his research areas, and in the minds of those who have had the pleasure of meeting him.

The book contains essays, research papers, historical accounts, and personal recollections. It covers most of Petr Hájek's professional career, including its circumstances, some glimpses into his family life, religion, opinions, and a good deal of storytelling. It spans over more than five decades, with contributions roughly arranged in a chronological order according to his research areas, ranging from set theory through data analysis, arithmetic, theory of vague information, to non-classical logic. The contributors are Petr Hájek's co-authors or colleagues, with a deep knowledge of both his research topics and of himself.

Authors responded with enthusiasm to our call; indeed, the desire to pay respects to Petr Hájek by presenting a familiar topic has been univocal. In order to achieve reasonable coverage of the research areas, we sometimes ventured to express our wish as to the desired topic for an author; in other cases, topics have been chosen by authors themselves. As to the form and nature of the contributions, we embraced the rare opportunity of a rather informal occasion to give the authors a free hand, apart from the initial hint of our desire for "scientific essays". Any degree of literary aspirations, of technical detail, of personal approach, has been acceptable. Subjectivity has been expected and accepted, with the important proviso that the responsibility for views expressed is fully that of the author. Our inclusive approach is reflected in the heterogeneous form of this volume, and rewarded by a large amount of warm recollections.

It will be noticed that contributions tend to be more numerous and more international as one moves forward in time. One of the reasons for this disproportion is the scientific isolation of Czechoslovakia during the Communist era. Nevertheless, those contributions that look back at that era are all the more interesting testimonies of scientific achievements and political obstacles that were characteristic of the times in this country.

Petr Hájek has been our teacher. There are many things to be learned from him, as a researcher and as a person. Among these are his scientific rigour, his inquisitiveness, his tolerance and generosity, and his fondness of simplicity. The acknowledgement of these qualities is a common thread in the contributions to this book, and their attainment has been our maxim in editing it. We thank Petr for being who he is, and we hope that he and other readers will enjoy the book.

We would like to thank all the people who helped at various stages of preparation of this book. First of all, we thank all the contributors to this book, for their effort and cooperation. Our thanks go also to the referees who, often at very short notice, undertook the task of assessment and improvement of the contributions. We thank our translators, who tried to preserve as much of the original flavour of the Czech contributions as possible.

The library of Institute of Computer Science has been recording Petr Hájek's bibliography for more than ten years, and also provided an initial and timely motivation for the preparation of this book. Dagmar Harmancová has been an invaluable source of information as to details of Petr Hájek's career. Petr Vopěnka has been our mentor, offering his encouragement and help. Marie Hájková, apart from contributing a lot of enthusiasm for the project, has obtained a lot of information that we could not ask for directly, as the volume was to a be a surprise. Karel Chvalovský has provided technical and typesetting support. Mirko Navara has been a meticulous proofreader, helping to improve the quality of the manuscript.

Our thanks go to Jane Spurr for her efficient help in the preparation process, and for her friendly attitude and prompt manner of communication.

This publication is financially supported by the Learned Society of the Czech Republic, in order to promote the Society and its members. Petr Hájek is one of them.

We thank the Institute of Computer Science, Academy of Sciences of the Czech Republic, for material support throughout. Petr Cintula and Zuzana Haniková have been partly supported by the grant ICC/08/E018 of the Grant Agency of the Czech Republic (a part of ESF Eurocores-LogICCC project FP006) and partly by the Institutional Research Plan AV0Z10300504. Vítězslav Švejdar has been supported by the research plan MSM 0021620839 that is financed by the Ministry of Education of the Czech Republic.

Petr Cintula
Institute of Computer Science
Academy of Sciences of the Czech Republic
Pod Vodárenskou věží 2
182 07 Prague 8, Czech Republic
Email: cintula@cs.cas.cz

Zuzana Haniková
Institute of Computer Science
Academy of Sciences of the Czech Republic
Pod Vodárenskou věží 2
182 07 Prague 8, Czech Republic
Email: zuzana@cs.cas.cz

Vítězslav Švejdar
Department of Logic, Faculty of Arts and Philosophy
Charles University in Prague
Celetná 20
116 38 Prague 1, Czech Republic
Email: vitezslav.svejdar@cuni.cz

Prague set theory seminar

PETR VOPĚNKA[1]

This text is dedicated to Petr Hájek
on the occasion of his 70th birthday

Czech mathematics between the world wars boasted advanced geometry: differential, projective, and descriptive. At Charles University in Prague, there was additionally good mathematical analysis. Meanwhile, modern set mathematics, namely topology, was being developed at Masaryk University in Brno, then the workplace of Eduard Čech. Among the excellent attendants of his topology seminar, one has to mention Bedřich Pospíšil, who died at the age of 32 after he had returned from a concentration camp. After the World War II, Eduard Čech moved to Prague, where another representative of modern Czech mathematics, Miroslav Katětov, was active. Mathematical logic was being underestimated by mathematicians at that time, as it was not supposed to yield any "hard" results. The only person who studied logic was Ladislav Rieger at the Czech Technical University in Prague.

I studied at Charles University in 1953 to 1958. The person who influenced me most was Eduard Čech, who supervised my master thesis in topology, namely, in the theory of dimension. At that time however, Čech was already seriously ill, which in fact made me his last student. Nevertheless, it was Eduard Čech once again who, shortly before his death, changed his opinion on mathematical logic, when he had learned about Gödel's proof of consistency of the continuum hypothesis. He then emphasized this result to me.

It was no easy task to get access to Gödel's work. During the 1950's, all contact between the East and the West had been broken and it was difficult for people in Czechoslovakia at that time to get in touch even with citizens of other countries of the Soviet block. After much toil, I managed to obtain a Russian translation of the Gödel's work I sought. At that time, I also began to frequent Rieger's seminar in mathematical logic, where I obtained the knowledge of Skolem's non-standard model of arithmetic of natural numbers. Thus armed, I created a non-standard model of Gödel-Bernays set theory (that is, a model with non-standard natural numbers) in 1961, using the ultraproduct method (instead of an ultrafilter, it uses a maximal ideal). This

---

[1] Translated by Zuzana Haniková

model was then publised in *Doklady akad. nauk SSSR, No 1 (1962)*. Shortly after that, Ladislav Rieger died after a severe illness.

Then, I decided to start a new seminar in axiomatic set theory, intended mainly for students. The students who enlisted were (in alphabetical order) Bohuslav Balcar, Tomáš Jech, Karel Hrbáček, Karel Příkrý, Antonín Sochor, Petr Štěpánek and some others. We were joined by Lev Bukovský from Bratislava, and, last but not least, Rieger's PhD student, Petr Hájek. The main target of the seminar was to study non-standard models of Gödel-Bernays set theory.

Set theory of the time was trapped in the cage of Gödel's constructibility axiom, and its biggest problem was to break its bars. Speaking loosely, the task was to create such a model of axiomatic set theory that contains a non-constructible subset of the set of natural numbers. At the same time, the model had to consist of constructible sets. There were two approaches. The non-constructible set could extend either a countable, or a non-standard model of set theory. I found the second option more hopeful, because an ultra-product on an uncountable set and its submodel consisting only of functions with at most countable ranges have the same natural numbers, but the former one has more subsets of the set of natural numbers. I spent two years trying in vain to extend the latter model with a suitable set from the former one.

During this period, the political situation in Czechoslovakia grew more open, and it was possible—after a considerable effort—to travel abroad, mainly to countries of the Soviet block. We enjoyed this opportunity in order to establish contacts with Polish mathematicians. Occasionally, we were able to travel to Warsaw, where Andrzej Mostowski worked.

In 1963, Paul Cohen created a countable model of axiomatic set theory that invalidated not only the constructibility axiom, but also the continuum hypothesis. I realized at once where I had been mistaken. The submodel of the ultraproduct to be extended had to be chosen as one consisting of those functions which are constant on sets out of a system of open sets in a chosen topology on the domain, such that their union belongs to the ultrafilter. Indeed, Cohen's conditions forcing a certain property form an open base of a particular, simple topological space. The transition to the open sets of an arbitrary topological space was elementary.

Thus, the so-called ∇-models depend on the topological space chosen, and allow for an interpretation of Gödel-Bernays set theory with an extra axiom in Gödel-Bernays set theory (in a manner similar to F. Klein's interpretation of non-Euclidean geometry in the Euclidean one). The extra axiom can be deduced from the characteristics of the chosen topological space. In the spring of 1964, I succeeded in proving this in the special case when the chosen topology was the usual topology on the set of real numbers. During that year I obtained a general theory of such models and in the autumn, I presented it on A. Mostowski's seminar in Warsaw. There, Andrzej Grzegorczyk pointed out to me that open sets of a topological space could be used as truth values in intuitionistic logic. Having returned to Prague, I revised the models in a corresponding way, and this version of the theory of these models was then published during 1965 in Bull. Acad. Polon. Sci., Sr. sci. math. et phys.

The models of axiomatic set theory described use open sets of a topological space as truth values of formulas. As a suitable ultrafilter must not contain a thin set, it is sufficient to consider the regular open sets as values of formulas, and these, as it is well known, form a complete Boolean algebra.

In 1967 I found out that Robert Solovay and Dana Scott were working on the so-called theory of Boolean-valued models of set theory. They were in essence the same models that I had created in 1964. Admittedly, their description of them was much more elegant than mine.

The breakout of set theory from the cage of Gödel's constructibility axiom brought a tumultuous development of the theory. With some exaggeration one may say that any statement that had resisted the efforts of mathematicians throughout the first half of the century turned out to be undecidable. Moreover, mathematicians began to learn non-trivial facts about the law governing set theory, which had been trivialized by the constructibility axiom. A study of such inaccessible cardinals the existence of which is inconsistent with the constructibility axiom was started, etc.

Each of the above mentioned participants of my seminar took an important part in this unprecedented development of set theory. In particular, Petr Hájek played a special part. He had, at the very beginning, learned everything crucial in mathematical logic of that time, and thus became a referee of soundness of all our endeavours.

After the occupation of Czechoslovakia by the Soviet Army in 1968, our seminar dispersed. Tomáš Jech and Karel Hrbáček emigrated to the United States. Likewise, Karel Příkrý continued his stay in the U.S., having left there in the middle of 1965 for a scientific stay (equipped with the knowledge of the latest version of models that I had created). Lev Bukovský left for Slovakia and Petr Štěpánek began his work in mathematical logic in computer science. In Prague, it was only Bohuslav Balcar who remained faithful to set theory.

Petr Hájek decided to tread the path of mathematical logic as such, which will no doubt be discussed in detail elsewhere in this volume.

As the solution of the problem of truth in Cantor's set theory (discussed in another contribution to this volume) turned out to be trivial, classical set theory lost much of its charm as far as I was concerned. I focused on the theory of infinity itself, in the study of which a significant role was played by the notion of semiset, introduced earlier for those subsets of sets of a non-standard model which are not sets in the model. After all, during those times when I was not allowed to travel even to Poland, and instead of letters,

I received only empty envelopes, there was little else left for me to do. Just a closing remark, I nevertheless travelled to Poland twice within those twenty years of darkness, when I illegally crossed the border in the mountains of Krkonoše.

BIBLIOGRAPHY

[Vop65a] P. Vopěnka. The limits of sheaves and applications on constructions of models. *Bull. Acad. Polon. Sci., Sér. Sci. Math. Astronom. Phys.*, 13(3):189–192, 1965.

[Vop65b] P. Vopěnka. On ∇-model of set theory. *Bull. Acad. Polon. Sci., Sér. Sci. Math. Astronom. Phys.*, 13(4):267–272, 1965.

[Vop65c] P. Vopěnka. Properties of ∇-model. *Bull. Acad. Polon. Sci., Sér. Sci. Math. Astronom. Phys.*, 13(7):441–444, 1965.

[Vop66]   P. Vopěnka. ∇-models in which GCH does not hold. *Bull. Acad. Polon. Sci., Sér. Sci. Math. Astronom. Phys.*, 14(3):95–99, 1966.

[Vop67]   P. Vopěnka. General theory of ∇-models. *Comm. Math. Univ. Carolinae*, 8:145–170, 1967.

Petr Vopěnka
Faculty of Philosophy and Arts
University of West Bohemia in Pilsen
Sedláčkova 38, 306 14 Plzeň
Czech Republic

My life with Petr, Petr's life with me

Marie Hájková[1]

Petr Hájek is my husband and it is from this viewpoint that my story about him, and also about myself, will be told. He is an excellent and renowned mathematical logician and I am the wife.

It was at least three times that Petr stepped onto a crossroads that could bring him closer to mathematical logic, or take him far away from it. The first crossroads: what to do after his maturity exam. Petr was seriously considering studying composition at the Academy of Performing Arts in Prague. The other option was mathematics. He enrolled in the Faculty of Mathematics and Physics, Charles University in Prague. The second crossroads: in his fourth year of study, he had a bicycle accident, occasioned by a fast ride which ended up in crashing into a truck. He suffered a fracture of the skull. He verged on the border of life and death, and survived thanks to his doctors and thanks to God. The third crossroads: Petr wrote his master thesis at the Faculty of Mathematics and Physics, and the subject was algebra. After the successful defense, it was the wish of professor Kořínek that Petr continued at his department of algebra as an assistant. But alas, the Communist Party made this impossible. They would not suffer a person who believed in God to be in touch with students. It was professor Kořínek, or someone else, who said it: Hájek must go to the Academy. Petr was accepted as a PhD student (then called "aspirant") in the Mathematical Institute of the Academy of Sciences and his advisor was the well-known logician, professor Ladislav Rieger; this initiated Petr's journey of a mathematical logician. Paradoxically, it was partly the doing of the Communist Party and its absurd ban.

I earned my master's degree in St. Petersburg, under the guidance of professor Šanin, having submitted my master thesis in 1964. Šanin was a constructivist, which made it difficult to learn anything substantial about classical logic and neighbouring areas, such as set theory, from him—many classical principles are not valid from a constructivist point of view. My work concerned automated theorem proving and involved a lot of programming in Basic. There were several computers at the Faculty, but the fastest computer available was a Ural 4, located in a bank where I managed to obtain access. My task was to find out, for a pair of given formulas, whether they were

---

[1]Translated by Zuzana Haniková

close in the given proof system (in particular, whether their distance was two proof steps). The work on this topic was quite painstaking. It was not until I returned to Prague that I realized how little I actually knew about logic.

Back in Prague in 1965, I entered a PhD (then called CSc) programme at Petr Vopěnka's Department of Logic, Faculty of Mathematics and Physics, in the Prague quarter of Karlín. Petr Hájek and I both frequented Vopěnka's set theory seminar, and this was where we met for the first time. Also Bohuslav Balcar, Tomáš Jech, Karel Hrbáček, Petr Štěpánek and some others frequented this seminar. During my PhDship I was also the secretary of Vopěnka's department.

I went away for a while to teach calculus to students of natural sciences. When I returned, Vopěnka's seminar in alternative set theory was already taking place in Karlín. Then, Petr Hájek took me under his wings. He made it his business to advise me respecting my reading list, and the topic of my thesis. The task was to prove that binumerations of arithmetic form a lattice. This was one of the best things that could happen, although sometimes I would find my cooperation with Petr a bit trying. The difficulty was especially to get Petr to agree with my ideas and to approve of the direction of research that they implied. I defended my thesis in 1969.

Petr and I are also co-authors. Our first joint paper was in arithmetic and my role consisted in proving a theorem that Petr needed.

In 1969, Petr spent one semester visiting professor Müller in Heidelberg. After his return, we got married, with Petr Vopěnka as the best man. Our daughter Marie was born in 1971 and consequently, I took a maternity leave. Soon after, I found myself diagnosed with multiple sclerosis. In the beginning I was just not feeling well and the doctors had no idea what was wrong with me, they suspected hypochondria. However, Petr did not underestimate the problems I had and provided a huge psychic support to me. These years were not easy for us as a family, as I often had to spend periods of time in a hospital, and during these, the care of our little daughter was entrusted to my mother.

Apart from mathematics, Petr also loves music. After our marriage he decided to commence a distant learning at the Academy of Performing Arts, with organ as his subject. Naturally this was very demanding, not only as regards time and scheduling, but also from the spatial point of view—even a small organ was clearly audible in a block of flats such as we lived in. However, music is a realm where Petr can relax, which is crucial for him. Another good thing is that our son, Jonáš, also became interested in music quite spontaneously, studied the conservatory in Prague with violoncello as his subject, and now is about to finish his studies at the Musicology Department of the Faculty of Arts and Philosophy, Charles University in Prague.

When thinking about this essay, I asked Petr about his other hobbies. He replied that he had me, was looking after me, and it gave him pleasure.

We own a small cottage in the Znojmo district, situated by the water reservoir of Vranov nad Dyjí. Long time ago, I persuaded Petr to take up windsurfing. I was glad he had a hobby that took him into the open, breathing fresh air. However, he already gave it up. It is quite interesting that Petr is in no way motivated to walk, to stroll in a forest, to gather mushrooms or berries, et cetera. For me, on the other hand, being in a forest or bathing in the open is a memorable experience. Another interesting thing about Petr is that if he finds himself in a forest after all, he insists on sticking to well-walked trails. Lately, the trend has been with Petr to bring his computer even to the cottage, and spend his time there working, trying not to waste a single minute.

During the communist period, if Petr was allowed to travel abroad to conferences, he was often afterwards invited for a hearing at the StB (State Security) as to the subject of his travel, which he did not take lightly. A hard blow came with the cancellation of the Logic Colloquium 1980, which should have taken place in Prague in the summer of that year and which he organized. The StB asked Petr Hájek to a meeting where he was told that the Colloquium had to be cancelled. There was an apprehension that the participants might stand up for Václav Benda, who was then imprisoned. Petr sent out letters of apology, stating that the Colloquium is cancelled for some reason that was fictitious. The cancellation hit him deeply, and consequently I helped him to seek therapy. Another blow was that after this experience, he was not allowed to travel to conferences.

After the Velvet Revolution, the situation turned the other way round. Petr continued at the Mathematical Institute, CSAS, until 1992, when he was elected director of the Institute of Computer Science, CSAS, after the decease of his predecessor, Tomáš Havránek. He remained at this post until 2000, and now continues in the Institute as a senior researcher.

Petr is a full professor, and also a honorary professor in Vienna. He is a member of the Learned Society of the Czech Republic, he has been awarded the medal *De scientiae et humanitate optime meritis* by the Academy of Sciences of the Czech Republic in 2006, and the Medal of Merit by the President of the Czech Republic in 2006. I do hope I have not omitted anything. Petr never mentions his honours and awards, but he is certainly pleased. To me, he said this much: "I am a good mathematician on an international scale, and a church organist. That's all." Petr's priorities in life are elsewhere. He is deeply religious, a member of the Evangelical Church of Czech Brethren. As a young girl in love with Petr, I accompanied him to church, understanding very little. University education did not help there, especially such as had been obtained in a totalitarian Communist country. Petr is a really tolerant

person, so I asked him directly—what is it that you believe in and what does
it offer you. A few words sufficed him to reply: "Jesus saith: I am the way,
the truth, and the life. I live according to this." This made me none the
wiser, I needed time. Nowadays, I am a member of the Evangelical Church,
too. I do not see myself as deeply religious—rather, I would quote from the
Bible again: Lord, I believe; help thou mine unbelief.

Marie Hájková

The problem of actualizability of the classical infinity

PETR VOPĚNKA[1]

ABSTRACT. The first part of this treatise is a brief summary of my lecture given in 1981 during a conference of the Union of Czech mathematicians commemorating the bicentenary of Bernard Bolzano's birth. My contribution was published in Polish in the journal *Wiadomosci Matematyczne XXII (1985)* under the title *Nieskonczonosc, zbiory i mozliwosc u B. Bolzana* and in a slightly modified form also in Russian as a second appendix to my book *Matematika v alternativnoj teorii mnozestv*, Izd. Mir, Moskva, 1983.

The subsequent parts of this treatise express that what was felt throughout the ages by many mathematicians and what they even managed to work with very successfully, and what I myself have purposefully worked with over some thirty years, but what appears so far to have eluded being distinctly and clearly exposed by anybody.

1 The problem of truth in Cantor's set theory

Consider a situation where there are some previously formed objects, and we single some of them out (regardless whether all at once or one after another). Then we obtain a *collection* consisting of these singled-out objects.

By *a set* we understand a collection that is sharply defined and which we interpret as an autonomous individual, or object.

When talking about people, we often have in mind not only people who are at that moment alive or have lived in the past, but also those who are yet to be born, and even those who have never been born and never will be. The extension of the concept 'people' is therefore no collection but a domain of all people.

A *domain* is no totality of some existing objects (regardless of the mode of their existence): it is the source and simultaneously a sort of a container encompassing suitable objects as they are emerging.

Naturally, every collection of some objects may be interpreted as a domain, albeit an exhausted one.

---

[1]Republished from *Pojednání o jevech povstávajících na množstvích*. OPS, Pilsen 2009, translated by Alena Vencovská.

When referring to an *actualization* of some domain, we mean the exhausting of it, effected by substituting for the domain the collection of all objects that fall or can fall in it.

In the case of a sharply defined domain the corresponding collection may be interpreted as a set.

It was via actualization of the domain of all natural numbers that set theory made its entry into mathematics. This actualization was accomplished by Bernard Bolzano and later extended by George Cantor to much wider infinite sharply defined domains. Set theory then went on to subsume mathematics of the 20th century as its own part via actualization of all subjects of mathematical inquiry.

However, already in the last years of the 19th century some limits in actualizability of sharply defined domains of mathematical objects began to emerge. In particular, the domain of all transfinite ordinal numbers cannot be actualized. For if there was a collection of all these numbers then it easily follows that the ordinal number that is the type of the well-ordering of these numbers according to size does not belong in this collection, which is absurd. Similarly of course in the case of the of the domain of all sets.

A ban on interpreting of the collection of all transfinite ordinal numbers as a set (along with a number of similar regulations concerning the forming of sets) did ward off contradictions from the theory of sets but such a purely pragmatic measure cannot be accepted as a solution of the problem of actualizability of sharply defined domains of mathematical objects. For there is nothing to prevent us from interpreting any sharply defined collection as an autonomous individual, that is as a set. The only genuine solution of the situation thus consists in admitting the non-actualizability of the domain of all transfinite ordinal numbers and hence of course also of the domain of all sets.

A mathematician working in mathematics based on set theory studies only sets belonging to some collection of sets. Elements of this collection are usually assumed to satisfy the usual rules of Cantor's intuitive (naive) set theory. They are the rules that served at the beginning of the 20th century as stimuli for axioms of axiomatic set theories such as that of Zermelo-Fraenkel. We shall call such collections Cantor's set universes, in short *CS-universes*.

Naturally, every CS-universe can be interpreted as a set which however is not – and cannot be – its own element.

For a CS-universe V, other sets beyond V itself must fail to be elements of V: some subsets of it and also for example the one-element set $\{V\}$ and any other set containing V as an element. On the other hand, a CS-universe V could be an element (or a subset) of some larger CS-universe.

In early mathematics it was the classical geometric space that decided about realizability of geometric objects, namely so that it either did or did not accommodate some such object. When, during the initial period of modern mathematical natural sciences, the whole of the real world had been placed in it, Isaac Newton used the name Sensorium Dei for this space. Based on a number of more or less obvious reasons we will now use this name in the following new sense:

By the term *Sensorium Dei* we understand that what decides about realizability of finite or infinite sets, namely so that it either does or does not accommodate a particular set.

It was during the development of classical geometry, probably sometimes at the beginning of modern times, that geometers started to transfer their interest from individual geometric objects to geometric space. Thus the space has also become an object of mathematical inquiry where forming various geometric objects served to reveal on the one hand possibilities offered by space and on the other hand necessities imposed by forming them.

Similarly, we shall now take Sensorium Dei to be our main subject of inquiry whilst forming of various CS-universes (and later also of other sets) will serve to reveal possibilities and necessities present in it.

Following Euclid, we shall use the term axiom for an assertion that we do not prove but that is worth accepting. In contrast, 'postulate' will stand for a primary task, that is a task which must be perceived as executable by anybody who wishes to study Sensorium Dei with us.

The domain of all ordinal numbers will be denoted On. The set of the ordinal numbers of a CS-universe V will be denoted $On(V)$. Obviously, $On(V) \subseteq V$, $On(V) \notin V$.

Axiom 1 *Ordinal numbers are absolute.*

That means that every CS-universe V satisfies:

(a) If $\alpha \in On(V)$ then α belongs to the domain On.

(b) Let α belong to the domain On and $\alpha \in V$. Then $\alpha \in On(V)$.

The following assertion is trivial.

Assertion 1 *Let V, W be CS-universes. Then either $On(V) \subseteq On(W)$ or $On(W) \subseteq On(V)$.*

If $On(V) \subseteq On(W)$ and $On(V) \neq On(W)$ then there exists an ordinal number $\alpha \in On(W)$ such that $On(V)$ is the set of all ordinal numbers smaller than α. Consequently $On(V) \in W$.

Axiom 2 *There exists the set $\mathbf{N}$ of all natural numbers.*

This axiom can be formulated also as the following primary task:

Postulate 1 *To actualize the domain of all natural numbers.*

Set theory-based mathematics takes both the above axioms to be true and hence consistent. Also the following assertion is trivial.

Assertion 2 *Let V be a CS-universe. Then we have*

(1) $\mathbf{N} \in V$

(2) *There exists an ordinal number ω such that $\omega \in On(V)$ and $\mathbf{N}$ is the set of all ordinal numbers smaller than ω.*

Sensorium Dei accommodates any set the existence of which is not contradictory. For there is no reason why such a set should not be realizable. Loosely speaking then, what is consistent is also realizable in Sensorium Dei.

This universal property of Sensorium Dei underlies, and in a sense also captures feasibility of the task described by the following postulate.

Postulate 2 *To form any set the existence of which is consistent with the two previously stated axioms.*

The difficulty with feasibility of this postulate lies predominantly in insuring consistency of the set that we intend to realize. For here we usually have to resign attempting a proof of consistency so that we cannot but rely on intuition obtained through forming sets and dealing with them. Such intuition also underlies our acceptance of realizability of a CS-universe and—if we are sufficiently daring—then also of a number of other CS-universes with longer ordinal numbers. Realizability of some CS-universes may also rest on their relative consistency with respect to some simpler one, the realizability of which we do not doubt.

Until mid 20th century mathematicians studying set theory have retained a view that could be expressed in saying that Sensorium Dei has only a single dimension in which actualizability of the domain of all sets is not achievable. The domain of all ordinal numbers is a pivotal axis and a representative of this dimension.

This means in particular that the domain of all subsets of any given set is actualizable. In other words, for every set X a set $P(X)$ can be formed of all subsets of the set X (that is, the powerset of X). In fact, an assertion according to which for every set X there exists its power-set $P(X)$ is one of the axioms of the Zermelo-Fraenkel system.

If it was the case that every set was constructible in Goedel's sense then it would naturally be true not only that the domain of all subsets of a given set was actualizable but also the above mentioned view concerning one-dimensionality of Sensorium Dei would be right.

As the 20th century entered its second half however, only a few set theorists remained who believed that non-constructible sets—in particular subsets of the set $\mathbf{N}$—cannot exist. However, it cost a lot of effort to show that this minority view is wrong—for example, by showing the relative consistency of the existence of non-constructible sets or even of the negation of the continuum hypothesis with respect to Zermelo Fraenkel axioms. This was achieved in 1963 by Paul Cohen.

What followed was a rapid development of set theory concentrating predominantly on finding various assertions such that neither they nor their negations are provable from Zermelo Fraenkel axioms. Many such assertions were found. With some exaggeration it can be said that all assertions that had previously resisted considerable efforts to prove or disprove them were shown to be independent of Zermelo Fraenkel axioms.

Mathematicians (excepting extreme formalists) thus found themselves faced with the pressing problem of truth in Cantor Set theory, that is with question which of these independent assertions were true. Hardly anybody now doubted that Goedel's axiom of constructibility was not true.

Of course, this problem has a trivial solution which all set theorists must have been subconsciously aware of but which they refrained from formulating clearly and explicitly, to the detriment of the further development of classical infinite mathematics. *Only questions like whether for example the continuum hypothesis is true in this or that CS-universe make sense, similarly as questions whether the commutativity axiom is true in this or that group make sense. On the other hand, asking whether the continuum hypothesis is true lacks sense just as asking whether the commutativity axiom is a true axiom of group theory lacks sense.*

Similarly as geometric space allows placing various intersecting geometric objects in it, various intersecting ZF universes may be placed in Sensorium Dei, even such that some independent assertion is true in one of them and not true in another. Placing of these CS-universes is subject only to axioms of absoluteness of the set of all ordinal numbers an of the set **N** of natural numbers.

Thus for example if Steinhaus-Mycielski's axiom of determinacy is consistent with Zermelo-Fraenkel axioms then it is possible to form simultaneously two CS-universes in Sensorium Dei such that in one of them the determinacy axiom holds and in the other one the axiom of choice holds.

Similarly, since the assertion that the set $P(\mathbf{N})$ has cardinality aleph three is consistent with Zermelo-Fraenkel axioms it is possible to form such a CS-universe V in Sensorium Dei, in which the set of all subsets of **N** (that is, of those subsets that belong to V) has cardinality aleph three. Just as well it is possible to form simultaneously a CS-universe W in which this set has cardinality aleph fifty. According to Axiom 1 these universes must satisfy either $On(V) \subseteq On(W)$ or $On(W) \subseteq On(V)$. We remark that it can be arranged so that $On(V) = On(W)$.

Clearly we can place whole bundles of CS-universes into Sensorium Dei.

The possibility of forming, for each CS-universe V a CS-universe W with more subsets of the set $\mathbf{N}$ of natural numbers leads us to the conclusion that *the domain of all subsets of the set $\mathbf{N}$ is not actualizable*. This domain thus represents a *second dimension of Sensorium Dei* in which actualization of the domain of all set is not achievable.

In this context we cannot fail to recall a statement, now almost forgotten, by René-Louis Baire who wrote in 1905[1]: *"If a set is given, for example the set of all natural numbers, then it is incorrect to consider parts of this set to be given, too."*

Taking account of principles of Sensorium Dei stated so far, we can reformulate results concerning collapses of cardinal numbers obtained through studying models of set theory as follows:

Let α be a transfinite ordinal number and let $[\alpha]$ denote the set of all ordinal numbers less than α. Let $\alpha \in V$, where V is a CS-universe. Then it is possible to form a CS-universe W for which $\alpha \in W$ and the set $[\alpha]$ is countable in W.

This fact and the universality of Sensorium Dei allow us to consider the task contained in the following postulate of the *general collapse* to be feasible:

Postulate 3 *To form a one-one mapping G of a given infinite set onto the set $\mathbf{N}$.*

If $X \in V$ is a set that is uncountable in the CS-universe V then the required mapping cannot be formed inside the CS-universe V (that is, $G \notin V$).

If α is an ordinal number and G a one one mapping of the set $[\alpha]$ onto the set $\mathbf{N}$ then we can easily form a relation $R \subseteq \mathbf{N} \times \mathbf{N}$ that is an ordering of the set $\mathbf{N}$ of the type α. It follows that in the direction of the above mentioned second dimension the depth of Sensorium Dei is just as bottomless as in the first direction

The general collapse has many interesting consequences since it means that even uncountable sets may – with a due care – be treated as if they were countable. Hence for example an isomorphism can be formed between any two infinite elementarily equivalent structures etc.

The general collapse could be glimpsed as early as 1851 in Bolzano's *Paradoxes of the Infinite* [2] and a suggestion of it it can be traced in the work of Leopold Löwenheim from 1915[3]. Backed by the mathematical results of the nineteen sixties it has resurfaced as a genuine principle for forming infinite sets.

Naturally, we also can form various other meaningful collections other than bundles of CS-universes in Sensorium Dei. Let us call them *set complexes*.

---

[1] *Cinque lettres sur la théorie des ensembles,* Bull de la Soc. Math., France, 1905
[2] *Paradoxien des Unendlichen (1851)*
[3] *Über Möglichkeiten im Relativkalkül,* Math. Ann.7, 1915

They can consist of various relational structures, algebraic groups, geometric spaces etc. We may then step out of such a set complex and investigate it, or form various additions to it, using the general collapse.

From this point of view CS-universes are merely special cases of set complexes and their advantage consists just in our being accustomed to work in them, knowing them well and believing in their consistency. Reasons other than these historical ones as to why we should focus on them would probably be hard to find.

The study of set complexes and relations between their parts is a new field that has opened to classical infinite mathematics.

In conclusion we remark that it is no other than the God of the medieval rational theology who accomplishes, in his mind, actualizing sets in Sensorium Dei. Bolzano and Cantor have admitted this openly. Other mathematicians, although claiming his faculties, try to suppress and ignore this. The title Sensorium Dei that we have chosen is therefore appropriate.

2 The problem of actualizability of the domain of all natural numbers

The problem of actualizability of classical infinity thus returns to its initial stage, which is the problem of actualizability of the domain of all natural numbers.

By natural numbers we understand numbers 0,1,2,3,... Those of them which we have written down form a negligibly small part of the domain of all natural numbers, and this would apply even if we wrote down the first hundred, million or trillion of them. It would not change even in the case of all numbers less than 10^{421} which, according to the old Indian script Lalitavistara, Budha could name.

The remaining infinitely many natural numbers, that is those which in the wildest dreams *Sensorium Humanum* cannot contain, hides in those three dots (or in the words 'and so on') always following whenever we write down or name some numbers wishing to explain natural numbers to somebody.

What is remarkable is that these three dots (this "and so on") suffice to bring about an idea of natural numbers, in some sense even very clearly, and to evoke understanding for their tending to infinity. Since we are unable to bring about the idea of natural numbers any better, there is no doubt that we are dealing with one of the archetypal ideas that mysteriously arise from the collective unconscious of the Euro-Indian mankind (in the sense of C.G.Jung)

Similar archetypes of this collective unconscious has formed those European gods that – thanks to their immortality and their unlimited faculty of seeing are able to attain natural numbers well beyond the reach of Sensorium Humanum. In the Greek antiquity they were the Olympian gods which were later replaced (as far as their faculties were concerned) by angels, including the fallen ones. Mathematicians need such superhumans. Not for these beings as such but in order that they could claim for themselves, mentally at least, the faculties impersonated by these divinities.

We shall call *accessible natural numbers* those natural numbers that may be reached by these divinities during their never ending lives, not only step by step but also by making big jumps which however only serve to ease the effort that they would need to expend when attaining these numbers stepwise. In other words, accessible are numbers of objects which *Sensorium Deorum* can contain.

For our purposes accessible natural numbers may be described as follows:

(0) Define a sequence $a_0, a_1, a_2, \ldots$ of natural numbers by recursively so that $a_0 = 1$, $a_{n+1} = 2^{a_n}$.

(1) The number 2 is accessible.

(2) If a number n is accessible and if $m < n$ then the number m is also accessible.

(3) If a number n is accessible then the number a_n is also accessible.

(4) There are no other accessible natural numbers.

We remark that the sequence $a_0, a_1, \ldots$ is not essential. Any other rapidly increasing sequence defined by a similar simple recursive rule would do.

The fundamental hypothesis of infinitary mathematics:

Every natural number is accessible.

How questionable this hypothesis is becomes apparent straight as we realize that it is an assertion according to which no larger natural numbers can exist than those attainable by Zeus.

We shall show that there is nothing at hand that we could use to support this hypothesis, let alone verify it.

In fact, mathematical formalism can cast serious doubts on this hypothesis as follows: Let T denote some axiomatic mathematical theory that contains at least implicitly the Peano arithmetic. We could for example take any strong axiomatic set theory. In this theory, neither Zeus nor any other more powerful being could write a formula $\phi(x)$ and prove that this theory defines exactly all the accessible natural numbers. That means, prove about any accessible natural number that it satisfies the formula $\phi(x)$ and also conversely, that any natural number x that satisfies $\phi(x)$ is necessarily accessible. For in a non-standard model of the theory T the numbers that satisfy this formula would be standard, but it is not possible in a non-standard model of T to define such natural numbers amongst which there exists no largest one by any formula. In other words, existence of natural numbers that are bigger than any accessible natural number cannot be shown as absurd along these lines.

The fact that there does not exist any formula $\phi(x)$ of the theory T that corresponds to the property 'x is an accessible natural number' means that the fundamental hypothesis of infinitary mathematics cannot be proved in this theory by induction as might be hoped prior to considering the matter seriously.

Hence arguments for the fundamental hypothesis of infinitary mathematics cannot be based on mathematical formalism: rather the contrary.

The answer to the question whether the fundamental hypothesis of infinitary mathematics is or is not true must be sought from that mysterious intuition about the tending of natural numbers to infinity. For this purpose we need to strip our intuition of all the additional interpretations that the tending of natural numbers to infinity has attracted.

When we carry out such a suspension of judgement or phenomenological *epoche* as E. Husserl would say, then we cannot but conclude that such a purified intuition for the tending of natural numbers to infinity contains no trace of the fundamental hypothesis of infinitary mathematics. We find no reason why we should shorten the domain of all natural numbers. On the contrary, the neat infinity contained in this domain excludes a priori all obstacles that could stand in the way of natural numbers in their expansion.

In summary, our unencumbered intuition for the tending of natural numbers to infinity does not admit the fundamental hypothesis of infinitary mathematics, it excludes it. This, supported by the above mentioned consistency of the negation of the fundamental hypothesis of infinitary mathematics leads us to consider the negation of this hypothesis to be the true assertion concerning the domain of all natural numbers. In other words, in the domain of all natural numbers there are latently present some inaccessible natural numbers.

It may seem that that the fundamental hypothesis of infinitary mathematics is useful since it relieves us of the obligation to consider such natural numbers that nobody needs. Hence it should appeal to mathematical pragmatism.

However, infinitely large (or inaccessible) natural numbers or more generally also real numbers, and especially their inverses, that is, infinitely small numbers, have proved to be extremely useful.

They were what allowed I.Newton and G.W.Leibnitz to start building the infinitesimal calculus (differential and integral calculus). The first textbook of it was probably written by G.F.A. de l'Hospital and appeared in 1696 under the title *Analyse des infinement petits*.

A foremost opponent of infinitely small numbers was George Cantor who tried hard to exclude them from mathematics. He even attempted to prove using transfinite numbers that infinitesimals do not exist.

The drive for exclusion of infinitely small numbers from mathematics joined by many has given birth to mathematical analysis. That is, a discipline the name of which arose from the title of the above mentioned book by crossing out what was being analyzed in it.

However, translating intuitively obvious notions and results from infinitesimal calculus into the somewhat sarcastically called ϵ, δ-analysis is, to put it mildly, quite awkward. For example, calculations with infinitely small quantities need to be replaced by assertions with proofs, numbers of alternating quantifiers in definitions of intuitively clear notions of infinitesimal calculus at least double when translated etc.

Trying to eliminate infinitesimals from mathematics thus turned out to be, to say the least, impractical, so there is little wonder that physicists (and others) stick with the original methods.

Hence the pragmatic approach clearly does not support the fundamental hypothesis of infinitary mathematics.

We have come to accept the fundamental hypothesis of infinitary mathematics as true merely because the ancient Greek geometers (Eudoxos, Archimedes) embedded it in the then discovered ideal geometric world.

*

Hence we start on a much wider and more open a journey towards a global investigation of natural numbers when we accept the principle below than was that chosen by mathematics at the beginning of the 20th century through the (albeit often subconscious) acceptance of the fundamental hypothesis of infinitary mathematics.

The First Principle *There exist inaccessible natural numbers.*

Wishing to conform to traditional terminology of infinitesimal calculus leads us to adopt the names *finite and infinite natural numbers*, for accessible and inaccessible natural numbers respectively.

In order to be able to study in some detail the position of the domain of all finite natural numbers within the domain of all natural numbers we do not need to require actualization of the domain of all natural numbers, that is, to assume the existence of the set **N** of all natural numbers. It suffices to accept the following task as feasible:

Postulate 4 (Of weak actualizability of the domain of natural numbers) *For each natural number α, to form the set $[\alpha]$ of all natural numbers smaller than α.*

Existence of at least one infinite natural number γ along with the fact that every finite natural number is smaller than γ implies that there exists a collection FN of all finite natural numbers. This is because the collection FN is a part of the set $[\gamma]$. In other words, the first principle and the postulate of weak actualizability of the domain of natural numbers imply actualizability of the domain of all finite natural numbers.

The collection FN cannot be interpreted as a set. If FN was a set we would have $FN \subseteq [\gamma]$ where γ is an infinite natural number and thus the set FN would have to have a largest element. However, that is not possible since as we can easily check, if n is a finite then $n + 1$ is also a finite natural number.

There is no reason why we should not consider the collection FN to be an autonomous individual, that is, interpret this collection as an object. On the contrary, such an interpretation is undoubtedly useful.

Hence we cannot but admit that the collection FN is not sharply defined. In order that also non-sharply defined collections can be treated as objects, we introduce the following concepts:

A semiset is a non-sharply defined collection of some objects singled out from some set, which we interpret as an autonomous individual, that is, as an object.

A class is any collection of objects, which we interpret as an autonomous individual, that is, as an object.

The class FN is thus a semiset.

The assertion about the class FN not being sharply defined is thus a logical consequence of the first principle and of the postulate of weak actualizability of the domain of natural numbers. Also the above mentioned non-existence of a formula $\phi(x)$ of the theory T representing the property 'x is a finite natural number' points to non-sharpness of the definition of the class FN. Even so, it is difficult to overcome doubts as to whether the class FN is not sharply defined after all – for the way in which we defined the domain of all finite natural numbers is transparent. However, the clear definition does not apply to the class FN but only to the way in which objects belonging to the domain of all finite natural numbers may be reached. Even Sensorium Deorum is limited by the horizon.

We can extend the domain of all finite natural numbers to various other farther-reaching domains of natural numbers. A prominent method for doing so is to choose some infinite natural number γ and to change the above definition of the domain of all accessible natural numbers by replacing the expression 'accessible number' with 'γ-number' and by writing the following instead of (1):

(1′) The number γ is a γ-number.

We shall use $(\gamma)\mathbf{N}$ for the class of natural numbers that would be obtained by actualization of the domain of all γ-numbers and through the decision to interpret the collection thus formed as an autonomous individual.

Now we could generalize the fundamental hypothesis of infinitary mathematics as follows:

> *There exists a natural number γ such that all natural numbers are*
> *γ-numbers.*

Similarly as it is in the case of finite natural numbers, also this generalized hypothesis shortens the domain of all natural numbers, moreover it does so quite openly without any pretence to being natural. Our understanding for tending the natural numbers to infinity, stripped of additional interpretations, now quite clearly does not contain it. Briefly speaking, there are more convincing reasons against this generalized hypothesis than against the original one. Hence we must reject it, that is, accept the following principle.

The second Principle *For any natural number γ there exists a natural number δ that is not a γ-number.*

We can easily see that the second principle along with the postulate of weak actualizability of the domain of natural numbers and the non-existence of a largest γ-number allows us to interpret the class $(\gamma)\mathbf{N}$ as a semiset.

As we have already pointed out, the neat infinity (that is the infinity obscured by no additional interpretations) present in the domain of all natural numbers excludes all obstacles that could hinder the expansion of natural numbers to infinity.

This is in accordance with the possibility of consecutive forming of semisets $(\gamma_1)\mathbf{N}$, $(\gamma_2)\mathbf{N}$, $(\gamma_3)\mathbf{N}$,...,$(\gamma_\alpha)\mathbf{N}$, $(\gamma_{\alpha+1})\mathbf{N}$, ... where

$$\gamma_2 \notin (\gamma_1)\mathbf{N}, \gamma_3 \notin (\gamma_2)\mathbf{N}, \quad ... \quad \gamma_{\alpha+1} \notin (\gamma_\alpha)\mathbf{N}$$

the growth of which is clearly just as unstoppable as the growth of the sequence of transfinite ordinal numbers, even though each of these two sequences has its own distinctive character.

Hence any effort to exhaust gradually the domain of all natural numbers is just as hopeless as an effort to exhaust gradually the domain of all transfinite ordinal numbers. The domain of all natural numbers is just as inexhaustible as the domain of all transfinite ordinal numbers.

In other words, the domain of all natural numbers is not actualizable.

Rejection of the fundamental hypothesis of infinitary mathematics thus opened the infinity present in the domain of all natural numbers in all its staggering depth and hence also a third dimension of Sensorium Dei in which actualizability of the domain of all sets is not achievable.

It may seem that the non-actualizability of the domain of all natural numbers contradicts what we have said in the first part of this treatise, namely the axiom of existence of the set **N** of all natural numbers. However, this is not a contradiction but a misunderstanding, as we will explain shortly.

3   Strictly finite natural numbers

It is also possible to shorten the domain of finite natural numbers, for example in the following way.

Let **strictly finite natural numbers** be those numbers that people can gradually reach. In other words, these count objects that can be contained in Sensorium Humanum.

We may describe the strictly finite natural numbers as follows.

(1″) The number 2 is strictly finite.

(2″) If a number n is strictly finite and if $m < n$ then the number m is also strictly finite.

(3″) If a number n is strictly finite then the number $n + 1$ is also strictly finite.

(4″) There are no other strictly finite natural numbers.

We remark that the forming of strictly finite natural numbers by adding the number one is limited by human faculties rather than those of Zeus.

The domain of all strictly finite natural numbers is an almost negligibly small part of the domain of all finite natural numbers. If we were interested in treating the collection SFN of all strictly finite natural numbers as an autonomous individual, the class SFN would be a semiset and there would exist a finite natural number n such that $SFN \subseteq [n]$.

The domain of all strictly finite natural numbers may be extended as necessary to various more inclusive domains which are also subdomains of the domain of all finite natural numbers. As an example we could use the domain of all natural numbers that some powerful computer can reach etc. Obviously, no such domain of natural numbers is sharply defined, that is, the natural numbers which we would obtain if such a domain were actualized and if we chose to interpret the collection of all these numbers as an autonomous individual, would form a semiset.

When we consider that the domain of all finite natural numbers is but some sort of an umbrella for all such domains of natural numbers, or, that Zeus is merely superhuman but not God we obtain another reason indicating non-sharpness of the definition of the domain of all natural numbers that can be gradually reached by members of the Greek Pantheon.

4   Possible directions for development of infinitary mathematics

The conception of mathematics under which we approach important domains of objects as if they were actualized has proved useful in many respects. Clarifying the position of the domain of all finite natural numbers in the domain of all natural numbers revealed the following paths along which infinitary mathematics may be developed within this conception.

The first path is described in the first part of this treatise. To clarify the above mentioned misunderstanding we just need to add that what has been called there the set $\mathbf{N}$ of all natural numbers really is the semiset FN. Otherwise, this is the path adopted by the classical set theory in which moreover the openness of both the first and the second dimension of sensorium Dei is used. The third dimension was closed to it by the fundamental hypothesis of infinitary mathematics which also obscures the fact that the class $\mathbf{N}=FN$ is not sharply defined.

The second path involves openness of both the first and the third dimension of Sensorium Dei but, unnecessarily, avoids the second dimension, in particular the possibility of exploiting the general collapse. What is called here the set $\mathbf{N}$ of all natural numbers is really the semiset FN. The set $\mathbf{N}^*$ (the nonstandard extension of the set $\mathbf{N}$) is in fact some semiset $(\gamma)\mathbf{N}$. This is the path adopted by nonstandard analysis.

The third path involves mainly openness of the third dimension of Sensorium Dei. This path takes the existence of the semiset FN into account, and of infinitely large natural numbers. When a class $\mathbf{N}$ is referred to, some semiset $(\gamma)\mathbf{N}$ is understood by it. This is the path followed by the alternative set theory, and prior to that trodden by the infinitesimal calculus up to the point when it was forced out of mathematics.

The fourth path is the path of natural infinitary mathematics that studies natural infinity, that is infinity present in non-sharpness and in non-definiteness. It focuses on small changeable subdomains of the domain of all finite natural numbers. The letter $\mathbf{N}$ stands here for the semiset FN. This fourth path is discussed in the work *Pojednání o jevech povstávajících na množstvích* (Treatise on Phenomena that Arise on Multitudes)

In conclusion we remark that the fourth of the paths mentioned above provides a phenomenological support for both the previous paths, in particular for the postulate of weak actualizability of the domain of all natural numbers that has a key role in our conception of mathematics.

Petr Vopěnka
Faculty of Philosophy and Arts, University of West Bohemia in Pilsen
Sedláčkova 38, 306 14 Plzeň
Czech Republic

Arithmetic in Prague in the 1970–80s

PAVEL PUDLÁK

ABSTRACT. In mid 1970s Petr Hájek decided to quit set theory and work in formal arithmetic. An important role in founding a group of researchers working in arithmetic was played by the seminar that he started in the Mathematical Institute. The seminar was soon accompanied by regular workshops in Alšovice. In this short note I will describe the atmosphere of those days, people involved and the research areas in which Hájek's group worked. This period of his scientific career was crowned with the successful monograph *Metamathematics of First Order Arithmetic* [HP93], which became the basic text in this field.

I was lucky to decide to write my master's thesis under the supervision of Petr Hájek. It was in 1974–75 when he was still active in set theory, but he was already becoming more interested in other fields. He has always been interested in applications of logic. Already in the mid 1960s he and several of his coworkers developed a system for *"automated generation of hypothesis"* from empirical data [HHC66, HH78]. Scientific terminology is developing and old names sound rather awkward today. What they were working on would be nowadays a part of the field called the *theory of databases*. While studying applications of logic, he has always been also interested in developing the theory of the particular fields. Sets of empirical data can be viewed as finite mathematical structures. We can study them using classical logics, eg., first order logic, but many theorems fail in the finite domain. For example, we cannot recursively axiomatize the first order sentences valid in all finite structures. Petr proposed to call the theory of finite structures *"the observational predicate calculus"*. With the advent of computerization, databases became very important and so their theory. Again the terminology is different nowadays, this field is called *finite model theory*. Petr wrote several papers about it [Háj76b, Háj76a] and proposed this area of research for my thesis.

The 1970s were exciting years of computational complexity. Stephen Cook stated the famous *P* vs. *NP* problem in 1971 and Richard Karp proved the *NP*-completeness of many combinatorial problems soon after. We also got hooked and started to think about *P* vs. *NP*. This and Petr's suggestion to work on the observational predicate calculus led me to proving that in finite structures *NP* sets are exactly the sets definable by second order existential quantifiers. This is the well-known *Fagin's Theorem*. Fagin, as I soon learned, proved this theorem before me, in his PhD thesis in 1973 [Fag74].[1] I did not continue to work in finite model theory and Petr also quit soon. It was probably a missed opportunity to establish this field in our country, because finite model theory has become a major field in computer-science logic.

Other exciting things were happening in the 1970s. Jeff Paris came up with the idea of using concepts from the large cardinal theory in models of arithmetic. This led him to the discovery of concrete sentences independent of Peano Arithmetic. The most striking result, the independence of a version of the finite Ramsey theorem, appeared in his seminal paper written jointly with Leo Harrington [PH77]. As is well-known, Gödel was the first one to prove the independence of true sentences on arithmetic and set theory. Gödel's sentences are, however, of a logical nature, far from the typical theorems in mathematics. Thus the first independent "mathematical" sentences appeared only in the work of Cohen about the Continuum Hypothesis. But Cohen's forcing method can only prove independence for sentences expressing fact about infinite sets. It was the Paris-Harrington result that opened the possibility to prove independence for statements about numbers and finite sets.

So why not proving the independence of some open problems, say the independence of $P = NP$, on Peano Arithmetic? After all, it has the same logical complexity, namely Π_2, as the Paris-Harrington sentence. I was young, ambitious and naive; Petr was more modest. Anyway, we agreed that models of arithmetic is the right field to work in. It was not a new field—Andrzej Mostowski worked on models of arithmetic already in 1950s, but the Paris-Harrington result gave a strong impetus to this area and many logicians started to work in this field. The two most important centers in the late 1970s and early 1980s were in Manchester and Warsaw, but there were many people working in this field in other countries. Although called 'models of arithmetic' the subject was understood in the broader sense: the study of formal systems of arithmetic. Petr had already published papers about interpretability, thus one subject that we started with was interpretability in arithmetical theories ([Háj79b, Háj81b, Háj93]). This was closely connected with selfreference and conservativity ([Háj80, Háj84, Háj87]), other subject in which he was an expert. He also worked in another area of pure logic, recursion theory, on problems related to arithmetic and computational complexity ([Háj79a, HK87, HK89]). Moreover, he still devoted a large part of his time to application.

We also had luck with students. Jan Krajíček started graduate study under my supervision and Vitězslav Švejdar was Petr's student. Thus we had

---

[1] My results were not completely subsumed by Fagin's, because I also proved a hierarchy theorem with respect to the number of quantifiers.

achieved "the critical mass" to be able to start a seminar. The main part of the Mathematical Institute is in an old building. At that time it was in essentially the same state as it was in the 19th century—no central heating, only a coal stove. The seminar typically started with the ritual of making fire in the stove and making cheap coffee. Yet it was the most active period of the seminar. There were so many new results, ours and results that we wanted to learn, that the standard two hours did not suffice. The seminar had two parts with a break in between. When Petr started writing the book about arithmetic, one part was reserved for him to lecture about what he was currently writing up.

Another important activity were workshops. We joined forces with our set theorists and recursion theorists to organize workshops in a small village Alšovice in the mountains Jizerské hory. Academy of Sciences owned a small house there, which was used for such events. Again, everything was very primitive there, especially, the lecture room was very small. But it did not matter—the workshop became very popular and we had many important participants from abroad. Our workshops used to be complemented by conferences in Karpacz, which is in Poland on the other side of the common border. Logicians from East Germany also organized workshops. Although they were interested in a different part of model theory, they also invited logicians working on models of arithmetic.

These personal contacts helped us to get closer ties with colleagues from abroad during the time when traveling abroad was very restricted for us. In one case the tie became more than a friendship—Jeff Paris met his future wife Alena Vencovská. As a result he visited Prague regularly, from which we mathematically profited very much. He gave lectures and courses, mainly about his new results obtained with Alex Wilkie. They were developing the theory of $I\Delta_0$, which is Peano Arithmetic with induction restricted to formulas with only bounded quantifiers, and of a slightly stronger theory $I\Delta_0+\Omega_1$, where Ω is the axiom stating that $x^{\lceil \log x \rceil}$ is a total function (see [WP87], a sample from a number of papers that they wrote). We were very impressed by their results and tried to catch up with them.

The theory $I\Delta_0$ as an interesting object of study was proposed by Rohit Parikh in his seminal paper [Par71]. He observed that Δ_0 predicates are easily decidable, in fact, decidable in linear space, hence the formal system has finitistic nature. The appeal of this theory for Paris and Wilkie, as well as for most logicians who studied the theory later, was rather the connection with computational complexity. The idea, now widely accepted in proof complexity, is that by studying theories axiomatized by induction for formulas that define a complexity class will help us more fully understand the class itself and, perhaps, someday even to prove independence results concerning the class. The class of sets decidable in linear space is an important complexity class, but there are more important classes, in particular, *P* and *NP*. A theory for *P* was defined by Cook in 1975 [Coo75]. The paper was published in a proceedings of a computer science conference, so it went almost unnoticed in the logic community. Many people, including me, realized only later that it was *the* founding paper of the new field which is now called *proof complexity*. Cook's theory PV is a theory in the language of equations; an extension PV1 introduced in that paper also used propositional connectives. First order theories for *P* and *NP* and other classes of the Polynomial Time Hierarchy were introduced by Buss in his thesis in 1985 [Bus86] and he coined the term *Bounded Arithmetic* for the union of these theories.

Buss's Bounded Arithmetic is equi-interpretable with $I\Delta_0+\Omega_1$, but he came up with a different way of viewing it and with new problems. It became more apparent that the problem of showing that different fragments of Bounded Arithmetic have different logical strength is very much related to the problem of showing that levels of the Polynomial time Hierarchy are distinct complexity classes, although Paris and Wilkie had asked an equivalent problem about finite axiomatizability of $I\Delta_0+\Omega_1$ before. The connection of Bounded Arithmetic with computational complexity attracted more logicians, including the famous proof-theorist Gaisi Takeuti.

In the late 1980s our interest started to diverge. Petr was working on higher fragments of Peano Theory, such as $I\Sigma_n$ (Peano Arithmetic with induction restricted to Σ_n formula), while I and Jan Krajíček were mainly interested in Bounded Arithmetic and other problems in the emerging field of proof complexity. At some point Petr was approached by members of the Ω-group with the proposal that he should write a book about formal arithmetic. He accepted and invited me to join. He wrote the chapters about higher fragments of Peano Arithmetic; my contribution was smaller—I wrote a chapter about Bounded Arithmetic. This book was published in 1993 and was quite a success, thus a paperback version was printed in 1998. The book is still used as the standard reference in formal arithmetic.

After finishing the book Petr changed his field of interest again and became a prominent figure in the new filed—as he had been in the previous ones.

BIBLIOGRAPHY

[Bus86]   S. R. Buss. *Bounded Arithmetic*. Bibliopolis, Napoli, 1986.
[Coo75]   S. A. Cook. Feasibly constructive proofs and the propositional calculus. In *Proc. Seventh Annual ACM Symposium on Theory of Computation*, pages 83–97, Albuquerque, 1975. ACM.
[Fag74]   R. Fagin. Generalized first-order spectra and polynomial-time recognizable sets. In R. Karp, editor, *Complexity of Computation*, SIAM-AMS Proceedings 7, pages 27–41. AMS, 1974.
[Háj66]   P. Hájek. Generalized interpretability in terms of models. *Časopis pro pěstování matematiky*, 91(3):352–357, 1966.

[Háj75]   P. Hájek. On logics of discovery. In J. Bečvář, editor, *Mathematical Foundations of Computer Science*, volume 32 of *LNCS*, pages 30–45, Berlin, 1975. Springer.

[Háj76a]   P. Hájek. Observationsfunktorenkalküle und die Logik der Automatisierten Forschung. *Elektronische Informationsverarbeitung und Kybernetik*, 12(4–5):181–186, 1976.

[Háj76b]   P. Hájek. Some remarks on observational model-theoretic languages. In *Set Theory and Hierarchy Theory. A Memorial Tribute to Andrzej Mostowski*, volume 537 of *Lecture Notes in Mathematics*, pages 335–345, Berlin, 1976. Springer.

[Háj77]   P. Hájek. Arithmetical complexity of some problems in computer science. In *Mathematical Foundations of Computer Science*, volume 53 of *LNCS*, pages 282–297, Berlin, 1977. Springer.

[Háj79a]   P. Hájek. Arithmetical hierarchy and complexity of computation. *Theoretical Comp. Sci.*, 8(2):227–237, 1979.

[Háj79b]   P. Hájek. On partially conservative extensions of arithmetic. In *Logic Colloquium '78. Proceedings of the Annual European Summer Meeting of the ASL*, pages 225–234, Amsterdam, 1979. North-Holland.

[Háj80]   P. Hájek. A note on partially conservative-extensions of arithmetic. *J. Symb. Logic*, 45(2):391–391, 1980.

[Háj81a]   P. Hájek. Completion closed algebras and models of Peano arithmetic. *Comm. Math. Univ. Carolinae*, 22(3):585–594, 1981.

[Háj81b]   P. Hájek. On interpretability in theories containing arithmetic II. *Comm. Math. Univ. Carolinae*, 22(4):667–688, 1981.

[Háj84]   P. Hájek. On a new motion of partial conservativity. In *Computation and Proof Theory*, volume 1104 of *Lecture Notes in Mathematics*, pages 217–232, Berlin, 1984. Springer.

[Háj87]   P. Hájek. Partial conservativity revisited. *Comm. Math. Univ. Carolinae*, 28(4):679–690, 1987.

[Háj93]   P. Hájek. Interpretability and fragments of arithmetic. In P. Clote and J. Krajíček, editors, *Arithmetic, Proof Theory and Computational Complexity*, pages 185–196, Oxford, 1993. Clarendon Press.

[HH78]   P. Hájek and T. Havránek. *Mechanizing Hypothesis Formation—Mathematical Foundations of a General Theory*. Springer, Berlin, 1978.

[HHC66]   P. Hájek, I. Havel, and M. Chytil. The GUHA-method of automatic hypotheses determination. *Computing*, 1(4):293–308, 1966.

[HK87]   P. Hájek and A. Kučera. A contribution to recursion theory in fragments of arithmetic. *J. Symb. Logic*, 52(3):888–889, 1987.

[HK89]   P. Hájek and A. Kučera. On recursion theory in $I\Sigma_1$. *J. Symb. Logic*, 54(2):576–589, 1989.

[HP93]   P. Hájek and P. Pudlák. *Metamathematics of First-Order Arithmetic*. Perspectives in Mathematical Logic. Springer, Berlin, 1993.

[Par71]   R. Parikh. Existence and feasibility in arithmetic. *J. Symb. Logic*, 36:494–508, 1971.

[PH77]   J. B. Paris and L. Harrington. A mathematical incompleteness in Peano arithmetic. In *Handbook of Mathematical Logic*, chapter D.8, pages 1133–1142. North-Holland, 1977.

[WP87]   A. J. Wilkie and J. B. Paris. On the scheme of induction for bounded arithmetical formulas. *Ann. Pure Appl. Logic*, 35:261–302, 1987.

Pavel Pudlák
Mathematical Institute, Academy of Sciences of the Czech Republic
Žitná 25 115 67 Praha 1, Czech Republic
Email: pudlak@math.cas.cz

Intermediate logics and the de Jongh property

DICK DE JONGH, RINEKE VERBRUGGE, AND ALBERT VISSER

Dedicated to Petr Hájek,
on the occasion of his 70th birthday

ABSTRACT. We prove that all extensions of Heyting Arithmetic with a logic that has the finite frame property possess the de Jongh property.

1 Preface

The three authors of this paper have enjoyed Petr Hájek's acquaintance since the late eighties, when a lively community interested in the metamathematics of arithmetic shared ideas and traveled among the beautiful cities of Prague, Moscow, Amsterdam, Utrecht, Siena, Oxford and Manchester. At that time, Petr Hájek and Pavel Pudlák were writing their landmark book *Metamathematics of First-Order Arithmetic* [HP93], which Petr Hájek tried out on a small group of eager graduate students in Siena in the months of February and March 1989.

Since then, Petr Hájek has been a role model to us in many ways. First of all, we have always been impressed by Petr's meticulous and clear use of correct notation, witness all his different types of dots and corners, for example in the Tarskian *'snowing'-snowing* lemmas [HP93]. But also as a human being, Petr has been a role model by his example of *living in truth*, even in averse circumstances [Hav89]. The tragic story of the Logic Colloquium 1980, which was planned to be held in Prague and of which Petr Hájek was the driving force, springs to mind [vDLS82]. Finally, we were moved by Petr's open-mindedness when coming to terms with a situation that turned out to look disconcertingly unlike the 'standard model'.[1]

Therefore, in this paper, we would like to pay homage to Petr Hájek. Unfortunately, we cannot hope to emulate his correct use of dots and corners. Instead, we do our best to provide some pleasing non-standard models and non-classical arithmetics.

---

[1] Rineke would also like to take the opportunity to thank the whole Hájek family—Petr, Marie, Marie, Jonáš and Jonáš—for their great hospitality extended to her during her long-time research stays in Prague in 1991, 1992 and 1993. The typical Sunday lunches at the Hájeks' flat are especially fondly remembered; these occasions could move from *svíčková na smetaně* with *knedlíky*, wonderfully cooked by Marie, through lively conversations over coffee, to an impressive private cello concert by the then eight-year old *střední* Jonáš.

2   Introduction

Consider a theory T in constructive predicate logic. A propositional formula φ is T-valid iff, for all substitutions σ of formulas of the language of T for propositional variables, we have $T \vdash \sigma(\varphi)$. The set of T-valid formulas is the propositional logic of T. We will call this logic Λ_T. The de Jongh property for T is the statement that the propositional logic Λ_T of T is precisely Intuitionistic Propositional Logic, in other words, $\Lambda_T = \mathsf{IPC}$.

The original theorem of de Jongh was that Heyting's Arithmetic HA has the de Jongh property. In the past many generalizations have been proved; we will give an enumeration below. In these generalizations the objective was mostly to prove (or in rare cases disprove) the de Jongh property for extensions of HA with some properties like e.g. Church's Thesis. In this paper we will go in a different direction. The idea is to strengthen the logic from the intuitionistic logic to an intermediate one. Intermediate logics are all those logics of strength between intuitionistic and classical logics. Define the de Jongh property for T with respect to an intermediate logic Λ as the statement that $\Lambda_T = \Lambda$. Our conjecture is the following.

CONJECTURE 1 *Let* HA(Λ) *be the result of extending* HA *with* Λ *for all formulas. Then* $\Lambda_{\mathsf{HA}(\Lambda)} = \Lambda$.

In this paper we will prove this conjecture for logics Λ with the *Finite Frame Property*, namely,

> *there exists a class of finite frames* $\mathfrak{F}$ *such that* Λ *is precisely the logic valid on all models on frames in* $\mathfrak{F}$. *For this class* $\mathfrak{F}$ *we then have:*

$$\Lambda = \{\varphi \mid \mathfrak{F} \models \varphi\} = \{\varphi \mid \text{ for all } \mathcal{M} \text{ on } \mathfrak{F}, \mathcal{M} \models \varphi\}.$$

For intermediate logics, as for normal modal logics, the finite frame property Λ is in fact equivalent to the ostensibly weaker finite model property (FMP), which expresses that there is a class of finite models $\mathfrak{M}$ such that Λ is precisely the logic valid on all models in $\mathfrak{M}$. (See Section 4 for more background.)

For logics Λ with the finite frame property, indeed HA(Λ) has the de Jongh property with respect to Λ. As we will discuss in the conclusion (Section 7), there seems to be little chance of generalizing the methods of this paper to a more extensive class of intermediate logics.

In our proof we will only employ substitutions of Π_2^0-sentences. From this it follows by quite general reasoning, that, assuming that our class of frames is recursive, we have a uniform version of the de Jongh property. This means that, for Λ with the finite frame property, there is a single substitution $\sigma^\star$ such that HA(Λ) $\vdash \sigma^\star(\varphi)$ iff $\Lambda \vdash \varphi$. Or, in a different formulation, there is an embedding of the Lindenbaum Heyting algebra of Λ into the Lindenbaum Heyting algebra of HA(Λ).

3   A Brief History of de Jongh's Theorem

The following brief overview of the history of de Jongh's Theorem for propositional logic is adapted from [Vis99].

1969 Dick de Jongh proves in an unpublished paper his original theorem that $\Lambda_{HA} = IPC$. He uses substitutions of formulas of a complicated form, namely $\forall x(\alpha(x) \vee \neg\alpha(x))$ with α almost negative. As a reminder, a formula is almost negative if it does not contain $\vee$, and $\exists$ only in front of an equation between terms (see [Tro73]). In fact he proves a much stronger result, namely that the logic of relative interpretations in HA is Intuitionistic Predicate Logic (see [J70]). de Jongh's argument uses an ingenious combination of Kripke models and realizability.

1973 Harvey Friedman in his paper [Fri73] gives another proof of de Jongh's theorem for HA. He provides a *single* substitution σ mapping the propositional variables to Π_2^0-sentences such that $HA \vdash \sigma(\varphi) \Leftrightarrow IPC \vdash \varphi$. Thus, Friedman shows that IPC is *uniformly complete* for Π_2^0-substitutions in HA. Friedman employs slash-theoretic methods as introduced by Kleene [Kle62].

1973 Craig Smoryński strengthens and extends de Jongh's work in a number of respects in his very readable paper [Smo73]. To state his results we need a few definitions. We write $D(\Pi_1)$ for the set of disjunctions of Π_1^0-sentences, $Prop(\Sigma_1)$ for propositional combinations of Σ_1^0-sentences. Let us remind the reader of some relevant principles (see [Tro73] for extensive discussions).

MP is *Markov's Principle* MP:

$$\forall x(A \vee \neg A) \wedge \neg\neg\exists x A \rightarrow \exists x A.$$

RFN_{HA} is the formalized uniform reflection principle for HA, where $\forall y Ay$ is closed:

$$Proof_{HA}(x, \ulcorner A\overline{y}\urcorner) \rightarrow Ay$$

$TI(\prec)$ is the transfinite induction scheme for a primitive recursive well-ordering $\prec$:

$$\forall x((\forall y \prec x)Ay \rightarrow Ax) \rightarrow \forall y Ay.$$

We have de Jongh's Theorem for the theories T: HA, HA+RFN(HA), HA+TI($\prec$), and HA+MP. For the first three theories we can take the range of our substitutions either Σ_1^0 or $D(\Pi_1)$. For HA+MP we can take the range of our substitutions $Prop(\Sigma_1)$. Smoryński uses Kripke models in combination with the Gödel-Rosser-Mostowski-Kripke-Myhill theorem to prove his results.

1975 Daniel Leivant in his PhD Thesis [Lei79] shows that the predicate logic of interpretations of predicate logic in HA is precisely intuitionistic predicate logic. Leivant's method is proof-theoretical. In fact Leivant shows that one can use as interpretation a fixed sequence of Π_2^0-predicates. Leivant's results yield another proof of Friedman's results described above.

1976 de Jongh and Smoryński in their paper [JS76] show de Jongh's Theorem for HAS. They also show uniform completeness for HAS with respect to a substitution with range among the Π_2^0-sentences.

1981 Yu.V. Gavrilenko in [Gav81] proves de Jongh's Theorem for HA+ECT$_0$, i.e., the theory of provable realizability over HA. The principle ECT$_0$ is Extended Church's Thesis (see [Tro73]):

$$\forall x(A \to \exists y By) \to \exists u \forall x(A \to \exists v(Tuxv \land B(Uv))),$$

where A is almost negative and u does not occur free in A and B and v not in B. In the formula ECT$_0$, T stands for Kleene's T-predicate and U for the corresponding result-extracting function [Kle43]. Gavrilenko proves this result as a corollary of the similar result of Smoryński for HA.

1981 As a reminder, the principle DNS stands for 'double negation shift' (see [Tro73]):

$$\forall x \neg\neg Ax \to \neg\neg\forall x Ax.$$

Albert Visser in his Ph.D. thesis [Vis81] provides an alternative proof of de Jongh's theorem for HA, HA+DNS, and HA+ECT$_0$ for Σ_1^0-substitutions adapting the method of Solovay's proof of the arithmetical completeness of Löb's logic for substitutions in PA [Sol76]. In fact, his proof extends to the same theories extended with appropriate reflection principles or transfinite induction over primitive recursive well-orderings.

1985 In his [Vis85], Albert Visser provides an alternative proof of de Jongh's uniform completeness theorem employing a single Σ_1^0-substitution. The proof is verifiable in HA+con(HA). Here, con(HA) formalizes the consistency of Heyting Arithmetic. Note that de Jongh's theorem *implies* con(HA), so the result is, in a sense, optimal. Visser's proof uses the NNIL-algorithm, an algorithm that is used to characterize the admissible rules for Σ_1^0-substitutions. See also [Vis02].

1991 Jaap van Oosten in his paper [vO91b] provides a more perspicuous version of de Jongh's semantical proof of de Jongh's theorem for (non-relativized) interpretations of predicate logic. Van Oosten uses Beth models and realizability. See also [vO91a].

1996 Using the methods developed by Visser in [Vis82] and by de Jongh and Visser in [JV96], one can prove uniform completeness with respect to Σ_1^0-substitutions for $\mathsf{HA+ECT_0}$, $\mathsf{HA+ECT_0+RFN(HA+ECT_0)}$, and $\mathsf{HA+TI(\prec)+ECT_0}$.

It is well known that the de Jongh property does not hold for $\mathsf{HA + MP + ECT_0}$. Consider the formulas χ and ρ, which are defined as follows.

- $\chi := (\neg p \vee \neg q)$,

- $\rho := [(\neg\neg\chi \rightarrow \chi) \rightarrow (\chi \vee \neg\chi)] \rightarrow (\neg\neg\chi \vee \neg\chi)$

Clearly, ρ is not provable in IPC. We use $\mathbf{r}$ for *Kleene realizability*. In his classical paper [Ros53], G.F. Rose showed that: $\exists e \ \forall\sigma \in \mathsf{sub_{HA}} \ \mathbb{N} \models e \, \mathbf{r} \, \sigma(\rho)$. Here $\mathsf{sub_{HA}}$ is the set of substitutions from the propositional variables to sentences of the arithmetical language. Thus, Rose refuted a conjecture of Kleene that a propositional formula is IPC-provable if all its arithmetical instances are (truly and classically) realizable. Note the remarkable fact that one and the same realizer realizes all instances! Inspecting the proof, one sees that only a small part of classical logic is involved in the verification of realizability: Markov's Principle. See David McCarthy's paper [McC91] for a detailed analysis. Thus we obtain:

$$\exists e \ \forall\sigma \in \mathsf{sub_{HA}} \ \mathsf{HA + MP} \vdash e \, \mathbf{r} \, \sigma(\rho).$$

Hence, a fortiori, $\rho \in \Lambda_{\mathsf{HA+MP+ECT_0}}$. See [Pli09] for an interesting recent survey of propositional realizability logic.

4   Intermediate Logics and the Finite Model Property

Intermediate logics (also called superintuitionistic logics) are the logics between IPC and CPC, classical propositional logic, i.e. the sets of of formulas closed under IPC-deduction and uniform substitution. Most anything one needs to know about these logics can be found in [CZ97]. We will enumerate the best known of these logics and mention a few basic facts, mainly concerning the finite model property. The well-known theorem for normal modal logics that the finite frame property and the finite model property for a model logic L coincide (see e.g. [BdRV02]), applies to intermediate logics as well [CZ97]. This means that every sentence not provable from the logic cannot only be refuted in a finite model of the logic, it can also be refuted in a frame validating the logic. So, it is appropriate to use the terminology that a logic has the finite model property (FMP) if there is a class of finite frames for which it is complete. As in the case of modal logics, not all intermediate logics are complete with respect to a class of frames, and not all those which are complete for such a class have the FMP. But all except one of the well-known intermediate logics which we will now discuss are known to be complete with respect to a class of finite frames.

LC, **Dummett's logic**, axiomatized by, e.g. $(\varphi \to \psi) \vee (\psi \to \varphi)$, is complete with respect to the finite linear frames.

KC, **Jankov's logic** (also called the logic of weak decidability, whereas real intuitionists might prefer, if anything, the logic of testability), axiomatized by $\neg \varphi \vee \neg\neg \varphi$, is complete with respect to the finite frames with a unique endpoint.

KP, **the logic of Kreisel and Putnam**, axiomatized by

$$(\neg\varphi \to \psi \vee \chi) \to (\neg\varphi \to \psi) \vee (\neg\varphi \to \chi).$$

It is complete with respect to the finite partial orderings satisfying the property: For each u and each set X of points succeeding u, there exists a v accessible from u such that all points of X are accessible from v and every endpoint above v is also above some point from X. KP was the first logic shown to have the *disjunction property*: If $\vdash \varphi \vee \psi$, then $\vdash \varphi$ or $\vdash \psi$, where $\vdash$ stands for provability in the given logic (here KP) [KP57].

T_n, the Gabbay-deJongh logics, which are complete with respect to the finite trees which have splittings of exactly n, i.e. each node has exactly n immediate successors. T_n is axiomatized by

$$\bigwedge_{k \leq n+1} ((\varphi_k \to \bigvee_{j \neq k} \varphi_j) \to \bigvee_{j \neq k} \varphi_j) \to \bigvee_{k \leq n+1} \varphi_k$$

T_1 coincides with LC. T_n-frames for $n > 1$ do not have a first order definition. Moreover, T_n for $n > 1$ is not canonical ([CZ97]). However, it is first-order definable on finite frames: it is characterized by the condition that every point has at most n immediate successors. Finally, T_n for $n > 1$ has the disjunction property [JG74].

BD_n, **the depth n logics** are complete with respect to the finite partial orderings (but also to the trees, or to the splitting trees) of depth n. BD_1 is classical logic axiomatized by Peirce's Law, $((\varphi \to \psi) \to \varphi) \to \varphi$. An axiomatization of BD_n for $n > 1$ is obtained by iteratedly substituting Peirce's Law into itself, e.g. BD_2 is axiomatized by $((\varphi \to (((\psi \to \chi) \to \psi) \to \psi)) \to \varphi) \to \varphi$.

Sc, **Scott's logic**, axiomatized by $((\neg\neg\varphi \to \varphi) \to (\varphi \vee \neg\varphi)) \to \neg\varphi \vee \neg\neg\varphi$. It is complete with respect to the finite partial orderings satisfying the property:

In each generated subframe all the endpoints are connected by an R, R^{-1}-chain containing only points of depth 0 or 1 (or equivalently the finite frames which do not have a p-morphism onto the asymmetric four-element tree of depth 3). This is not equivalent to a first-order definition. Like T_n and KP, Scott's logic also has the disjunction property [KP57, CZ97].

ML, Medvedev's logic of finite problems [Med66] has the finite frame property (it is complete with respect to the finite boolean algebras without their top element), but it has no known axiomatization and is not known to be decidable. It contains Scott's logic and KP and is contained in Jankov's logic. It is not finitely axiomatizable and its infinite axiom systems will need an infinity of atoms [MSS79]. ML coincides with the set of formulas all whose essentially negative substitution instances are provable in KP, or stated in another form, it is the logic of the valid schemata obtained by adding $\neg\neg p \to p$ for atoms only to KP (due to [Lev69], see also [Cia09]).

The **Propositional Logic of Realizability**. Plisko [Pli09] discusses several variants out of which we choose the logic of the effectively realizable formulas, although for most purposes it makes little difference. The logic has no known axiomatization, it is not known whether it has the finite model property or is decidable. It does have the disjunction property. From our earlier remarks it follows that it contains Scott's logic with $\neg\psi \vee \neg\theta$ substituted for the sole variable in Scott's axiom. Just like ML, the Propositional Logic of Realizability is contained in KC, but it has been shown to neither contain nor be contained in ML.

All logics using only $\to, \wedge, \neg$ in their axioms have the finite model property, but none but IPC itself have the disjunction property (see [CZ97]). The same is true for logics with axioms using only *NNIL*-formulas (No Nesting of Implications on the Left); in fact, these logics are the same as the ones using only $\to$ (see [Yan08]).

5   Arithmetics and Finite Frames

In this section, we prove our main result. We follow Smoryński's classical paper [Smo73]. In that paper Smoryński gives two proofs. We use the second more complicated one, which is more flexible for applications. But first we will sketch in a few lines the first proof which gives us a partial result, and indicate why it does not generalize. The idea (of both proofs) is, given a propositional Kripke-model, to construct a Kripke model for HA on the same frame, and then to exhibit arithmetical sentences with exactly the same forcing behavior as the propositional variables on the propositional model.

First, one should realize that any node k in a model $\mathcal{K}$ is attached to a classical structure $\mathcal{M}_k$ (not necessarily a model of HA [Bus93]). The simple proof now starts with the basic lemma:

LEMMA 2 *Given a (non-rooted) model of* HA *(or equivalently a set of models for* HA*), one can obtain a new model of* HA *by equipping this non-rooted model with a root to which the standard model of the natural numbers is attached.*

Proof Assume $\mathcal{K}$, $\mathcal{K}'$, r as in the statement of the lemma. Assume also that $r \Vdash A(0) \wedge \forall x\, (A(x) \to A(x+1))$. It is sufficient to show that $r \Vdash \forall x\, A(x)$, since the $\mathcal{K}$-part of the model already satisfies (internal) induction. We have that $r \Vdash A(0) \wedge (A(n) \to A(n+1))$ for each natural number n. This enables us, by applying external induction, to conclude that the new root forces $A(n)$ for each n, from which $r \Vdash \forall x\, A(x)$ follows. $\qquad\square$

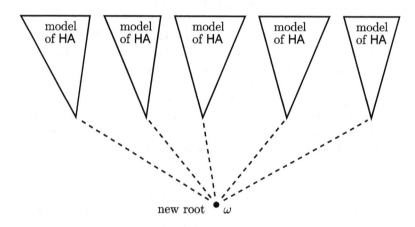

Figure 1. Scheme of Lemma 2: A new model of HA is constructed by equipping a set of models of HA with a new root and attaching the standard model ω to it.

The simple proof now continues using the fact that IPC is complete with respect to the finite trees that are everywhere splitting (each node has at least two immediate successors). One builds a propositional Kripke-model on the same tree by starting with non-standard models for PA (and thus for HA!) on the endpoints, and attaching the standard model to each of the other nodes. By the basic Lemma 2, using induction on the depth of the nodes, this is a model of HA.

The non-standard models at the endpoints are chosen to be incomparable in the following sense: each of them has its own Σ^0_1-sentence, which, in PA, contradicts all the others. Then, because one has a finite everywhere splitting tree, each node is completely determined by the endpoints which are accessible from it. This is sufficient to create for each node a sentence true on exactly that node (and its successors). Finally, disjunctions of those sentences will characterize arbitrary upward closed subsets of the model, in particular those determined by the valuations of the propositional variables in the propositional model.

To those intermediate logics that are complete with respect to a set of finite splitting trees, this method will immediately apply. The point is that we do not only get a model of HA but, because of the form of the frame, also of Λ. This is immediately clear for $\varphi(\alpha_1, \ldots, \alpha_n)$ if $\varphi(p_1, \ldots, p_n)$ is a member of Λ and $\alpha_1, \ldots, \alpha_n$ are arithmetic sentences with constants for the elements of the appropriate domains of the models. If in such a case $\alpha_1, \ldots \alpha_n$ have free variables, then the truth value of the universally quantified version of $\varphi(\alpha_1, \ldots, \alpha_n)$ depends only on the value of sentences arising by substitution in $\varphi(p_1, \ldots, p_n)$, and, since we declared these to be true already, the universally quantified forms have to be true as well. In particular, one obtains in this manner the de Jongh property for HA with respect to the logics T_n (for $n > 1$) and BD_n, but not for KC, KP, and ML. Also not for $\mathsf{T}_1 = \mathsf{LC}$, because linear frames do not split everywhere.

Especially, the need to characterize each node exactly by the endpoints accessible from it seems essential in this setup of the proof. So, for the general case we will need a different way to proceed. One basic idea remains the same. We want to fit a root to a number of models. Smoryński's second method does this in a much more sophisticated manner than the first, and with this method one can under certain strict conditions have a root with a non-standard model attached to it. To set the stage, we need a sequence of lemmas. We start with a few well-known lemmas on Kripke models of HA. For definitions of forcing in such Kripke models, see [Smo73] and [TvD88].

LEMMA 3 *Suppose $\mathcal{K}$ is a Kripke model of* HA. *Let $A\vec{x}$ be a Σ_1^0-formula. Then, for any node k and any $\vec{d}$ in D_k, we have $k \Vdash A\vec{d}$ iff $\mathcal{M}_k \models A\vec{d}$.*

Proof We first prove, by induction on the complexity of A, that, for any Δ_0^0-formula $A\vec{x}$, any node k, and any $\vec{d}$ in D_k, we have $k \Vdash A\vec{d}$ iff $\mathcal{M}_k \models A\vec{d}$. The proof uses the fact that HA proves the decidability of Δ_1^0-formulas. We treat the case of restricted universal quantification.

Suppose we have the desired property for $By\vec{x}$ in Δ_0^0. If $k \Vdash \forall y < eBy\vec{d}$, then evidently, for all $d' <_{\mathcal{M}_k} e$, $k \Vdash Bd'\vec{d}$. Hence, by the Induction Hypothesis, for all $d' <_{\mathcal{M}_k} e$, $\mathcal{M}_k \models Bd'\vec{d}$. Hence, $\mathcal{M}_k \models \forall y < e\, By\vec{d}$.

Conversely, suppose (a) $\mathcal{M}_k \models \forall y < e\, By\vec{d}$. In order to derive a contradiction, suppose (b) $k \Vdash \exists y < e \neg By\vec{d}$. It follows by the Induction Hypothesis and writing out the negation and restricted existential quantification, that $\mathcal{M}_k \models \exists y < e\, \neg By\vec{d}$, contradicting (a). So, canceling (b), we conclude $k \nVdash \exists y < x\, \neg By\vec{d}$. Then, by decidability, $k \Vdash \neg\exists y < e\, \neg By\vec{d}$. So, by intuitionistic predicate logic, $k \Vdash \forall y < e\, \neg\neg By\vec{d}$. By decidability again, $k \Vdash \forall y < e\, By\vec{d}$.

The step from Δ_0^0 to Σ_1^0 is easy. $\hfill\square$

The following lemma is due to Wim Ruitenburg.

LEMMA 4 *Let $\mathcal{K}$ be a model of* HA. *Let $A\vec{x}$ be a Π_2^0-formula. Then, for any node k and for any $\vec{d}$ in D_k, we have: $k \Vdash A\vec{d}$ iff, for all $k' \geq k$, $\mathcal{M}_{k'} \models A\vec{d}$.*

Proof Suppose $A\vec{x}$ is $\forall\vec{y}\, S\vec{y}\vec{x}$, where S is Σ_1^0. We have:

$$
\begin{aligned}
k \Vdash \forall\vec{y}\, S\vec{y}\vec{d} \quad &\Leftrightarrow\quad \forall k' \geq k\, \forall\vec{e} \in D_{k'}\ k' \Vdash S\vec{e}\vec{d} \\
&\Leftrightarrow\quad \forall k' \geq k\, \forall\vec{e} \in D_{k'}\ \mathcal{M}_{k'} \models S\vec{e}\vec{d} \\
&\Leftrightarrow\quad \forall k' \geq k\ \mathcal{M}_{k'} \models \forall\vec{y}\, S\vec{y}\vec{d}
\end{aligned}
$$

This gives us the desired property. $\hfill\square$

Next, we need some basic insights from the theory of interpretations.

LEMMA 5 *Let T be any RE theory. Then* $\mathsf{Q} + \mathsf{con}(T)$ *interprets T.*

Here Q is Robinson's Arithmetic, a very weak arithmetical theory introduced by Tarski, Mostowski and Robinson in their book [TMR53].

This is a general version of a fundamental theorem due to Hilbert & Bernays, which was worked out by Wang. The proof was simplified by Feferman. The methods that lead to the general version are due to among others Solovay, Friedman, and Pudlák. See [Vis09], for context, explanation and references. A very crude explanation of the result is as follows. The Model Existence Lemma tells us that, if a theory T is consistent, there is a model $\mathcal{N}$ of T. A moment's reflection shows that the construction of this model, via the Henkin construction, is syntactical in nature. By fine-tuning the argument, we can see that inside a theory that contains the consistency statement of T we can construct an interpretation of T. Here the heuristic is that interpretations are something like uniformly defined internal models.

Thus, inside any model $\mathcal{M}$ of $\mathsf{Q} + \mathsf{con}(T)$ we can construct a model $\mathcal{M}'$ of T. Moreover, if $\mathcal{M}$ is a model of PA, and if T is an arithmetical theory extending Q, then we can easily prove, using an argument that is essentially due to Dedekind, that $\mathcal{M}'$ is an *end-extension* of $\mathcal{M}$ modulo a definable embedding. The idea is that inside $\mathcal{M}$ we can define the obvious function that maps the numbers of the theory into the numbers of the internal model $\mathcal{M}'$ by recursion. We can prove by induction that the numbers of $\mathcal{M}$ are mapped to an initial segment of $\mathcal{M}'$. We formulate this in a lemma. For precise discussions of this and similar lemmas, see [HP93].

LEMMA 6 *Suppose T is an arithmetical theory extending* Q. *Then every model $\mathcal{M}$ of* PA $+$ con(T) *contains an internally definable end-extension (modulo a definable embedding) satisfying T.*

Finally, we formulate Smoryński's fundamental lemma. Consider a non-rooted Kripke model $\mathcal{K}$ of HA and a model $\mathcal{M}$ of PA. Suppose $\mathcal{K}$ is definable in $\mathcal{M}$. This means that the set of nodes K of $\mathcal{K}$ is definable, that the ordering on the nodes is definable, that there is a formula $\delta(k, d)$ giving the domain elements of the node k, and that there are arithmetical formulas $A_P(k, \vec{d})$ representing $k \Vdash P\vec{d}$. Moreover, we ask that $\mathcal{M}$ verifies basic properties like the fact that the relation between the nodes is a partial ordering. We have:

LEMMA 7 *There is a definable embedding of $\mathcal{M}$ as an initial segment of the model associated with each node of $\mathcal{K}$. This embedding is unique in the strong sense that there can be only one definable embedding that commutes with 0 and successor. Thus we can form a rooted model $\mathcal{K}^+$ by adding $\mathcal{M}$ as a new root to $\mathcal{K}$. We have: $\mathcal{M}^+$ is a model of* HA.

The proof of this result is in Smoryński's classical paper [Smo73]. The idea is a simple extension of the idea of the basic Lemma 2. Given a nonrooted model of HA, we cannot just add an arbitrary non-standard model as the root, since such a model does have induction for its own language but not for the enriched language in which we can also talk about $\mathcal{K}$. However, if $\mathcal{K}$ is internally definable this problem disappears and we can add the non-standard model as a root. One could say that the non-standard model internally thinks it is standard and that as soon as it can talk about $\mathcal{K}$ the earlier external argument can be internalized.

We are now ready for the main construction, which diverges from Smoryński's proof [Smo73].

THEOREM 8 *Suppose that Λ is the logic of a given class of finite frames $\mathfrak{F}$. Let* HA(Λ) *be the result of extending* HA *with Λ for all formulas. Then*

$$\Lambda_{\mathsf{HA}(\Lambda)} = \Lambda,$$

i.e. the propositional logic of HA(Λ) *is Λ. Our result works both when we consider logics of substitutions of formulas and when we consider the logics of substitutions of sentences.*

Proof Let Λ be the logic of a given class of finite frames $\mathfrak{F}$. Suppose $\Lambda \nvdash \varphi$. There is a finite model $\mathcal{K}$ with frame in $\mathfrak{F}$, such that $\mathcal{K} \nVdash \varphi$. Let the ordering of our model be $\preceq$. We can arrange it so that the nodes of $\mathcal{K}$ are $0, \ldots, n-1$ and, if $i \preceq j$, then $i \leq j$. We define:

- $\mathsf{incon}^0(\mathsf{PA}) := \bot$, $\mathsf{incon}^{k+1}(\mathsf{PA}) := \mathsf{prov}_{\mathsf{PA}}\ (\mathsf{incon}^k(\mathsf{PA}))$,

- $C_k := \neg\mathsf{incon}^k(\mathsf{PA}) \wedge \mathsf{incon}^{k+1}(\mathsf{PA})$,

- $\overline{C}_0(x) := \mathsf{incon}(\mathsf{PA})$,

- $\overline{C}_{k+1}(x) := \neg\,\mathsf{proof}_{\mathsf{PA}}(x, \mathsf{incon}^k(\mathsf{PA})) \wedge \mathsf{incon}^{k+2}(\mathsf{PA})$.

We note that the C_i are mutually exclusive. We define T_i, for $0 \leq i < n$, by $T_i := \mathsf{PA} + C_{n-i}$. By Löb's Theorem and Σ^0_1-soundness, each of the T_i is consistent. Moreover, again by Löb's Theorem, for $i < n-1$, we have $T_i \vdash \mathsf{con}(T_{i+1})$. Let $\mathcal{N}_0$ be any model of T_0. By Lemma 5, we can construct an internal model, say $\mathcal{N}_1$ of T_1, in $\mathcal{N}_0$. The model $\mathcal{N}_1$ will be an end-extension of $\mathcal{N}_0$. We iterate this construction, obtaining an internal model $\mathcal{N}_{i+1}$ of T_{i+1} in $\mathcal{N}_i$. Since 'being an iternal model of' and 'being an end-extension of' are transitive relations, we find that if $i \leq j \leq n-1$, then $\mathcal{N}_j$ is an internally defined end-extension of $\mathcal{N}_i$.

We now construct a Kripke model $\mathcal{S}$ by making $\mathcal{M}_k$ the model associated to the node k and taking over the ordering of $\mathcal{K}$. By Smoryński's Lemma 7, we find that $\mathcal{S}$ is a model of HA. Since the frame of our model is in $\mathfrak{F}$, we see that $\mathcal{S}$ satisfies $\mathsf{HA}(\Lambda)$.

Consider any proposition p in $\mathcal{K}$. We want to find an arithmetical formula $\sigma(p)$ such that, for all $k < n$, $\mathcal{S}, k \Vdash \sigma(p)$ iff $\mathcal{K}, k \Vdash p$. A first choice to consider would be $\sigma(p) := \bigvee_{\mathcal{K}, k \Vdash p} C_k$. However, it is easy to see this won't wash. Consider, for example, $i \prec j$ and suppose $\mathcal{S}, i \Vdash \sigma(p)$. Then $\mathcal{S}, i \Vdash C_m$, for some m with $\mathcal{K}, m \Vdash p$. It follows that $\mathcal{N}_i \models C_m$, and hence that $m = i$. By persistence, we find that $\mathcal{S}, j \Vdash C_i$, but then $\mathcal{N}_j \models C_i$. A contradiction.

So this does not work. In the light of Ruitenburg's Lemma 4, we can diagnose the problem as follows: the sentence $\sigma(p)$ is not *constructively* equivalent to a Π^0_2-sentence or some other kind of sentence that gives us a transfer from classical satisfaction and constructive forcing. However, our $\sigma(p)$ *is classically* equivalent to a Π^0_2-sentence, and therein lies the simple solution. Suppose $\{k \mid \mathcal{K}, k \Vdash p\} = \{k_0, \ldots, k_{m-1}\}$. We define:

- $\sigma^\star(p) := \forall x_0, \ldots x_{m-1}\ (\overline{C}_{k_0}(x_0) \vee \ldots \vee \overline{C}_{k_{m-1}}(x_{m-1}))$.

It is easy to see that *constructively* $\sigma^\star(p)$ is equivalent to a Π^0_2-sentence and that *classically* $\sigma^\star(p)$ is equivalent to $\sigma(p)$.

By Ruitenburg's Lemma 4, we find:

$$\mathcal{S}, i \Vdash \sigma^\star(p) \quad \Leftrightarrow \quad \mathcal{K}, i \Vdash p.$$

Hence, by induction, we find that, for any ψ:

$$\mathcal{S}, i \Vdash \sigma^\star(\psi) \;\;\Leftrightarrow\;\; \mathcal{K}, i \Vdash \psi.$$

So, $\mathcal{S}, 0 \nVdash \sigma^\star(\varphi)$. $\qquad\qquad\qquad\qquad\qquad\qquad\qquad\qquad\qquad\square$

REMARK 9 *A remarkable aspect of the above proof is that, where Smoryński used Rosser-style self-reference, it only employs Gödelean self-reference— since it uses Löb's Theorem.*

REMARK 10 *We note that we can employ a fixed series of models in our main argument that can be chosen independently of the finite frames. This uses an argument originally due to Harvey Friedman [Fri78]. We define by Carnap's version of the Fixed Point Lemma a formula $A(x)$ such that:*

$$\mathsf{PA} \vdash \forall x \, (A(x) \leftrightarrow (\mathsf{con}(\mathsf{PA} + A(x+1)) \vee \exists y {\leq} x \, \mathsf{proof}_{\mathsf{PA}}(y, \neg A(0)))).$$

Let $T_i := \mathsf{PA} + A(i) + \mathsf{incon}(\mathsf{PA} + A(i))$. Let $C_i := \mathsf{con}(T_{i+1}) \wedge \mathsf{incon}(T_i)$. We easily see that the T_i are consistent and that T_i is equivalent to $\mathsf{PA} + C_i$. Moreover, the C_i are incompatible. We note that, by Löb's Theorem, $T_i \vdash \mathsf{con}(T_{i+1})$. We now start with a model $\mathcal{N}_0$ of T_0 in which we construct an internally definable end-extension $\mathcal{N}_1$ that is a model of T_1, etc. Thus we obtain a sequence of models $\mathcal{N}_i$, where each next model is a definable end-extension of the previous one. Moreover, $\mathcal{N}_i \models C_i$.

Can we extend the above proof to other arithmetical theories besides HA as basis? We note that we can extend it to weaker theories like i-$I\Sigma_1$, the intuitionistic version of $I\Sigma_1$, since the construction of an end-extension from a consistency statement already works in the classical theory $I\Sigma_1$ [HP93]. Secondly, the proof also works for stronger theories like HA plus uniform reflection, or HA extended with an RE set of negations that are Π_2^0-conservative over PA.

We have one somewhat more interesting extension of our result, for intermediate logics with what we will call the endpoint replacement property:

DEFINITION 11 *Consider an intermediate logic Λ.*

1. *Let $\mathfrak{F}$ be a class of frames for the logic. We say that Λ has the endpoint replacement property with respect to $\mathfrak{F}$ if the following holds. Take any $\mathcal{F}$ in $\mathfrak{F}$, a set of endpoints $\{u_i \,|\, i \in I\}$ of $\mathcal{F}$, and a set $\{\mathcal{M}_i \,|\, i \in I\}$ of (possibly infinite) models of Λ with roots $\{r_i \,|\, i \in I\}$. Let $\mathcal{G}$ be the frame that is the result of replacing the endpoints $\{u_i \,|\, i \in I\}$ by the frames of $\{\mathcal{M}_i \,|\, i \in I\}$. Then any model $\mathcal{M}$ on $\mathcal{G}$ which is such that its submodels generated by $\{r_i \,|\, i \in I\}$ are exactly the $\{\mathcal{M}_i \,|\, i \in I\}$ will validate Λ.*

2. Λ *has the* endpoint replacement property *if* Λ *has the endpoint replacement property with respect to its class of frames.*

Of the logics we mentioned, the following have the endpoint replacement property: IPC, KC, LC, and T_n. The logics KP, Sc, ML, and BD_n do not have the property with respect to a class of frames for which they are complete. On the negative side: for KP and Sc this is obvious from the fact that the simple 3-element fork which is a frame for both is no longer a frame for either when attached to an endpoint of itself. On the positive side, we will just show the most complicated case: the T_n.

PROPOSITION 12 T_n *has the endpoint replacement property.*

Proof Let $\mathcal{F}$ be a frame for T_n, let $\{u_i \mid i \in I\}$ be a set of endpoints of $\mathcal{F}$, and let $\{\mathcal{M}_i \mid i \in I\}$ be a set of models of T_n with roots $\{r_i \mid i \in I\}$. Let $\mathcal{M}$ be a model on the frame $\mathcal{G}$ that is the result of replacing the endpoints $\{u_i \mid i \in I\}$ by the frames of $\{\mathcal{M}_i \mid i \in I\}$ such that the models generated by the r_i are exactly the $\mathcal{M}_i$. We have to show that the assumption that $T_n(\varphi_1, \ldots, \varphi_{n+1})$ is falsified in $\mathcal{M}$ will lead to a contradiction. We can then assume without loss of generality that the root satisfies the antecedent of T_n for $\varphi_1, \ldots, \varphi_{n+1}$ and falsifies each of $\varphi_1, \ldots, \varphi_{n+1}$. Let us define N as the set of nodes in $\mathcal{M}$ that falsify each of $\varphi_1, \ldots, \varphi_{n+1}$. This will be a downward closed set containing the root, falsifying $T_n(\varphi_1, \ldots, \varphi_{n+1})$ everywhere. Since for each $k \leq n+1$, $\bigvee_{j \neq k} \varphi_j$ is falsified at each $v \in N$, also $\varphi_k \to \bigvee_{j \neq k} \varphi_j$ is false at v for some $k \leq n+1$. This means that for each $v \in N$, there is a $k \leq n+1$ and a w_k with $v \leq w_k$ such that w_k makes φ_k true but no φ_j for $j \neq k$. Note that such nodes w_k and w_j for different k and j are always incomparable. Note also that $x \Vdash T_n$ iff $x \in N$.

Let us first consider the case that one of the r_i is in N. That is impossible, since then $r_i \not\Vdash T_n(\varphi_1, \ldots, \varphi_{n+1})$ whereas the model $\mathcal{M}_i$ is supposed to be a model for T_n. The second possibility is that none of the r_i are in N. Note that then, if $\mathcal{M}_i$ contains any w_k-node as described above, its root r_i will have to be a w_k-node, and there cannot be w_j-nodes for any $j \neq k$ in $\mathcal{M}_i$. If we now restore the endpoints $\{u_i \mid i \in I\}$ in $\mathcal{M}$ in place of the models $\{\mathcal{M}_i \mid i \in I\}$ and define

$$V'(p_m) = \{w \in \mathcal{F} - \{u_i \mid i \in I\} \mid w \Vdash \varphi_m\} \cup \{u_i \mid r_i \Vdash \varphi_m\},$$

then it is obvious that for this new valuation, for each $v \in N$ (and in particular for the root of $\mathcal{F}$), $v \not\Vdash' T_n(p_1, \ldots, p_{n+1})$, since one of the $p_k \to \bigvee_{j \neq k} p_j$ will be falsified at the appropriate w_k. Thus, the resulting model will falsify $T_n(p_1, \ldots, p_{n+1})$. This is again a contradiction, because $\mathcal{F}$ is a frame for T_n. $\qquad\square$

THEOREM 13 *Suppose that Λ has both the finite model property and the endpoint replacement property with respect to a class $\mathfrak{F}$ of finite frames. Suppose that U is a consistent extension of $\mathsf{HA}(\Lambda)$ with an RE set of negations N. Then U has the de Jongh property for Λ.*

Before we turn to the proof of our theorem, we remind the reader of some basic facts concerning Rosser sentences. Let V be any consistent RE extension of i-EA, the intuitionistic version of Elementary Arithmetic. Using Craig's trick, we can arrange that the axiom set of V is given by an elementary predicate α. This predicate can be taken to be a $\Delta_1^0(i\text{-EA})$-formula. This means that both αx and $\neg\, \alpha x$ are i-EA-provably equivalent to a Σ_1^0-formula and that αx is i-EA-provably decidable. Using this predicate, we can find a reasonable arithmetization proof_V, the proof-predicate for V, such that $\mathsf{proof}_V(x,y)$ is $\Delta_1^0(i\text{-EA})$. Let $\mathsf{prov}_V(x)$ be $\exists y\, \mathsf{proof}_V(y,x)$.

Consider formulas A and B. Suppose $A = \exists x\, A_0 x$ and $B = \exists y\, B_0 y$. We define:

- $A \leq B :\leftrightarrow \exists x\, (A_0 x \wedge \forall y{<}x\, \neg B_0 y)$,

- $A < B :\leftrightarrow \exists x\, (A_0 x \wedge \forall y{\leq}x\, \neg B_0 y)$,

- $(A < B)^\perp :\leftrightarrow B \leq A$,

- $(A \leq B)^\perp :\leftrightarrow B < A$.

Using Gödel's Fixed Point Lemma, we can find a Rosser sentence R for V such that i-EA $\vdash R \leftrightarrow \mathsf{prov}_V(\neg R) \leq \mathsf{prov}_V(R)$. We clearly have i-EA $\vdash \neg\,(R \wedge R^\perp)$. Using Rosser's argument, we can show that both $V + R$ and $V + \neg R$ are consistent. We note that both R and $R^\perp$ can be rewritten over i-EA plus Σ_1^0-collection to the strict Σ_1^0 form. Since we work in extensions of HA, we may assume that R and $R^\perp$ are Σ_1^0.

We turn to the proof of Theorem 13.

Proof Let Λ and U be as required for Theorem 13. Consider any φ such that $\Lambda \nvdash \varphi$. Let $\mathcal{K}$ be a counter-model in $\mathfrak{F}$. Suppose the non-endpoints of $\mathcal{K}$ are $a_0, \ldots, a_{n-1}$ and that we have chosen our enumeration in such a way that $i \leq j$ implies $a_i \preceq a_j$. We repeat our construction from the proof of Theorem 8 of a sequence of models with $\mathcal{N}_i$ for $0, \ldots, n-1$, with the modification that we take as our base theory $W := \mathsf{PA} + \mathsf{con}(U)$. So $\mathcal{N}_{n-1}$ will satisfy $\mathsf{con}(U)$ and $\mathsf{incon}(\mathsf{PA} + \mathsf{con}(U))$.

We enumerate the endpoints as $b_0, \ldots, b_{r-1}$. We construct a sequence of Σ_1^0-sentences $S_0, \ldots, S_{r-1}$, with the following properties, for $i < j \leq r - 1$:

$$W \vdash \mathsf{con}(U + S_i) \text{ and } U \vdash \neg\,(S_i \wedge S_j).$$

The easiest way to construct such a sequence is by induction on r. If $r = 1$, we take $S_0 := \top$. Suppose we have constructed $S'_0, \ldots S'_{s-1}$ with the desired properties. Let R and $R^\perp$ be the pair consisting of the Σ_1^0 Rosser sentence for $U + S'_{s-1}$ and its Σ_1^0 opposite. We take $S_i := S'_i$, for $i < r - 1$, $S_{r-1} := (S'_{r-1} \wedge R)$, and $S_r := (S'_{r-1} \wedge R^\perp)$. We now construct in $\mathcal{N}_{n-1}$ inner Kripke models of $U + S_i$. The construction is analogous to the classical case. It is essential that the base theory in which we do the construction is classical! See Appendix A of [Vis98] for a description of how to do an internal Kripke model construction. This gives us models $\mathcal{M}_j$ of $U + S_j$, internally definable in each $\mathcal{N}_i$.

We build a Kripke model for arithmetic by associating $\mathcal{N}_i$ with a_i and $\mathcal{M}_j$ with b_j. This will be a model of $\mathsf{HA} + N$ by the internal version of Smoryński's Lemma 7, noting that the additional negations in N are always downwards preserved. It will be a model of Λ by the endpoint replacement property. We associate to each propositional atom p in $\mathcal{K}$ the sentence:

$$E_p := \bigvee_{a_i \Vdash p} (C_i \wedge \mathsf{con}(U)) \vee \bigvee_{b_j \Vdash p} S_j.$$

As before in the proof of Theorem 8, we transform E_p to a classically equivalent Π_2^0-sentence, say E_p^+. Let σ be the substitution $p \mapsto E_p^+$. By Ruitenburg's Lemma 4, we find that our new model does not force $\sigma(\varphi)$. $\square$

So, for example, for $\Lambda = \mathsf{IPC}, \mathsf{LC}, \mathsf{KC}, \mathsf{T_n}$, we do have the de Jongh property for $\mathsf{HA}(\Lambda)$ plus the negation of the sentence expressing the Primitive Recursive Markov's Principle, and for $\mathsf{HA}(\Lambda)$ plus the negation of the sentence expressing the decidability of the Halting Problem. (See [Tro73] for definitions of these properties.)

6   A Uniformization Result

In this section we show how to prove uniformization using recursion-theoretic arguments. This style of result is originally due to Franco Montagna [Mon79] and, independently, Albert Visser [Vis81].

THEOREM 14 *Consider any theory T with elementary axiom set and any interpretation N, such that N interprets i-EA, the constructive version of EA (which is also known as $I\Delta_0 + \exp$), in T. Let A_i be an elementary decidable sequence of sentences in the language of T.*

Suppose, for every i, $T \nvdash A_i$. Then there is a Σ_1^0-formula $R(x)$ such that:

a. i-$\mathsf{EA} \vdash (R(x) \wedge R(y)) \to x = y$,

b. for any i, $T \nvdash R^N(i) \to A_i$.

Proof Let $R(x)$ be such that:

$$i\text{-EA} \vdash R(x) \quad \leftrightarrow \quad \exists p \, (\text{proof}_T(p, R^N(x) \to A_x) \wedge$$
$$\forall q \leq p \, \forall y \leq q \, \neg \text{proof}_T(q, R^N(y) \to A_y))).$$

We work with the reasonable assumption that the code of the numeral of y is larger than y, that the code of a formula in which a numeral occurs is larger than the code of that numeral, and that the code of a proof is larger than the code of any formula occurring in it. Also we assume that proofs have single conclusions. (See [HP93] for discussions of such reasonable assumptions.)

We verify the uniqueness clause (a). Reason in i-EA. Suppose $R(x)$ and $R(y)$ with witnesses p and q, respectively. In case $p = q$, we are done. Suppose that $q < p$. From $R(x)$, we have: $\forall y' \leq q \, \neg \text{proof}_T(q, R^N(y') \to A_{y'})$. Because, by our assumptions on coding, we must have $y < q$, we find $\neg \text{proof}_T(q, R^N(y) \to A_y)$. This contradicts $R(y)$.

We verify (b). Suppose, in order to derive a contradiction, that

$$T \vdash R^N(i) \to A_i.$$

Let π be a proof that witnesses this fact. Via a finite search among proofs with Gödel number smaller or equal to the Gödel number of π, we may obtain the T-proof π^* with smallest Gödel number with a conclusion of the form $R^N(j) \to A_j$. Say we have: $\pi^* : T \vdash R^N(i^*) \to A_{i^*}$. By the definition of R, it follows that $R(i^*)$ is true. So, by Σ_1^0-completeness, $T \vdash R^N(i^*)$. Hence, $T \vdash A_{i^*}$, contradicting the assumption. We may conclude that $T_i \nvdash R(i) \to A_i$. $\square$

Let Λ be any intermediate logic with the finite frame property. Suppose the set of finite frames $\mathfrak{F}$ corresponding to Λ is decidable. We can easily find an elementary enumeration $(\varphi_k)_{k \in \omega}$ of all Λ-underivable formulas, such that we can find a counter-model $\mathcal{K}_k$ with frame in $\mathfrak{F}$ of φ_k in an elementary way from k. In our main theorem, we have shown that we can transform $\mathcal{K}_k$ in an elementary way into a Π_2^0-substitution σ_k such that $\text{HA}(\Lambda) \nvdash \sigma_k(\varphi_k)$.

THEOREM 15 *Under the circumstances described above, we have the uniform de Jongh property for* $\text{HA}(\Lambda)$ *with respect to* Λ.

Proof Let $A_i := \sigma_i(\varphi_i)$. Applying Theorem 14, with $\text{HA}(\Lambda)$ in the role of T, we find $R(x)$ with the promised properties. Take

$$\tau(p) := \exists y \, (R(y) \wedge \text{true}_{\Pi_2^0}(\ulcorner \sigma_y(p) \urcorner)).$$

Here $\text{true}_{\Pi_2^0}$ is the Π_2^0-truth predicate and $\ulcorner \sigma_y(p) \urcorner$ is a arithmetization that sends y to the Gödel number of $\sigma_y(p)$.

Suppose $\text{HA}(\Lambda) \vdash \tau(\varphi_j)$. It follows that $\text{HA}(\Lambda) \vdash R(j) \to \sigma_j(\varphi_j)$, contradicting the assumption. $\square$

7   Cautionary Afterword

We have proved our results with pleasant, pedestrian methods. The reason for this luxurious situation is the simplicity and power of the two basic ideas on which everything rests: Smoryński's idea of the preservation of HA under adding ω as a root and Smoryński's extension of the idea to non-standard models, provided that the rest of the Kripke model is definable in the new root. These ideas allow us to construct Kripke models step-by-step with a lot of control over their properties.

However, as Johan Cruijff said *elk voordeel heb z'n nadeel*, i.e., *every advantage comes with a disadvantage*. The disadvantage is that the essential dependence on these ideas makes the results not easily extendible.

A first possible direction of extension would be *more frame classes*. We note that there is not much hope to extend our results to frame classes with infinite models —at least not with the methods at hand. The reason is that the preservation of HA under adding a suitable root is essentially a tool for constructing finite objects. Of course, we may extend the construction to *non-standardly finite models*, but these are rather artificial from the point of view of frame classes of intermediate logics. The good news here is that almost all known natural logics do have the finite frame property. A possible exception is the Propositional Logic of Realizability. In this connection it is worth pointing out that de Jongh's original proof could handle infinite frames but that it did need trees.

The second possible extension would be *more arithmetical theories*. There seems to be no hope to extend our results to extensions of HA that lack the Smoryński property. For example, we do know that $HA + ECT_0$ has the de Jongh property of IPC, but the extension of $HA + ECT_0$ with intermediate logics is a complete *terra incognita*. In this case it could be interesting to look for *counterexamples* to the extension of our results.

BIBLIOGRAPHY

[BdRV02]   P. Blackburn, M. de Rijke, and Y. Venema. *Modal Logic*, volume 53 of *Cambridge Tracts in Theoretical Computer Science*. Cambridge University Press, 2002.

[Bus93]   S. R. Buss. Intuitionistic validity in T-normal Kripke structures. *Ann. Pure Appl. Logic*, 59:159–173, 1993.

[Cia09]   I. Ciardelli. Inquisitive semantics and intermediate logics. Master's thesis, ILLC, University of Amsterdam, Amsterdam, 2009. MOL-2009-11.

[CZ97]   A. Chagrov and M. Zakharyashev. *Modal Logic*. Oxford Logic Guides. Oxford University Press, Oxford, 1997.

[J70]   D. H. J. de Jongh. The maximality of the intuitionistic predicate calculus with respect to Heyting's Arithmetic. *J. Symb. Logic*, 36:606, 1970.

[JG74]   D. H. J. de Jongh and D. Gabbay. A sequence of decidable finitely axiomatizable intermediate logics with the disjunction property. *J. Symb. Logic*, 39:67–78, 1974.

[JS76]   D. H. J. de Jongh and C. Smoryński. Kripke models and the intuitionistic
           theory of species. *Ann. Math. Logic*, 9:157–186, 1976.
[JV96]   D. H. J. de Jongh and A. Visser. Embeddings of Heyting algebras. In
           [HHST96], pages 187–213, 1996.
[Fri73]   H. Friedman. Some applications of Kleene's methods for intuitionistic systems.
           In A. R. D. Mathias and H. Rogers, editors, *Cambridge Summer School in
           Mathematical Logic*, pages 113–170. Springer, 1973.
[Fri78]   H. Friedman. Classically and intuitionistically provably recursive functions.
           In G. Müller and D. Scott, editors, *Higher Set Theory*, pages 21–27. Springer,
           1978.
[Gav81]   Y. V. Gavrilenko. Recursive realizability from the inuitionistic point of view.
           Soviet Mathematics Doklady, 23:9–14, 1981.
[Hav89]   V. Havel. *Living in Truth*. Faber and Faber, London, 1989.
[HHST96]   W. Hodges, M. Hyland, C. Steinhorn, and J. Truss, editors. *Logic: From
           Foundations to Applications*, Oxford, 1996. Clarendon Press.
[HP93]   P. Hájek and P. Pudlák. *Metamathematics of First-Order Arithmetic*. Per-
           spectives in Mathematical Logic. Springer, Berlin, 1993.
[Kle43]   S. C. Kleene. Recursive predicates and quantifiers. *Trans. Amer. Math. Soc.*,
           53:41–73, 1943.
[Kle62]   S. C. Kleene. Disjunction and existence under implication in elementary intu-
           itionistic formalisms. *J. Symb. Logic*, 27:11–18, 1962.
[KP57]   G. Kreisel and H. Putnam. Eine Unableitbarkeitsbeweismethode für den Intu-
           itionistischen Aussagenkalkül. *Zeitschr. Math. Logik Grundlagen Math.*, 3:74–
           78, 1957.
[Lei79]   D. Leivant. *Absoluteness in Intuitionistic Logic*, volume 73. Mathematical
           Centre Tract, Amsterdam, 1979. The Thesis was originally published in 1975.
[Lev69]   V. A. Levin. Some syntactic theorems on the calculus of finite problems of Yu.
           T. Medvedev. *Soviet Mathematics Doklady*, 10:288–290, 1969.
[McC91]   D. C. McCarty. Incompleteness in intuitionistic metamathematics. *Notre
           Dame J. Formal Logic*, 32:323–358, 1991.
[Med66]   Y. Medvedev. Interpretation of logical formulas by means of finite problems.
           Soviet Mathematics Doklady, 7:180–183, 1966.
[Mon79]   F. Montagna. On the diagonalizable algebra of Peano Arithmetic. *Bollettino
           Unione Matematica Italiana*, 16–B:795–813, 1979.
[MSS79]   L. L. Maksimova, V. Shehtman, and D. Skvortsov. The impossibility of a finite
           axiomatization of Medvedev's logic of finitary problems. *Soviet Mathematics
           Doklady*, 20:394–398, 1979.
[Pli09]   V. E. Plisko. A survey of propositional realizability logic. *Bull. Symb. Logic*,
           15:1–42, 2009.
[Ros53]   G. F. Rose. Propositional calculus and realizability. *Trans. Amer. Math. Soc.*,
           75:1–19, 1953.
[Smo73]   C. Smoryński. Applications of Kripke models. In A. S. Troelstra, editor,
           Metamathematical Investigations of Intuitionistic Arithmetic and Analysis,
           volume 344 of *Lecture Notes in Mathematics*, pages 324–391. Springer, Berlin,
           1973.
[Sol76]   R. Solovay. Provability interpretations of modal logic. *Israel J. Math.*, 25:287–
           304, 1976.
[TMR53]   A. Tarski, A. Mostowski, and R. M. Robinson. *Undecidable Theories*. North-
           Holland, Amsterdam, 1953.
[Tro73]   A. S. Troelstra, editor. *Metamathematical Investigations of Intuitionistic
           Arithmetic and Analysis*. Number 344 in Lecture Notes in Mathematics.
           Springer, Berlin, 1973.
[TvD88]   A. S. Troelstra and D. van Dalen. *Constructivism in Mathematics, vol 2*,
           volume 123 of *Studies in Logic and the Foundations of Mathematics*. North-
           Holland, Amsterdam, 1988.

[vDLS82]   D. van Dalen, D. Lascar, and T. Smiley, editors. *Logic Colloquium '80: Papers Intended for the European Summer Meeting of the Association for Symbolic Logic*, volume 108 of *Studies in Logic and the Foundations of Mathematics*, Amsterdam, 1982. North-Holland.

[Vis81]   A. Visser. *Aspects of Diagonalization and Provability*. PhD thesis, Department of Philosophy, Utrecht University, Utrecht, 1981.

[Vis82]   A. Visser. On the completeness principle. *Ann. Math. Logic*, 22:263–295, 1982.

[Vis85]   A. Visser. Evaluation, provably deductive equivalence in Heyting's Arithmetic of substitution instances of propositional formulas. Logic Group Preprint Series 4, Department of Philosophy, Utrecht University, Utrecht, 1985.

[Vis98]   A. Visser. Interpretations over Heyting's Arithmetic. In E. Orłowska, editor, *Logic at Work*, Stud. Fuzziness Soft Comput., pages 255–284. Physica Verlag, Heidelberg, New York, 1998.

[Vis99]   A. Visser. Rules and arithmetics. *Notre Dame J. Formal Logic*, 40(1):116–140, 1999.

[Vis02]   A. Visser. Substitutions of Σ_1^0-sentences: Explorations between intuitionistic propositional logic and intuitionistic arithmetic. *Ann. Pure Appl. Logic*, 114:227–271, 2002.

[Vis09]   A. Visser. Can we make the Second Incompleteness Theorem coordinate free? *J. Logic Comput.*, 2009. DOI 10.1093/logcom/exp048.

[vO91a]   J. van Oosten. *Exercises in Realizability*. PhD thesis, Department of Mathematics and Computer Science, University of Amsterdam, Amsterdam, 1991.

[vO91b]   J. van Oosten. A semantical proof of de Jongh's theorem. *Arch. Math. Logic*, 31:105–114, 1991.

[Yan08]   F. Yang. Intuitionistic subframe formulas, NNIL-formulas and n-universal models. Master's thesis, ILLC, University of Amsterdam, Amsterdam, 2008. MOL-2008-12.

Dick de Jongh
ILLC, University of Amsterdam,
Science Park 904, 1098 XH Amsterdam,
The Netherlands
Email: d.h.j.dejongh@uva.nl

Rineke Verbrugge
Artificial Intelligence, University of Groningen,
PO Box 407, 9700 AK Groningen,
The Netherlands
Email: rineke@ai.rug.nl

Albert Visser
Department of Philosophy, Utrecht University,
Heidelberglaan 8, 3584 CS Utrecht,
The Netherlands
Email: Albert.Visser@phil.uu.nl

My cooperation with Petr Hájek

ZOFIA ADAMOWICZ

I first met Petr Hájek in 1973, when Andrzej Mostowski organized a Logical Semester in the Banach Center in Warsaw. We both worked then in set theory and I was still an undergraduate student then. After having graduated I worked on degrees of constructibility in set theory. This was the subject of my PhD, written under the supervision of Andrzej Mostowski. I completed the PhD in 1975 with a thesis on degrees of constructibility of reals. Later I continued working on degrees of constructibility. I constructed a model of the set theory ZFC, where degrees of constructibility formed an arbitrary constructible well founded upper semi-lattice with countable initial segments ([Ada76]). Next time, as I remember, I met Petr at a conference in Bierutowice in 1976, again in Poland. We talked about my result. Petr told me that he had a model of ZFC with a decreasing chain of degrees of constructibility ([BH78], [Háj77b], [Háj77a]). I was very much interested since this was complementary to my result.

In the eighties we both, me and Petr, moved from set theory to the study of models of arithmetic. We were both inspired by the famous results of Paris, Kirby and Harrington ([PH77]) on combinatorial sentences independent of Peano Arithmetic. I met Petr at many conferences and we exchanged news about models of arithmetic. We had a lot of inspiring discussions. I visited Prague several times in the eighties and also Petr visited Poland. We both were interested a lot in, so called, fragments of first order Peano Arithmetic. An extensive study of such fragments can be found in the Hájek-Pudlák work [HP93]. The results Petr got on the fragments were later put by him into the book. The fragments consist mainly in partialization of either the induction scheme or the collection scheme. By "partialization" I mean here restriction to the class of Σ_n arithmetical formulas, for a fixed n—see [HP93] for all the related notions. In particular, we were both interested in collection schemes $B\Sigma_n$:

$$\forall a\big(\forall x{\leq}a\exists y\psi(x,a,y) \Leftrightarrow \exists w\forall x{\leq}a\exists y{\leq}w\psi(x,a,y)\big),$$

where ψ runs over Σ_n formulas.

In particular the above scheme can be treated as a translation principle of translating formulas of the form $\forall x{\leq}a\exists y\psi(x,a,y)$, i.e. having a bounded quantifier in front of a strict Σ_n formula, to strict Σ_n formulas.

In the late eighties Petr asked me an interesting question on the scheme $B\Sigma_n$. Then the ways of our research met again very closely. Namely Petr suggested to formulate a translation principle very generally, without indicating the way how the translation is done. He suggested the following scheme $\hat{B}\Sigma_n$.

Let $\sigma_n(x, y)$ denote the usual Σ_n formula which, provably in $I\Delta_0 + \text{Exp}$, is universal for Σ_n formulas with one variable. Then $\hat{B}\Sigma_n$ is the following scheme:

$$\exists s \forall a \big(\forall x \leq a \exists y \psi(x, a, y) \Leftrightarrow \sigma_n(s, a) \big),$$

where ψ runs over Σ_n formulas.

The intuitive meaning of this scheme is that every formula of the form $\forall x \leq a \exists y \psi(x, a, y)$, (i.e. having a bounded quantifier in front of a strict Σ_n formula) is equivalent to some strict Σ_n formula, i.e. to an instance of the universal Σ_n formula corresponding to some parameter s (depending on the formula ψ, i.e. on the formula which we are translating).

Petr asked me what was the strength of his scheme and in particular whether it was weaker than the usual scheme $B\Sigma_n$. For instance, the question was whether $\hat{B}\Sigma_n$ was provable in a weak Σ_n induction scheme—induction for parameter-free Σ_n formulas—in which $B\Sigma_n$ itself is not provable.

In a joint work with Roman Kossak I succeeded to answer Petr's question. We showed ([AK88]) that $\hat{B}\Sigma_n$ is equivalent to the usual $B\Sigma_n$ (over $I\Delta_0 + \text{Exp}$), whence, in particular, it is not provable in induction for parameter free Σ_n formulas.

Later, in the nineties, I was very impressed by the book of Hájek and Pudlák "Metamathematics of First Order Arithmetic", ([HP93]), an extensive and excellent monograph on models of arithmetic. Since then, I have been using this book permanently. Our scientific ways split when Petr left arithmetic for fuzzy logic, and I stayed in arithmetic. Anyway we remained friends.

BIBLIOGRAPHY

[Ada76]   Z. Adamowicz. Constructible semi-lattices of degrees of constructibility. In A. Lachlan, M. Srebrny, and A. Zarach, editors, *Set Theory and Hierarchy Theory V*, pages 1–45, Bierutowice, 1976. Springer, 1977.

[AK88]   Z. Adamowicz and R. Kossak. A note on $B\Sigma_n$ and an intermediate induction scheme. *Zeitschr. Math. Logik Grundlagen Math.*, 34:261–264, 1988.

[BH78]   B. Balcar and P. Hájek. On sequences of degrees of constructibility (solution of Friedman's Problem 75). *Zeitschr. Math. Logik Grundlagen Math.*, 24(4):291–296, 1978.

[Háj77a] P. Hájek. Another sequence of degrees of constructability. *Notices Amer. Math. Soc.*, 24(3), 1977.

[Háj77b] P. Hájek. Some results on degrees of constructibility. In G. H. Müller and D. S. Scott, editors, *Higher Set Theory*, page 285, Oberwolfach, April 13–23 1977. Springer, 1978.

[HP93]   P. Hájek and P. Pudlák. *Metamathematics of First-Order Arithmetic.* Perspectives in Mathematical Logic. Springer, Berlin, 1993.
[PH77]   J. B. Paris and L. Harrington. A mathematical incompleteness in Peano arithmetic. In *Handbook of Mathematical Logic*, chapter D.8, pages 1133–1142. North-Holland, 1977.

Zofia Adamowicz
Institute of Mathematics of the Polish Academy of Sciences,
Śniadeckich 8
00-956 Warszawa, Polska (Poland)
Email: zosiaa@impan.pl

Recollections of a non-contradictory logician

MATTHIAS BAAZ

Meeting Petr Hájek

In summer 1986 I tried to cross the English Channel using the Calais Dover ferry to attend the Logic Colloquium in Hull. The ferry got stranded and I reached the University of Hull at midnight with enormous delay. I was unable to locate my dormitory on the campus without a map. Fortunately, an anti-rape unit, complete with German shepherd dogs, stopped me from roaming and escorted me in the right direction. For me, however, the most memorable event at this Logic Colloquium—besides the presentation of Bounded Arithmetic by Sam Buss in full Sioux war paint—was my meeting Petr Hájek for the first time. During our first conversation he invited me to visit him and his students Pavel Pudlák and Jan Krajíček in Prague, students to become prominent themselves in the world of logic. Jan Krajíček and I worked at that time on problems related to Kreisel's conjecture. Petr Hájek told me that he was unable to provide any kind of support for my visit, except for his unlimited time. Only many years later I became fully aware of the generosity of this offer. Our conversation was in German: Petr Hájek has always spoken grammatically perfect German with a Czech accent, so familiar to Viennese ears. He impressed me then (as he impresses me still) by his benevolent seriousness and honesty, disappointingly rare in science (and in the world in general).

A City of Wisdom

Travelling for the first time to Prague I had to cross the border near the small Austrian town of Gmünd. One coach was detached from the local train and pushed over the border. The coach had exactly two passengers besides me. The meeting with Petr Hájek was of clandestine nature as I was not supposed to enter the Mathematical Institute of the Academy of Sciences, but that was to change soon during my later visits. Prague at this time gave the impression of loneliness, its inner town being somehow derelict. Even the street lights were switched off some hours after midnight. I remember that

I once crossed Charles Bridge on a Sunday afternoon in the early spring being completely alone on the bridge. I was fascinated by the total silence in the heart of a legendary European town. For me however Prague became a city of wisdom, because never in my life had I been part of a perpetual seminar without fixed schedules. The wide scope of this seminar comprising all fields of logic can be guessed already from the mere fact, that Pavel Pudlák and Jan Krajíček turned to proof theory (and to complexity theory), considering the strong model theoretic background of Petr Hájek himself. I started to visit Prague more and more often.

A World Divided

When I started to visit Prague Petr Hájek was not in full grace of the authorities because he had transported the printout of a banal computer program over the iron border (and probably for other reasons as well). He has always been politically aware: he told me that even his small son Jonáš was able to distinguish East and West since his LEGO stones from the German Democratic Republic looked exactly as his LEGO stones from Denmark, but, unlike those from Denmark, did not really fit together. He told me several times about Józef Maria Bocheński who had expressed to him once the hope of being able to cross again unmolested the borders within Europe, like in the time of the Austrian-Hungarian Monarchy. I know that Petr Hájek shared Bocheński's dream. Petr Hájek was however always convinced that the only way to deal with such circumstances is not to try to change others (or the world) but to remain righteous oneself. (As we know, the world did change after all.)

A Proof of the Existence of God

Petr Hájek is a member of the not so numerous Evangelical Church of Czech Brethren having deep roots in the Chech reformation aka the Hussite Church and the Moravian Church. He is theologically well educated. (By the way, it is a pleasure to follow Petr Hájek's organ play in his church.) Therefore, he was very thrilled by the possibility to obtain a copy of Gödel's ontological proof of the existence of God, which I brought to the CSSR long before it had been published in the west [Daw97, Göd95]. He became however very critical to this proof, which he considers as a mildly formal variant of Saint Anselm's proof (via *summum bonum*), cf. [Háj02]. This critical disposition is maybe the best example of Petr Hájek's conviction that formalism should not be misused to pretend things to be simple which are not simple. (Unfortunately, stock market analysts do not seem to share this conviction.) Petr Hájek's colleague and former advisor Petr Vopěnka announced an official lecture on Gödel's proof of the existence of God, which was duly cancelled by the authorities.

The lecture took place nevertheless, but in private flats of dissidents. (A personal remark: I have been told that Georg Kreisel commented on Gödel, with respect to the draft of the ontological proof, that all assumptions do not only hold for G denoting God but also for G denoting Gödel and that therefore the existence of Gödel is established beyond reasonable doubts. This is another way of criticism: we should recall Socrates and Diogenes.)

A Shift of Emphasis

The division of the world shifted and the neighbours of Austria got the opportunity to accustom themselves to the stupidities of *our* political systems just after getting rid of the stupidities of their own. At this time, a shift of emphasis occurred in Petr Hájek's scientific work: he became interested in the mathematics of fuzzy logic. (This shift of emphasis was accompanied by increased travel and lecture activities on European level). The change considered as fundamental by many of his colleagues is for me no change of principle: It constitutes a scientist's travel to a scientifically virgin jungle together with the farmer's pleasures to cultivate new land by old means. It is the *application* of abilities and knowledge gained in Petr Hájek's scientific life. After years of hard work he could justly and proudly summarize [Háj98, p. 3]: *The reader coming from a logician's perspective will (hopefully) be pleased by seeing that the logical systems of many-valued logic relevant to fuzzy logic have a depth and beauty comparable with that of classical logic.*

The Beauty of Mathematics

Let me describe Petr Hájek's approach to mathematics using two small mistaken suggestions. (Such problematic suggestions remain extremely rare in his scientific life.) He originally suggested to omit axiom

$$(\phi \,\&\, (\phi \to \psi)) \to (\psi \,\&\, (\psi \to \phi))$$

from his axiom system for Basic Logic, hoping that it was derivable. He was wrong in the sense that this axiom turned out to be independent of the others, but right in the sense that the system without this axiom (a fragment of MTL) is beautiful in all respects of deduction, cf. [BCM04]. (It is not a small achievement to be right even when one is wrong!)

Secondly, he offered a bottle of Cognac to anybody who could show that the axiom

$$\forall x(\phi(x) \vee \psi) \to (\forall x \phi(x) \vee \psi)$$

was dependent, as this axiom is the obstacle to a beautiful axiomatic compositionality of predicate fuzzy logics. (As to be expected, the award condition did not gain me a bottle of Cognac when I hinted at the independence of the axiom using Kripke semantics: this remark did not satisfy the desired standard of beauty).

A Proof of the Existence of God, continued

Let me conclude my short recollections by a speculative paragraph. The main difference of Petr Hájek from many of us is that we live several lives as single persons. I am for example sometimes mathematician, manager, explorer etc; and when I am one of those, I cease to be all the others. Petr Hájek lives only one life. In this life, mathematics is for him the hard work doing bits and pieces and solving problems ridiculous in themselves. All this is done for the large picture to emerge, but there is no guarantee and maybe not even a hope that this happens. But the large picture becomes visible and one might think that this is a proof of the existence of God for Petr Hájek.

BIBLIOGRAPHY

[BCM04] M. Baaz, A. Ciabattoni, and F. Montagna. Analytic calculi for monoidal t-norm based logic. *Fund. Informaticae*, 59(4):315–332, 2004.

[Daw97] J. W. Dawson Jr. *Logical Dilemmas: The Life and Work of Kurt Gödel*. AK Peters Ltd., 1997. This book contains the German original of the ontological proof.

[Göd95] K. Gödel. Ontological proof. In *Collected Works: Unpublished Essays & Lectures III*, pages 403–404. Oxford University Press, 1995.

[Háj98] P. Hájek. *Metamathematics of Fuzzy Logic*, volume 4 of *Trends in Logic*. Kluwer, Dordrecht, 1998.

[Háj02] P. Hájek. A new small emendation of Gödel's ontological proof. *Studia Logica*, 71(2):149–164, 2002.

Matthias Baaz
Institute for Discrete Mathematics and Geometry (E104)
Vienna University of Technology
Wiedner Hauptstrasse 8-10 A-1040 Wien, Austria
Email: baaz@logic.at

LC'80, or else how we wanted to organise a conference

Kamila Bendová[1]

The Logic Colloquium is an annual conference in logic, organised under the auspices of the Association for Symbolic Logic (ASL). It takes place in the summer in some European country, whose logicians consider this a great honour. The conference attracts researches from many European and non-European countries.

Czechoslovakia, Prague had been chosen to become the host in 1980; Petr Hájek was elected the chairman of the Organizing Committee. The hosting institutions were the Institute of Mathematics of the Czechoslovak Academy of Sciences (CSAS), where I worked at that time, and the Faculty of Mathematics and Physics of the Charles University. The topics were similar as in analogous earlier conferences: set theory, automata theory, recursive functions. Petr Vopěnka was nominated the chair of the Programme Committee.

At the time of the conference preparations, the Czech dissident movement suffered a heavy blow, as five important members of the Committee for the Defence of the Unjustly Persecuted (CDUP) were imprisoned: Václav Havel (the future president), Jiří Dienstbier (the future minister of foreign affairs), Petr Uhl, Otta Bednářová and Václav Benda. Their "crime" consisted in standing up for citizens who had been unjustly prosecuted, even by the laws then in force (the existing legal order included the Law 120 on human and citizen's rights): they were sending letters to courts, disseminating information about unlawful convictions through typewritten leaflets, and informing the foreign public. They announced this intention already in April 1979 and after a year of the above mentioned activities were sentenced to several years' imprisonment. A significant portion of European artists and university instructors stood up in their support; e.g., representatives of the French cultural and scientific elite, director Patrick Chereau and mathematician Jean Dieudonné, tried to attend the trial. (They were not allowed into the courtroom and were expelled from Czechoslovakia on the night after the trial.)

Next heavy blow to the dissident movement came with the wave of forced emigration which followed the trial of 1980.

Sometime in the spring of 1980, Zdena Tominová, a spokeswoman of Charter 77, approached me with a question of her friend and English university professor Kathleen Wilkes, whether the western participants should not boycott the LC conference in protest to the imprisonment of the members of the CDUP.

---

[1] Translated by Anna Lauschmannová

International conferences in Eastern countries were of two kinds: some were supported by the state, i.e., the Communist Party, which meant that only those Czech and Slovak specialists who satisfied the criteria of the Communist Party were allowed to take part; these conferences were exploited to propagate the socialist ideology and depict Czechoslovakia as a normal democratic country. I did not object against boycotting such large propagandist conferences.

On the other hand, there were also other conferences, tolerated rather than supported by the Communist regime, in particular in the less controlled area of mathematics and in computer science in general, that allowed even the scientists who would not be allowed to travel abroad (not even to other communist countries) to meet their colleagues from the Western Europe and United States.

Through Zdena Tominová, I tried to explain to professor Wilkes that LC falls into the second category and that it would indeed be very meritable to take part. Then I went to Petr Hájek, and delightedly announced I have "saved" the LC.

Every conference has its organising committee, but in the communist countries this committee was subject to the so-called nomenclatural consultation, i.e., a responsible party institution checked up on the members with respect to ideology and security. In Prague, almost all logicians from the Institute of Mathematics and from the Faculty of Mathematics and Physics were in the committee, as they organised not only the scientific and social programme, but also put together the address list, sent out the invitations (put them into envelopes and glued the addresses on them), and organised accommodation and boarding. All personal correspondence with the participants was carried on by Petr Hájek, who was thus responsible for the conference.

In communist countries, the Organising committee had almost no discretionary trust. The conference was organised by the Institute of Mathematics, a division of the Czechoslovak Academy of Sciences, and thus the President of the Academy was fully in charge of the course of events of the conference—and he was, as we shall see, following the orders of the State Security.

Surprisingly, no problem occurred with me as the wife of Václav Benda (no-one seems to have noticed, possibly since I was the only woman in the committee, and so they regarded me as a secretary), but with the so-called crossed-out and excluded communists.

After the Russian occupation in 1968, a massive purge occurred in the Czechoslovak Communist Party. "Crossed-out" were the people who, after the occupation, left the party of their own accord, whilst "exluded" were the people who were expelled by a party meeting as a punishment for "inapt" behaviour during the Prague spring or after the occupation. Quite often, the crossed-out members were additionally excluded.

The committee was subjected to this control, some members voluntarily resigned their membership to avoid complications, and the rest passed. Another difficulty was averted.

Czechoslovakia imposed visa requirements on other countries, and again this facilitated opportunities for screening and granting permission to enter the country only to some. Most importantly, it was a big struggle to obtain visas for the citizens of Israel (without whom a logical conference is inconceivable), and in particular for Mrs. Hanna Gaifman, a lady of Czech origins, but married to professor Haim Gaifman. (The State Security forewarned in the entry of December 18th, 1979: *Information concerning the mathematical conference with Israeli attendance... Under the heading of a mathematical conference, a visit of foreign spies is likely.*) Even this hurdle was finally overcome and ten Israeli logicians accompanied by seven family members obtained the entry visa for Prague.

Tomáš Jech, a well known Czech mathematician and logician, had emigrated to the United States after the Russian occupation. He was keeping track of the news from his motherland and cooperating with Amnesty International, an organisation which demands justice for the unjustly prosecuted everywhere in the world. Moreover, he believed that the solidarity among professionals should stretch beyond the boundaries of a given state, and if an imprisoned mathematician could not be helped by his colleagues from the communist countries, then his colleagues from the democratic countries might be more successful. He sent the following letter to these colleagues:

Dear ...   February 21, 1980

I understand that you have been invited by the organizing cormmittee to deliver a lecture at the European Summer Meeting of the Association for Symbolic Logic to be held in Prague, Czechoslovakia in August 1980. I hope that you will be able to attend the conference and contribute to its success.

The reason I am sending you this letter is to bring to your attention the case of a Czech mathematician Dr. Vaclav BENDA. Benda is one of several people arrested in April 1979 and sentenced in October for subversion. He was sent to prison for four (4) years. Before his arrest, Benda was a spokesman for the civil rights group ''Charter 77'' and a member of the Committee for Defense of the Unjustly Persecuted. Benda's wife is a mathematical logician; they have five small children. Dr. Vaclav Benda has been adopted been adopted by Amnesty International as a ''prisoner of conscience".

There is not much we mathematicians can do when one of our colleagues is persecuted by his government. What might help, and what has occasionally been successful before in similar cases, is when sufficently many people of stature write to the authorites to express their concern for the prisoner. Although

the East European governments rarely pay attention to public
opinion, they have been known on occasions to prefer releasing
individual to the the ''nuissance'' of continuing attention of
the outside world.
If you are willing to help I suggest that you (as well as the
other speakers at the Prague meeting) write a letter to the
President of the Czechoslovak Academy of Sciences (address:
Narodni 3, Praha 1, Czechoslovakia) to express your concern for
the well-being of Dr. Vaclav Benda and to protest his
incarceration. You should mention that you are an invited
speaker at the European Summer Meeting and might point out that
Benda's detention is a breach of the International Covenant on
Civil and Political Rights as well as of the Czechoslovak
Constitution.
I hope you will give a serious consideration to my request and
I am looking forward to seeing you in Prague in August.
With best regards,
Thomas Jech

Tomáš Jech kept only one of the replies, that of Azriel Levy:

Dear Thomas,   9th April, 1980
I have received your letter concerning Vaclav Benda. Before
that I talked with Ken McAloon about the matter. He felt we
should do something for the 77 group. My advice was to
concentrate on Benda because our weight is not sufficient to
help all of them and we may be more effective if we try to help
Dr. Benda alone. I discussed this matter with Professor Gert
Müller, who is now on a short visit here. He is afraid that if
we make much noise now we may give more trouble to Vopenka and
Hajek than help Benda. Cancelling the Prague conference is no
threat to the Czech regime since I see no reason why they
should care so much about this conference; this is not the
Olympic Games.
Müller suggested that we shall get the A.S.L. and the German
Union of History and philosophy of Science to intervene on
Benda's behalf. We now also discuss the matter in Prague and
maybe arrange for a letter of all the participants of the
conference to the Czech authorities. In any case I definitely
agree we should not let the matter be forgotten and we shall
discuss in Prague how is the best way to proceed.
Sincerely yours,
Azriel Levy

My husband Václav Benda signed the Charter 77 in January 1977, in April he became one of the founding members of the Committee for the Defence of the Unjustly Persecuted (CDUP) and in 1979 he even became the spokesman of the Charter. Since then, a wiretap was installed in our flat and everything that was said in the study was recorded. After my husband had been imprisoned, the wiretap should have been removed, because I had not signed the Charter 77 and I had not been branded as a hostile person. However, the wiretap stayed where it was; only the records were now labelled as an agent's accounts, because the State Security, well aware of the immense power that its employers could accumulate, inspected compliance with certain rules. The files of the hostile persons have been destroyed soon after the events of November 1989. My file had been kept only during my husband's imprisonment, and hence was not destroyed as one of the "active" files, and I got its copy sometime in 1990. It serves me as an invaluable source of information, not only about the operation of the State Security, but also as a private diary, because almost anything that happened at our place is documented there. From this file I also draw my information on the year 1980. Analogously, a file of Ivan Havel, the brother of Václav Havel, still exists. (In the Czech Republic, any citizen may view the documents which are outcome of the operation of the State Security until 1989. However, he or she must be aware of the reason why these records were taken: to control the country through information.)

A record from May 20th, 1980:

... Bendová informed that events in support of the mathematicians in Charter 77 are planned during the mathematical conference (August 24th – August 30th, 1980). Most probably, Czech participants will not be asked for their signatures.

Next time we talked about this with Zdena Tominová on May 27th, 1980:

... K. Bendová informed about the mathematical conference under prepa-ration, during which foreign mathematicians plan events of support for the mathematicians among the charterers. A large number of foreing materials (participants?) are enrolled and the greatest difficulties appar-ently arise with the Israelis—the official institutions do not approve of their attendance. The accommodation of the participants will be provided in the campus of the Agricultural University in Suchdol.

She also indicated that foreign mathematicians consider boycotting the conference as a sign of their disapproval and support for the mathemati-cians among CH77.

Tominová is familiar with the whole affair and indicated that a boycott would not help, quite to the contrary, a large number of participants are necessary and then some action.

According to B., the conditions of the mathematicians at her Institute are poor and these are not allowed to visit foreign conferences or travel abroad. For this, she blames the regime and the lack of freedom, as always.

Consequently, a complice is to concentrate on eliciting the names of the foreign persons engaged in this operation and the degree of provocation arrangements.

This report is to be used for the preparation of measures in co-ordination with the S-StB Praha, X. S-SNB and XI. S-SNB.[1]

This happened immediately:

27. 5. 1980 S-SNB of the capital city Prague and Central Bohemia Region, S-StB

Comrade Gen. Mjr. Bohumír Molnár, Chief of XI. S-SNB

Dear Comrade Chief,

I share information on an attempt to exploit the international conference Logic Colloquium (LC), which is about to take place from August 20th to August 31st 1980 in Prague, for provocation actions. The organiser is the Institute of Mathematics of the Czechoslovak Academy of Sciences (IM CSAS), which falls under the competence of XI. S-SNB.

In view of the fact that foreign centres are involved in this event, I present the report for further use. Consequently, I advise co-operation with X. and XI. S-SNB in applying precautions to prevent the exploitation of the conference for provocations against the Czechoslovak Socialist Republic.

The Chief of S-StB Prague, Colonel Zdeněk Němec

The attachment of the letter included the following notice:

Information on the preparation of a provocative exploitation of conference LC held on 24.-30. 8. 1980

From 24. 8. until 30. 8. 1980, an international conference Logic Colloquium 80 is about to take place on the campus of the Agricultural University in Suchdol, out of the initiative of CSAS. The conference should pass as a European meeting of mathematicians with the participation of the international non-governmental Association for Symbolic Logic – ASL.

According to the preliminary applications, 398 mathematicians are planning to take part. Out of this number, 220 come from the capitalist countries, 105 from other socialist countries, 62 from CSSR and 11 from the

---

[1]S-StB stands for the Directorate of the State Security, S-SNB for the Directorate of the National Security Corps. The Directorate of the State Security was in charge of the capital, X. and XI. S-SNB were nicknames for the directorate of the counterintelligence for the struggle against the internal adversary (dissent etc.) and the directorate of the counterintelligence for the protection of the economy (the supervision of industries etc.).

developing countries. In addition, large numbers of accompanying persons have to be taken into account. E.g., 10 mathematicians and 7 accompanying persons from Izrael should take part; according to a pronouncement of the members of the organising committee, there is no scientific ground for this.

Abroad, a co-ordination board of French, British and Polish mathematicians has been established. The board is preparing provocative actions in support of the rightist and anti-socialist elements in CSSR among the signatories of Charter 77.

Currently, members of the board have contacted some charterers who are professional mathematicians and are in touch with the organising committee of the conference. These are: Kamila Bendová (wife of the spokesman of CH77 and secretary of the CDUP Václav Benda, presently serving a sentence), who works as a secretary of the organising committee of the conference; and Ivan Havel, a mathematician and the brother of the spokesman of CH77 Václav Havel, who is also serving a sentence. Through them, they even intend to exploit for provocative purposes the member of the organising committee responsible for the scientific programme – Petr Vopěnka.

Mathematicians from France, Great Britain and the People's Republic of Poland, who intend to organise these provocative actions, are planning to arrive to CSSR early, in order to gain time to prepare everything together with Bendová, Havel and other involved persons.

The provocative actions would take place during the conference, both in the form of personal speeches of several participants and with use of written materials.

According to Bendová and Havel, these actions should be of large extent and, abroad, they should serve for seditious campaigns against CSSR and socialist countries.

How much could be deduced from my pronouncement that some mathematicians could not travel abroad! As far as I know, Ivan Havel did not know anything about the whole issue, he used to organise conferences of slightly different kind—the Mathematical Foundations of Computer Science.

Nonetheless, nothing was happening so far. The worst happened at the moment when an attempt was made at finding a compromise solution. In my file, I found the following report of agent Vladimír, dated June 12th, 1980:

On June 11th, 1980, Vl. Landa, a corresponding member of the CSAS, informed about a conversation with prof. Gert Müller from the Federal Republic of Germany (FRG) on the logic colloquium in preparation. Dr.

Petr Hájek, an employer of the IM CSAS and the chairman of the organising committee of the colloquium, was present at the conversation.

Müller stated openly that actions in support of V. Benda and V. Havel are in preparation. He indicated that he himself will strive to lessen the extent of these, but that he cannot give any guarantee for the behaviour of younger mathematicians from the FRG.

... in view of the possibility of exploitation of the congress, to provide the conditions for its cancellation. In case that this cannot be accomplished, negotiate changes in the structure of the organising committee, which does not guarantee the representation of the CSAS and CSSR.

In Ivan Havel's file, I found the following document from June 19th, 1980:

In the view of the setup of a mathematical colloquium, to take place in August 1980 in Prague, in the executive board of which right-oriented persons with bounds to the members of the CDUP and CH77 are engaged, lists of persons of Jewish origin from Israel who should take part in the conference were established.

As can be seen from the attached documentation, the list of Israeli delegates was taken from the lists and address lists of participants of the 1977 colloquium in Wroclaw, the 1976 logic colloquium in Oxford and the address list of members of the International Mathematical Union, published in The Journal of Symbolic Logic, Volume 41, Number 4.

... and, on the side of X. S-SNB, control the measures negotiated with the governing body of the CSAS on June 14th, 1980, so that the colloquium does not take place in the CSSR.

The increasing restlessness of the State Security is witnessed by the fact that a separate file was set up for the Institute of Mathematics of the CSAS on July 8th, 1980, just because of the colloquium. Regrettably, the file does not exist anymore.

After that, the State Security attempted to press upon the conference chairman Petr Hájek to disavow the provocations of the foreign mathematicians in favour of Václav Benda; he refused, fully aware that this prestigious conference for which he had been striving for a long time, with all its importance and profit for the Czech logical community, was in danger of cancellation.

I found one more extraordinary text in the file of Ivan Havel: on June 25th, 1980, the academic Josef Říman asks the Deputy Foreign Minister for his standpoint on the following fact:

During the negotiations at the CSAS Presidium, Professor Müller announced facts which substantially change the so-far favourable conditions for the successful progress of the colloquium. According to his findings, some participants from the capitalist countries are preparing protest actions criticising the imprisonment of Dr. Václav Benda as one of the spokespeople of the "charterers" and the breaches against human rights in CSSR.

On the grounds of this information, the President of the CSAS Jaroslav Kožešník decided to cancel the colloquium for technical reasons.

In view of the importance of the whole matter, I ask you, Comrade Deputy, to communicate your viewpoint.

Evidently, the Deputy did not object very strongly, as six weeks before the conference, a telegram was sent to all registered participants, with an excuse that the conference had been cancelled for technical reasons.

A sequel

Just before the Prague conference, another conference took place in Patras in Greece, so many American mathematicians, once they already were in Europe and had the visa for Czechoslovakia, wanted to visit Prague or the Czech mathematicians privately. However, the Czechoslovak Embassy claimed that the departure of the plane which should had left after the arrival of the bus from Patras had been shifted to an earlier hour. This turned out to be a lie: the plane, which usually lands in the Athens, flied from Beirut directly to Prague. The attempts to get a chance to fly to Prague on the subsequent days were not successful, so the American mathematicians gave up. The fact that, just because of possible protests, the secret police of Prague goes as far as to change the flight of a regular line, shattered even the down-to-earth professors.

Postscripts

I could visit my husband in the prison for one hour four times a year. During my visit in the spring 1980, at the time when a conference on human rights (where the commitments from Helsinki were to be evaluated) was taking place in Madrid, I had been warned by the "educator" (the prison warden) that I was not allowed to talk about the conference. As I had mistaken the Madrid conference with the Prague LC'80 conference, I told him that I can hardly obey as the conference is my job. The warden had been totally disconcerted by this and he interrupted our conversation ("only private matters") all the more.

In 1982, the Proceedings of the Logic Colloquium 1980 were published by North-Holland Pub. Co., thanks to D. van Dalen, D. Lascar and T. J. Smiley, consisting of the materials from the cancelled conference.

Tomáš Jech did not get the visa to Czechoslovakia since then until 1989.

The epilogue

Naturally, it remains an open question whether it was a good idea to connect the conference with the defence of a colleague mathematician, imprisoned for defending other unjustly prosecuted people. However, it remains a fact that 10 years later the communist regime fell and that Charter 77 had its worth in this fall. In 1998, the Logic Colloquium finally took place in Prague, under the chairmanship of Petr Hájek; Tomáš Jech held an invited lecture. Petr Vopěnka has finally been appointed Professor, he became the head of the Department of Mathematical Logic and the Philosophy of Mathematics, he even was the Minister for Education; Petr Hájek was also appointed Professor and for eight years he was the Director of the Institute of Computer Science. Both were appraised by the President not only for their scientific work, but also for their civic attitudes.

Most importantly, the people involved in the story need not feel ashamed that they have not stood up against despotism and manifest injustice.

Kamila Bendová
Email: kamila.bendova@uoou.cz

Seeing numbers

IVAN M. HAVEL

This essay is dedicated to Petr Hájek,
my teacher, colleague, and friend

1 Introduction

The ability to represent time and space and
number is a precondition for having any experience whatsoever.

Randy Gallistel

In his influential book *The principles of Psychology* (1890) the well-known
psychologist and philosopher William James listed seven "elementary mental
categories" that he postulated as having a *natural* origin [Jam07, p. 629]. In
an alleged order of genesis he listed, on the third place, together with ideas
of time and space, the idea of number.

His mentioning the idea of number along with the ideas of time and space
as something natural is truly interesting both philosophically and from the
viewpoint of cognitive science. In this respect it is worth noting that one of
the symptomatic features of the mainstream cognitive science is the tendency
not to talk so much about ideas as about their *representations,* either in the
mind or, even better, in the brain. Just notice the maxim by Randy Gallistel
in the above motto (quoted in [Cal09]). The hypothesis that the human sense
of number and the capacity for arithmetic finds its ultimate roots in a basic
cerebral system has been frequently proposed and elaborated, for instance by
the neuroscientist Stanislas Dehaene [Deh97, Deh01, Deh02].

In this essay I am going to take up somewhat different perspective.[1] Instead of following up with various extraneous concerns offered by brain research and behavioral and/or developmental psychology I will base my considerations on the way things, in our case numbers,[2] are *actually experienced* (a
digression towards animal "arithmetic" in Section 3 is an exception). Hence
the nature of my interest is more phenomenological (in the philosophical sense
of the word) than scientific and, I dare say, my claims are speculative, rather

---

[1] In this paper I expand on some of the ideas briefly presented in my earlier essays,
e.g., [Hav08, Hav09b]. The work on this paper was funded by the Research Program CTS
MSM 021620845.

[2] Throughout this paper the term "number" always means "natural number".

than conforming to empirical or deductive knowledge. Correspondingly, I use the term "experience" in its philosophical sense, when referring to any mental state associated with one's conscious "living through" a certain event or situation—"conscious" in the sense that there is something "it is like" to be in that state of the mind.

First of all we should make clear what kind of "thing" numbers are, i.e., what is the nature of entities that humans may experience as numbers? The way mathematicians formulate, introduce, or conceptually represent *their* idea of a number is quite a different question. This, however, does not mean that the relation between the mathematicians' concept and human experience should be ignored as irrelevant.

Now, our first observation is that the very word "number" (in English as well as its equivalents in some other languages) is rather multivalent. This fact may lead to difficulties, especially in the context of our study— but, at the same time, it may hint at certain interesting hypotheses about the origin of this very multivalence. Thus I shall distinguish here at least three different senses of the word: (1) "numbers" as *numerical quantities*, i.e. counts *of* something, (2) "numbers" as *abstract entities* emerging in human mind and allowing mathematical formalization in the framework of one or another formal theory,[3] (3) "numbers" as natural or conventional *symbols* or *numeral signs* that represent numbers in either one of the previous two senses. In view of the importance of this conceptual distinction in our study, I shall henceforth distinguish the above three senses by using, respectively, the terms (1) *count,* (2) *number,* and (3) *numeral.* However, when there is no danger of confusion, I shall often use the term *number* in the general, more encompassing sense.

2   Human sense of numerical quantities

Are we, people, endowed with anything like a *sense of counts*? Here the term "sense" (another multivocal word) vaguely refers to human ability or disposition to recognize some (not too large) counts of items that are perceived, remembered or imagined. The allusion to perceptual "senses" (vision, hearing, touch, etc.) may be intuitively apt but it is worth elaborating a little (cf. [Hav09a, Chapter 5]).

Let us start with the intended meaning of the general notion of having (or being endowed with) *the sense of X* where X is a certain predetermined quality (in our case, somewhat oddly, the term "quality" refers to numerical "quantity"). First, such "sense" is something to be attributed to a person who is, so to speak, the possessor or owner of the said sense, and second, "having

---

[3]What is characteristic of abstract numbers is that they do not emerge (or they are not constructed) individually but always together with some (or all) other numbers.

the sense of" refers to one's *disposition,* rather than to factual *employing* such disposition in a concrete situation (analogically to, say, the sense of humor, sense of responsibility, etc.). This subtle distinction is not always properly taken care of in the scientific literature but it is usually implied by the context. It may turn out to be particularly relevant to cognitive science which attributes various "senses of ..." or "feelings for ..." to conscious subjects. I shall call such subjective dispositions, or "senses of something", simply *inner senses* (take it just as a technical term).

Most of such inner senses are implicit, pre-reflective features of our everyday experience even if sometimes we may subject them to conscious reflection, especially when they are actually being employed. This holds, mutatis mutandis, for the usual perceptual senses as well as for the sense of counts. All of them are subjectively experienced as well as objectively inferable dispositions. Unlike perceptual senses, the sense of counts does not have its own dedicated physiological organ; rather it indirectly utilizes various perceptual organs (as for its dedicated brain area there exist various conjectures).

We may tentatively put forth the idea of a minimal, pre-reflective inner sense of counts as something innate, already built into the very structure of experience. For this, however, we would have to distinguish various subcategories of counts: depending on whether they correspond to small, moderately large, or very large numbers (prematurely said).

Indeed, we directly perceive, *without* counting, very small counts.[4] For instance, we normally "see at a glance" the triplicity in triangles or tripods, quaternity in squares, quintuplicity in five-point stars—all that without any actual process of counting angles (or legs or vertices or tips)—but when the group becomes larger we gradually become wrong in a direct grasp of the count. This may happen around seven, eight, or more items in the group (depending on the individual and context). For larger groups we cannot but resort to a slower but more reliable actual counting procedure. Let us refer to this transition phase as to the *first horizon* of number apprehension. Then the *second horizon* of number apprehension might vaguely delimit what can be *conceivably* counted in practice (possibly in thought only); finally, numbers that are beyond the second horizon and stretch towards the potential infinity can only be grasped through indirect theoretical tools.

The sense of counts should be distinguished from another, perhaps originally independent *inner sense of numerosity* comprising the ability to notice that a certain group of entities either swells or shrinks, or that it is either larger or smaller than another group of entities. The sense of numerosity NEED not entail, in general, the ability of counting or the idea of a count.

---

[4]Let us live with this little ambiguity in terms—by the verbal form (*to count*) I refer to an active temporal process aimed at determining *the count.*

Let us quote W. James' account of (the sense of) counts, namely the ideas of number, of the increasing number-series, and of the emergence of arithmetic [Jam07, pp. 653-654]:

> Number *seems to signify primarily the strokes of our attention in dis-criminating things. These strokes remain in the memory in groups, large or small, and the groups can be compared. The discrimination is, as we know, psychologically facilitated by the mobility of the thing as a total. But within each thing we discriminate parts; so that the number of things which any one given phenomenon may be depends in the last instance on our way of taking it. [...] A sand-heap is one thing, or twenty thou-sand things, as we may choose to count it. We amuse ourselves by the counting of mere strokes, to form rhythms, and these we compare and name. Little by little in our minds the number-series is formed. This, like all lists of terms in which there is a direction of serial increase, car-ries with it the sense of those mediate relations between its terms which we expressed by the axiom "the more than the more is more than the less." That axiom seems, in fact, only a way of stating that the terms do form an increasing series. But, in addition to this, we are aware of certain other relations among our strokes of counting. We may interrupt them where we like, and go on again. [...] We thus distinguish between our acts of counting and those of interrupting or grouping, as between an unchanged matter and an operation of mere shuffling performed on it. [...] The principle of constancy in our meanings, when applied to strokes of counting, also gives rise to the axiom that the same number, operated on (interrupted, grouped) in the same way will always give the same result or be the same.*

Some authors use the term *number sense* for "our ability to quickly under-stand, approximate, and manipulate numerical quantities" [Deh01]. In this study I dare to claim, probably despite James, that our ability to perform actual counting (above the rudimentary sense of counts) may *not* be a nec-essary prerequisite for the number sense. In my view, two inner senses—the direct sense of (small) counts and the sense of numerosity—may be more essential.

3   Can animals count?

There is a growing number of studies with animals exhibiting certain limited abilities to count and, as it is often claimed, to perform elementary arithmetic operations [Cal09]. We may be enticed to immediate hypotheses about the evolutionary origin of such abilities. This, under the prevailing Darwinian paradigm, would make us look for one or another survival advantage of such abilities, analogous to the advantages of having, say, the sense of colors, of shapes, of spatial directions, etc.

Perhaps we may conjecture that what appeared relatively early in the animal world might have been abilities that are not based on the process of counting. Such abilities may be primarily two: (1) the ability to identify small counts *at a glance*—a count of eggs, offspring, wolves in a pack, etc.—and (2) the ability to distinguish, without counting, between a smaller amount and a larger amount—of grain, leaves of grass, ants in a colony, etc.—(as James puts it, "the more than the more is more than the less.") Only much later, perhaps among humans, actual counting procedures came to be used, and after that there emerged the abstract concept of a number together with the idea of the "number line" endowed with various arithmetical operations.

Various experiments show that there are certain reasons to attribute the sense of (small) counts to animals. There are reports on primates, elephants, salamanders, chimpanzees, birds, even fish and bees, which can reliably recognize small counts of presented objects [Cal09]. The favorites are four-day-old chicks (thus no training could be assumed) that reportedly are able to correctly determine that $1 + 2$ is greater than $4 - 2$, that $0 + 3$ exceeds $5 - 3$, and that $4 - 1$ is more than $1 + 1$. However, one has to be careful with interpreting such experiments. Researchers frequently use certain appealing phrases like "number recognition" or even "arithmetic skills" when talking about animals. My own small survey of the literature has revealed, first, that such experiments with animals involved counts of some specific objects, usually objects with survival importance to the animal. Second, in most cases the tested abilities could be explained simply by the ability to discriminate between larger and smaller amounts, without any need to do actual counting. (The rash claims about "arithmetic" skills of animals could be enticed by the habitual tendency of us, numerate humans, to do counting, adding, subtracting, etc. even when we deal with relatively small counts.)

There are two ideas for suitable that may support various hypotheses about evolutionary origins of the number sense. One idea was already mentioned: looking for obvious survival advantages. The second idea is more logical. For instance, the sense of counts already presumes other inner senses, namely the sense of sameness and difference, or more specifically, the sense of individuality of elements of a group and the sense of that very type of similarity which characterizes the group. (Surprisingly enough, James placed "ideas of difference and resemblance, and of their degrees" into a later, fourth position in his list of natural mental categories.)

4 Representing numbers

How did we, the human species, develop the abstract *concept of a number*, as something implicitly related to counts but without reference to particular entities counted? Talking and thinking in human way about abstract numbers is only possible with their appropriate symbolic representations.[5] Here comes the third meaning of the word "number", viz. that of a *numeral*. Let us use this word in the most general sense, including not only word numerals (like "one" "two", "three", ...) and their combinations ("thirty six") but also other kinds of symbol, or better sets of symbols, that unambiguously represent (abstract) numbers. You can think of the usual graphical signs (1, 2, 3, 4, ..., or 01, 10, 11, 100, ..., or I, II, III, IV, ...). Let us allow, for the purposes of the present study, even more general representations, for instance geometrical shapes, whether drawn, written or merely imagined.

To avoid misunderstanding: I do not assume that the concept of numeral has to be derived from, or dependent on, a prior concept of (abstract) number. We could equally well associate numerals directly with counts of some entities, real or imagined. The simplest idea is the *analog representation* of counts (or numbers). Think, for instance, of scribbling down or imagining groups of some concrete objects like dots, strokes, tokens, marks, knots on a rope etc.[6] Each such group can be also viewed as a symbolical numeral directly representing the count of items in the same group. In mathematics we speak about the unary numeral system for representing numbers.

As a matter of fact, many cultures use analog notation (signs) for three smallest numbers, sometimes even for four or five; see Fig. 1. Obviously, for slightly larger numbers (near the first horizon of number apprehension) a danger of misinterpretation may increase. Probably because of this and for the sake of compression specific signs have outplaced the analog signs. Moreover, and more importantly, arbitrarily large numbers can be represented by sequential juxtaposition of figures.

It is worth noting that some languages use different words for counts of different categories of object. This fact may suggest that the concept of count may be more natural than the concept of number.

It is interesting to note, in passing, that some languages use *object-specific* numeral systems which depend on the kind of objects counted [BB08]. For instance, on one of the island groups in Polynesia, tools, sugar cane, pandanus, breadfruit, and octopus are counted with different sequences of numerals. There is a current dispute about whether such object-specific counting systems were predecessors of the abstract conception of numbers and number line (Beller and Bender argue in the opposite).

---

[5] In this point I differ from the view of some scientists like Dehaene (quoted above) and Jean-Pierre Changeux. They associate the number sense with activation of specialized neuronal circuits in the brain (this applies also to non-human animals) so that human language and consciousness are not assumed to be a prerequisite for dealing with numerical quantities.

[6] Perhaps even temporal sequences of events like sounds of a tolling bell.

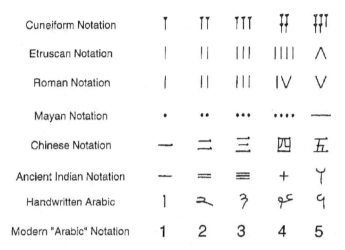

Figure 1. Numeral signs in various cultures (redrawn from [Ifr94])

5   Numbers dancing in our heads

We can follow up the above considerations in various directions. We have already observed that there are three different conceptualizations of the pre-arithmetic idea of number: (1) number as a *count* of some identifiable (maybe visible or tangible) items, (2) (invisible, intangible) *abstract number,* and (3) number as a *numeral* of certain type (visible, speakable etc.). In theory we can easily point to inherent, indeed even necessary, interrelationships between these three conceptualizations. For instance, we may be interested in comparing three respective roads to infinity: (1') the intuition of gradual but unbounded swelling the group of items toward larger and larger counts (passing over the first two to the third of the above mentioned horizons of number apprehension), (2') developing a formal (axiomatic) theory of natural numbers, and (3') assuming such numerical representations that allow for depiction of arbitrarily large counts (or of arbitrarily large abstract numbers).

Here I am not going to discuss the theoretical issue of infinity. Instead I am going to pose a different question, which may be more important if human cognition is at issue. Let us look "inside our minds", so to speak, and ask whether and how far we (humans) could mentally grasp the idea of a number.[7] If we are able to "see at a glance" (i.e., without counting) small counts of things, why not to venture into imagining an analogously direct access to much larger counts, or perhaps even to abstract numbers.

---

[7]For philosophical reasons I am not comfortable with the term "introspection" but the reader may happily make do with it.

Ivan M. Havel

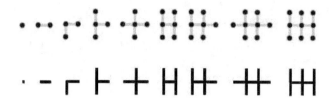

Figure 2. Turning dots into figures

True, in our culture we are too captive in the framework of words (numerals), symbols (number signs), arithmetic operations, and indeed, of the whole number line. Speaking for myself, whenever I hear, say, the sound "thirty six" I immediately happen to hear "six time six" (as a leftover of memorizing the multiplication table in the primary school), or alternatively, I could imagine the formula "$36 = 2^2 \cdot 3^2$" (provided I were obsessed with prime number representations). Surprisingly enough, I never imagine a rectangle of six rows and six columns of dots (or something), or a prism made of small cubes, four horizontal, three vertical, and three backwards. Why not? Wouldn't it be easier to form mental images of various geometrical shapes, to remember them, and perhaps to manipulate them in various exciting ways?

Consider for example Fig. 2. In its upper part there are nine groups of dots; each group corresponds to a different count of dots (a singleton, pair, triplet, quartet, etc.). Let us fix the relative position of dots in each group by shaded lines. In the lower part of Fig. 2 only the lines are depicted. Why not treat these line figures as numeral signs, easy to be imagined and remembered? Notice that most of these figures are formed by adjoining previous figures in the sequence; this immediately leads to the tentative idea of using such adjoining process for pictorial representing much larger numbers. Incidentally, each such numeral can be viewed as both analog and symbolic representation of a certain number.[8] (In Section 8 this representation will be used to develop more expressive pictorial representation of numbers.)

There is a plausible hypothesis that autistic savants with extraordinary numerical powers can mentally grasp numbers in some synesthetic way (pictorial, auditory, tactile), or maybe even in a form of some dynamic objects, rather than in the ordinary numeral representation.[9] In this respect it is worth mentioning the case of Daniel Tammet (known for his record of reciting 22 514 digits of π from memory). Let me quote some of Tammet's own

---

[8]The adjoining procedure may lead even to 2D or 3D figures.

[9]The term synesthesia refers to perceptual experience in multiple modalities in response to stimulation in one modality. For some people, for example, letters or numbers (numerals) evoke vivid color sensations [RH03].

reflections (from recently published conversation with him in Scientific American [Leh09]):

I have always thought of abstract information—numbers for example—in visual, dynamic form. Numbers assume complex, multi-dimensional shapes in my head that I manipulate to form the solution to sums, or compare when determining whether they are primes or not. [...] In my mind, numbers and words are far more than squiggles of ink on a page. They have form, color, texture and so on. They come alive to me, which is why as a young child I thought of them as my "friends." [...] I do not crunch numbers (like a computer). Rather, I dance with them. [...] What I do find surprising is that other people do not think in the same way. I find it hard to imagine a world where numbers and words are not how I experience them! [... My] number shapes are semantically meaningful, which is to say that I am able to visualize their relationship to other numbers. A simple example would be the number 37, which is lumpy like oatmeal, and 111 which is similarly lumpy but also round like the number [numeral figure] three (being 37·3). Where you might see an endless string of random digits when looking at the decimals of π, my mind is able to "chunk" groups of these numbers [figures] spontaneously into meaningful visual images that constitute their own hierarchy of associations.

This is a rare case of a savant able to report on his exceptional inner experience. No wonder such a report generates more questions than answers. As I already mentioned, contemporary discussions mostly seek solutions from the brain research, which is expectable in view of the tremendous recent progress in various brain imagining techniques. My opinion is, however, that we could hardly expect from the brain scientist direct and valuable answers to phenomenally formulated questions about subjective experience. When Tammet says, for instance, that numbers "assume complex, multi-dimensional shapes," or that he "dances with them," it is of little help to the neuroscientist in his quest for adequate phenomena in the brain. By no means I want to say that Tammet's reports are meaningless—quite conversely, I believe that his semi-metaphorical statements say more, in a sense, about human mind than the empirical scientist could formulate in the language of neuronal dynamics.

6 The riddle of prime twins

Here "prime twins" is a little pun: I do not mean twin primes (i.e. prime numbers of the form p, $p+2$), but John and Michel, the famous autistic, severely retarded twins, studied in 1966 by the neurologist Oliver Sacks [Sac85]. I am not going to discuss here their significance for scientific (neurological or psychological) research but merely use them as a valuable source of, or motivation for, wild hypotheses about the way human mind can deal with numbers in the extreme. Let us quote Sacks' own report on one of his encounter with the twins [Sac85, p. 191]:

> [They] were seated in a corner together, with a mysterious, secret smile on their faces, a smile I had never seen before, enjoying the strange pleasure and peace they now seemed to have. I crept up quietly, so as not to disturb them. They seemed to be locked in a singular, purely numerical, converse. John would say a number—a six-figure number. Michael would catch the number, nod, smile and seem to savour it. Then he, in turn, would say another six-figure number, and now it was John who received, and appreciated it richly. They looked, at first, like two connoisseurs wine-tasting, sharing rare tastes, rare appreciations. I sat still, unseen by them, mesmerized, bewildered.

Sacks wisely wrote down their numbers and later, back at home, he consulted a book of numerical tables and what he found was that all the six-figure numbers were *primes*! The next day he dared to surprise the twins and ventured his own eight-figure prime. Twins paused a little time and then both at the same time smiled. The exchange of primes between Sacks and the twins continued during the following days, with gradual increase of the length of the numeral up to one of twenty figures brought out by the twins, for which, however, Sacks had no way to check its primality. For generating and recognizing larger primes the twins needed more time, typically several minutes.

Let us point out four conspicuous aspects of the twins' performance: (i) their striking emotional fondness for the primes, even if (ii) "they [could] not do simple addition or subtraction with any accuracy, and [could] not even comprehend what multiplication or division means" [Sac85, p. 187], (iii) there were delays in their reactions—the larger the longer numerical lengths of the primes processed, and (iv) there were certain indications that visualization was at play, as Sacks writes about another of the twins' abilities (calendar calculation) [Sac85, p. 187]:

> [Their] eyes move and fix in a peculiar way as they do this—as if they were unrolling, or scrutinizing, an inner landscape, a mental calendar. They have the look of 'seeing', or intense visualization [...]

One of the above listed aspects (the third one, about delays) may suggest that production or recognition of a prime was not an instantaneous act but involved a certain internal procedure (what we would call "procedure") that consumed a certain time—a nontrivial time compared to the usually instantaneous responses of number prodigies. Well, what may such a procedure be?

It is not my intention to put serious effort into trying to explain the particular extraordinary powers of Sacks' twins[10] (or other number prodigies). I rather take up the twin case as an incentive to deal with numbers in alternative ways, not grounded in the ordinary arithmetic.

7 Primality sans arithmetic

A simple characterization of primes which is not based on arithmetical operations may be as follows: A number is a prime if and only if it corresponds to the count of items that cannot be arranged into a regular rectangle (except for a row or line). Imagine, for instance, a platoon of soldiers that cannot be drawn up into a rectangular formation (file by three, file by four, etc., except for a single row or line). Then we can easily design an effective finitary procedure P which, given a group X of individual elements (let us call them tokens), would systematically arrange it into various rectangular formations, always checking whether some of the tokens are left over or not. If the only possible arrangement without a leftover is a single row (or line) then the count of all tokens in X is a prime number.

The following four observations about such primality test P seem to be particularly relevant to our approach: (1) the procedure P may be carried out, in principle, with the help of visual imagery only, (2) feasibility of P in imagination could reach a certain (biological or psychological) limit if X happens to be too large, (3) the larger X the more steps may P require (hence more time would be needed for realizing P), and, last but not least, (4) in order to execute P there is no need to know the precise (numerical) count of tokens in X. What is only needed is a presentation of group X as a whole and allowing for simple mobility of its tokens. (Incidentally, one can imagine a mechanical apparatus that would realize the procedure).

Now, the oddity of prime number prodigies may be primarily related to point (2). With the current relatively moderate (albeit rapidly growing) scientific knowledge of the structure and dynamics of the brain, and without any knowledge at all about inner experiences of the prodigies, we can only conjecture that, for the prodigies, the feasibility horizon of the procedure P must be substantially farther than for normal people.

Of course, there may exist (and probably do exist) some unimaginable, entirely different solutions for dealing with primes. We are still far from understanding even the particular case of Sacks' prime twins. How, in addition to some marvelous mental imagery, were they able to communicate about

---

[10]In response to Sacks' report, several researchers published various theoretical speculations. However, most of them (if not all) are based on ordinary arithmetical properties of primes (cf., e.g., [Yam09].

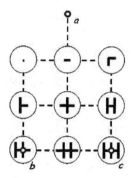

Figure 3. First level of number representation.

numbers also in the ordinary form of (Arabic) numerals or (English) words? And why did they sense precisely primes as something joyful?

8   Numbers turned into curves

Following up with the first question (about communication), I shall develop here an illustrative example of a relatively simple direct translation between two types of number representation: on the one hand, numerals in the usual form of sequences of decimal digits (or expressions formed from number names), and on the other hand, made-up pictorial "numerals" representing even huge numbers, preferably in a way that would allow "seeing them at a glance".

Let us discuss the latter issue, that of pictorial representation. To be more specific, consider the numeral system introduced in Section 5 (cf. Fig. 2). There I mentioned the possibility of extending such system to arbitrarily large numbers. However, the presumed miraculous imagery of even the most prodigious savants cannot be unbounded. Under the theory of embodied mind, whatever the (yet unknown) nature of pictorial number representation may be, there have to be certain limits to it. Such limits could somehow resemble the first horizon of human direct number apprehension (as mentioned in Section 2), except that they may be located considerably far away ("far away" on *our* number line).

My speculative idea for a more powerful savant number representation (assume that it can be still called "representation") is based on the notion that it may have a *hierarchical nature.*[11] I shall illustrate it with a concrete example (admittedly fabricated and most likely without any relation to reality). In

---

[11]In fact, the positional numeral system is hierarchical too, since each position represents different power of 10. In our case the word "hierarchical" should point to different representational strategy associated with each level.

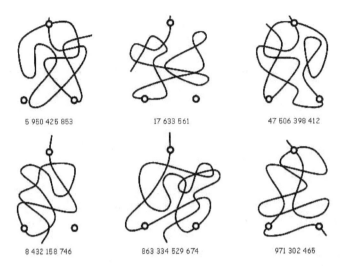

Figure 4. Second level of number representation.

the first step choose a certain, perhaps small set N_0 of elementary numerals, for instance the nine figures as in Fig. 2. This determines the first level of number representation. In the next step arrange elements of N_0 into a predetermined fixed formation, let us say into a 3×3 regular grid $G(9)$ with one additional point (for zero), as shown in Fig. 3. Obviously, in the Euclidean plane this grid can be uniquely determined by three reference points (a, b, c), also depicted in Fig. 3.

Imagine now a certain number n presented in the form of a decimal numeral (sequence of digits), say, $n = 5\,950\,425\,853$. Draw a smooth curve through the grid, starting in point a and successively passing through all and only those vertices that represent digits 3, 5, 8, 5, 2, 4, 0, 5, 9, 5 in that order (for technical reasons in the reverse order comparing to n). There may be infinitely many such curves but each would represent precisely number n (one of them is the upper left curve in Fig. 4, the other curves in the figure represent other numbers; I included them for those who like pictures).

For the inverse procedure, consider a given planary curve (for instance one of those in Fig. 4) together with three reference points, a, b, and c. These points are sufficient to draw a unique regular grid (as in Fig. 3) that lays over the curve. Now proceed along the curve, starting in the upper point a, and list all the digits associated with vertices you pass through (you should list them from the right to the left). Eventually the generated numeral expresses the unique number represented by the curve.

I admit that the procedure is somewhat artificial and strange. Moreover,

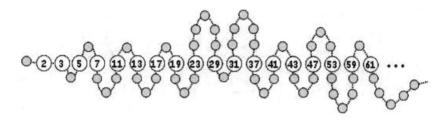

Figure 5. Prime number line and non-prime number line

it has a serious drawback (for our approach): even if it thinkable that the curve itself can be recognized at a glance (analogously to, e.g., human face recognition), the snag is the requirement of knowing its exact position with respect to three fixed reference points. Yet the proposed procedure has certain noticeable properties: first, one can in principle get along with visual imagery, second, there is no need of counting anything, and third, no standard arithmetical operations are used. Yet the procedure allows for handling considerably large numbers (as seen in Fig. 4) with only modest increase of complexity.

9   Seeing primes all at once

As a matter of fact, quite a different alternative track of thought may be pursued, too. It is conceivable that number prodigies do not care so much about concrete numbers and their individual properties like primality. Let me quote in this respect a passage about a synesthetic subject from [RH03, pp. 56–57]:

> If asked to visualize numbers the subject finds that they are arranged in a continuous line extending from one point in the visual field to another remote point—say from the top left corner to bottom right. The line does not have to be straight—it is sometimes curved or convoluted or even doubles back on itself. In one of our subjects the number line is centered around "world centered" coordinates—he can wander around the 3D landscape of numbers and "inspect" the numbers from novel noncanonical vantage points. Usually the earlier numbers are more crowded together on the line and often they are also coloured.

Obviously a modified version of such a process may be hypothesized for the case of prime number savants like Sacks' twins. They may be miraculously gifted with a sort of direct access to a certain *whole set* of primes, small or large (certainly not to the set of *all* primes in the mathematical sense). They can perhaps visualize the entire set as a single, very complex geometrical object in which primes play a salient role, while nonprimes fill the space between or around them. A very simple illustrative example of such an idea is depicted in Fig. 5 where the ordinary number line is curved around a straight "prime-number line".

Now is a good time to comment on the question at the end of Section 7 concerning Sacks' prime twins who, even though severely retarded, paid special attention precisely to primes. Why was it just primes, out of all numbers, that made them so happy? A weird but plausible idea may be that the geometrically prominent positions of primes on or inside the complex geometrical object mentioned above would charge them with some emotionally strong, perhaps esthetic quality.

Another, easier guess may proceed from our earlier characteristic of primes (see Section 7): a prime number corresponds to such a count of elements of a group that wriggles out of all (nontrivial) rectangular formations. With a grain of aphorism: an entity that resists something as banal as a rectangular formation deserves a joyful welcome.

10 Conclusion

I outlined several purposely simple but highly speculative ideas about mental number processing, motivated by some extraordinary cases of numerical savants. I admit that the ideas are far from being mutually compatible. The reader may have noticed, for instance, that there is no clear way how to combine the curved-line pictorial representation of numbers in Section 8 with the procedure of primality testing presented in Section 7. The remarkable powers of Sacks' twins and of other number prodigies remain, and probably will remain for a long time, a mystery.

However, my motivation for seeking various number representations and procedures did not consist in trying to suggest some explanatory tools and even less did it consist in proposing something of practical use. The only purpose was heuristic: to open possibilities of quite unusual lines of reasoning about cognition and arithmetic.

BIBLIOGRAPHY

[BB08]   S. Beller and A. Bender. The limits of counting: Numerical cognition between evolution and culture. *Science*, 319:213–215, 2008.

[Cal09]   E. Callaway. Animals that count: How numeracy evolved. *New Scientist*, 2009.

[Deh97]   S. Dehaene. *The Number Sense. How the Mind Creates Mathematics*. Oxford University Press, New York, 1997.

[Deh01]   S. Dehaene. Précis of the number sense. *Mind and Language*, 16(1):16–36, 2001.

[Deh02]   S. Dehaene. Single-neuron arithmetic. *Science*, 297:1652–1653, 2002.

[Hav08]   I. M. Havel. Zjitřená mysl a kouzelný svět. *Vesmír*, 87:810, 2008. In Czech.

[Hav09a] I. M. Havel. Uniqueness of episodic experience. In *Kognice 2009*, pages 5–24, Hradec Králové, 2009. Gaudeamus, 2009.

[Hav09b]   I. M. Havel. Vidět počty a čísla. *Vesmír*, 88:810, 2009. In Czech.

[Ifr94]   G. Ifrah. *Histoire Universelle des Chiffres*, volume I et II. Robert Lafont, Paris, 1994.

[Jam07]   W. James. *The Principles of Psychology*, volume II. Cosimo Inc., New York, 2007. Originally published in 1890 in Boston.

[Leh09]   J. Lehrer. Inside the savant mind: Tips for thinking from an extraordinary thinker. *Scientific American*, 2009. http://www.scientificamerican.com/article.cfm?id=savants-cognition-thinking.

[RH03]   V. S. Ramachandran and E. M. Hubbard. The phenomenology of synaesthesia. *J. of Consciousness Studies*, 10(8):49–57, 2003.

[Sac85]   O. Sacks. The twins. In *The Man Who Mistook His Wife for a Hat*, pages 185–203. Pan Books, London, 1985.

[Yam09]   M. Yamaguchi. On the savant syndrome and prime numbers. *Dynamical Psychology*, 2009. Electronic journal, http://www.goertzel.org/dynapsyc/yamaguchi.htm.

Ivan M. Havel
Center for Theoretical Study
Jilská 1, 110 00 Prague 1
Czech Republic
Email: office@cts.cuni.cz

Where are hypotheses being born? On exploratory and confirmatory faces of the GUHA method

DAN POKORNÝ

*Dedicated to Petr Hájek,
on the occasion of his 70th birthday*

1 Beginning

At the Beginning was the Hypothesis. Lead by the Hypothesis, the Empirical Data were collected. The Statistical Test came as a witness to testify the value of the Hypothesis, so that through him all men might believe. The Statistical Test weighted the Hypothesis with the Empirical Data and decided eventually.

Many textbooks explain this basic confirmatory principle of the statistical inference. One issue remains open: How hypotheses are being born. When ongoing researchers in medical sciences or in our particular area of psychotherapy research ask this secret they are told: The *Hypotheses* should be founded with the *Theory*. A closer look reviews that the *Theory* usually consists of series of journal papers reporting results of previous empirical studies in the field. Hence, a repetition of previous studies seems to be a safe way. The question remains where the hypotheses of these studies have been born. Rules of statistical inference do not demand a theoretical grounding of hypotheses: A question may be asked no matter where it comes from. It may even arise fully unconsciously and be more precisely formulated later.

The exploration of existing empirical data is one of the resources for the generation of hypotheses. The procedures of exploratory data analysis support the researcher by offering comprehensible representations of the data structure. Only a few of exploratory methods have an explicit goal to generate hypotheses for the following research. Hájek, Havel, and Chytil [HHC66] proposed and implemented probably the first practically applicable procedure. The aim of the GUHA procedure ASSOC was to generate and test hypotheses on multivariate categorical data [HHC83]. The output of the exploratory procedure consists of satisfied hypotheses that can be optionally used as an input for the following confirmatory study. Hájek's and Havránek's [HH78] logic of observational calculi thus focuses on the usually considered a dark side of the discovery process.

The ASSOC procedure generates hypotheses according to syntactical rules:

a generalized quantifier connects two elementary conjunctions—called antecedent and succedent—of literals representing dichotomously scaled variables. Exact definitions can be found in the literature cited; some examples will be shown below. The exhaustiveness is related to a problem often tackled in the theory and practice of the GUHA method: The requirement to present all satisfied hypotheses (resp. metaschemes representing them) can quickly lead to huge and hardly comprehensible result lists. We will present experiments with possible modifications of the procedure ASSOC.

a. The sorting of presented hypotheses has approved its usefulness in the practice. On top of that, we will propose a meaningful sorting of literals *within* a presented hypothesis.

b. The selection rule focusing on the statistically highly relevant hypotheses will be proposed which can considerably reduce the volume of presented results.

c. Exploratory and confirmatory processes will be discussed; some alternative perspectives will be compared. A procedure will be proposed enabling to perform exploratory and confirmatory analysis within a single step.

Empirical data from a study concerning psychosocial counseling for university students will serve as an experimental object of these proposals. Proposals will be presented in a semi-formal way and confined to the following basic options of the ASSOC procedure. The information in the data matrix is supposed to be complete and the sets of antecedent and succedent variables are supposed to be mutually disjoint. Our favorite exact Fisher test is cast in the role of the quantifier. The data in our example are two-valued; generalizations to the nominal case could be done easily.

2   Study and a general question

The Students Counseling Center of the University of Ulm offers a support in solving personal problems and in critical situations. Problems include learning disabilities, examination anxieties, social difficulties, family or partnership difficulties, and many other topics. Students' worries are discussed in one or more initial interviews including the test-diagnostic procedures. The counselor estimates the main problems of the student, and recommends the following steps. Problems can sometimes be solved already during the initial interview. The individual counseling or group courses in the Center are recommended in most cases. Some problems need diagnostic or therapeutic treatment in psychotherapy, psychiatry, somatic medicine or other fields.

The complete sample of students who contacted the Center during the period 2005–2008 for the first time was investigated in a study [APGP09]. The sample data were collected by the Center's head Renate Aschoff-Pluta and her collaborators.

We will focus here on the relationship between student's main complaints and counselor's recommendations. Both questions were completely answered by 427 students (out of 489) who constitute our observational sample. The list of main problems and the simplified list or counselor's recommendations—both sorted by descending category frequencies—are presented in Table 1.

problem	abs.	rel.
working problems	233	4%
family problems	229	47%
depressive moods	206	42%
relationship problems	128	26%
low self-esteem	126	26%
anxiety feelings	113	23%
emotional instability	112	23%
fear of examinations	98	20%
psychosomatic complaints	60	12%
other problems	52	11%
obsessive thoughts	48	10%
eating disorders	27	6%
addiction	23	5%
psychiatric disorders	21	4%
out-of-control aggressions	15	3%
sexual problems	10	2%
suicidal thoughts	9	2%
sample	**427**	**100%**

RECOMMENDATION	abs.	rel.
COUNSELING	351	72%
PSYCHOTHERAPY	218	45%
OTHER RECOMMEND.	67	14%
PSYCHIATRY	65	13%
SOMATIC TREATMENT	38	8%
NO RECOMMENDATION	29	6%
sample	427	100%

Table 1.

The problem (task) for the ASSOC procedure was formulated as follows: We are interested in associations between 17 main problems (antecedent) and 6 possible recommendations (succedent; the list was simplified here). We are interested in consequences of the concurrent occurrence of problems. Hence, the elementary conjunctions of 1 to 3 literals in a positive form should be generated. The observational association between problems and recommendations is operationalized by the one-sided exact Fisher's test on the significance level $\alpha_{\text{crit}} = .05$.

3 Literal sorting within a hypothesis

The ASSOC procedure generated 4998 hypotheses; 206 of them were satisfied by the data using the criterion of the Fisher's exact test. One of satisfied hypotheses states that there is an association between the co-occurrence of emotional instability and depressive moods on one hand and the recommendation of the psychotherapy treatment on the other hand (Table 2). The observed frequency 57 is considerably higher than the expected one 29.10. The reached significance level of the Fisher's test 0.152×10^{-14} suggested that the association hypothesis can be accepted with a high security. The Yule's association quotient .80 estimated that the association is a very close one.

hypothesis A1&A2 ~ S1		S1 psychotherapy recommended		
		yes	no	total
	yes	57	8	65
		88%	12%	100%
		a	b	r
A1&A2 emotional instability & depressive moods	no	161	261	422
		38%	62%	100%
		c	d	s
	total	218	269	487
		45%	55%	100%
		k	l	m

hypothesis	$A1\&A2 \sim S1$	Fisher's test:	$p = .0000000000000152$
		Yule's $Q =$	$(ad - bc)/(ad + bc) = .84$
hypothesis	$A1 \sim S1$	Fisher's test:	$p = .00000000000387$
hypothesis	$A2 \sim S2$	Fisher's test:	$p = .00000000349$
model	$(A1, A2),$ $(A1, S1),$ $(A2, S1)$	exact test:	$p = .0355$

Table 2.

By definition, the set of satisfied hypotheses is a solution of the given problem. Hypotheses can appear in the presented solution in any order. The first GUHA procedures used the lexicographical ordering. However, it was recognized that the ordering of presented hypotheses has a great psychological impact on the beneficiary of the solution: the user. Procedures for sorting and selecting satisfied hypotheses with combined lexicographical and numerical criteria have been proposed (cf. our procedure COLLAPS in [HHC83] or early procedures for sorting of ASSOC results). For instance, satisfied hypotheses can be sorted by the increasing antecedent and succedent lengths first; and within these segments by the significance level reached with the Fisher test. The basic principle of the GUHA method "find everything interesting" was enhanced by "show the most interesting first".

This idea can be extended to the sorting of syntactical parts *within* a hypothesis. Literals within the antecedent $\varphi \equiv A_1, \ldots, A_u$ as well as literals within the succedent $\psi \equiv S_1, \ldots, S_v$ can be presented in any order; the resulting sentences $\varphi \sim \psi$ remain equivalent. The elementary conjunction φ of antecedent literals $A_1, \ldots, A_u$ is associated with the succedent formula ψ. For each sentence $\varphi_1 \equiv A_1 \sim \psi, \ldots, \varphi_u \equiv A_u \sim \psi$, we can evaluate the appropriate test statistic like the significance level reached with the Fisher's exact test. The antecedent literals can be sorted by this statistic now; the antecedent literal with the strongest association to the succedent formula is presented as the first one. Succedent literals can be sorted analogously.

The antecedent of the satisfied hypothesis in Table 2 consists of two literals. The significance level reached with the Fisher test for the association with the recommended psychotherapy was 0.387×10^{-12} for the emotional instability and 0.349×10^{-9} for depressive moods. Hence, the emotional instability—likely to have a greater impact on the chosen recommendation—is listed on the first place.

This presentation seems to be close to the natural reasoning. In medicine, symptoms that are more relevant for a certain disease are usually referred to as first. As another example, the second sentence out of the following ones seems to be formulated more luckily

1. "cloudy day & winter season $\sim$ snowfall",

2. "winter season & cloudy day $\sim$ snowfall".

The snowfall is likely to be more strongly associated with winter than with clouds.

4   Statistically motivated hypothesis selection

The amount of "all interesting hypotheses" like 206 satisfied out of 4998 generated hypotheses still represents a high challenge for the user's patience. In our example, just 20 hypotheses are of the minimal length (1–1); longer hypotheses constitute a great deal of the presented solution (see Table 3). Some longer hypotheses can bring new substantial messages. However, many of them can occur just like trivial extensions of some statistically strong short hypothesis.

DEFINITION 1 *Let us have two sentences* $\varphi \sim \psi$ *and* $\varphi \& \alpha \sim \psi \& \beta$, *where* $\varphi, \psi, \alpha, \beta$ *are elementary conjunctions; and, both* φ, ψ, *and least one of* φ, β

are non-empty. Then we say that the second sentence is an extension *of the first one; and that the first one is a* reduction *of the second one.*

hypotheses	antecedent-succedent length			all lengths
	1-1	2-1	3-1	
all investigated hypotheses	102	816	4080	4998
all satisfied (Fisher test: p ≤ .05)	20	88	98	206
using the conservative improvement	20	88	91	199
using the strict improvement	20	88	88	196
using multivariate statistical criteria	20	23	9	52

Table 3.

One of the short satisfied hypotheses stated that "working problems~counseling". Seventeen extensions of this hypothesis were found, too, having the form "working problems&something~counseling" where "something" is a conjunction of some other problems. With the increasing lengths of antecedent and succedent, the user's mind can be overflowed by a huge amount of mutually similar hypotheses.

The improving rules of Petr Hájek were designed with the sake to reduce the amount of logically redundant information, cf. [HH78, HHC83]. The solution can contain metasentences containing a satisfied core sentence $\varphi \sim \psi$ and two lists of literals improving the antecedent, resp. succedent formula. Any extension of the core sentence by improving literals is also satisfied by the investigated data. Statistics like reached significance level, have the same value for all extended sentences (conservative improvement) or they can be even better (strict improvement). In this way, metaschemes of sentences can reduce the amount of reported results and concurrently preserve the exhaustiveness of the solution.

The number or presented hypotheses in our example decreased from 206 to 199 using the conservative improvement rule or to 196 using the strict improvement rule (see Table 3). It seems that only such a moderate saving is a common price for the requirement of the exhaustiveness.

An alternative approach can be the selection of highly interesting hypotheses due to some statistically motivated criteria. Such a criterion can be based on exact tests for log-linear models. Let us consider the hypothesis

$$\varphi \sim \psi \equiv (A_1 \& \ldots \& A_u) \sim (S_1 \& \ldots \& S_v)$$

The evaluation of contained literals yields a multiway contingency table 2^{u+v}. We assume all the effects within the antecedent as well within the succedent are fixed. Moreover, we assume all effects expressed by any reduced hypothesis to be fixed. The resulting model of all interactions of the order $u + v - 1$ differs from the saturated model only by the removal of the highest-order effect. This effect of the order $u + v$ can be examined directly by the one-sided exact test; no iterative proportion fitting is needed for the exact procedure.

Using this test, we can establish the following selection criterion:

A satisfied hypothesis is highly interesting if

  1. *no reduced hypothesis is satisfied OR*

  2. *the association is statistically stronger than it would deduced from the set of reduced hypothesis.*

Let us illustrate this criterion at the satisfied hypothesis $A1\&A2 \sim S1$ in Table 2. Both reduced hypotheses $A1 \sim S1$ and $A2 \sim S1$ are satisfied, too; hence the condition (1) does not apply. The within-antecedent effect is $(A1, A2)$, two effects corresponding to reduced hypotheses are $(A1, S1)$ and $(A2, S1)$. Hence, we like to investigate the model of all second inter-actions $((A1, A2), (A1, S1), (A2, S1))$ applying the exact test of the effect $(A1, A2, S1)$. The result $p = 0.0355$ indicated that the effect of co-occurring emotional instability and depressive moods is significantly higher than the expected additive effect. Indeed, the observed frequency 57 still exceeds the frequency 52.84 expected by the all-second-interaction model. The hypothesis is "highly interesting" because of the condition (2). The tested effect should express the "novelty" gained by the extension.

In general, there was no hypothesis found having the succedent "no recommendation". The selected hypotheses for the recommendations of counseling, psychotherapy, or psychiatry are shown in Table 4. The result volume was reduced considerably (cf. Table 3): Only 52 out of 206 satisfied hypotheses were selected. Clearly, the proposed criterion does not guarantee the exhaustiveness. It rather tries to reduce the redundancy and to focus on hypotheses potentially extending the knowledge already presented.

5   Inference in the simultaneous case

The GUHA procedure ASSOC generates and tests hypotheses; their number is "very high" in the rule. From the confirmatory point of view, this represents a problem. The statistical tests—used as quantifiers within the procedure—were theoretically developed for the "one question, one experiment, one answer" situation in order to guarantee a clearly limited probability of the falsely positive finding. The simultaneous inference can cause the increase of this probability to a hardly controllable extent—should each satisfied hypothesis be believed as a unique result. Nevertheless, formal significance levels obtained during the exploratory search can be interpreted in

Hypothesis	Fisher's exact test, p		
	single	simulation	Bonferroni
working problems ~ COUNSELING	<.000001	<.000001	.000011
fear of examinations ~ COUNSELING	.000003	.00065	.013
family problems ~ COUNSELING	.044	>.99	1
emotional instability ~ PSYCHOTHERAPY	<.000001	<.000001	<.000001
depressive moods ~ PSYCHOTHERAPY	<.000001	<.000001	.000018
eating disorders ~ PSYCHOTHERAPY	.000011	.0032	.053
psychiatric disorders ~ PSYCHOTHERAPY	.0030	.67	1
sexual problems ~ PSYCHOTHERAPY	.026	>.99	1
psychosomatic complaints ~ PSYCHOTHERAPY	.034	>.99	1
emotional instability & depressive moods ~ PSYCHOTHERAPY	<.000001	<.000001	<.000001
psychiatric disorders & working problems ~ PSYCHOTHERAPY	.0020	.52	1
family problems & working problems ~ PSYCHOTHERAPY	.021	>.99	1
psychiatric disorders ~ PSYCHIATRY	<.000001	.000002	.000042
suicidal thoughts ~ PSYCHIATRY	.000017	.0046	.081
addiction (alcohol, drugs) ~ PSYCHIATRY	.00026	.082	1
anxiety feelings ~ PSYCHIATRY	.025	>.99	1
anxiety feelings & working problems ~ PSYCHIATRY	.0025	.59	1
anxiety feelings & family problems ~ PSYCHIATRY	.0031	.68	1
working problems & obsessive thoughts ~ PSYCHIATRY	.0085	.97	1
working problems & depressive moods ~ PSYCHIATRY	.034	>.99	1
emotional instability & fear of examinations ~ PSYCHIATRY	.038	>.99	1
emotional instability & psychosomatic complaints ~ PSYCHIATRY	.038	>.99	1
anxiety feelings & working problems & family problems ~ PSYCHIATRY	.00098	.30	1
depressive moods & family problems & fear of examinations ~ PSYCHIATRY	.0054	.87	1
emotional instability & depressive moods & fear of examinations ~ PSYCHIATRY	.022	>.99	1
working problems & relationship problems & depressive moods ~ PSYCHIATRY	.030	>.99	1

Table 4. $N = 487$

different ways. In the context of logic of discovery, different approaches were carefully discussed by Tomáš Havránek (cf. [HH78, Chapter VIII]).

Hypothesis generation

The GUHA method generates and "offers" formally satisfied hypotheses. The researcher selects the relevant ones and examines them in a following confirmatory study. A hypothesis proposed by the procedure maybe one which was not on the researcher's mind before. The GUHA method fulfilled a heuristic function in this case. The historically first hypothesis found was surprisingly unsurprising: a male is not a female.

Explorative description

The goal is to describe the dependency structure in the multidimensional categorical data. Standard methods for multiway contingency tables can successfully at the same time manage data up to approximately seven dimensions. The number of variables concurrently investigated by the GUHA method can go considerably beyond this limit. The set of the presented hypotheses is understood as a global pattern; its interpretation does not rely so substantially on each local finding.

For instance, the "high-level" hypothesis of our study example was as follows: There is an association between the problems of a student and the recommendations proposed by the counselor. The solution—partly reproduced in Table 4—enables to gain a general impression about the character of problems suitable for various treatments. A limited number of falsely positive finding does not considerably change the general impression. At least the error rate falsely positive findings can be estimated.

Simultaneous inference

When an issue is the simultaneous correctness of all inferences, procedures for simultaneous inference have to be applied. Let n_h be the number of simultaneous tests, α_{crit} the considered critical significance level and α the significance level reached for a certain hypothesis. Instead of testing $\alpha \leq \alpha_{\text{crit}}$ in a single-hypothesis case, the principle of Bonferroni tests all hypotheses using a considerably more conservative level $\alpha \leq \alpha_{\text{crit}}/n_h$ in the sake to preserve the global error-probability level α_{crit}.

A general question of data analysis concerns the scope of hypotheses that should be considered as "simultaneously" tested. Paradoxes can be found when comparing the generally shared opinions to various situations in the data analysis. It is not clear why the simultaneous inference is generally recommended in some situations (like for the pairwise group comparison in the variance analysis) and not recommended in other cases (like for the inspection of the correlation matrix).

Hájek and Havránek (in [HH78], 8.2.3 and ff.) consider the solution set—

or even its subset selected by the researcher—as one possible universe of discourse. This makes a good sense especially in the above mentioned case of hypothesis generation where the solution of one study serves as a theoretical input for the following one. In the presented example it would mean to use the corrected critical value $.05/206 \approx .00024$.

The simultaneous confirmation of results *in the current study* would demand a much more rigorous approach. The set of all hypotheses H to be examined has a cardinality $n_h = \mathrm{card}(H)$. In the presented example, the corrected critical significance level would be $.05/4998 \approx .000010$. Hypotheses where marginal zeros in the frequency table occur can be excluded from the testing a-priori; the critical level in the example is then $.05/2790 \approx .000017$.

In practical applications of the GUHA method, the corrected significance levels are mostly much too rigorous. The Bonferroni procedure as well as its various improvements are based on the worst-case analysis. In the following section, we will propose a method of more powerful simultaneous inference for the ASSOC procedure.

6 Concurrent exploration and confirmation: a one-step procedure

Let us consider a problem to be solved by the ASSOC procedure. Limitations mentioned above are essential now: The information in the data matrix M is supposed to be complete and the sets of variables used for antecedent and succedent literals in the problem definition are mutually disjoint. The goal is to find the correction of the significance level enabling the simultaneous inference for the whole set of investigated hypotheses.

Two submatrices of the data matrix M should be investigated corresponding to vectors of antecedent, respectively succedent as stated in the problem definition (see Table 5 ignoring the symbol π at first). In our example, the most significant level among 4998 investigated hypotheses was $\alpha_{\min} \approx 0.152 \times 10^{-14}$ and it was reached with the hypothesis shown in the Table 2. We will assume both the antecedent and succedent vectors as fixed. As a general null hypothesis, we assume that both vectors are mutually independent. We ask what is the global probability that the best observed value $\alpha_{\min}$ is reached with at least one investigated hypothesis under these random conditions. The global probability will be used as a criterion both for the global test and for the simultaneous local tests of association.

Consider a permutation π on the set of m cases constituting a one-one mapping between antecedent and succedent vectors (see Table 5). In the data matrix $M(\pi)$, fixed antecedent and fixed succedent parts are newly recombined together. All hypotheses $h \in H$ of the original problem can be evaluated; let us denote $\alpha = t(h, \pi)$ the significance level reached for the hypothesis h and let us denote $\alpha_{\min}(\pi) = \min\{t(h, \pi) \mid h \in H\}$ the most significant level reached. Let Π be a set of all possible $m!$ permutations. Let α be a significance level used in a single-hypothesis inference; like the critical value or the significance level reached. Then we define the corrected significance level $u(\alpha)$ as follows:

$$\alpha_{\text{simult}} = u(\alpha) = P(\alpha_{\min}(\pi) \leq \alpha \mid \pi \in \Pi)$$

Let α_{crit} be a critical value considered. Let $H_1 \subseteq H_0$ be a subset of true null hypotheses and $H_2 = H_0 - H_1$ a set of remaining hypotheses; as Petr Hájek uses to say, "only the God knows" $H1$ and $H2$. Nevertheless, it holds for any $H_1 \subseteq H_0$:

$$P(\min\{t(h, \pi) \mid h \in H_1\} \leq \alpha_{\text{simult}} \mid \pi \in \Pi)$$
$$\leq P(\min\{t(h, \pi) \mid h \in H_0\} \leq \alpha_{\text{simult}} \mid \pi \in \Pi) \leq \alpha_{\text{crit}}$$

Hence, if we use α_{simult} as a more strict criterion for individual hypotheses then the global probability to obtain any falsely positive results remains limited by the desired level α_{crit}.

This approach—two fixed vectors and all permutations on the set of cases—is a generalization of the simple one-sided Fisher exact test.

"A little computational problem" is caused by the cardinality of the permutation set $\text{card}(\Pi) = m!$. Therefore, a subset $\Pi_S \subset \Pi$ can be used instead. The simultaneous significance levels can be estimated by using a large number of random permutations.

In our example, 10^7 random permutations were generated. Each permutation connected two submatrices of antecedent and succedent vectors in a new way (Table 5) and the most significant hypothesis was found. Resulting simultaneous critical value was $\alpha_{\text{simult}} = u(.05) = .000174$. This value constitutes a criterion that is about 17 times more advantageous than the Bonferroni procedure. In comparison, this is like if one would test just 288 hypotheses instead of 4998 ones.

Simultaneously confirmed hypotheses can be compared in Table 4. The Bonferroni procedure points to 6 hypotheses: "working problems" and "examination fears" indicate counseling; "emotional instability" and "depressive moods" as well as their concurrent occurrence indicate psychotherapy; "psychiatric disorders" indicate psychiatry. The proposed simultaneous procedure points to 9 hypotheses; it offers thus 3 additional hypotheses. One hypothesis regards the recommendation of the somatic treatment. Remaining two hypotheses are important from the clinical point of view: "eating disorders indicate psychotherapy" and "suicidal thoughts indicate psychiatry".

The solution can be seen as two nested sets of selected satisfied hypotheses. The inner one represents the simultaneously confirmed core of the solution

i	antecedent	π	$\pi(i)$	succedent
1	01100010001001000	$\leftrightarrow$	$\pi(1)$	110000
2	00101100000000000	$\leftrightarrow$	$\pi(2)$	100000
3	10011001010000000	$\leftrightarrow$	$\pi(3)$	110000
4	00000000001000000	$\leftrightarrow$	$\pi(4)$	010000
5	01011000000000000	$\leftrightarrow$	$\pi(5)$	010000
$\vdots$	$\vdots$	$\vdots$	$\vdots$	$\vdots$
485	11101010000000000	$\leftrightarrow$	$\pi(485)$	110000
486	00100000000000000	$\leftrightarrow$	$\pi(486)$	100000
487	10000100010000000	$\leftrightarrow$	$\pi(487)$	100000

Table 5.

consisting of nine mentioned hypotheses. The confirmation can be seen as secure as for an a-priori postulated single hypothesis.

The outer set contains remaining hypotheses that are interpreted in an exploratory way. Sets of antecedents associated to four selected succendents can be compared structurally. Let us order the recommendations in the following way:

no recommendation—counseling—psychotherapy—psychiatry.

(a) No particular problems were associated with "no recommendation"; rather common and daily problems of the studies were associated with counseling; more severe ones were associated with psychotherapy; and the most severe disorders like addiction or suicidal thoughts were associated with psychiatry. (b) W.r.t. this ordering, the number of hypotheses is increasing: 0, 3, 9 and 14 hypotheses. (c) W.r.t. this ordering, the antecedent length is increasing: from single literals for counseling to the conjunctions of three literals for psychiatry.

Conclusion: Ordering the four recommended treatments as shown, we can see that the latter ones are indicated by (a) more severe problems, (b) by a greater number of different problems, and (c) by problem combinations co-occurring concurrently.

7   Yin and Yang of the data analysis

The GUHA method generating hypotheses is anchored in the exploratory data analysis. One can imagine the presented set of exploratory hypotheses like a larger white surface in the Figure 1 and the confirmatory core like a small black circle within it. Analogously, classical confirmatory studies can be imagined like a large black area containing at least a small white exploratory part. In the real data analysis, both processes of exploration and confirmation follow and complement each other. The denial of the discovery part of the research processes would bring the data analysis into an unfavorable situation:

The light shines in the darkness, but the darkness has not understood it.

Figure 1. Exploratory (white, yang) and confirmatory (black, yin) data analysis.

ACKNOWLEDGEMENT

This text on the GUHA method within the frame of exploratory data analysis is a contribution to the celebration of Petr Hájek's anniversary. A heuristic search for names performed by the GUHA community is remembered in Appendix A; the title was inspired by the GUHA-style sci-fi story [Cla53]. Petr Hájek's work in logic and his interest for logical paradoxes [HPS00] are addressed in Appendix B.

BIBLIOGRAPHY

[APGP09]   R. Aschoff-Pluta, A. Gutbord, and D. Pokorný. Psychosoziale Beratungsstelle für Studierende, Universität Ulm. Jahresstatistik 2008 mit Vergleich zu den Vorjahren 2005–2007. Technical report, University of Ulm, 2009.

[Cla53]   A. C. Clarke. The nine billion names of God. In Frederik Pohl, editor, *Star Science Fiction Stories No. 1*. Ballatine Books, USA, 1953.

[HH78]   P. Hájek and T. Havránek. *Mechanizing Hypothesis Formation— Mathematical Foundations of a General Theory.* Springer, Berlin, 1978.

[HHC66]   P. Hájek, I. Havel, and M. Chytil. The GUHA-method of automatic hypotheses determination. *Computing*, 1(4):293–308, 1966.

[HHC83]   P. Hájek, T. Havránek, and M. K. Chytil. *Metoda GUHA – Automatická Tvorba Hypotéz.* Academia, Prague, 1983.

[HPS00]   Petr Hájek, Jeff Paris, and John C. Shepherdson. The liar paradox and fuzzy logic. *J. Symb. Logic*, 65(1):339–346, 2000.

Appendix A

The nine hundred names of GUHA

When the GUHA method was developed in the 1960's, the inventors checked its name General Unary Hypothesis Automaton very carefully. The abbreviation should not have any meaning in any existing language. Metoděj K. Chytil confirmed this by searching in numerous dictionaries and by asking many people.

The Czech language creates new words mostly by adding suffices to the already existing words. There is a great treasure of suffices. These suffices are a-priori meaningless. Nevertheless, there is some common feeling about their character. In the 1970's, the GUHA method became to be popular in broad circles of enthusiastic and creative people. A small continuing game has been developed: To build new words based on the name GUHA and to guess and dispute about their meaning. As a result, a fairy-tale-like GUHA world was created. Some examples—how I can remember them—follow.

GUHÁJEK, GUHAVRÁNEK Two highest ministers of king GUHA.

METODODĚJ The prophet of the GUHA method.

GUHAVEL The first GUHA machine constructor.

GUHOMIL, GUHOMÍR, GUHOSLAV Three brave sons of the old king GUHA.

GUHANA His clever daughter.

GUHATÝR A great hero fighting for the GUHA method. Not to be confused with German *Guhatier* = a GUHA animal.

GUHÁNEK A journeyman going with → GUHADLO to villages to find local hypotheses there.

GUHADLO A simple transportable wooden instrument for generating hypotheses.

GUHÁRNA (1) A manufacture where hypotheses are being produced, a very loud place. (2) A shop where sweets, coffee, and hypotheses can be consumed.

GUHINET A mechanized music generation automaton, being used at smaller castles.

GUHIŠTĚ A place (1) ... for recycling old hypotheses, (2) ... for impressive solution performances, (3) ... where children can play with low-level hypotheses.

GUHIVERSITA A place for relatively advanced GUHA studies.

One day, a guest from India visited the Centre of Biomathematics in Prague, and the surprise was on both sides. He saw the name GUHA written on numerous document boxes, and he asked, "WHO IS MISTER GUHA?" It turned out that Guha is one of the most common names in India. The inventors of GUHA did not manage to check it before.

Clearly, there was no GOOHAGLE in those times.

Appendix B

Logician and the miraculous fish

The fisherman was sitting on the riverbank. He was observing his fishing rod over the dark water and thinking about the barber. The barber on a ship who shaves those and only those men on board who do not shave themselves. The question is then: Who shaves the barber? If he does not shave himself then he does. And, if he does then he does not. And so forth. A good question for long hours of waiting by the river.

The fisherman loved such famous paradoxes. He was a logician by his heart and soul as well as profession. He had investigated plenty of logical axiomatic systems, and developed some himself. These systems enable to formulate logically sound questions that can, in some lucky cases, even lead to sound answers. The question about the barber was not a sound one. Where there is no question, there can be no answer; one can just be amused with this fascinating hook of words. The fisherman was in love with the beautiful secure world of logic as much as with this beautiful morning by the river.

The hot sun was climbing up the blue sky and the fisherman was thinking of his grandson. He had been asking so many wonderful questions recently. What if he asked who shaves the barber on the ship? Should the fisherman explain that this is not a good question? That it is not a question at all? Would the boy understand why reasonable adults forbid him to ask in such a way? Well, there are some secrets that should better stay unanswered. For instance, what does the little Child Jesus—bringing Christmas gifts to the Czech children—look like. However, the quite practical question of the barber's shaving does not belong to this mysterious circle. Suddenly, the fisherman did not find a logical solution of the barber's paradox so satisfactory.

The fisherman's contemplation was interrupted by something very unexpected for him: a fish on the hook. He took it carefully into his hands and looked at it, a nice grey little fish. The dialogue was opened by the fish: "Let me stay alive, please. I am a miraculous fish. If and only if you set me free I will fulfill any wish you ask." Normally, the fisherman was catching fish just accidentally and set them free every time. However, this seemed like *the* chance of his life. He asked carefully: "Did you say *any wish*?" and "Did you say *if-and-only-if*?" The fish confirmed: "Yes, if-and-only-if" and the fisherman asked: "Let me think a little bit." After a moment of silence, the fisherman said: "My wish is to eat you for dinner tonight."

The fish loved its life and the freedom to swim in all the waters. However, it was not its main problem at this moment. A miraculous fish has to be truthful and honest in any case. Miraculous fish can never ever lie because this would be in a contradiction with their miraculous nature, worth more than life. (The contrary behavior was observed by their near relatives, *liar fish* that are never allowed to tell the truth. Anatomically, the liar fish is very similar to the miraculous fish except for its very bright colors.) The current problem of the miraculous fish was obvious. If the fisherman will eat it then he cannot give the freedom to it; and if he gives it its freedom, he cannot eat it. The fisherman's wish was not satisfiable. The miraculous fish was facing a problem of such kind for the first time in its life.

To have the miraculous fish for dinner was by *no* means a wish of the fisherman. In fact, he was already looking forward to the promised cheese with fried potatoes. His true, hidden and great desire was: to hear the answer of the miraculous fish. The fisherman was absolutely sure about one issue. A fully satisfactory answer to a paradoxical question of this kind can be found ... by a miracle only.

Dan Pokorný
University of Ulm
Department of Psychosomatic Medicine and Psychotherapy
Am Hochsträß 8, D-89081 Ulm
Email: dan.pokorny@uni-ulm.de

The GUHA method and observational calculi: tools for data mining

JAN RAUCH[1]

Dedicated to Petr Hájek,
on the occasion of his 70th birthday

ABSTRACT. The paper presents the GUHA method and observation calculi, the development of which was started by Petr Hájek in 1960's, as important tools for data mining—today's vibrant discipline of informatics. The main features of the GUHA method and observational calculi are summarized and a survey of selected important related results is given. Experience with applications of GUHA procedures are presented and the importance of dealing with domain knowledge and analytical reports is emphasized. The possibilities of applying the theoretical results on observational calculi in dealing with domain knowledge are discussed.

1   Introduction

Data mining is a discipline of informatics the development of which started in the 1990's when lots of owners of databases realized the huge potential hidden in their data. Uncovering the information and knowledge hidden in the databases can remarkably enhance the possibilities of their owners. The book [Fay96] illustrates the state of the art in the mid-1990's. Today, data mining is both an established scientific discipline and important application area; see e.g. [QX06, kdn]. Note that one of the top ten algorithms in data mining is the apriori algorithm [Xin08] that mines for association rules [Agg96].

The GUHA method is an original Czech method of exploratory data analysis. Its aim is to provide all interesting facts from the analyzed data for the given problem. The method uses GUHA-procedures. A GUHA-procedure is a computer program, the input of which consists of the analyzed data and a simple definition of a large set of relevant patterns. The GUHA procedure automatically generates each particular pattern and tests if it is true for the analyzed data. The output consists of all prime patterns. The pattern is prime if it is true for the analyzed data and does not immediately follow from the other more simple output patterns.

[1] The work described here has been supported by the project 201/08/0802 of the Czech Science Foundation.

The most important GUHA procedure is the ASSOC procedure. It mines for patterns that can be seen as a generalization of association rules described in [Agg96]. However, the ASSOC procedure is much older than the apriori algorithm mining for association rules. The prehistory of the GUHA method goes back to 1954 [Háj04]. The first GUHA - related publication is [HHC66a]. There are lots of publications on the GUHA method, most of them are written or co-written by Petr Hájek, see e.g. [HHC66a, HHC66b, Háj68, HBR71, Háj73, HH77, Háj78, HH78, Háj81, HHC83, HH82, HI82, Háj04, HSZ95]. The book [HH78] and two special numbers of the International Journal of Man Machine Studies [Háj78, Háj81] are very important. The book [HH78] starts with two questions:

Q_1: Can computers formulate and justify *scientific* hypotheses?

Q_2: Can they comprehend empirical data and process it rationally, using the apparatus of modern mathematical logic and statistics to try to produce a rational image of the observed empirical world?

A theory developed in [HH78] to answer these questions is based on a scheme of inductive inference:

$$\frac{\text{theoretical assumptions, observational statement}}{\text{theoretical statement}} \, .$$

This means that if we accept theoretical assumptions and verify a particular statement about the observed data, we accept the conclusion - a theoretical statement. It is crucial that an intelligent statement about the observed data leads to theoretical conclusions, not the data itself.

Observational calculi formulas which correspond to suitable statements about the observed data are defined and studied in [HH78]. The association rules the GUHA procedure ASSOC mines for are understood as formulas of special observational calculi. Theoretically interesting and practically important results about association rules - formulas of observational calculi - are achieved. The observational problem and the GUHA method are introduced and investigated in [HH78] as tools to study possible answers to questions Q_1 and Q_2.

The ASSOC procedure has been implemented several times and applied many times, see e.g. [Háj78, Háj81, HHC83, HSZ95]. The intention of application of the GUHA method to databases led to a deeper study of association rules and new results were achieved in [Rau86]. The boom of association rules in the 1990's was the start of a new effort in the study of association rules as formulas of observational calculi, see e.g. [Rau97, Rau05, Rau08a, Rau10]. The results can be understood as the logic of association rules.

The enhanced GUHA procedure ASSOC as well as additional GUHA procedures were implemented and applied [HHR10, LDRŠ04, RŠ05a, RŠ05b,

RŠL05]. Experience of applying these procedures exists in various areas, e.g. in medical data analysis [RŠ07, RŠ08, RŠ09b, RT07].

This experience led to a strong belief in the importance of dealing with domain knowledge in particular steps of applications of GUHA procedures. Also the importance of the careful presentation of results to "problem owners" was recognized together with possibilities of the automatic preparation of analytical reports presenting the results of analyses and disseminating them through the Semantic web. Note: this is in accordance with the current research trends in data mining [QX06].

The goal of the paper is to introduce selected important results on observational calculi and the GUHA method, to summarize the experience from applying newly implemented GUHA procedures and to outline new related research topics. The principles used in [HH78] to answer questions Q_1 and Q_2 are outlined in Section 2 together with selected important results. An overview of the latest results on observational calculi is given in Section 3. Experience of applying GUHA procedures is summarized in Section 4. The possibilities of applying the results on observational calculi in dealing with domain knowledge are outlined in Section 5.

2   Principles

2.1   Logic of Discovery

A logic of discovery is developed in [HH78] to answer questions Q_1 and Q_2. The scheme of inductive inference inspired additional questions L0 - L4:

(L0) In what languages does one formulate observational and theoretical statements? (What is the syntax and semantics of these languages? What is their relation to the classical first order predicate calculus?)

(L1) What are rational inductive inference rules bridging the gap between observational and theoretical sentences? (What does it mean that a theoretical statement is justified?)

(L2) Are there rational methods for deciding whether a theoretical statement is justified (on the basis of given theoretical assumptions and observational statements)?

(L3) What are the conditions for a theoretical statement or a set of theoretical statements to be of interest (importance) with respect to the task of scientific cognition?

(L4) Are there methods for suggesting such a set of statements which is as interesting (important) as possible?

Answering questions (L0)–(L2) leads to a logic of induction, answers to questions (L3) and (L4) opens the way to a logic of suggestion. Answers to questions (L0)–(L4) constitute a logic of discovery. The rational inductive inference rules bridging the gap between observational and theoretical sentences are based on statistical approaches, i.e. estimates of various parameters or statistical hypothesis tests are used. This leads to theoretical statements about state dependent systems. A theoretical statement is justified if a condition concerning estimate of a used parameter or a statement given by a statistical hypothesis test are satisfied in the analyzed observed data.

Very detailed answers to questions (L0)–(L4) are given in the book [HH78] and a logic of discovery is developed. This paper involves two topics related to the logic of discovery, i.e. GUHA method and observational calculi. The GUHA method is defined as a solution of an observational research problem, see Section 2.2.

Question L0 leads, among other things, to defining observational predicate calculi. Observational predicate calculi are modifications of classical predicate calculi—only finite models are allowed and generalized quantifiers are added. The finite models correspond to analyzed data and generalized quantifiers correspond to statements on the observed data given by used statistical methods. Observational predicate calculi are introduced in Section 2.3.

Many results about observational predicate calculi can be seen as results about association rules. An overview of them is given in Section 2.4. Note: logical calculi of association rules are introduced in Section 3 together with an overview of new results. Some of the new results were achieved in [Rau86] as results about observational predicate calculi.

2.2   Observational Research Problem and the GUHA Method

Using induction rules based on statistical methods usually means there is 1:1 correspondence between observational and theoretical statements. The conceptual frame of scientific research is given by observational language and theoretical language and by collected data - an observational model. Observational sentences can be then classified as relevant or irrelevant questions. An observational sentence Φ is a *relevant question* if a decision whether Φ is true fot the given data or not is valuable because

- we do not know whether Φ is true

- if Φ is true then it leads via the inference rule, to an interesting theoretical hypothesis which is justified since Φ is true.

We call Φ a *relevant observational truth* if it is a relevant observational question and if it is true for the data.

Question L3 i.e. *What are the conditions for a theoretical statement or a set of theoretical statements to be of interest (importance) with respect to the task of scientific cognition?* can be converted into a similar question about observational statements. We are interested in a set RT of all relevant observational truths to get a set of theoretical statements interesting with regard to the given task of scientific cognition.

However, the set RT could be formidably large and it is reasonable to search for its concise representation. The notion of immediate conclusion is used and actually we are interested in a set X of truths such that each relevant truth is an immediate conclusion from X. The immediate conclusion is represented by a transparent inference rule I.

The *observational research problem* $\mathcal{P}$ is given by a set RQ of relevant questions and by inference rule I. The solution of a given observational research problem $\mathcal{P}$ is each set X of observational sentences where each relevant truth $\Phi \in RT$ can be inferred by I from X. The definition of the observational research problem (i.e. *r-problem*) leads to modifying **(L4)** to

(L4′) Are there methods for constructing good solutions to r-problems?

The GUHA method is defined in [HH78] as a method for solving of observational research problems. Several general aspects of the GUHA method are emphasized, i.e. the possibilities of choosing an appropriate type of relevant question and a satisfactory variable syntactical descriptions of sets of relevant questions. Note: this is only a very brief and informal sketch of both formal and informal large considerations given in [HH78].

The GUHA method is carried out using GUHA procedures. A GUHA procedure is a computer program the input of which consists of analyzed data and a set of parameters defining the large set of relevant questions. The output of the GUHA procedure is the set X representing the set of all relevant truths. It must be emphasized that the most used GUHA procedure ASSOC does not use the well known apriori algorithm [Agg96]. It is based on representing analyzed data by strings of bits [Rau78].

2.3 Monadic Observational Predicate Calculi

Observational predicate calculi are modifications of classical predicate calculi—only finite models are allowed and generalized quantifiers are added. The finite models correspond to the analyzed data and the generalized quantifiers correspond to statements on the observed data. The observed data is usually in the form of a data matrix. Predicates correspond to columns of data matrices - thus we are particularly interested in monadic observational predicate calculi.

We are dealing with predicate observational calculi, thus only two values i.e. *true* and *false* can be used. An example is the data matrix $\mathcal{M}$ in Fig. 1.

We assume the data matrix $\mathcal{M}$ is the result of observing objects $o_1, \ldots, o_n$ and their predicates. The values of functions $f_1, \ldots, f_K$ are results of observation. The functions correspond to the columns of the data matrix. 1 stands for *true* and 0 stands for *false*. We see that the value of function f_1 for object o_1 is 1 i.e. *true*, etc.

Monadic observational predicate calculus $\mathcal{MOPC}$ is a language for talking about the results of such observations. The predicates are symbols that correspond to the observed functions i.e. to the columns of the data matrices. Thus if the $\mathcal{MOPC}$ is intended to talk about observations such as the data matrix $\mathcal{M}$, it has K predicates usually named $P_1, \ldots, P_K$.

object	f_1	f_2	$\ldots$	f_K
o_1	1	0	$\ldots$	0
$\vdots$	$\vdots$	$\vdots$	$\vdots$	$\vdots$
o_n	0	1	$\ldots$	1

Figure 1. Data matrix $\mathcal{M}$

The formulas $(\forall x)P_1(x)$, $(\exists x)(P_1(x) \wedge P_2(x))$ and $P_3(x) \vee P_K(y)$ are examples of formulas of a $\mathcal{MOPC}$, the first two formulas are closed and have only classical quantifiers $\forall$, and $\exists$. These closed $\mathcal{MOPC}$ formulas can be evaluated using suitable contingency tables. The values of closed formulas $(\forall x)P_1(x)$ and $(\exists x)(P_1(x) \wedge P_2(x))$ can be defined using contingency tables T_1 and $T_{1\wedge 2}$, see Fig. 2.

$\mathcal{M}$	objects
P_1	u_1
$\neg P_1$	v_1

Table T_1

$\mathcal{M}$	objects
$P_1 \wedge P_2$	u_2
$\neg(P_1 \wedge P_2)$	v_2

Table $T_{1\wedge 2}$

Figure 2. Tables T_1 and $T_{1\wedge 2}$

Here u_1 is the number of objects described in the data matrix $\mathcal{M}$ satisfying P_1 (i.e. the number of rows for which $f_1(o) = 1$) and v_1 is the number of objects not satisfying P_1 (i.e. the number of rows for which $f_1(o) = 0$), similarly for u_2, v_2 and $P_1 \wedge P_2$. The formula $(\forall x)P_1(x)$ is true in $\mathcal{M}$ if and only if $v_1 = 0$, the formula $(\exists x)(P_1(x) \wedge P_2(x))$ is true in $\mathcal{M}$ if and only if $u_2 > 0$.

The $\mathcal{MOPC}$ formulas with generalized quantifiers expressing statements on the observed data corresponding to statistical hypothesis tests are also evaluated using contingency tables. We outline how a simple closed formula with a generalized quantifier is evaluated.

An example of a statistical hypothesis test is Fisher's test (on the level α) of the null hypothesis of the independence of P_1 and P_2 against the alternative one of the positive dependence. Fisher's quantifier $\sim_\alpha$ where $0 < \alpha \leq 0.5$ and the (closed) formula $(\sim_\alpha x)(P_1(x), P_2(x))$ is used to express the corresponding fact about P_1 and P_2 in the given data matrix $\mathcal{M}$.

The associated function $F_{\sim_\alpha}$ of Fisher's quantifier is used. This is a $\{0, 1\}$-valued function defined for all quadruples $\langle a, b, c, d \rangle$ of natural numbers satisfying $a + b + c + d > 0$ so that

$$
F_{\sim_\alpha}(a, b, c, d) = \begin{cases} 1 & \text{if } \sum_{i=a}^{\min(r,k)} \frac{\binom{k}{i}\binom{n-k}{r-i}}{\binom{r}{n}} \leq \alpha \wedge ad > bc \\ \\ 0 & \text{otherwise} \end{cases}
$$

where $r = a + b$, $k = a + c$, and $n = a + b + c + d$.

The formula $(\sim_\alpha x)(P_1(x), P_2(x))$ is evaluated in the data matrix $\mathcal{M}$ so that the associated function $F_{\sim_\alpha}$ of Fisher's quantifier $\sim_\alpha$ is applied to the four fold contingency table $4ft(P_1, P_2, \mathcal{M}) = \langle a, b, c, d \rangle$ of P_1 and P_2 in $\mathcal{M}$, see Fig. 3. Here a is the number of objects satisfying both P_1 and P_2 (i.e.

$\mathcal{M}$	P_2	$\neg P_2$
P_1	a	b
$\neg P_1$	c	d

$4ft(P_1, P_2, \mathcal{M})$

$\mathcal{M}$	ψ	$\neg\psi$
φ	a	b
$\neg\varphi$	c	d

$4ft(\varphi, \psi, \mathcal{M})$

Figure 3. Four fold contingency tables $4ft(P_1, P_2, \mathcal{M}$ and $4ft(\varphi, \psi, \mathcal{M})$

the number of rows in which both $f_1(o) = 1$ and $f_2(o) = 1$), b is the number of objects satisfying P_1 and not satisfying P_2, etc.

We say that the formula $(\sim_\alpha x)(P_1(x), P_2(x))$ is true in the data matrix $\mathcal{M}$ if $F_{\sim_\alpha}(a, b, c, d) = 1$, otherwise it is false. An exact description of the Fisher's quantifier is given in [HH78].

Each formula $(\sim_\alpha x)(\varphi(x), \psi(x))$, where both $\varphi(x)$ and $\psi(x)$ are quantifier free formulas, is evaluated in the same way i.e. the associated function $F_{\sim_\alpha}$ is applied to the contingency table $4ft(\varphi, \psi, \mathcal{M})$ of φ and ψ in $\mathcal{M}$, see Fig. 3. An example of this type of formula is the formula $(\sim_\alpha x)(P_1(x) \wedge P_2(x), P_3(x) \vee P_4(x))$.

The formula $(\sim_\alpha x)(\varphi(x), \psi(x))$ is closed, i.e. it has no free variables. However, $(\sim_\alpha x)(\varphi(x), \psi(y))$ has one free variable y. General definitions of atomic formulas, formulas with quantifiers of various types, formulas with free and bound variables, and values of all formulas are given in [HH78].

2.4   Results on Observational Predicate Calculi

Many both theoretically interesting and practically important results were achieved in [HH78], they cover detailed answers to questions L0 - L4. Let us mention some results involving formulas of the form $(\approx x)(\varphi(x),\ \psi(y))$ introduced above. These formulas can be understood as a generalization of association rules introduced in [Agg96], see below.

The Quantifier of Likely p-implication

First we introduce an additional generalized quantifier - the quantifier $\Rightarrow^{!}_{p,\alpha}$ of *likely p-implication* for $0 < p < 1$ and $0 < \alpha \leq 0.5$. It is also called the *lower critical implication*. The formula $(\Rightarrow^{!}_{p,\alpha}, x)(P_1(x), P_2(x))$ corresponds to a statistical binomial test on the significance level of α of the null hypothesis H0: $P(P_2|P_1) \leq p$ against the alternative one H1: $P(P_2|P_1) > p$. $P(P_2|P_1)$ means the conditional probability of P_2 given P_1.

The formula $(\Rightarrow^{!}_{p,\alpha}, x)(P_1(x), P_2(x))$ is evaluated using an associated function $F_{\Rightarrow^{!}_{p,\alpha}}$ of the quantifier $\Rightarrow^{!}_{p,\alpha}$. This is a $\{0,1\}$-valued function defined for all quadruples $\langle a, b, c, d \rangle$ of natural numbers satisfying $a + b + c + d > 0$:

$$
F_{\Rightarrow^{!}_{p,\alpha}}(a,b,c,d) = \begin{cases} 1 & \text{if } \sum_{i=a}^{a+b} \binom{a+b}{i} i (1-p)^{a+b-i} \leq \alpha \\ \\ 0 & \text{otherwise.} \end{cases}
$$

The formula $(\Rightarrow^{!}_{p,\alpha} x)(P_1(x), P_2(x))$ is true in the data matrix $\mathcal{M}$ if it is $F_{\Rightarrow^{!}_{p,\alpha}}(a,b,c,d) = 1$ where $\langle a,b,c,d \rangle = 4ft(P_1, P_2, \mathcal{M})$; see Fig. 3.

Quantifiers of Type $\langle 1, 1 \rangle$ and Association Rules

Both Fisher's quantifier $\sim_\alpha$ and the quantifier $\Rightarrow^{!}_{p,\alpha}$ of likely p-implication are examples of generalized quantifiers of the type $\langle 1, 1 \rangle$. Each quantifier $\approx$ of the type $\langle 1, 1 \rangle$ is used to express the relations of two predicates. Examples of closed formulas with such a quantifier are $(\approx x)(P_1(x),\ P_2(x))$ and $(\approx x)(P_1(x) \wedge P_2(x), P_3(x) \vee P_4(x))$.

A $\{0,1\}$-valued associated function $F_\approx(a,b,c,d)$ is defined for each quantifier $\approx$ of the type $\langle 1, 1 \rangle$ for all quadruples $\langle a, b, c, d \rangle$ of natural numbers satisfying $a+b+c+d > 0$. The functions $F_{\sim_\alpha}$ and $F_{\Rightarrow^{!}_{p,\alpha}}$ are examples of this associated function. The function $F_\approx(a,b,c,d)$ is used to evaluate formulas $(\approx x)(\varphi(x),\ \psi(x))$ where both $\varphi(x)$ and $\psi(x)$ are quantifier free formulas. The formula $(\approx x)(\varphi(x),\ \psi(x))$ is true in the data matrix $\mathcal{M}$ if and only if $F_\approx(a,b,c,d) = 1$ where $\langle a,b,c,d \rangle = 4ft(\varphi,\psi, \mathcal{M})$; see Fig. 3.

Let us define a generalized quantifier $\to_{C,S}$ with associated function $F_{\to_{C,S}}$ satisfying

$$
F_{\to_{C,S}}(a,b,c,d) = 1 \text{ if and only if } \frac{a}{a+b} \geq C \wedge \frac{a}{a+b+c+d} \geq S\ .
$$

Then, for example the rule $(\to_{C,S} x)(P_1(x) \wedge P_2(x), P_3(x) \wedge P_4(x))$ corresponds to the association rule $P_1 \wedge P_2 \to P_3 \wedge P_4$ with confidence C and support S, see [Agg96].

We can conclude that the closed formulas $(\approx x)(\varphi(x), \psi(x))$ where both $\varphi(x)$ and $\psi(x)$ are quantifier free formulas, and $\approx$ is a generalized quantifier of the type $\langle 1,1 \rangle$ can be understood as a generalization of association rules defined in [Agg96].

Classes of Quantifiers of Type $\langle 1,1 \rangle$

Two important classes of generalized quantifiers of the type $\langle 1,1 \rangle$ are defined using associated functions:

The quantifier $\approx$ is *associational* if the following holds for all quadruples of natural numbers $\langle a, b, c, d \rangle$, $\langle a', b', c', d' \rangle$ satisfying $a + b + c + d > 0$ and $a' + b' + c' + d' > 0$: if $F_{\approx}(a, b, c, d) = 1$ and $a' \geq a \wedge b' \leq b \wedge c' \leq c \wedge d' \geq d$ then also $F_{\approx}(a', b', c', d') = 1$.

The quantifier $\approx$ is *implicational* if the following holds for all quadruples of natural numbers $\langle a, b, c, d \rangle$ and $\langle a', b', c', d' \rangle$ satisfying $a + b + c + d > 0$ and $a' + b' + c' + d' > 0$: if $F_{\approx}(a, b, c, d) = 1$ and $a' \geq a \wedge b' \leq b$ then also $F_{\approx}(a', b', c', d') = 1$.

Both Fisher's quantifier $\sim_{\alpha}$ and the quantifier $\Rightarrow^!_{p,\alpha}$ of lower critical implication are associational, the quantifier $\Rightarrow^!_{p,\alpha}$ is also implicational. Fisher's quantifier $\sim_{\alpha}$ is not implicational. It is easy to show that each implicational quantifier is also associational. In addition, if the quantifier $\Rightarrow^*$ is implicational, then the value of its association function $F_{\Rightarrow^*}(a, b, c, d)$ depends neither on c nor on d and we can write only $F_{\Rightarrow^*}(a, b)$. It is however easy to prove that the generalized quantifier $\to_{Conf,Sup}$ corresponding to association rules defined in [Agg96] is not associational [Rau98b].

Additional important particular quantifiers of the type $\langle 1,1 \rangle$ are defined and studied in [HH78]. We introduce two simple examples of them: The quantifier $\Rightarrow_{p,Base}$ of *founded p-implication* with parameters $0 < p \leq 1$ and $Base > 0$ natural is defined by a $\{0,1\}$-valued associated function $F_{\Rightarrow_{p,Base}}$ satisfying

$$F_{\Rightarrow_{p,Base}}(a, b, c, d) = 1 \text{ if and only if } \frac{a}{a+b} \geq p \wedge a \geq Base .$$

It is easy to prove that the quantifier $\Rightarrow_{p,Base}$ is implicational.

The quantifier $\sim$ of *simple association* is defined by a $\{0,1\}$-valued associated function $F_{\sim}$ satisfying

$$F_{\sim}(a, b, c, d) = 1 \text{ if and only if } ad > bc .$$

The quantifier $\sim$ is associational, but it is not implicational.

Deduction Rules

Both the Fisher's quantifier $\sim_\alpha$ and the quantifier $\sim$ of simple association are *symmetric*. The quantifier $\approx$ is symmetric if its associated function $F_\approx$ satisfies $F_\approx(a,b,c,d) = F_\approx(a,c,b,d)$.

The deduction rule of symmetry involves *designated formulas*. The formula is designated if it is open (i.e. quantifier free) and if it contains only the designated variable x. The formulas $\varphi(x) = P_1(x) \wedge P_2(x)$ and $\psi(x) = P_3(x) \vee P_4(x)$ are examples of designated formulas. If φ and ψ are designated formulas and $\approx$ is a quantifier of the type $\langle 1,1 \rangle$ then we can write $\varphi \approx \psi$ instead of $(\approx x)(\varphi, \psi)$. The *deduction rule* SYM of *symmetry* is

$$\text{SYM} = \frac{\varphi \approx \psi}{\psi \approx \varphi} \text{ where } \varphi, \psi \text{ designated.}$$

Note: if $4ft(\varphi,\psi, \mathcal{M}) = \langle a,b,c,d \rangle$ then $4ft(\psi,\varphi, \mathcal{M}) = \langle a,c,b,d \rangle$. This means that the deduction rule SYM of symmetry is sound for symmetric quantifiers. An example: if the formula $P_1(x) \wedge P_2(x) \sim_\alpha P_3(x) \vee P_4(x)$ is true in the data matrix $\mathcal{M}$, then also $P_3(x) \vee P_4(x) \sim_\alpha P_1(x) \wedge P_2(x)$ is true in the data matrix $\mathcal{M}$, here $\sim_\alpha$ is Fisher's quantifier.

There is an additional important deduction rule SpRd that is sound for implicational quantifiers. The rule SpRd is called the *despecifying-dereducing rule*. It involves *elementary associations* i.e. formulas $\kappa \approx \delta$ where κ is an elementary conjunction and δ is an elementary disjunction. The *elementary conjunction* is a designated formula $\epsilon_1 P_{i_1}(x) \wedge \cdots \wedge \epsilon_k P_{i_k}(x)$ where each ϵ_j is either negation symbol $\neg$ or it is empty. The *elementary disjunction* is defined similarly as $\epsilon_1 P_{i_1}(x) \vee \cdots \vee \epsilon_k P_{i_k}(x)$.

The SpRd rule is defined on the set of all elementary associations as

$$\text{SpRd} = \frac{\kappa_1 \approx \delta_1}{\kappa_2 \approx \delta_2}$$

where $\kappa_1 \approx \delta_1$ results from $\kappa_2 \approx \delta_2$ by successive reduction and specification. We are not going to describe this process formally; we are only going to give an example. This means that we find δ_3 so that $\kappa_1 \approx \delta_1$ despecifies to $\kappa_2 \approx \delta_3$ and $\kappa_2 \approx \delta_3$ dereduces to $\kappa_2 \approx \delta_2$.

We start with the elementary association $P_1(x) \wedge P_2(x) \approx P_3(x)$. This despecifies to $P_1(x) \approx \neg P_2(x) \vee P_3(x)$. Despecification means that one element of the conjunction κ_1 (in our case $P_2(x)$) is moved from κ_1 to δ_1 and a negation sign is added (or removed). Then $P_1(x) \approx \neg P_2(x) \vee P_3(x)$ dereduces to $P_1(x) \approx \neg P_2(x) \vee P_3(x) \vee P_4(x)$. Dereduction means that a new element is added to the disjunction on the right side. Thus the formula $P_1(x) \approx \neg P_2(x) \vee P_3(x) \vee P_4(x)$ results from $P_1(x) \wedge P_2(x) \approx P_3(x)$ by successive dereduction and despecification.

The deduction rule SpRd is sound for implicational quantifiers, e.g. if the rule $P_1(x) \wedge P_2(x) \Rightarrow^!_{p,\alpha} P_3(x)$ is true in the data matrix $\mathcal{M}$ then the rule $P_1(x) \Rightarrow^!_{p,\alpha} \neg P_2(x) \vee P_3(x) \vee P_4(x)$ is also true in $\mathcal{M}$.

Some Additional Results

Additional results are achieved in [HH78], many of them are related to association rules. Special interest must be paid to results on logical calculi with missing information. There are also results on the definability of generalized quantifiers by the classical quantifiers $\forall$ and $\exists$. Calculi with qualitative values that are closely related to calculi of association rules are very important; see Section 3.

Intention to apply the GUHA method to databases led to new results on monadic predicate calculi [Rau86]. They involve new and more general deduction rules among association rules and deeper criteria for the definability of association rules in classical predicate observational calculi. Some of them are mentioned in Section 3. Many-sorted observational predicate calculi were also defined and studied in [Rau86], see also [Rau06b].

3   Logic of Association Rules

The boom of association rules in the 1990's was the start of a new effort in the study of association rules as formulas of observational calculi [Rau97, Rau98a, Rau98b, Rau05, Rau06a, Rau07, Rau08a, Rau08b, Rau09, Rau10]. The new results can be understood as the logic of association rules [Rau05]. The syntax of used formulas of observational calculi has been significantly simplified and special logical calculi formulas of which correspond to association rules in a very natural way were defined. These calculi are called logical calculi of association rules. An association rule is understood as a general relation of two Boolean attributes; see Section 3.1.

New classes of association rules are defined and studied, an overview of them is in Section 3.2. Important new results on deduction rules among association rules were achieved; they are listed in Section 3.3. There are also additional new results about dealing with missing information and the definability of association rules in classical predicate calculi [Rau08a, Rau08b].

3.1   Logical Calculi of Association Rules

Experience of teaching the GUHA method and discussions with data mining specialists with no or little experience of mathematical logic led to simplifying the language of observational predicate calculi. We are interested in association rules.

Predicate Calculus of Association Rules

Closed formulas $(\approx x)(\varphi(x), \psi(x))$ where both $\varphi(x)$ and $\psi(x)$ are quantifier free formulas of monadic predicate observational calculus and $\approx$ is a generalized quantifier of the type $\langle 1, 1 \rangle$, can be understood as a generalization of the association rules defined in [Agg96], see Section 2.4.

An example is the formula $(\approx x)(P_1(x) \vee P_2(x), P_3(x) \wedge P_4(x))$. This formula can be written as $P_1 \vee P_2 \approx P_3 \wedge P_4$. There is a simple one–one mapping between closed formulas $(\approx x)(\varphi(x), \psi(x))$ and formulas $\varphi \approx \psi$ where both $\varphi(x)$ and $\psi(x)$ are quantifier free formulas and φ and ψ are created as Boolean combinations of predicates $P_1, \ldots, P_K$. The mapping is carried out so that each $\varphi(x)$ is transformed to φ by omitting all substrings "(x)" and φ is transformed to $\varphi(x)$ by expanding each predicate P_i into $P_i(x)$.

A logical calculus with formulas $\varphi \approx \psi$ where both φ and ψ are Boolean combinations of predicates $P_1, \ldots, P_K$ is called *predicate calculus of association rules*. Modules of this calculus are $\{0, 1\}$ - data matrices, an example is the data matrix $\mathcal{M}$ in Fig. 1. The association rule $\varphi \approx \psi$ is true or false in the given data matrix $\mathcal{M}$. It is true if and only if $F_{\approx}(a, b, c, d) = 1$ where $F_{\approx}$ is an associated function of $\approx$ and $\langle a, b, c, d \rangle$ is a four-fold table $4ft(\varphi, \psi, \mathcal{M})$ of φ and ψ in $\mathcal{M}$, see Fig. 3.

Calculus of Association Rules

Association rules - formulas of predicate calculus of association rules concern $\{0, 1\}$ - data matrices. However most of the analyzed data matrices are not $\{0, 1\}$ - valued. There is a final set of possible values that can occur in each particular column of the data matrix. The columns of these data matrices are usually called *attributes*, their possible values are called *categories*.

These data matrices can be seen as data matrices of the type $T = \langle t_1, \ldots, t_K \rangle$ where $t_i \geq 2, i = 1, \ldots, K$ are natural numbers. The columns of data matrices are $f_1, \ldots, f_K$, the column f_i can only have the values $1, \ldots, t_i$. An example is data matrix $\mathcal{M}_T$; see Fig. 4.

object	f_1	f_2	$\ldots$	f_K	$A_1(5)$	$A_2(1, 3)$
o_1	4	1	$\ldots$	3	0	1
o_2	7	3	$\ldots$	9	0	1
$\vdots$	$\vdots$	$\vdots$	$\vdots$ $\vdots$ $\vdots$	$\vdots$	$\vdots$	$\vdots$
o_n	5	2	$\ldots$	7	1	0

Figure 4. Data matrix $\mathcal{M}_T$ of the type $\mathbf{T} = \langle t_1, \ldots, t_K \rangle$

Our goal is to define logical calculus formulas that correspond to general Boolean attributes derived from the columns of these data matrices. It can be done using basic Boolean attributes instead of predicates in the formulas of predicate calculus of association rules.

We assume there are attributes $A_1, \ldots, A_K$ corresponding to the columns $f_1, \ldots, f_K$ of matrices of the type $T = \langle t_1, \ldots, t_K \rangle$. A basic Boolean attribute is an expression of the form $A_i(\alpha)$ where $\alpha \subset \{1, \ldots, t_i\}$ for $i = 1, \ldots, K$. The basic Boolean attribute $A_i(\alpha)$ is true in a row o_j of the data matrix if and only if the value $v_{j,i}$ that is in the row o_j and column f_i satisfies $v_{j,i} \in \alpha$. We say that these Boolean attributes are of the type $\mathcal{T} = \langle t_1, \ldots, t_K \rangle$.

Fig. 4 contains examples $A_1(5)$ and $A_2(1,3)$ of basic Boolean attributes. In Fig. 4 we use 1 for *true* and 0 for *false*. Please note that we use $A_1(1,3)$ instead of $A_1(\{1,3\})$, similarly for additional basic Boolean attributes.

We use basic Boolean attributes instead of predicates in formulas of predicate calculus of association rules, thus

$$A_1(5) \sim_\alpha A_4(2) \quad \text{and} \quad A_1(5) \wedge A_2(1,3) \Rightarrow^!_{p,\alpha} A_4(2) \vee A_3(1,2)$$

are examples of association rules.

A logical calculus with formulas $\varphi \approx \psi$ where both φ and ψ are Boolean combinations of basic Boolean attributes of the type $\mathcal{T} = \langle t_1, \ldots, t_K \rangle$ is called *calculus of association rules*. The models of this calculus are data matrices of the type $\mathcal{T} = \langle t_1, \ldots, t_K \rangle$, an example is the data matrix $\mathcal{M}_{\mathcal{T}}$ in Fig. 4. The association rules $\varphi \approx \psi$ is true or false in the given data matrix $\mathcal{M}$. It is true if and only if $F_\approx(a,b,c,d) = 1$ where $F_\approx$ is an associated function of $\approx$ and $\langle a, b, c, d \rangle$ is a four-fold table $4ft(\varphi,\psi, \mathcal{M})$ of φ and ψ in $\mathcal{M}$, see Fig. 3. The generalized quantifiers used in (predicate) calculus of association rules are called 4ft-quantifiers.

Please note that we can also get logical calculus of association rules from calculi with qualitative values defined in [HH78] in a similar way to the way we get predicate calculus of association rules from observational monadic predicate calculi.

3.2   Classes of Association Rules

Two important classes of 4ft-quantifiers (i.e. generalized quantifiers of the type $\langle 1, 1 \rangle$) are defined in [HH78] - associational and implicational, see section 2.4 Additional classes are defined and studied in [Rau98a, Rau98b, Rau05, Rau07]. Each class of 4ft-quantifiers defines a class of association rules. If the 4ft-quantifier $\approx$ is implicational, then the association rule $\varphi \approx \psi$ is also implicational. The same is true for the additional classes of association rules.

The notion *Truth Preservation Condition* (i.e. TCF) is introduced [Rau97]. TCF is a condition concerning two four-fold contingency tables $\langle a, b, c, d \rangle$ and $\langle a', b', c', d' \rangle$. The conditions $a' \geq a \wedge b' \leq b \wedge c' \leq c \wedge d' \geq d$ and $a' \geq a \wedge b' \leq b$ used in [HH78] to define associational and implicational quantifiers can be also seen as TCF.

Most of additional classes are defined using particular TCF this way: The 4ft-quantifier $\approx$ is of the class Ω if $F_\approx(a,b,c,d) = 1 \wedge TPC_\Omega$ implies $F_\approx(a',b',c',d') = 1$ for all couples of 4ft tables $\langle a,b,c,d \rangle, \langle a',b',c',d' \rangle$ satisfying TPC_Ω. TPC_Ω is called a *true preservation condition for the class* Ω.

Important truth preservation conditions are listed in Tab. 1 together with the classes of 4ft-quantifiers they define.

Class		Truth preservation condition
Implicational	$TPC_{\Rightarrow}$	$a' \geq a \ \wedge \ b' \leq b$
Double implicational	$TPC_{\Leftrightarrow}$	$a' \geq a \ \wedge \ b' \leq b \ \wedge \ c' \leq c$
Σ-double implicational	$TPC_{\Sigma,\Leftrightarrow}$	$a' \geq a \ \wedge \ b' + c' \leq b + c$
Associational (i.e. Equivalence)	$TPC_{\equiv}$	$a' \geq a \ \wedge \ b' \leq b \ \wedge \ c' \leq c \ \wedge \ d' \geq d$
Σ-equivalence	$TPC_{\Sigma,\equiv}$	$a' + d' \geq a + d \ \wedge \ b' + c' \leq b + c$

Table 1. Classes of Quantifiers Defined by Truth Preservation Conditions

The class of associational quantifiers is also known as the class of equivalence quantifiers. One of reasons is that the 4ft-quantifier $\rightarrow_{C,S}$ corresponding to the association rule defined in [Agg96] is not associational. All implicational, double implicational, Σ-double implicational and Σ-equivalence quantifiers are associational [Rau98b].

The class of *quantifiers with the F-property* is an additional important class of association quantifiers. The 4ft-quantifier $\approx$ has the *F-property* if the following is true when $F_{\approx}(a, b, c, d) = 1$: if $b \geq c - 1 \geq 0$ then $F_{\approx}(a, b+1, c-1, d) = 1$ and if $c \geq b - 1 \geq 0$ then $F_{\approx}(a, b-1, c+1, d) = 1$. Note: there are various additional classes of 4ft-quantifiers [Rau98b, Rau10].

The 4ft-quantifiers $\Rightarrow_{p,Base}$ and $\Rightarrow^{!}_{p,\alpha}$ are implicational. The 4ft-quantifier $\Leftrightarrow_{p,Base}$ of *founded double implication* is for $0 < p \leq 1$, $Base > 0$ defined in [HHC83] so that $F_{\Leftrightarrow_{p,Base}} = 1$ if and only if $\frac{a}{a+b+c} \geq p \wedge a \geq Base$. This is an example of a double implicational quantifier, it is also Σ-double implicational. The 4ft-quantifier $\equiv_{p,Base}$ of *founded equivalence* is for $0 < p \leq 1$, $Base > 0$ defined in [HHC83] so that $F_{\equiv_{p,Base}} = 1$ if and only if $\frac{a+d}{a+b+c+d} \geq p \wedge a \geq Base$. This is an example of an Σ-equivalence quantifier.

Both Fisher's quantifier $\sim_{\alpha}$ and the quantifier $\sim$ of *simple association* are 4ft-quantifiers with the F-property, they are also associational. An additional important quantifier with F-property is the 4ft-quantifier $\Rightarrow^{+}_{p,Base}$ of *above average dependence* defined for $0 < p$ and $Base > 0$ in [Rau05] so that $F_{\Rightarrow^{+}_{p,Base}} = 1$ if and only if $\frac{a}{a+b} \geq (1+p)\frac{a+c}{a+b+c+d} \wedge a \geq Base$. The 4ft-quantifier $\sim^{+}_{p,Base}$ is however not associational.

3.3 Deduction Rules for Association Rules

Section 2.4 contains examples of important deduction rules. New important results on deduction rules concerning classes of association rules defined in

Section 3.2 were achieved [Rau98b]. The results involve the soundness of deduction rules of the form $\frac{\varphi \approx \psi}{\varphi' \approx \psi'}$ where both $\varphi \approx \psi$ and $\varphi' \approx \psi'$ are association rules - formulas of calculus of association rules. Reasonable criteria of the soundness of these deduction rules for important subclasses of 4ft-quantifiers have been proved.

We introduce these results for the class of *interesting implicational quantifiers* that is an important subclass of implicational quantifiers.

The implicational quantifier $\Rightarrow^*$ is interesting if it is a-dependent, b-dependent and if $F_{\Rightarrow^*}(0,0) = 0$; bear in mind that for implicational quantifiers we can only write $F_{\Rightarrow^*}(a, b)$ instead of $F_{\Rightarrow^*}(a, b, c, d)$, see section 2.4. The 4ft quantifier $\approx$ is *a-dependent* if there are non-negative integers a, a', b, c, d so that $\approx (a, b, c, d) \neq \approx (a', b, c, d)$ and analogously for b-dependent 4ft-quantifier. The following criterion of soundness is proved [Rau97, Rau98a, Rau05]:

If $\Rightarrow^*$ is an interesting implicational quantifier then there are formulas $\omega_{1A}, \omega_{1B}, \omega_2$ of propositional calculus created from φ, ψ, φ', ψ' so that the deduction rule $\frac{\varphi \Rightarrow^* \psi}{\varphi' \Rightarrow^* \psi'}$ is sound if and only if at least one of the following conditions are satisfied:

- Both ω_{1A} and ω_{1B} are tautologies.

- ω_2 is a tautology.

It is proved in [Rau86] that the important implicational quantifiers (e.g $\Rightarrow_{p, Base}$ of founded implication and $\Rightarrow^!_{p, \alpha, Base}$ of lower critical implication) are interesting implicational quantifiers. Similar theorems are proved for the classes of Σ-double implicational quantifiers, Σ-equivalence quantifiers and for the quantifiers with F-property, [Rau86, Rau98b, Rau05].

4   Applying the GUHA method

The most used implementation of the GUHA procedure ASSOC is the procedure 4ft-Miner that is part of the LISp-Miner system. It has very fine tools to define the set of relevant association rules to be verified, see section 4.1.

It is important that the association rules are simple enough for non-specialists in data mining. However the particular rules must be carefully presented and explained, a simple list of rules true in the given data is unacceptable. The true rules must be related to a reasonable analytical question. It is very important that the consequences of known domain knowledge are filtered out. To fulfill this requirement domain knowledge must be stored and maintained, see Section 4.2.

The best tool for presenting results to domain experts appears to be an analytical report. The analytical report answers a given reasonable analytical question and is structured according to the needs of the domain expert, see

Section 4.3. The idea of disseminating these analytical reports through the Semantic web is very challenging, Section 4.4 contains some comments.

Automatically producing reasonable analytical reports is a great challenge. It requires, among other things, automatically filtering the consequences of domain knowledge. Deduction rules in calculi of association rules can be used to solve this task. The principles are shown in Section 5.

Most of the presented experience is related to the procedure 4ft-Miner. There are however six additional GUHA procedures involved in the LISp-Miner system but their description is out of the scope of this paper, for more details see e.g. [HHR10, RŠ05b, RŠL05, RŠ09b, RŠ09a].

4.1   Defining Set of Relevant Association Rules

The procedure 4ft-Miner mines for association rules $\varphi \approx \psi$ [RŠ05a]. It also mines for conditional association rules, but we will not consider them here. The input of the 4ft-Miner consists of the analyzed data matrix and the parameters defining the set of association rules to be verified. The Boolean attribute φ is called *antecedent*, ψ is called *succedent* [HH78]. The definition of the set of association rules to be verified consists of a definition of a *set Φ of relevant antecedents*, a definition of a *set Ψ of relevant succedents* and a specification of the 4ft-quantifier $\approx$.

There are 18 possible 4ft-quantifiers and also conjunctions of these quantifiers can be used. All association rules $\varphi \approx \psi$ where $\varphi \in \Phi$, $\psi \in \Psi$, and φ and ψ have not common attributes are verified. We only outline the way in which the set Φ is defined, the set Ψ is defined analogously, for details see [RŠ05a].

The antecedent is a conjunction of partial antecedents $\varphi_1, \ldots \varphi_k$. There are definitions $\Phi_1, \ldots \Phi_k$ of sets of partial antecedents. Each antecedent φ is a conjunction $\varphi = \varphi_1 \wedge \cdots \wedge \varphi_k$ where $\varphi_i \in \Phi_i$ for $i = 1, \ldots, k$. Particular sets of relevant partial antecedents can contain an empty conjunction that is identically true. The minimum and maximum number of attributes in the antecedent φ can also be set.

Each particular set Φ_i of relevant partial antecedents is a set of conjunctions or a set of disjunctions of literals. The literal is a basic Boolean attribute $A(\alpha)$ or its negation $\neg A(\alpha)$. The set Φ_i is given by the choice of conjunctions/disjunctions, the minimum and maximum number of literals in each $\varphi_i \in \Phi_i$ and the specifications of the sets $\mathcal{B}(A'_1), \ldots, \mathcal{B}(A'_q)$ of relevant literals that are automatically created from the attributes $A'_1, \ldots, A'_q$. The set of attributes $\{A'_1, \ldots, A'_q\}$ is a subset of the set $\{A_1, \ldots, A_k\}$ of all attributes given by the analyzed data matrix. Each $\varphi_i \in \Phi_i$ is then created as conjunctions/disjunctions of literals, the number of used literals must be within the specified limits. Some additional detailed limitations can be used for the common occurrence of attributes.

We show how the sets of relevant literals are defined. We use attribute A with categories $\{1, 2, 3, 4, 5\}$. The set α in the literal $A(\alpha)$ is called the *coefficient* of $A(\alpha)$. The *length of the literal* is the number of categories in its coefficient.

The set of all literals to be generated for a particular attribute is given by: (i) the *type of coefficient* - there are six types (ii) the minimum and maximum length of the literal, (iii) one of the possible options: generate only positive literals $A(\alpha)$—generate only negative literals $\neg A(\alpha)$—generate both positive and negative literals. Examples of important types of coefficients are:

Subsets: definition of subsets of length 2-3 gives literals $A(1,2)$, $A(1,3)$, $A(1,4)$, $A(1,5)$, $A(2,3)$,   ..., $A(4,5)$, $A(1,2,3)$, $A(1,2,4)$, $A(1,2,5)$, $A(2,3,4)$, ..., $A(3,4,5)$.

Intervals: definition of intervals of length 2-3 gives literals $A(1,2)$, $A(2,3)$, $A(3,4)$, $A(4,5)$, $A(1,2,3)$, $A(2,3,4)$ and $A(3,4,5)$. If we have attribute *BMI* (Body Mass Index) with categories $15, 16, 17, \ldots, 34, 35$, then coefficients of the length 5-5 define a "sliding window" of the length 5 i.e. intervals $\langle 15, 19 \rangle$, $\langle 16, 20 \rangle$, ..., $\langle 30, 34 \rangle$, $\langle 31, 35 \rangle$.

Left cuts: a definition of left cuts with a maximum length of 3 defines literals $A(1)$, $A(1,2)$ and $A(1,2,3)$. Left cuts can be used to define intervals corresponding to the minimum values of the given attribute.

Right cuts: a definition of right cuts with a maximum length of 3 defines literals $A(5)$, $A(5,4)$, and $A(5,4,3)$. Right cuts can be used to define intervals corresponding to the maximum values of the given attribute.

The type *Cuts* means both left cuts and right cuts. Cuts can be used to investigate the extreme values of attributes.

Note: the ability to deal with basic Boolean attributes with automatically generated coefficients is an important feature of mining for association rules by 4ft-Miner. The usual apriori algorithm does not have this feature. The applications of the 4ft-Miner procedure confirm the usefulness of the wide possibilities to specify the set of relevant rules, see e.g. [RŠ07, RT07].

4.2   Dealing with Domain Knowledge

The fine possibilities of setting the set of relevant association rules can be used to investigate various relations among ordinal attributes. We can, for example, get rules stating that a high consumption of beer corresponds to high BMI.

There are also wide possibilities of transforming the original data to new attributes. We can, for example, define a new attribute *Weight of a patient* in kg with categories: intervals $\langle 20, 25 \rangle, \langle 25, 30 \rangle, \ldots, \langle 115, 120 \rangle$ and then by choosing a coefficient of the *Cuts* type with length 1–5 investigate patients both with low weight i.e. *Left cuts*: $\langle 20, 25 \rangle, \langle 20, 30 \rangle, \ldots, \langle 20, 45 \rangle$ and with high weight i.e. *Right cuts*: $\langle 95, 120 \rangle, \langle 100, 120 \rangle, \ldots, \langle 115, 120 \rangle$. Note: a left cut of the length 2 consists of the two most left categories i.e. intervals $\langle 20, 25 \rangle$ and $\langle 25, 30 \rangle$ and thus this left cut can be written as an interval $\langle 20, 30 \rangle$, the same applies for additional cuts.

These possibilities can be used to tune the set of relevant association rules very finely i.e. we can use the 4ft-Miner procedure to answer sophisticated questions and to deal with rules related to various items of domain knowledge. Thus it is natural to try to store and maintain reasonable items of domain knowledge in a formalized way and to use them in particular steps of data mining. There are two software systems for solving this task: *Knowledge base* - an inherent part of the LISp-Miner system [RŠ09b] and an independent web based software BKEF [Kli09]. Items of three types of domain knowledge can be maintained:

Details of particular attributes - name, list of possible values for nominal and ordinal attributes (e.g. *single/married/divorced/widowed* for the attribute *Marital status*), minimum and maximum values and details of suitable discretization for rational attributes etc.

Groups of attributes - an example is the group *Personal characteristics* containing the attributes *Sex, Age, Marital status, Beer consumption* etc., an additional example is the group *Cardiovascular risk factors* that contains the attributes *Hypertension, Obesity* (i.e. high BMI), *Smoking* etc.

Mutual influence between attributes - there are several types of influence among attributes, most of them are relevant to specified types of attributes. Two examples of dependencies between ordinal or rational-valued attributes A and B are $A \uparrow\uparrow B$ and $A \uparrow\downarrow B$. Item $A \uparrow\uparrow B$ of domain knowledge means that if attribute A increases then attribute B also increases. Item $A \uparrow\downarrow B$ of domain knowledge means that if attribute A increases then attribute B decreases. An example of the mutual influence between attributes is the item of domain knowledge *Beer consumption* $\uparrow\uparrow$ *BMI* stating that if beer consumption increases then BMI also increases.

The stored knowledge can be used in various ways, namely to input transformation of the analyzed data; to formulate reasonable analytical questions, to filter out the consequences of known items of domain knowledge and to arrange resulting analytical reports [Rau09, RŠ07, RŠ08].

A simple example of an analytical question created using groups of attributes is *"Are there any strong unknown relations between the Boolean characteristics of the group* Personal characteristics *and the Boolean characteristics of the group* Cardiovascular risk factors *in the given data set?"*. This question can be formally written as

$$\mathcal{B}(\textit{Personal characteristics}) \approx^? \mathcal{B}(\textit{Cardiovascular risk factors}) \ .$$

It can be solved by several sophisticated applications of the 4ft-Miner procedure. The items of domain knowledge such as *Beer consumption* $\uparrow\uparrow$ *BMI* can be used to filter out the association rules that are consequences of known facts. The results related to the logical calculi of association rules can be used, see also Section 5.

An additional example of an analytical question formulated using items of domain knowledge is the question: Are there exceptions to the item *Beer consumption*↑↑ *BMI*? This exception can be an association rule

$$Beer\ consumption(high)\ \land\ Physical\ activity\ (high) \Rightarrow_{0.9,50} BMI\ (normal)$$

where *Beer consumption(high)*, *Physical activity (high)*, and *BMI (normal)* are basic Boolean attributes with suitable coefficients - intervals found on the basis of an exhaustive search through all the possible intervals. This is, however, only a simple example, more sophisticated approaches can be used.

4.3   Analytical Reports

The analytical report answers a given analytical question. It is a textual document written with regard to the needs of the domain expert who is usually not a data mining expert. A sketch of an analytical report follows. The names of chapters are in `typewriter` and brief characterizations of particular chapters are given. The report answers the above formulated question $\mathcal{B}(Personal\ characteristics) \approx^? \mathcal{B}(Cardiovascular\ risk\ factors)$.

1. `Introduction`: Formulating the analytical question, the principles of solving it, description of the structure of the report.

2. `Analyzed data`: List and basic characteristics of the used attributes of both groups.

3. `Principles of analysis`: Association rules are explained in a form suitable for a non-specialist in data mining. Principles of 4ft-Miner are introduced together with the settings of its input parameters.

4. `Domain knowledge`: Items of domain knowledge that are considered to be known are listed, e.g. *Beer consumption*↑↑ *BMI*.

5. `Results of the analysis`. Lists of found true rules arranged in a suitable way is the core of this chapter. Suitable statements "of the second order" can be used, e.g. "There is no unknown relation involving beer consumption".

6. `Conclusions`
Summary and suggestions of further analysis.

Various specifications of the given question can be required, e.g. to find all patterns true in the analyzed data that do not follow from the given items of domain knowledge or to find only exceptions to the given items of domain knowledge.

4.4   Analytical Reports and the Semantic Web

One of ways of disseminating analytical reports is to present them using web technologies. There is already good experience in this field. Thus there is a natural challenge to prepare an analytical report using only analytical reports already presented on the web. The SEWEBAR research project is trying to answer this challenge [RŠ07].

SEWEBAR (i.e. *SEmantic WEB and Analytical Reports*) is a research project the goal of which is to study the possibilities of producing readable analytical reports answering reasonable analytical questions. An analytical question involving the given particular data set is called a *local analytical question*, the corresponding analytical report is *local analytical report*. There are also *global analytical reports* synthesizing the results presented in several additional analytical reports (local or global). Global analytical reports give answers to suitable *global analytical questions*. We assume that all we can use to answer global analytical questions are local and/or additional global analytical reports and that there is no possibility of getting the original local data and mining in it.

Let us consider a global analytical question "*What is the difference between Hospital$_1$ and Hospital$_2$ with regard to the relation of* Personal characteristics *and* Cardiovascular risk factors*?* ". We assume this analytical question can only be answered if we have local analytical reports answering the local questions *What interesting relations of* Personal characteristics *and* Cardiovascular risk factors *are true in* Hospital*?* both for Hospital$_1$ and for Hospital$_2$.

There are still many ways of answering these local analytical questions and preparing the local analytical report. In the first step we limit ourselves to local analytical reports that are prepared in two main steps:

1. All interesting relations of *Personal characteristics* and *Cardiovascular risk factors* described by association rules are found using the GUHA procedure 4ft-Miner.

2. The local analytical report is arranged on the basis of the found association rules.

Filtering the consequences of given items of domain knowledge can of course be used. A research activity is directed at this [Kli09, Rau09].

5   Deduction Rules and Domain Knowledge

An important task of dealing with domain knowledge is to remove all rules from the result of the 4ft-Miner procedure that can be understood as consequences of a given set of items of domain knowledge; see Section 4.2.

Bear in mind the notion of the *observational research problem* introduced in Section 2.2. The observational research problem $\mathcal{P}$ is given by a set RQ of relevant questions and inference rule I. The solution of a given observational research problem $\mathcal{P}$ is each set X of observational sentences so that each relevant truth $\Phi \in RT$ (RT is a set of all relevant observational truths) can be inferred by I from X. The GUHA method is a method for solving observational research problems, the output of the GUHA procedure is the set X.

We have the set of rules i.e. a solution X produced by the 4ft-Miner procedure. Thus each $\Phi \in X$ is an association rule of the form $\varphi \approx \psi$. Our task is to decide for each rule $\varphi \approx \psi \in X$ if it can be considered a consequence of a given set of items of domain knowledge. We give an outline of a solution to this problem.

We assume that we have only one item $\mathcal{IDK}$ of domain knowledge. If we have a set of items of domain knowledge, we can use the outlined approach for each particular item. We define a set $\mathcal{C}_{\mathcal{IDK}}$ of simple rules $\rho \approx \sigma$ so that the set $\mathcal{C}_{\mathcal{IDK}}$ can be naturally considered as a set of all the consequences of the item $\mathcal{IDK}$ in the form $\varphi \approx \psi$ that are as simple as possible. The set $\mathcal{C}_{\mathcal{IDK}}$ is called the *set of atomic consequences of* $\mathcal{IDK}$.

We consider a given rule $\varphi \approx \psi$ to be a consequence of $\mathcal{IDK}$ if either $\varphi \approx \psi$ is identical with a $\rho \approx \sigma \in \mathcal{C}_{\mathcal{IDK}}$ or if there is a $\rho \approx \sigma \in \mathcal{C}_{\mathcal{IDK}}$ so that $\varphi \approx \psi$ logically follows from $\rho \approx \sigma$ i.e. if the deduction rule $\frac{\rho \approx \sigma}{\varphi \approx \psi}$ is sound.

We give an example of the definition of a set of atomic consequences for a mutual influence between attributes. An example of such an item of domain knowledge is the expression *Beer consumption* $\uparrow\uparrow$ *BMI* saying that patients with higher beer consumption have higher BMI. More generally we have an item $A \uparrow\uparrow B$ of domain knowledge where A is an attribute with categories $1, \ldots, u$ and B is an attribute with categories $1, \ldots, v$.

Let us assume that we use the quantifier $\Rightarrow_{p,Base}$ of founded p-implication. This means that we have to define a set of rules in the form $\rho \Rightarrow_{p,Base} \sigma$ that can be naturally considered as a set of all the consequences of the item $A \uparrow\uparrow B$ and that are as simple as possible. In this case, we consider the simplest rules to be in the form $A(\alpha) \Rightarrow_{p,Base} B(\beta)$, we assume $\alpha \subset \{1, \ldots, u\}$ and $\beta \subset \{1, \ldots, v\}$.

The rule $A(low) \Rightarrow_{p,Base} B(low)$ stating that "If A is low then B is low" can be understood as a natural consequence of $A \uparrow\uparrow B$. The only problem is to define the coefficients α and β that can be understood as "low". This can be done so that we choose a natural A_{low}, $1 < A_{low} < u$ and B_{low}, $1 < B_{low} < v$ and then we consider α as small if and only if $\alpha \subset \{1, \ldots, A_{low}\}$ and β as small if and only if $\beta \subset \{1, \ldots, B_{low}\}$.

Also the rule $A(high) \Rightarrow_{p,Base} B(high)$ stating that "If A is high then B is high" can be understood as a natural consequence of $A \uparrow\uparrow B$. The coefficients

α and β can be defined as "high" in the following way. We choose a natural A_{high}, $1 < A_{low} < A_{high} < u$ and B_{high}, $1 < B_{low} < B_{high} < v$ and then we consider α as high if and only if $\alpha \subset \{A_{high}, \ldots, v\}$ and β as high if and only if $\beta \subset \{B_{high}, \ldots, v\}$.

The set of all rules $A(low) \Rightarrow_{p,Base} B(low)$ and $A(high) \Rightarrow_{p,Base} B(high)$ can be considered as the set $\mathcal{C}_{A\uparrow\uparrow B}$ of atomic consequences of $A \uparrow\uparrow B$. Note: the set $\mathcal{C}_{A\uparrow\uparrow B}$ can be defined in a finer way, e.g. by adding rules of the form $A(medium) \Rightarrow_{p,Base} B(medium)$ with a suitable definition of "medium". Then rules $A(low, medium) \Rightarrow_{p,Base} B(medium)$, $A(low, medium) \Rightarrow_{p,Base} B(medium, high)$, and $A(medium) \Rightarrow_{p,Base} B(medium, high)$ can also be added. We can even use rules $A(very\ low) \Rightarrow_{p,Base} B(very\ low)$ and $A(very\ high) \Rightarrow_{p,Base} B(very\ high)$ etc.

This gives us a reasonable set $\mathcal{C}_{A\uparrow\uparrow B}$ of rules $A(\alpha) \Rightarrow_{p,Base} B(\beta)$ that can be considered as a set of atomic consequences of the item $A \uparrow\uparrow B$. Note: the set $\mathcal{C}_{A\uparrow\uparrow B}$ depends on the 4ft-quantifier $\Rightarrow_{p,Base}$ and thus we should write $\mathcal{C}_{A\uparrow\uparrow B,\Rightarrow_{p,Base}}$ and call it a set of atomic consequences of $A \uparrow\uparrow B$ for $\Rightarrow_{p,Base}$.

The set $\mathcal{C}_{A\uparrow\uparrow B,\Rightarrow_{p,Base}}$ of atomic consequences of $A \uparrow\uparrow B$ can be used to filter out all rules $\varphi \Rightarrow_{p,Base} \psi$ that can be considered as a consequence of $A \uparrow\uparrow B$. The rule $\varphi \Rightarrow_{p,Base} \psi$ is considered as a consequence $A \uparrow\uparrow B$ if it is in the form $A(\alpha) \Rightarrow_{p,Base} B(\beta)$ and $A(\alpha) \Rightarrow_{p,Base} B(\beta) \in \mathcal{C}_{A\uparrow\uparrow B,\Rightarrow_{p,Base}}$ or if there is a $A(\alpha) \Rightarrow_{p,Base} B(\beta) \in \mathcal{C}_{A\uparrow\uparrow B,\Rightarrow_{p,Base}}$ so that the deduction rule $\frac{A(\alpha)\Rightarrow_{p,Base}B(\beta)}{\varphi\approx\psi}$ is sound according to criterion given in section 3.3.

The principle of defining the set of atomic consequences for five additional items of mutual influence between attributes and for the quantifiers $\Rightarrow_{p,Base}$ of founded p-implication, $\Leftrightarrow_{p,Base}$ of founded double implication, $\equiv_{p,Base}$ of *founded equivalence* and $\Rightarrow^{+}_{p,Base}$ of above average dependence introduced in Section 3.2 is outlined in [Rau09].

Note: there is a natural requirement on the reasonable consistency of the set $\mathcal{C}_{\mathcal{IDK}}$ of atomic consequences of the item of domain knowledge $\mathcal{IDK}$ i.e. there cannot be two atomic consequences $\rho_1 \approx \sigma_1$ and $\rho_2 \approx \sigma_2$ that contradict each other. A detailed discussion of this topic is however without the scope of this paper.

6   Conclusions

We have introduced the GUHA method and observational calculi the research of which was started by Petr Hájek in the 1960's. We have presented their main principles and results achieved in the book [HH78]. In addition we have introduced new theoretical results related to the GUHA method and observational calculi. We have argued for their usefulness to data mining research, namely to association rules.

We have presented the main experience with applications of the 4ft-Miner GUHA procedure that is the current implementation of the ASSOC procedure defined in [HH78]. We can conclude that its fine possibilities of tuning the set of relevant association rules means that various, even sophisticated, analytical questions can be solved. The presentation of results to domain experts is very important. Analytical reports seem to be a reasonable way of presenting the results. We consider reports that substantially use the fact that the GUHA procedure offers *all* the relevant facts to the given question. This leads to the possibility of disseminating analytical reports according to the principles of the Semantic web. Formalized domain knowledge is very important. It can be used in various ways in the data mining process—from formulating of reasonable analytical questions to arranging analytical reports.

We have also outlined some current research topics related to the GUHA method and observational calculi and have shown that they are relevant to the most challenging data mining problems.

BIBLIOGRAPHY

[Agg96]   R. Aggraval et al. Fast discovery of association rules. In U. M. Fayyad et al., editor, *Advances in Knowledge Discovery and Data Mining*, pages 307–328. AAAI Press, Menlo Park, 1996.

[Fay96]   U. M. Fayyad et al., editor. *Advances in Knowledge Discovery and Data Mining*, Menlo Park, 1996. AAAI Press.

[Háj68]   P. Hájek. Problém obecného pojetí metody GUHA. *Kybernetika*, 4(4):505–515, 1968.

[Háj73]   P. Hájek. Automatic listing of important observational statements I–II. *Kybernetika*, 9:187–205, 251–271, 1973.

[Háj78]   P. Hájek, editor. *GUHA*, volume 10 of *Int. J. Man-Machine Stud.*, 1978. Special issue.

[Háj81]   P. Hájek, editor. *GUHA II*, volume 15 of *Int. J. Man-Machine Stud.*, 1981. Special issue.

[Háj04]   P. Hájek. Metoda GUHA v minulém století a dnes. In V. Snášel, editor, *Znalosti 2004*, pages 10–20, Ostrava, 2004. FEI VŠB.

[HBR71]   P. Hájek, K. Bendová, and Z. Renc. The GUHA method and the three valued logic. *Kybernetika*, 7(6):421–435, 1971.

[HH77]   P. Hájek and T. Havránek. On generation of inductive hypotheses. *Int. J. Man-Machine Stud.*, 9(4):415–438, 1977.

[HH78]   P. Hájek and T. Havránek. *Mechanizing Hypothesis Formation—Mathematical Foundations of a General Theory*. Springer, Berlin, 1978.

[HH82]   P. Hájek and T. Havránek. GUHA 80: An application of artificial intelligence to data analysis. *Computers and Artificial Intelligence*, 1(2):107–134, 1982.

[HHC66a]   P. Hájek, I. Havel, and M. Chytil. The GUHA-method of automatic hypotheses determination. *Computing*, 1(4):293–308, 1966.

[HHC66b]   P. Hájek, I. Havel, and M. Chytil. GUHA-metoda automatického zjišťování hypotéz. *Kybernetika*, 2(1):31–47, 1966.

[HHC83]   P. Hájek, T. Havránek, and M. K. Chytil. *Metoda GUHA – Automatická Tvorba Hypotéz*. Academia, Prague, 1983.

[HHR10]   P. Hájek, M. Holeňa, and J. Rauch. The GUHA method and its meaning for data mining. *J. of Computer and System Sci.*, 76(1):34–48, 2010.

[HI82]   P. Hájek and J. Ivánek. Artificial intelligence and data analysis. In H. Causs-
            inus, P. Ettinger, and R. Tomassone, editors, *COMPSTAT '82*, pages 54–60,
            Wien, 1982. Physica Verlag.
[HSZ95]   P. Hájek, A. Sochorová, and J. Zvárová. GUHA for personal computers. *Com-
            putational Statistics and Data Analysis*, 19(2):149–153, 1995.
[kdn]   http://www.kdnuggets.com/.
[Kli09]   T. Kliegr et al. Semantic analytical reports: A framework for post-processing
            data mining results. In J. Rauch et al., editor, *Foundations of Intelligent
            Systems. Proceedings of ISMIS 2009*, page 8898. Springer, 2009.
[LDRŠ04]   V. Lín, P. Dolejší, J. Rauch, and M. Šimůnek. The *KL-Miner* procedure for
            datamining. *Neural Network World*, 5(4):411–420, 2004.
[QX06]   Y. Qiang and W. Xindong. 10 challenging problems in data mining research.
            Int. J. of Information Technology & Decision Making, 5(4):597–604, 2006.
[Rau78]   J. Rauch. Some remarks on computer realisations of GUHA procedures. *Int.
            J. Man-Machine Stud.*, 10:23–28, 1978.
[Rau86]   J. Rauch. *Logical Foundations of Hypothesis Formation from Databases*.
            PhD thesis, Mathematical Institute of the Czechoslovak Academy of Sciences,
            Prague, 1986. In Czech.
[Rau97]   J. Rauch. Logical calculi for knowledge discovery in databases. In J. Ko-
            morowski and J. M. Żytkow, editors, *Principles of Data Mining and Knowl-
            edge Discovery*, volume 1263 of *LNAI*, pages 47–57. Springer, 1997.
[Rau98a]   J. Rauch. Classes of four-fold table quantifiers. In J. M. Żytkow and
            M. Quafafaou, editors, *Principles of Data Mining and Knowledge Discovery*,
            volume 1510 of *LNAI*, pages 203–211. Springer, 1998.
[Rau98b]   J. Rauch. Contribution to logical foundations of KDD, 1998. Assoc. Prof. The-
            sis at Faculty of Informatics and Statistics. University of Economics, Prague.
            In Czech.
[Rau05]   J. Rauch. Logic of association rules. *Applied Intelligence*, 22:9–28, 2005.
[Rau06a]   J. Rauch. Definability of association rules in predicate calculus. In T. Y. Lin
            et al., editor, *Foundations and Novel Approaches in Data Mining*, volume 9
            of *Studies in Computational Intelligence*, pages 23–40. Springer, 2006.
[Rau06b]   J. Rauch. Many sorted observational calculi for multi-relational data mining.
            In *Data Mining Workshops*, pages 417–422. IEEE Computer Society Press,
            2006.
[Rau07]   J. Rauch. Observational calculi, classes of association rules and F-property.
            In T. Y. Lin et al., editor, *Granular Computing 2007*, pages 287–293, Los
            Alamitos, 2007. IEEE Computer Society Press.
[Rau08a]   J. Rauch. Classes of association rules—an overview. In T. Y. Lin et al., editor,
            Datamining: Foundations and Practice, volume 118 of *Studies in Computa-
            tional Intelligence*, pages 283–297. Springer, 2008.
[Rau08b]   J. Rauch. Definability of association rules and tables of critical frequencies. In
            T. Y. Lin et al., editor, *Datamining: Foundations and Practice*, volume 118
            of *Studies in Computational Intelligence*, pages 299–321. Springer, 2008.
[Rau09]   J. Rauch. Considerations on logical calculi for dealing with knowledge in
            data mining. In Z. W. Ras and A. Dardzinska, editors, *Advances in Data
            Management*, pages 177–202. Springer, 2009.
[Rau10]   J. Rauch. Logical aspects of the measures of interestingness of association
            rules. In J. Koronacki et al., editor, *Recent Advances in Machine Learning
            (Dedicated to the Memory of Ryszard S. Michalski)*, Studies in Computational
            Intelligence, pages 175–203. Springer, 2010.
[RŠ05a]   J. Rauch and M. Šimůnek. An alternative approach to mining association
            rules. In T. Y. Lin, S. Ohsuga, C. J. Liau, and S. Tsumoto, editors, *Data
            Mining, Foundations, Methods, and Applications*, pages 219–238. Springer,
            2005.

[RŠ05b]   J. Rauch and M. Šimůnek. GUHA method and granular computing. In X. Hu et al., editor, *Proceedings of IEEE conference Granular Computing 2005*, pages 630–635. IEEE Computer Society Press, 2005.

[RŠ07]   J. Rauch and M. Šimůnek. Semantic web presentation of analytical reports from data mining - preliminary considerations. In *Web Intelligence*, pages 3–7. IEEE Computer Society Press, 2007.

[RŠ08]   J. Rauch and M. Šimůnek. LAREDAM - considerations on system of local analytical reports from data mining. In A. An et al., editor, *Foundations of Intelligent Systems. Proceedings of ISMIS 2008*, volume 4994 of *LNAI*, pages 143–149. Springer, 2008.

[RŠ09a]   J. Rauch and M. Šimůnek. Action rules and the GUHA method: Preliminary considerations and results. In J. Rauch et al., editor, *Foundations of Intelligent Systems. Proceedings of ISMIS 2009*, pages 76–87. Springer, Berlin, 2009.

[RŠ09b]   J. Rauch and M. Šimůnek. Dealing with background knowledge in the SEWE-BAR project. In B. Berendt et al., editor, *Knowledge Discovery Enhanced with Semantic and Social Information*, pages 89–106. Springer, 2009.

[RŠL05]   J. Rauch, M. Šimůnek, and V. Lín. Mining for patterns based on contingency tables by KL-Miner—first experience. In T. Y. Lin et al., editor, *Foundations and Novel Approaches in Data Mining*, pages 155–167. Springer, 2005.

[RT07]   J. Rauch and M. Tomečková. System of analytical questions and reports on mining in health data - a case study. In J. Roth et al., editor, *Proceedings of IADIS European Conference Data Mining 2007*. IADIS Press, 2007.

[Xin08]   W. Xindong et al. Top 10 algorithms in data mining. *Knowledge and Information Systems*, 14:1–37, 2008.

Jan Rauch
Faculty of Informatics and Statistics
University of Economics, Prague
nám. W. Churchilla 4
130 67 Prague, Czech Republic
Email: rauch@vse.cz

Fuzzification of some statistical principles of GUHA

MARTIN HOLEŇA

Dedicated to Petr Hájek,
on the occasion of his 70th birthday

ABSTRACT. The probably best known parts of Petr Hájek's legacy are the knowledge discovery method GUHA and formal fuzzy logic. There is an obvious connection between them: fuzzy logic generalizes the classical Boolean logic, a variant of which forms the logical principles of GUHA. However, there is actually also another, less known connection of fuzzy logic to the GUHA method: its connection to hypotheses testing, which forms one of the key components of the statistical principles of GUHA. An overview of research into fuzzy hypotheses testing in the context of the GUHA method is the topic of the present chapter. A motivation for that research is explained and a survey of the published results is given.

1   Introduction

From mid 1960s till early 1990s, Petr Hájek paid much attention to the *method for knowledge discovery GUHA*. From time to time, GUHA was even his main concern, for example when writing the monograph [HH78]. Since mid 1990s, on the other hand, his research clearly focused on *formal fuzzy logic*, for which he elaborated theoretical fundamentals in his monograph [Háj98]. Needless to say, GUHA and fuzzy logic are not unrelated: fuzzy logic is a generalization of the classical Boolean logic, and a variant of Boolean logic—observational logic—forms the logical fundamentals of the GUHA method. However, it would be incorrect to simplify the relationship between GUHA and fuzzy logic only to the relationship between Boolean and fuzzy logic. The reason is that the GUHA method relies not only on logical, but also on statistical principles. In my opinion, the close connection between logic and statistics is actually the most remarkable feature of GUHA. I am not aware of any other knowledge discovery method in which that connection is so rigorously elaborated as in the GUHA method.

I learned about GUHA not from Petr Hájek, but from the other principal author of the method, responsible for its statistical fundamentals, the late Tomáš Havránek. Moreover, I even first personally met Petr through Tomáš

Havránek. I recall here all this to express my gratitude to Tomáš for the influence he had on me during my PhD studies, and also to explain why my research into fuzzy generalization of GUHA did not concern its logical principles, but the statistical principles. More precisely, it concerned *testing of fuzzy hypotheses*, which has been, independently of the GUHA method, investigated for more than 20 years [Arn96, Arn98, Buc04, Buc05, Cas93, DVV85, GH97, Ngu06, SS90, TB01, WI93].

In this chapter, the motivation for that research is outlined, and its main results are summarized. The next section reviews the statistical fundamentals of the generalized quantifiers of observational logic used in GUHA. The relevance of fuzzy hypotheses testing to GUHA, which motivated my research into this subject, is discussed in Section 3. Finally, Section 4 surveys the key results of the research, which have been published in the papers [Hol98] and [Hol04] in the journal Fuzzy Sets and Systems.

2 Review of statistical fundamentals of GUHA quantifiers

To understand the context of fuzzification of statistical properties of GUHA, it is necessary to recall that an overwhelming majority of generalized quantifiers used in the GUHA method have a statistical motivation. GUHA adopts the statistical point of view of data about a set of objects as a random sample from some probability distribution. Then also the evaluations $\|\varphi\|, \|\psi\|, \ldots$ of Boolean predicates $\varphi, \psi, \ldots$ on those objects are a random sample, but this time a random sample from a probability distribution on $\{0, 1\}^m$, where m denotes the arity of the quantifier (typically, binary quantifiers are considered, thus $m = 2$). The truth functions of generalized quantifiers are then defined in such a way that they correspond to the application of some statistical method to such random samples (cf. the examples below). In particular, their definitions correspond to two important types of statistical methods: parameter estimation and hypotheses testing.

1. Generalized quantifiers based on parameter estimation: the truth function states that some estimator of some parameter of the probability distribution of the random sample $\|\varphi\|, \|\psi\|, \ldots$ fulfills some prescribed condition. Actually, only estimators based on the contingency table of that random sample are used in GUHA. For binary quantifiers, this is the four-fold table

$$
(1) \quad
\begin{array}{c|cc|c}
 & \psi & \neg\psi & \sum \\
\hline
\varphi & a & b & r \\
\neg\varphi & c & d & s \\
\hline
\sum & k & l & n \\
\end{array}
$$

Consequently, the truth function simplifies to a function on quadruples of non-negative integers, with values $\mathrm{Tf}(a, b, c, d) \in \{0, 1\}$. Moreover, the sentence $(Qx)(\varphi(x), \psi(x))$, for simplicity written $\varphi Q \psi$, is usually required to have at least a prescribed minimal support in data, i.e., $a \geq A$, respectively $\frac{a}{n} \geq S$ for some constant A, respectively S.

The best known quantifier of that kind is the *founded implication* $\to_\theta$, for which the conditional probability $P(\|\psi\| = 1 \mid \|\varphi\| = 1)$ that a predicate ψ is true conditioned on a predicate φ being true for the same object is estimated using the unbiased estimator $\frac{a}{a+b}$, and the prescribed condition is to reach at least a given value $\theta \in (0, 1)$. Hence, the definition of the truth function $\mathrm{Tf}_{\to_\theta}$ including a requirement for minimal support in data is

$$(2) \quad \mathrm{Tf}_{\to_\theta}(a, b, c, d) = \begin{cases} 1 & \text{if } \frac{a}{a+b} \geq \theta \ \& \ a \geq A, \\ 0 & \text{else.} \end{cases}$$

2. Generalized quantifiers based on hypotheses testing: the truth function corresponds to the result of some statistical test of some null hypothesis H_0 concerning the probability distribution of the random sample $\|\varphi\|, \|\psi\|, \ldots$ against some alternative H_1. Similarly to quantifiers based on parameter estimation, only hypotheses concerning the contingency table of that random sample are tested (for example, the hypothesis that the probability distributions of the columns and the rows of the table are independent). If t denotes the test statistics of a test and $C_\alpha \in (0, 1)$ defines its critical region on a significance level $\alpha \in (0, 1)$, and if Q is a binary quantifier with the table (1), its truth function Tf_Q including the requirement $a \geq A$ has to be defined in one of the following ways:

$$(3) \quad \mathrm{Tf}_Q(a, b, c, d) = \begin{cases} 1 & \text{if } t \in C_\alpha \ \& \ a \geq A, \\ 0 & \text{else;} \end{cases}$$

$$(4) \quad \mathrm{Tf}_Q(a, b, c, d) = \begin{cases} 1 & \text{if } t \notin C_\alpha \ \& \ a \geq A, \\ 0 & \text{else.} \end{cases}$$

An advantage of the definition (3) compared to the definition (4) is that if the condition for the minimal support is fulfilled, the validity of the sentence $\varphi Q \psi$ coincides with rejecting the H_0 in favour of H_1. Consequently, $\varphi Q \psi$ can be interpreted as "H_1 holds with probability at least $1 - \alpha$". Therefore, most of the GUHA quantifiers based on hypotheses testing are defined in accordance with (3).

Most frequently used among the generalized quantifiers of this kind are the quantifiers *lower critical implication*, sometimes also called *likely implication*, $\to_{\theta, \alpha}^{!}$, *Fisher* $\to_\alpha^F$, and $\chi^2 \to_\alpha^{\chi^2}$ (the index α always denotes the significance level of the corresponding test). The first of them is also closely related to the above mentioned publications [Hol98, Hol04]. The quantifier $\rightarrow^{!}_{\theta,\alpha}$, in which $\theta \in (\frac{1}{2}, 1)$ is a given constant, is defined in accordance with (3) for testing the hypothesis $H_0 : P(\|\psi\| = 1 \mid \|\varphi\| = 1) \in \langle 0, \theta \rangle$ against the alternative $H_1 : P(\|\psi\| = 1 \mid \|\varphi\| = 1) \in (\theta, 1)$ by means of the binomial test. That test has the test statistics $\sum_{i=a}^{a+b} \binom{a+b}{i} \theta^i (1-\theta)^{a+b-i}$ and the critical region $C_\alpha = \langle 0, \alpha \rangle$, which leads to the truth function

$$(5) \quad \mathrm{Tf}_{\rightarrow^{!}_{\theta,\alpha}}(a, b, c, d) = \begin{cases} 1 & \text{if } \sum_{i=a}^{a+b} \binom{a+b}{i} \theta^i (1-\theta)^{a+b-i} \leq \alpha \ \& \ a \geq A, \\ 0 & \text{else.} \end{cases}$$

3 Motivation for testing fuzzy hypotheses in the context of GUHA

For which φ and ψ the sentence $\varphi \rightarrow^{!}_{\theta,\alpha} \psi$ is valid, depends according to (5) on the choice of the constant $\theta > \frac{1}{2}$ (e.g., $\theta = 0.8$, $\theta = 0.9$). However, data mining is typically performed in situations when there is only very little knowledge available about the probability distribution governing the data, thus no clue for the choice of the constant θ in (5). Therefore, using the lower critical implication (as well as several other GUHA quantifiers) entails a large amount of subjectivity. It was an effort to decrease this subjectivity that motivated the research reported in the publications [Hol98, Hol04]. The basic idea of that research is to replace hypotheses described with traditional, crisp sets by hypotheses described with fuzzy sets (therefore called *fuzzy hypotheses*). In the particular case of the lower critical implication, the interval $(\theta, 1)$, $\theta > \frac{1}{2}$ for the conditional probability $P(\|\psi\| = 1 \mid \|\varphi\| = 1)$ in H_1 is replaced by a fuzzy set with the intended meaning "high probability". The complementary interval $\langle 0, \theta \rangle$ for $P(\|\psi\| = 1 \mid \|\varphi\| = 1)$ in H_0 is then replaced with a fuzzy set with the intended meaning "probability that is not high". Consequently, the sentence $\varphi \rightarrow^{!}_{\theta,\alpha} \psi$ can be, due to (3), interpreted "with probability at least $1 - \alpha$, the conditional probability $P(\|\psi\| = 1 \mid \|\varphi\| = 1)$ is high". The publication [Hol04] replaces even the significance level α by a fuzzy set. If that fuzzy set has the intended meaning "high significance level", the sentence $\varphi \rightarrow^{!}_{\theta,\alpha} \psi$ can be finally interpreted "at a high significance level, the conditional probability $P(\|\psi\| = 1 \mid \|\varphi\| = 1)$ is high".

It is worth noticing that the intended meaning of a fuzzy set does not uniquely determine its definition—cf. different definitions of fuzzy sets with the intended meaning "high probability" in Figure 1. Hence, even fuzzy hypotheses do not remove all subjectivity from GUHA quantifiers. Compared to hypotheses described with crisp sets, this subjectivity is nevertheless restricted, in the following respects:

a) The definition of a fuzzy set must not allow an interpretation that would contradict the intended meaning.

b) All fuzzy sets with the same intended meaning usually have to fulfil certain requirements. For example, for the fuzzy sets μ on $(0,1)$ with the intended meaning "high probability", the following requirements have been proposed in [Hol98]:

(i) μ is nondecreasing,

(ii) $\lim\limits_{p \to 0+} \mu(p) = 0$,

(iii) $\lim\limits_{p \to 1-} \mu(p) = 1$.

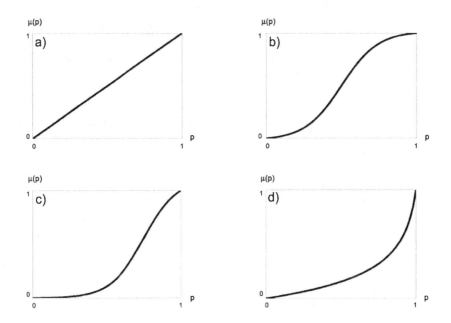

Figure 1. Examples of fuzzy sets with the intended meaning "high probability"

4   Survey of GUHA-related results about testing fuzzy hypotheses

The difference between the publications [Hol98] and [Hol04] consists in the formal framework in which fuzzy hypotheses testing is studied:

- In the older publication [Hol98], this is the framework of fuzzy set theory. The result of testing a fuzzy hypothesis H_0 is viewed as a fuzzy set on the pair of possible outcomes {"H_0 is rejected","H_0 is not rejected"}. The membership grades of those outcomes are in the case of binary quantifiers the values of two functions that generalize the truth functions (3)–(4).

- In the more recent paper [Hol04], fuzzy hypotheses testing is studied in the framework of fuzzy logic. The result of testing H_0 is viewed as an evaluation of the sentence $R_{FL\forall}$ that states the rejection of H_0 in an appropriate fuzzy predicate logic. This sentence is evaluated in a model that comprises crisp sets evaluating predicates of traditional observational logic on data, together with fuzzy sets evaluating the fuzzy hypotheses H_0, H_1 and a fuzzy significance level.

In connection with the different frameworks considered in the publication [Hol98] and [Hol04], it is worth recalling that fuzzy logic as a formal logic system is very young—it has been fully developed mainly in the monographs [Háj98], published 1998, and [Got01], published 2001. Before, the term "fuzzy logic" was (and even nowadays still frequently is) actually used for *fuzzy set theory*. To distinguish both meanings of that term, fuzzy set theory is sometimes called *fuzzy logic in broader sense*, whereas the formal logic system is called *fuzzy logic in narrow sense* (cf. [NPM00]). The paper [Hol98] reports final results of a research performed in the years 1995–97 (some preliminary results have been reported in [Hol95, Hol96]), i.e., before the monographs [Háj98, Got01] appeared. Therefore, it approaches fuzzy hypotheses testing from the point of view of fuzzy logic in broader sense. On the other hand, the paper [Hol04] has been strongly influenced by the monograph [Háj98], and it attempts to approach the topic of fuzzy hypotheses testing from the point of view of fuzzy logic in narrow sense.

Main results reported in the paper [Hol98] can be summarized as follows:

1. A generalization of the truth functions (3)–(4) to two $[0,1]$-valued functions: *accepting fuzzy associated function* and *rejecting fuzzy associated function* (presented in the paper [Hol98] as Definition 4.1).

2. A proof that the accepting fuzzy associated function of the quantifier $\to_{\theta,\alpha}^{!}$ as well as the rejecting fuzzy associated function of another generalized quantifier, $\to_{\theta,\alpha}^{?}$, are constant and equal 1, and derivation of equations allowing to compute the remaining fuzzy association functions of those quantifiers (presented in Theorem 4.4 of the paper). In particular, the rejecting fuzzy associated function of the lower critical implication is

$$(6) \quad \mathrm{Faf}_{\to_{\theta,\alpha}^{!}}^{\mathrm{reject}}(a,b,c,d) = 1 - \lim_{p \to p(a,\alpha)+} \mu_0(p),$$

where μ_0 denotes the membership function of H_0, $a > 0$, and $p_{(j,\alpha)}$ for $j = 1, \ldots, r$ is a solution of the equation

(7) $\qquad \sum\limits_{i=j}^{r} \binom{r}{i} p_{(j,\alpha)}^{i} (1 - p_{(j,\alpha)})^{r-i} = \alpha.$

3. A proof that if in particular H_0 is described with the crisp interval $\langle 0, \theta \rangle$, the fuzzy associated function $\mathrm{Faf}^{\mathrm{reject}}_{\to^{!}_{\theta,\alpha}}$ coincides with the truth function $\mathrm{Tf}_{\to^{!}_{\theta,\alpha}}$, and a proof of a similar property for $\mathrm{Faf}^{\mathrm{accept}}_{\to^{?}_{\theta,\alpha}}$ (in Theorem 4.6).

4. A generalization of the *power function* of a test, i.e., of the function that assigns to values of parameters of the probability distribution of the test statistic corresponding probabilities of rejecting the null hypothesis. The power function has been generalized by means of two concepts: a *fuzzy power function*, the values of which are fuzzy sets on $\langle 0, 1 \rangle$ (i.e., on the possible values of probabilities of rejecting the null hypothesis), and a *minimal power function* for a level $\delta \in \langle 0, 1 \rangle$, the values of which are infima of the δ-cuts of the values of the fuzzy power function (in Definition 5.1).

5. A derivation of several properties of minimal power functions that hold for all *fuzzy-implicational quantifiers*, i.e., for all quantifiers $\to$ that fulfil $a' \geq a$ & $b' \leq b \Rightarrow \mathrm{Faf}^{\mathrm{accept}}_{\to}(a',b',c',d') \geq \mathrm{Faf}^{\mathrm{accept}}_{\to}(a,b,c,d)$ & $\mathrm{Faf}^{\mathrm{reject}}_{\to}(a',b',c',d') \geq \mathrm{Faf}^{\mathrm{reject}}_{\to}(a,b,c,d)$ (in Theorem 5.2(a)).

6. A derivation of equations allowing to compute the fuzzy power function and the minimal power function for the quantifiers $\to^{!}_{\theta,\alpha}$ and $\to^{?}_{\theta,\alpha}$ (in Theorem 5.2(b)-(c) and Theorem 5.6, respectively). In this case, the parameter of the probability distribution of the test statistic is the conditional probability $P(\|\psi\| = 1 \mid \|\varphi\| = 1)$. In particular, for a value $P(\|\psi\| = 1 \mid \|\varphi\| = 1) = \pi \in \langle 0, 1 \rangle$ and the lower critical implication, its fuzzy power function $\mathrm{Fpf}_{\to^{!}_{\theta,\alpha}}(\pi)$ has the membership function

(8) $\quad \nu(y) = \begin{cases} \lim\limits_{p \to p_{(j,\alpha)}+} \mu_0(p) & \text{if } (\exists j < r) \ y = \sum\limits_{i=j+1}^{r} \binom{r}{i} \pi^i (1 - \pi)^{r-i}, \\ 0 & \text{else,} \end{cases}$

where μ_0 is the membership function of H_0, and $p_{(0,\alpha)} = 0$ completes solutions $p_{(1,\alpha)}, \ldots, p_{(r,\alpha)}$ of (7). Its minimal power function, $\mathrm{Mpf}_{\to^{!}_{\theta,\alpha}}$, for a level δ has the value

(9) $\quad \mathrm{Mpf}_{\to^{!}_{\theta,\alpha}}(\pi) = \begin{cases} \sum\limits_{i=j}^{r} \binom{r}{i} \pi^i (1 - \pi)^{r-i} & \text{if } \lim\limits_{p \to p_{(j,\alpha)}+} \mu_0(p) < \delta \leq \\ & \qquad \lim\limits_{p \to p_{(j-1,\alpha)}+} \mu_0(p), j \in \{1, \ldots, r\} \\ 0 & \text{if } \delta \in (0, \lim_{p \to p_{(r,\alpha)}+} \mu_0(p)\rangle. \end{cases}$

Further, main results reported in the more recent paper [Hol04] can be summarized as follows:

1. A formal definition of a *fuzzy predicate logic suitable for hypotheses testing*, FL∀, which includes Boolean predicates $\varphi_1(x), \ldots, \varphi_k(x)$ from the observational logic, fuzzy predicates H_0, H_1 for the hypotheses and S for a significance level, and a generalized quantifier $\nabla()$ for testing H_0 against H_1 (presented in paper [Hol04] in Definitions 3.1 and 3.2).

2. A formal definition of the *representation of a statistical test of H_0 against H_1 with the quantifier* $\nabla()$ in a model of FL∀ that includes evaluations of the predicates of the observational logic FL∀ on data, and fuzzy sets evaluating H_0, H_1 and S, as well as a formal definition of the *degree of rejection of H_0* as the evaluation of a particular sentence, $R_{\mathrm{FL}\forall}$, of FL∀ in that model (presented as Definition 3.4). Since the interpretation of fuzzy predicates in any model of FL∀ are fuzzy sets, each such model is actually a bridge between the approach to fuzzy hypotheses testing in [Hol04], based on fuzzy logic in narrow sense, and the traditional approach used in [Hol98], which relies on the fuzzy set theory.

3. For the case of a crisp significance level, [Hol04] proves the equivalence

$$(10) \quad R_{\mathrm{FL}\forall} \equiv (\forall x)((c \preceq x \;\&\; c \neq x) \rightarrow \neg H_0(x)),$$

provided the language of the logic FL∀ can be extended with a binary predicate $\preceq$ that is interpreted as a linear order at the set Θ of possible values of some parameter of the probability distribution of the test statistics, and with a constant c that is interpreted as a particular element $\theta_c \in \Theta$ such that for $\theta \in \Theta$, the considered test rejects H_0 if and only if $\theta \leq \theta_c$ (presented in Proposition 3.8 of the paper). According to (10), the degree of rejection of H_0 can be then obtained also as the evaluation $\|(\forall x)((c \preceq x \;\&\; c \neq x) \rightarrow \neg H_0(x))\|$, which is frequently computed in a similar way as in the traditional, fuzzy-set based approach. In particular for the lower critical implication, $\Theta = (0, 1)$, $\theta_c = p_{(a,\alpha)}$, with $a > 0$ from the contingency table (1) and $p_{(1,\alpha)}, \ldots, p_{(r,\alpha)}$ defined in (7), the evaluation $\|(\forall x)((c \preceq x \;\&\; c \neq x) \rightarrow \neg H_0(x))\|$ yields

$$(11) \quad \|(\forall x)((c \preceq x \;\&\; c \neq x) \rightarrow \neg H_0(x))\| = \inf_{p \in (p_{(a,\alpha)}, 1)} \ominus \mu_0(p),$$

where μ_0 is the membership function of the fuzzy set interpreting H_0, and $\ominus$ is a precomplement (interpreting the negation). For example, if FL∀ extends the Lukasiewicz logic (hence, $\ominus x = 1 - x$), $H_0 = \neg H_1$, and the fuzzy set evaluating H_1 has a non-decreasing membership function μ_1 (cf.

the requirement (i) from page 133), (11) turns to

(12) $\|(\forall x)((c \preceq x \,\&\, c \neq x) \rightarrow \neg H_0(x))\| = 1 - \lim_{p \rightarrow p_{(a,\alpha)}+} \mu_0(p),$

i.e., to the result (6) obtained by means of the approach relying on fuzzy sets.

4. A simplification of the equivalence (10) for a crisp significance level to

(13) $R_{\text{FL}\forall} \equiv \neg H_0(c),$

if the predicate $\preceq$ fulfills the axiom $H_0(x) \equiv (\exists y)x \preceq y \,\&\, x \neq y \,\&\, H_0(y),$ that corresponds to the requirement that the membership function of the fuzzy set evaluating H_0 is non-decreasing right-continuous. The evaluation $\|\neg H_0(c)\|$ for the lower critical implication yields

(14) $\|\neg H_0(c)\| = \ominus\mu_0(p_{(a,\alpha)}),$

in particular for Lukasiewicz logic and $H_0 = \neg H_1,$

(15) $\|\neg H_0(c)\| = 1 - \mu_0(p_{(a,\alpha)}).$

5. An analogy of the results 3. and 4. for general significance levels. More precisely, the deducibility of the equivalence

(16) $R_{\text{FL}\forall} \equiv [H_0(c) \rightarrow (\exists z)(S(z) \,\&\, (\nabla(c,z)x)(D_1(x),\ldots,D_k(x)))].$

from a specific additional assumption A_c, which depends on the above mentioned constant c extending FL$\forall$ (in Proposition 3.9). This equivalence can be used to alternatively obtain the degree of rejection of H_0, as the evaluation $\|H_0(c) \rightarrow (\exists z)(S(z) \,\&\, (\nabla(c,z)x)(D_1(x),\ldots,D_k(x)))\|$. In particular for the lower critical implication, Lukasiewicz logic and the significance level interpreted with a fuzzy set with the membership function $\sigma_1(\alpha) = 1 - \alpha$, it is shown in Example 3.10 in the paper

(17) $\|H_0(c) \rightarrow (\exists z)(S(z) \,\&\, (\nabla(c,z)x)(D_1(x),\ldots D_k(x)))\| = \min_{p \in (0,1)} f_{\sigma_1}(p)$

where f_{σ_1} is a function defined as

(18) $f_{\sigma_1}(p) = 1 - \mu_0(p) + \sum_{k=0}^{a-1} \binom{a+b}{k}p^k(1-p)^{a+b-k}.$

In particular, to $H_0 = \neg H_1$ and H_1 interpreted with fuzzy sets from Figure 1 correspond the functions f_{σ_1} in Figure 2.

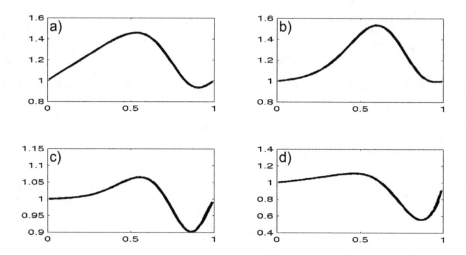

Figure 2. Graphs of the function f_{σ_1} defined in (18) corresponding to the interpretation of H_1 with the four fuzzy sets from Figure 1, which have the intended meaning "high probability"

6. An elaboration of the result 5 for the lower critical implication, which had a key role in the paper [Hol98], in combination with all three fundamental fuzzy logics (Łukasiewicz, product and Gödel) and two kinds of fuzzy significance level. Besides the significance level interpreted with a fuzzy set that has the membership function $\sigma_1(\alpha) = 1 - \alpha$, which in the case of the Łukasiewicz logic leads to (17)–(18), the significance level "approximately α^*" has been considered. This is interpreted with a fuzzy set the membership function of which, e.g., for $\alpha^* = 0.05$, can be defined

$$(19) \quad \sigma_2(p) = \begin{cases} 20\alpha & |\alpha \in (0, 0.05\rangle \\ 20(0.1 - \alpha) & |\alpha \in \langle 0.05, 0.1\rangle \\ 0 & |\alpha \in \langle 0.1, 1). \end{cases}$$

For example, in the case of the product logic $\min_{p \in \Theta_{\text{supp}}} f_{\sigma_2}(p) =$

$$(20) \quad \|H_0(c) \rightarrow (\exists z)(S(z) \ \& \ (\nabla(c, z)x)(D_1(x), \ldots, D_k(x)))\|,$$

where the function f_{σ_2} and the set Θ_{supp} are defined

$$(21)\quad f_{\sigma_2}(p) = \frac{20(0.1 - \sum_{k=a}^{a+b} \binom{a+b}{k} p^k (1-p)^{a+b-k})}{\tilde{H}_0(p)},$$

$$\Theta_{\mathrm{supp}} = \{p \in (p_{(a,.05)}, p_{(a,.1)}) \mid \tilde{H}_0(p) > 20(.1 - \sum_{k=a}^{a+b} \binom{a+b}{k} p^k (1-p)^{a+b-k})\}$$

with $a > 0$ again from the contingency table (1), and $p_{(j,\alpha)}$, $j = 1, \ldots, r$, $\alpha \in (0,1)$ defined in (7).

5   Conclusion

This chapter summarized my research into fuzzy hypotheses testing in the context of the GUHA method. Fuzzy hypotheses testing represents a kind of connection between two key areas of Petr Hájek's legacy: GUHA, and formal fuzzy logic. However, it is not the direct and straightforward connection betweeen fuzzy logic and the variant of Boolean logic underlying the logical principles of the GUHA method. It is an alternative connection between fuzzy logic and the statistical principles of GUHA. Though statistical principles of the method used to be not the domain of Peter, but of Tomáš Havránek, Petr nevertheless showed a vivid interest in this research and contributed to it through interesting discussions and valuable comments. As he read the first version of the paper [Hol04], he proposed to replace my cumbersome proofs of some key results with more elegant and shorter ones. But most importantly, it is Peter's monograph [Háj98] that accounts for the most fundamental shift in the reported research—a shift from dealing with fuzzy hypotheses testing within the framework of fuzzy set theory (or equivalently, fuzzy logic in broad sense) to dealing with it within the framework of formal fuzzy logic (or equivalently, fuzzy logic in narrow sense). The paper [Hol98] approaches fuzzy hypotheses testing from the point of view of fuzzy set theory, similarly to all preceding literature about that topic (e.g., [GH97, SS90, WI93]), but also to the more recent publications [Buc04, Buc05, Ngu06, TB01]. The paper [Hol04], on the other hand, is actually the first attempt (except for some preliminary results reported in [Hol01a, Hol01b]) to approach the topic of fuzzy hypotheses testing from the point of view of formal fuzzy logic.

BIBLIOGRAPHY

[Arn96]   B. F. Arnold. An appraoch to fuzzy hypotheses testing. *Metrika*, 42:119–126, 1996.

[Arn98]   B. F. Arnold. Testing fuzzy hypotheses with crisp data. *Fuzzy Sets Syst.*, 94:232–333, 1998.

[Buc04]   J. J. Buckley. *Fuzzy Statistics*. Springer, 2004.

[Buc05]   J. J. Buckley. Fuzzy statistics: Hypothesis testing. *Soft Computing*, 9:512–518, 2005.

[Cas93]   M. R. Casals. Bayesian testing of fuzzy parametric hypotheses from fuzzy information. *Recherche Opérationnelle / Operations Research*, 27:189–199, 1993.

[DVV85]   M. Delgado, J. Verdegay, and M. Vila. Testing fuzzy hypotheses. a Bayesian approach. In M. M. Gupta, A. Kandel, W. Bandler, and J. B. Kiszka, editors, *Approximate Reasoning in Expert Systems*, pages 307–316. North-Holland, 1985.

[GH97]   P. Grzegorzewski and O. Hryniewicz. Testing statistical hypotheses in fuzzy environment. *Mathware & Soft Computing*, 4:203–217, 1997.

[Got01]   S. Gottwald. *A Treatise on Many-Valued Logics*, volume 9 of *Studies in Logic and Computation*. Research Studies Press, Baldock, 2001.

[Háj98]   P. Hájek. *Metamathematics of Fuzzy Logic*, volume 4 of *Trends in Logic*. Kluwer, Dordrecht, 1998.

[HH78]   P. Hájek and T. Havránek. *Mechanizing Hypothesis Formation—Mathematical Foundations of a General Theory*. Springer, Berlin, 1978.

[Hol95]   M. Holeňa. Fuzzy hypotheses testing and Guha implicational quantifiers. *Bull. for Studies and Exchanges on Fuzziness and Its Appl.*, 63:10–15, 1995.

[Hol96]   M. Holeňa. Exploratory data processing using a fuzzy generalization of the Guha approach. In J. F. Baldwin, editor, *Fuzzy Logic*, pages 213–229. John Wiley & Sons, 1996.

[Hol98]   M. Holeňa. Fuzzy hypotheses for Guha implications. *Fuzzy Sets Syst.*, 98:101–125, 1998.

[Hol01a] M. Holeňa. A fuzzy logic framework for testing vague hypotheses with empirical data. In *Proceedings of the Fourth International ICSC Symposium on Soft Computing and Intelligent Systems for Industry*, pages 401–407. IOS Press, 2001.

[Hol01b] M. Holeňa. A fuzzy logic generalization of a data mining approach. *Neural Network World*, 11:595–610, 2001.

[Hol04]   M. Holeňa. Fuzzy hypotheses testing in the framework of fuzzy logic. *Fuzzy Sets Syst.*, 145:229–252, 2004.

[Ngu06]   H. T. Nguyen. *Fundamentals of Statistics with Fuzzy Data*. Springer, 2006.

[NPM00]   V. Novák, I. Perfilieva, and J. Močkoř. *Mathematical Principles of Fuzzy Logic*. Kluwer, Dordrecht, 2000.

[SS90]   J. J. Saade and H. Schwarzlander. Fuzzy hypotheses testing with hybrid data. *Fuzzy Sets Syst.*, 35:213–217, 1990.

[TB01]   S. M. Taheri and J. Behboodian. A Bayesian approach to fuzzy hypotheses testing. *Fuzzy Sets Syst.*, 123:39–48, 2001.

[WI93]   N. Watanabe and T. Imaizumi. A fuzzy statistical test of fuzzy hypotheses. *Fuzzy Sets Syst.*, 53:167–178, 1993.

Martin Holeňa
Institute of Computer Science
Academy of Sciences of the Czech Republic
Pod Vodárenskou věží 2
182 07 Prague 8, Czech Republic
Email: martin@cs.cas.cz

From relative frequences to lattice-
-valued possibilistic measures

IVAN KRAMOSIL

Dedicated to Professor Petr Hájek
on the occasion of his 70th birthday

1 Introduction

Professor Petr Hájek is an excellent Czech mathematician and logician of the
world top scientific level and with the world-wide known prestige and repu-
tation. It is not the aim of this short essay to list and evaluate prof. Hájek's
extremely rich and extended scientific work and I hope this will be done in
this volume at much higher and better level than when relying just to my
very limited personal abilities and powers. Just a few facts being in the close
connection with the contents of this essay following below are re-called.

As far as I know, Prof. Hájek's earliest works in mathematics were focused
towards non-classical logics and set theory. The alternative ways of processing the truth values in non-classical logics involved a nonstandard way of
understanding these truth values, hence, within the set-theoretic framework,
to an alternative understanding of the notion of the membership of an element
in some subset of the basic space under consideration. Prof. Hájek, together
with Prof. Vopěnka, introduced, for the sake of mathematical formalization
of these matters, the notion of semiset and they published an excellent detailed analysis of semisets and related notions in their book "The Theory
of Semisets" [VH72]. From the point of view of mathematical elegance and
metamathematical correctness this approach to the "third type of value of
membership relation in set theory" should be appreciated much higher than
the notion of fuzzy set introduced by L. A. Zadeh ([Zad65]). Hence, semisets
may be understood as (meta)mathematically perfect introduction, in a qualitative sense, of the phenomenon of uncertainty (in the ontological sense)
as far as the membership relation is concerned. Fuzzy sets may be taken
as an over-simplified (from the methodological point of view) quantitative
diversification of the standard membership relation.

As his answer to the notion of fuzzy sets, Prof. Hájek oriented his scientific
aim to the notion of fuzzy logics as a method for describing fuzzy sets, taken
as truth values distributions on a set of formulas, by the tools of deduction

rules applied to some sets of axioms, as this approach offers more general results not immediately related to a particular distribution of fuzzy truth values. So, the great part of prof. Hájek's works in this field is related to the notion of uncertainty in the sense of fuzziness and vagueness.

My own scientific work has been always more or less closely related to uncertainty quantification and processing beginning with uncertainty in the sense of randomness, but since several years my investigations have been oriented rather to non-numerical, in particular, to lattice-valued possibilistic distributions and measures. This is why I dare hope that this essay, sketching very briefly and informally the way from relative frequencies to non-numerical possibilistic measures as mathematical tools for uncertainty quantification and processing, could be kindly accepted as thematically appropriate and very modest contribution to the volume celebrating Prof. Hájek's important life anniversary.

2 Classical Real-Valued Probability Measure

Perhaps one of very soon demonstration of human intelligence was the ability of man to distinguish the two qualitatively different classes of phenomena occurring in the surrounding world—the phenomena which can be called deterministic (with respect to given collection of conditions) and those which are nondeterministic in the sense of randomness. A phenomenon deterministic under some conditions occurs whenever these conditions hold and this relation may be used if we want either to involve the occurrence of this phenomenon or, contrary, to defend its realization. For nondeterministic phenomena, even when the same conditions are repeatedly realized, the phenomenon sometimes occurs, but sometimes not.

However, even if the occurrence of the phenomenon, let us introduce the term random event, in a particular sample cannot be predicted, under certain meta-conditions, later on called independent and identical distribution, the resulting sequence of samples proves certain regularity, even if it is rather difficult, at the beginning of the research, to describe this regularity in an explicit way. The experimenting subject may observe, e.g., that the ratio of the number of occurrences and non-occurrences of the tested event proves certain stability. When denoting by $\langle i_1, i_2, \ldots, i_n \rangle, i_* = 1$, if the event E occurs, $i_* = 0$ otherwise, the sequence encoding the results on n repeated samples, the stability of the value $e(n) = n^{-1} \sum_{j=1}^{n} i_j$, in a sense to be more correctly specified, may be taken as the first non-trivial empirical observation concerning the sequence of samples in question and the numerical characteristic $e(n)$ of this sequence, called *relative frequency* of successes in the sequence of the length n of samples can be taken as the first empirical mathematical tool applied when processing the random sequence in question.

Applying the methodological principle of duality between an ideal formalized Platonist description of the surrounding us world and the empirical data and observations as the only phenomena accessible to human beings and reflecting only partially and incompletely the ideas from the Platonist world, the observed stability of relative frequencies of successes (occurrences of a fixed random event) can be explained as follows: there exists a real number $p \in [0, 1]$ such that the relative frequencies of occurrences of the event in question are incomplete and not quite precise projections of the value p into the unit interval $[0, 1]$. This value p is called the probability of the random event E and denoted by $p(E)$. Hence, in this most elementary approach probability values are real numbers from the unit interval of reals in order to be directly comparable with the values of certain relative frequencies and modified according to them if necessary.

It was why the Kolmogorov axiomatic approach to probability theory has been based on the idea of normalized set function-measure—defined on (certain, resp.) subsets of an elementary space Ω. The uncertainty whether a fixed phenomenon E occurs in the given sample is described in such a way that it depends on the actual value of a parameter ω taking its values in Ω, but this actual value not being known in the moment. So, the only what we may process is the subset of Ω defined by those $\omega \in \Omega$ for which E occurs and from the mathematical point of view we may directly identify the phenomenon or random event E directly with this set. Given a σ-field $\mathcal{S}$ of subsets of Ω such that $E \in \mathcal{S}$ holds and a σ-additive measure P on $\mathcal{S}$, we may define probability of the random event E as the value $P(E)$ ascribed to the set E, hence, as a numerically evaluated size of the subset of those values $\omega \in \Omega$, for which E occurs, hence, as the size of (those values $\omega \in \Omega$ which are in) E. Considering a sequence of independent and identically distributed random samples with E as possible result, we obtain as a consequence of our model that the relative frequencies of the number of occurrences of E in the sequences of samples in question proves stability in the sense that it tends to $P(E)$ with probability one. Hence, in this sense the classical Kolmogorov probability theory is compatible with the intuitive idea of informal stability of relative frequencies of occurrences of the random event E taken as the primary point of output of our reasonings as outlined above.

It is perhaps worth being re-called explicitly, that also under the classical axiomatic probability theory probabilities take their values in the unit interval of real numbers and relative frequencies can be seen as imperfect images or shadows of probability values, but this time the real-valued nature of probability values is declared axiomatically and stability of relative frequencies in various forms follows as the law of large numbers.

Taking a real-valued normalized probability space $\langle \Omega, \mathcal{S}, P \rangle$ with P defined as σ-additive normalized mapping which takes $\mathcal{S} \subseteq \mathcal{P}(\Omega)$ into $[0, 1]$, we obtain that for each $E_1, E_2 \in \mathcal{S}$ either $P(E_1) \leq P(E_2)$ or $P(E_2) \leq P(E_1)$ hold, in other terms, each two random events are comparable according to the sizes of their probability values. As relative frequencies of occurrences of random events E_1, E_2 in a finite sequence of samples are real numbers from $[0,1]$, they may be compared according to their sizes, however, in general we cannot be sure that if $e^1(n) \leq e^2(n)$ is the case for relative number $e^1(n)$ of occurrences of E_1 in n samples and relative number $e^2(n)$ of occurrences of E_2 in the same n-tuple of samples then the same inequality will hold for limit values $\lim_{n \to \infty} e^1(n) \leq \lim_{n \to \infty} e^2(n)$ in the sense of strong law of large numbers (convergence with probability one). Hence, up to the trivial case when for $E_1, E_2 \in \mathcal{S}$ we have $E_1 \subseteq E_2 \subseteq \Omega$, the relations like $P(E_1) \leq P(E_2)$ must be implemented into the model in question as explicit additional assumptions.

3   Boolean-Valued Probability Measures

Let us mention, very briefly, the idea of abandoning the real-valued nature of probability measure in favour of some non-numerical probability measures. In the classical probability theory what is primary is the notion of random event as a measurable subset of a basic space Ω of elementary random events, and to this set, secondary, a real number obeying certain axiomatic demands is ascribed. From the computational point of view, this ascription of probability values suffers from certain disadvantages, e.g., it is not extensional in the sense that there do not exist, in general, functions f and g from $[0,1] \times [0,1]$ to $[0,1]$ such that $P(E_1 \cup E_2) = f(P(E_1), P(E_2))$ and $P(E_1 \cap E_2) = g(P(E_1), P(E_2))$ would be the case for each $E_1, E_2 \in \mathcal{S}$.

The power-set of all subsets of the space Ω obviously meets the conditions imposed on complete Boolean algebra with respect to set inclusion as the relation of partial ordering on $\mathcal{P}(\Omega)$ and with set complement as Boolean complement. So, the identical mapping $I \colon \mathcal{P}(\Omega) \to \mathcal{P}(\Omega)$ may be taken as trivial set-valued, i.e., Boolean-valued probability measure on the system $\mathcal{S}$ of all random events, probability of a random event being identical with the random event in question. This Boolean-valued probability measure is obviously extensional, as both the functions f and g mentioned a few lines above also coincide with identity on $\mathcal{P}(\Omega)$: $A \cap B = A \cap B$ and $A \cup B = A \cup B$. What is more important, however, is the fact that this trivial Boolean-valued probability measure may be taken just as the starting point of a sequence of Boolean-valued measures more and more distinguishing and more and more less likely to the identical measure I on $\mathcal{P}(\Omega)$, but still conserving some Boolean-like properties of the original set function I and so keeping some information on the original sets from $\mathcal{S}$, being lost when ascribing to random events as probability values real numbers from the unit interval.

4   Uncertainty in the Sense of Vagueness-Fuzziness

Let us begin with the following example considering a coin being tossed at random. Before the tossing, or even after the tossing supposing that the coin was immediately hidden (by a hand, say), we do not know the result of the toss till the moment when the coin becomes accessible to our sight we do not know what the result of the toss might be, but we are sure that it is just one of the two possible cases—head or tail. Hence, the uncertainty in the sense of randomness illustrated by this example is an uncertainty of the gnoseological type as it is the lack of knowledge of the state of things as far as the tossed coin is concerned but the state as such is, from the ontological point of view, fully determined—head or tail.

Let a coin be digged up during an archeological research. However, during its several centuries continuing lying in the soil with its destroying erosive influences the coin is damaged in such a way that the digging archeologist is not able to decide which side of the coin *used to be*, long time ago, the head and which used to be a tail, at this moment both the sides of the coin have lost some aspects of their former properties and the digger cannot say that the side of the coin just observed possesses all the properties defining the attribute of being the head (the tail, resp.) of the coin in question. What remains is the question whether the property of being the head, say, of the coin under consideration may be quantified by a degree of uncertainty taken as a real number from the unit interval of real numbers. This may be done, e.g., in the following way. Let the archeological research has arrived at the conclusion that there are, for the head of an ancient coin, ten details possessed by each head of such coin. If the digged coin does not possess, on its head, these ten characteristic details but only, say, seven of them, the digged coin is not a full valid case of the coin under consideration, but we could say that it is the coin under consideration to the degree 0.7 (seven characteristic details are present). The three remaining details are not present, are missing, it is not the case that the coin is partially hidden or covered somehow, so that we are not able to decide about the presence or absence of all the ten details decisive for standard 0-1 classification of the digged coin.

Let us recall explicitly the legislative rules according to which reimbursements for damaged or partially destroyed banknotes are calculated and payed are motivated by a similar idea: what matters is the relative ratio of the size of the damaged banknote. Namely, if at most 25% of the size of the banknote is damaged or destroyed, the banknote is refunded in its full value, e.g., a 100-crown banknote damaged on at most 25% of its surface is refunded by 100 crowns, i.e., it is taken as the 100-crown banknote to the degree one. On the other side, if more than 75% of the surface of the banknote is damaged, no reimbursement is paid, hence, such a damaged banknote is taken as a

100-crown banknote to the degree 0. For the degree of destruction (damage) between 25%–75% the damaged banknote is taken as a 100-crown banknote to the degree identical with the relative portion of the non-damaged (saved) size of the banknote, i.e., the banknote damaged on 30% of its surface is re-funded by the amount of 70 crowns.

The following shift in the interpretation of the used terms is important. Instead of saying that the digged object *used to be*, eight centuries ago, a perfect denarius, in the fuzzy terms, this object belonged in the degree 1 to the set of Prague denarii, but this conclusion is not quite reliable (it is reliable to the degree 0.7), the same object is now, at the moment when being digged, not a perfect denarius from 12th century, but it is just an object belonging to the fuzzy set of denarii only to the degree 0.7.

So, we have arrived at the uncertainty measure, in the sense of vagueness, fuzziness or similarity degree, mathematically defined by a mapping μ from Ω to $[0,1]$. Supposing that three exists at least one ω_0 such that $\mu(\omega_0) = 1$ (perhaps weakened to the demand $\sup\{\mu(\omega) : \omega \in \Omega\} = 1$ in the case when Ω is infinite), we have arrived at the notion of real-valued fuzzy subset of the basic set (space) Ω, as introduced in 1965 by L. A. Zadeh.

5 Real-Valued Fuzzy Sets

Hence, to recall once more, real-valued fuzzy subset of a nonempty space Ω is a mapping $E: \Omega \to [0,1]$ such that $\sup\{E(\omega): \omega \in \Omega\} = 1$. If Ω is finite, the last condition is equivalent with the demand that there exists $\omega_0 \in \Omega$ such that $E(\omega) = 1$. Wanting to define a set function Π induced by the fuzzy set E, ascribing to each $A \subseteq \Omega$ a real number $\Pi(A) \in [0,1]$ and possessing at least some reasonable properties like extensionality and monotonicity ($\Pi(A) \leq \Pi(B)$ for each $A \subseteq B \subseteq \Omega$), we arrive at the supremum function, so setting $\Pi(A) = \sup\{E(\omega): \omega \in A\}$ for each $A \subseteq \Omega$. From the theoretical point of view, the two approaches to the definition of the function Π are equivalent: either we define $\Pi(A)$ as just introduced, easily arriving at the consequence that for each nonempty system $\mathcal{A}$ of subsets of Ω the relation $\Pi(\bigcup_{A\in\mathcal{A}} A) = \bigvee_{A\in\mathcal{A}} \Pi(A)$ (complete maxitivity) holds, or we may take complete maxitivity as the axiomatic demand imposed on the mapping $\Pi: \mathcal{P}(\Omega) \to [0,1]$, arriving at the definition of $\Pi(A)$ when applying the complete maxitivity condition to singletons of the space Ω.

In what follows, we will not introduce some results concerning the elementary properties of real-valued fuzzy sets, but we will rather analyze positive and negative consequences following from the choice of real numbers from the unit interval as degrees of fuzziness. First of all, in the case of fuzzy sets there is nothing what could play the role of relative frequencies in probability theory and mathematical statistics, hence, fuzziness degrees cannot be taken as idealized values in the Platonic sense, more or less good projections of which being empirically observable and accessible. If something like this can be introduced, e.g., the proportion of conserved characteristic details on the head or tail of ancient coin discovered in the soil or the proportion of the damaged/saved size of a banknote, the applied rules and relations may be reasonably justified within just a very narrow field of application and even in this case a non-negligible portion of particular knowledge concerning this field of application is necessary.

The main advantage of real-valued fuzzy sets consists in the general riches of various structures and operations which may be introduced and implemented on $[0, 1]$ together with the fact that operations of supremum and infimum are applicable and their results defined for each subset of the unit interval, in spite of the conditions of measurability and integrability which must be introduced and checked when applying σ-additive measures. Moreover, the set function Π is obviously extensional, as we have proved above when showing the complete maxitivity of Π.

However, the problem remains that the real numbers in $[0, 1]$ are completely ordered, i.e., each two different values of fuzzy degrees are comparable by the relation "... is greater than...", but with the role of relative frequencies disappearing the arguments in favour of this and that relation become to be still more weak and vague. In other words, the more and more portion of fuzzy sets (i.e., fuzziness degrees) in the mathematical model in question is implemented into the model according to the subjective feelings of the human subject proposing the model and keeping the full responsibility for the consequences of her/his choice. Consequently, the subject prefers the models with as much as possibly weakened assumptions imposed on the particular parts of the model and containing also considerations about the numerical character of the fuzziness degrees.

We will discuss this approach in more detail below, but before doing so let us slightly modify the used terms according to the alternatives proposed by L. A. Zadeh in 1978 ([Zad78]). Namely, instead of the term "fuzzy subset of the set (space) Ω" we will say "(real-valued) possibilistic distribution π on Ω", hence, $\pi\colon \Omega \to [0, 1]$ and $\sup\{\pi(\omega)\colon \omega \in \Omega\} = 1$ holds, and instead of the tern "set function Π induced by μ or E" we will say "possibility or possibilistic measure Π on $\mathcal{P}(\Omega)$ induced by the possibility or possibilistic distribution π on Ω".

6 Lattice-Valued Possibilistic Distributions and Measures

After a slightly more detailed analysis of the notions and relations applied and processed when investigating probability theory and related parts of mathematical statistics we observe easily that the operations of addition, series computation, integration and a number of related mathematical tools prevail when applying probability theory to uncertainty quantification and processing. Contrary to this case, the main operations when introducing and processing degrees of fuzziness (possibilistic distributions and measures) are the relations of (partial) orderings and the operations of suprema and infima, which may be defined in more general and non-numerical sets and structures over them, above all, in partially ordered sets (p.o.sets). It is why, quite naturally, partially ordered sets in which suprema and infima of all subsets of the support set of the p.o.set in question are defined have been the most often used candidates for non-numerical fuzziness degrees; these structures are called *complete lattices*. The first paper on lattice-valued fuzzy sets ($\mathcal{L}$-fuzzy sets) were published as soon as in 1967 (two years after the Zadeh's pioneering paper on fuzzy sets) by J. A. Goguen ([Gog67]) and it contains also the first discussion dealing with the motivation for non-numerical (in general) and lattice-valued (in particular) uncertainty degrees and measures. The first perfect and detailed mathematical setting of the theory of lattice-valued uncertainty quantification and processing was presented by G. De Cooman in his paper "Possibility Theory" ([DC97]). De Cooman's analysis tries to develop a systematical parallel way to the standard theory of measure and integration, but now within the framework of lattice-valued possibilistic distributions and measures. As could be expected, some notions and results of standard measure theory can be more or less closely and intuitively translated into the possibilistic terms and the related results are at least syntactically similar to the classical ones, say, just with additions, series and integrals replaced by suprema and products by infima, other results achieved in possibility theory are more or less close to some measure-theoretic results, but a syntactical similarity is missing, and some results achievable in standard measure and integration theory do not possess any reasonable counterparts in the possibility theory.

De Cooman's analysis is conceived for lattice-valued possibilistic measures in general and particular complete lattices as support sets for fuzziness degrees in some specific cases are introduced just as examples to illustrate more generally defined and analyzed theses. From the theoretical point of view the conditions imposed on complete lattices, according to which supremum and infimum values are defined for each subset of the support set on which the poset in question is constructed, substantially simplify the case. Indeed, the closedness of all subsets of this support set to suprema and infima not being supposed, we had to analyze, in particular, each supremum and infimum coming to scene during our investigations and to distinguish between the three possibilities. Either to prove that the particular supremum (infimum) is defined so that we may go on in our reasonings, or to add to our premises a new item declaring the existence of just the particular extreme value in question, or to abandon our construction and proofs, replacing them by a weaker model, if it is possible and reasonable as an approximation of the originally pursued way of construction and reasoning.

It may be rather difficult to introduce a more or less serious prognosis as far as the further development of non-numerical uncertainty degrees and related theories is concerned, but let us try to explicate at least some ideas. Continuing the investigation in the field of lattice-valued possibilistic measures, the effort could be shifted from general results to a deeper analysis of lattice-valued possibilistic measures over particular complete lattices fitted for uncertainty and fuzziness quantification and processing over particular structures important from the point of view of their actual or in a close future promising applications. Immediately following problems may be those of optimization of the chosen structure of non-numerical uncertainty degrees at least in the two following directions: the uncertainty degrees and structures over them should be as simple as possible when defining and verifying the preliminary uncertainty description and, second, as simple as possible from the point of view of computation complexity of computing and deductive procedures necessary to process the uncertainty values within the framework of the theoretical models in question. Within the traditional scale of uncertainty values (defined by the unit interval of real numbers and the standard structures over them) replaced by a wider space of competing models and structures over non-numerical support sets the problems of comparing different approaches to uncertainty quantification and processing can be obviously seen to be worth a more detailed analysis and investigation.

BIBLIOGRAPHY

[DC97]   G. De Cooman. Possibility theory I, II, III. *Int. J. General Systems*, 25:291–323, 325–351, 353–371, 1997.
[Gog67]   J. A. Goguen. *L*-fuzzy sets. *J. Math. Anal. Appl.*, 18:145–174, 1967.
[VH72]   P. Vopěnka and P. Hájek. *The Theory of Semisets*. Academia, Prague, 1972.
[Zad65]   L. A. Zadeh. Fuzzy sets. *Information and Control*, 8:338–353, 1965.
[Zad78]   L. A. Zadeh. Fuzzy sets as the basis for a theory of possibility. *Fuzzy Sets Syst.*, 1:3–28, 1978.

Ivan Kramosil
Institute of Computer Science
Academy of Sciences of the Czech Republic
Pod Vodárenskou věží 2
182 07 Prague 8, Czech Republic
Email: kramosil@cs.cas.cz

A note on simple factorization in Dempster-Shafer theory of evidence

RADIM JIROUŠEK

In honour of Petr Hájek
on the occasion of his 70th birthday

ABSTRACT. The paper introduces (in our knowledge first) attempt to define the concept of factorization within the Dempster-Shafer theory of evidence. In the same way as in probability theory, the presented concept can support procedures for efficient multidimensional model construction and processing. The main result of this paper is a *factorization lemma* describing, in the same way as in probability theory, the relationship between factorization and conditional independence.

1 Introduction

It was at the beginning of 80's of the last century when I first heard about Dempster-Shafer theory of Evidence, and it was Petr Hájek who was lecturing on this topic. In fact it was him who organized (and still is organizing) a series of working seminars (so called Hájek's seminars) at the Institute of Mathematics of the Czechoslovak Academy of Sciences, where not only me but many regular attendees from several research and university institutions heard about some interesting topics first time.

To tell the truth, in those days, when mainframe computers with several hundreds of kilobytes of memory represented the best available computational technique, I considered Dempster-Shafer theory to be a rather academic topic. At that time I could not imagine that a model with super-exponential space complexity could ever be applied to problems of practice. And yet, in the second half of 80's, Petr Hájek became an enthusiastic booster of this technique [Háj87, Háj92, HH92b, HH92a, Háj93, Háj94]. And as it has appeared later he was right. Since that time the computational tools have empowered in such a way that these models have become computationally tractable.

Naturally, before Dempster-Shafer models started being employed to problems of practice, there was a great flowering of probabilistic (and also possibilistic) models with their exponential computational complexity. This was because a substantial decrease of computational complexity was achieved with the help of models taking advantage of the concept of conditional independence. So, representing multidimensional probability distributions in a form of Graphical Markov Models (GMMs) (e.g. Bayesian networks) made it possible to store in a computer memory distributions of hundreds (or even thousands) of dimensions. However, studying properly probabilistic GMMs one can realize that it is not the notion of *conditional independence* that makes it possible to represent these models efficiently. The efficiency is based on the notion of *factorization*, which in probability theory (due to factorization lemma presented here as Lemma 1) coincide with conditional independence. Going into details, one can notice that the notion of factorization has been introduced in several different ways in probability theory, and some others can still be studied.

This is why in this paper we shall first briefly analyze the notion of factorization in probability theory and only afterwards will generalize the simplest definition for the Dempster-Shafer theory of evidence.

2   Probabilistic Factorization

In this section we recall several notions from probability theory, which served as an inspiration for the considerations presented in the further parts of this paper. Here, we will consider a probability measure π (or ν) on a finite space

$$\mathbf{X}_N = \mathbf{X}_1 \times \mathbf{X}_2 \times \ldots \times \mathbf{X}_n,$$

i.e. an additive set function

$$\pi : \mathcal{P}(\mathbf{X}_N) \longrightarrow [0, 1],$$

for which $\pi(\mathbf{X}_N) = 1$. For any $K \subseteq N$, symbol $\pi^{\downarrow K}$ will denote its respective marginal measure (for each $B \subseteq \mathbf{X}_K$):

$$\pi^{\downarrow K}(B) = \sum_{\substack{A \subseteq \mathbf{X}_N \\ A^{\downarrow K} = B}} \pi(A),$$

which is a probability measure on subspace

$$\mathbf{X}_K = \underset{i \in K}{\times} \mathbf{X}_i.$$

Let us remark that for $K = \emptyset$ we get $\pi^{\downarrow \emptyset} = 1$.

An analogous notation will be used also for projections of points and sets. For a point $x = (x_1, x_2, \ldots, x_n) \in \mathbf{X}_N$ its projection into subspace $\mathbf{X}_K$ will be denoted

$$x^{\downarrow K} = (x_{i, i \in K}),$$

and for $A \subseteq \mathbf{X}_N$

$$A^{\downarrow K} = \{y \in \mathbf{X}_K : \exists x \in A, x^{\downarrow K} = y\}.$$

Consider a probability measure π and three disjoint groups of variables $X_K = \{X_i\}_{i \in K}$, $X_L = \{X_i\}_{i \in L}$ and $X_M = \{X_i\}_{i \in M}$ ($K, L, M \subset N$, $K \neq \emptyset \neq L$) having their values in $\mathbf{X}_K$, $\mathbf{X}_L$ and $\mathbf{X}_M$, respectively. We say that X_K and X_L are *conditionally independent given* X_M (with respect to probability measure π) if for all $x \in \mathbf{X}_{K \cup L \cup M}$

$$\pi^{\downarrow K \cup L \cup M}(x) \cdot \pi^{\downarrow M}(x^{\downarrow M}) = \pi^{\downarrow K \cup M}(x^{\downarrow K \cup M}) \cdot \pi^{\downarrow L \cup M}(x^{\downarrow L \cup M}).$$

This property will be denoted by the symbol $K \perp\!\!\!\perp L \mid M \,[\pi]$. In case that $M = \emptyset$ then we say that groups of variables X_K and X_L are (*unconditionally*[1]) *independent*, which is usually denoted by a simplified notation: $K \perp\!\!\!\perp L \,[\pi]$.

As already mentioned in Introduction, the notion of factorization is introduced in probability theory in several different ways, and therefore we will use some adjectives to distinguish them from each other. The properties presented further in this section as lemmata and corollaries can be either found in [Lau96] or can be directly deduced as trivial consequences of properties presented there.

Simple factorization

Consider two nonempty sets $K, L \subseteq N$. We say that π *factorizes with respect to* (K, L) if there exist two nonnegative functions

$$\phi : \mathbf{X}_K \longrightarrow [0, +\infty) \quad \text{and} \quad \psi : \mathbf{X}_L \longrightarrow [0, +\infty),$$

such that for each $x \in \mathbf{X}_{K \cup L}$ the equality[2]

$$\pi^{\downarrow K \cup L}(x) = \phi(x^{\downarrow K}) \cdot \psi(x^{\downarrow L})$$

holds true.

LEMMA 1 (Factorization lemma) *Let $K, L \subseteq N$ be nonempty. π factorizes with respect to (K, L) if and only if $K \perp\!\!\!\perp L \,[\pi]$.*

COROLLARY 2 π *factorizes with respect to (K, L) if and only if*

$$\pi^{\downarrow K \cup L}(x) = \pi^{\downarrow K}(x^{\downarrow K}) \cdot \pi^{\downarrow K \cup L}(x^{\downarrow L \backslash K} \mid x^{\downarrow K \cap L}),$$

for all $x \in \mathbf{X}_{K \cup L}$.

COROLLARY 3 *Let $\pi_1, \pi_2, \pi_3, \dots$ be a sequence of probability measures each of them factorizing with respect to (K, L). If this sequence is convergent then also the limit measure $\lim_{j \to +\infty} \pi_j$ factorizes with respect to (K, L).*

---

[1]Some author call it marginal independence.

[2]As usually, we do not distinguish between a singleton set and its element, so x stands also for $\{x\}$, and $x^{\downarrow K}$ is the only element from $\{x\}^{\downarrow K}$.

Multiple factorization

Consider a finite system of nonempty subsets $K_1, K_2, \ldots, K_p$ of a set N. We say that π *factorizes with respect to* $(K_1, K_2, \ldots, K_p)$ if there exist p functions $(i = 1, 2, \ldots, p)$

$$\phi_i : \mathbf{X}_{K_i} \longrightarrow [0, +\infty),$$

such that for all $x \in \mathbf{X}_{K_1 \cup \ldots \cup K_p}$

$$\pi^{\downarrow K_1 \cup \ldots \cup K_p}(x) = \prod_{i=1}^{p} \phi_i(x^{\downarrow K_i}).$$

REMARK 4 *In this general case one can (using Lemma 1) derive a system of conditional independence relations valid for a measure π factorizing with respect to $(K_1, K_2, \ldots, K_p)$ but no assertion that could be considered a direct analogy to any of the preceding Corollaries hold true. This is why the following type of factorization is often considered.*

Recursive factorization

Consider a finite system of nonempty subsets $K_1, K_2, \ldots, K_p$ of a set N. We say that π *recursively factorizes with respect to* $(K_1, K_2, \ldots, K_p)$ if for each $i = 2, \ldots, p$ π (simply) factorizes with respect to the pair

$$(K_1 \cup \ldots \cup K_{i-1}, K_i).$$

REMARK 5 *Using Corollary 2 iteratively one can get a formula expressing the multidimensional measure $\pi^{\downarrow K_1 \cup \ldots \cup K_p}$ with the help of its respective marginals*[3]

$$\pi^{\downarrow K_1 \cup \ldots \cup K_p}(x) = \prod_{i=1}^{p} \pi^{\downarrow K_i}\left(x^{\downarrow K_i \setminus (K_1 \cup \ldots \cup K_{i-1})} \big| x^{\downarrow K_i \cap (K_1 \cup \ldots \cup K_{i-1})}\right).$$

So we are getting a trivial assertion saying that if π recursively factorizes with respect to $K_1, K_2, \ldots, K_p$ then it also factorizes with respect to this system of subsets. Let us stress that recursive factorization is much stronger than multiple factorization. For example, for recursive factorization an analogy to Corollary 3 holds true.

---

[3]Read $(K_1 \cup \ldots \cup K_0)$ as $\emptyset$.

Marginal factorization

Marginal factorization, which is introduced in this paragraph is usually not considered by other authors. It is weaker than the recursive factorization (and in a sense stronger than factorization).

Consider a finite system of nonempty subsets $K_1, K_2, \ldots, K_p$ of a set N. We say that π *marginally factorizes with respect to* $(K_1, K_2, \ldots, K_p)$ if $\pi^{\downarrow K_1 \cup \ldots \cup K_p}$ is uniquely given by its marginals $\pi^{\downarrow K_1}, \pi^{\downarrow K_2}, \ldots, \pi^{\downarrow K_p}$. More precisely, for this type of factorization we assume that there exists a function $\mathcal{F}$ such that

$$\pi^{\downarrow K_1 \cup \ldots \cup K_p} = \mathcal{F}(\pi^{\downarrow K_1}, \pi^{\downarrow K_2}, \ldots, \pi^{\downarrow K_p}).$$

As an interesting example of this type of factorization may serve the following function[4] $\mathcal{F}$

$$\mathcal{F}(\pi^{\downarrow K_1}, \pi^{\downarrow K_2}, \ldots, \pi^{\downarrow K_p}) = \arg\max\{\mathbf{H}(\nu)\},$$

where the maximization is performed over all measures ν on $\mathbf{X}_{K_1 \cup \ldots \cup K_p}$ for which $\nu^{\downarrow K_1} = \pi^{\downarrow K_1}, \nu^{\downarrow K_2} = \pi^{\downarrow K_2}, \ldots, \nu^{\downarrow K_p} = \pi^{\downarrow K_p}$.

REMARK 6 *It is obvious that this type of factorization strongly depends on the function* $\mathcal{F}$. *Generally, no factorization lemma holds true for marginal factorization. However, if one considers a "reasonable" function* $\mathcal{F}$ *(for example if it is continuous in a sense) then Corollary 3 holds true.*

Decomposition

We say that a sequence $K_1, K_2, \ldots, K_p$ meets the *running intersection property* (RIP) if for all $i = 2, \ldots, p$ there exists $j, 1 \leq j < i$, such that

$$K_i \cap (K_1 \cup \ldots \cup K_{i-1}) \subseteq K_j.$$

LEMMA 7 (Decomposition lemma) *If* $(K_1, K_2, \ldots, K_p)$ *meets RIP, then* π *factorizes with respect to* $(K_1, K_2, \ldots, K_p)$ *if and only if it recursively factorizes with respect to* $(K_1, K_2, \ldots, K_p)$.

In the literature, measures factorizing with respect to systems of sets meeting (after a possible reordering) RIP are usually called *decomposable measures*.

---

[4] $\mathbf{H}$ denotes the classical Shannon entropy

$$\mathbf{H}(\nu) = - \sum_{x:\nu(x)>0} \nu(x) \log(\nu(x)).$$

REMARK 8 *It is interesting to notice that all these definitions coincide if one considers only two sets of indices: π factorizes with respect to (K, L) if and only if it marginally factorizes with respect to (K, L). Notice also that a two-element sequence meets always RIP. This is why not many authors distinguish different types of factorization.*

3 D-S Theory—Notation

As in the previous section, we consider a finite *multidimensional frame of discernment*

$$\mathbf{X}_N = \mathbf{X}_1 \times \mathbf{X}_2 \times \ldots \times \mathbf{X}_n,$$

and its *subframes* $\mathbf{X}_K$. Consider $K, L \subseteq N$ and $M \subseteq K$. In addition to a projection of a set A we will need also an opposite operation, which will be called a join. By a *join* of two sets $A \subseteq \mathbf{X}_K$ and $B \subseteq \mathbf{X}_L$ we will understand a set

$$A \otimes B = \{x \in \mathbf{X}_{K \cup L} : x^{\downarrow K} \in A \ \& \ x^{\downarrow L} \in B\}.$$

Let us note that if K and L are disjoint, then $A \otimes B = A \times B$, if $K = L$ then $A \otimes B = A \cap B$.

In view of this paper it is important to realize that if $x \in C \subseteq \mathbf{X}_{K \cup L}$, then $x^{\downarrow K} \in C^{\downarrow K}$ and $x^{\downarrow L} \in C^{\downarrow L}$, which means that always

$$C \subseteq C^{\downarrow K} \otimes C^{\downarrow L}.$$

However, and it is important to keep this in mind, it does not mean that $C = C^{\downarrow K} \otimes C^{\downarrow L}$. For example, considering 3-dimensional frame of discernment $\mathbf{X}_{\{1,2,3\}}$ with $\mathbf{X}_i = \{a_i, \bar{a}_i\}$ for all three $i = 1, 2, 3$, and $C = \{a_1 a_2 a_3, \bar{a}_1 a_2 a_3, a_1 a_2 \bar{a}_3\}$ one gets

$$\begin{aligned} C^{\downarrow\{1,2\}} \otimes C^{\downarrow\{2,3\}} &= \{a_1 a_2, \bar{a}_1 a_2\} \otimes \{a_2 a_3, a_2 \bar{a}_3\} \\ &= \{a_1 a_2 a_3, \bar{a}_1 a_2 a_3, a_1 a_2 \bar{a}_3, \bar{a}_1 a_2 \bar{a}_3\} \supsetneq C. \end{aligned}$$

In Dempster-Shafer theory of evidence several measures are used to model the uncertainty (belief, plausibility and commonality measures). All of them can be defined with the help of another set function called a *basic (probability or belief) assignment* m on $\mathbf{X}_K$, i.e.

$$m : \mathcal{P}(\mathbf{X}_K) \longrightarrow [0, 1]$$

for which $\sum_{A \subseteq \mathbf{X}_K} m(A) = 1$. Since we will consider in this paper only normalized basic assignments we will assume that $m(\emptyset) = 0$. Set $A \subseteq \mathbf{X}_K$ is said to be a *focal element* of m if $m(A) > 0$.

Analogously to marginal probability measures we consider also marginal basic assignment of m defined on $\mathbf{X}_N$. For each $K \subseteq N$ a *marginal basic assignment* of m is defined (for each $B \subseteq \mathbf{X}_K$):

$$m^{\downarrow K}(B) = \sum_{\substack{A \subseteq \mathbf{X}_N \\ A^{\downarrow K} = B}} m(A).$$

Considering two basic assignments m_1 and m_2 defined on $\mathbf{X}_K$ and $\mathbf{X}_L$, respectively, we say that they are *projective* if

$$m_1^{\downarrow K \cap L} = m_1^{\downarrow K \cap L}.$$

4   Independence and factorization in D-S Theory

Let us now present a generally accepted notion of independence ([YSM02a, She94, Stu93]).

DEFINITION 9 Let m be a basic assignment on $\mathbf{X}_N$ and $K, L \subset N$ be nonempty disjoint. We say that groups of variables X_K and X_L are *independent*[5] with respect to basic assignment m (in notation $K \perp\!\!\!\perp L \ [m]$) if for all $A \subseteq \mathbf{X}_{K \cup L}$

$$m^{\downarrow K \cup L}(A) = \begin{cases} m^{\downarrow K}(A^{\downarrow K}) \cdot m^{\downarrow L}(A^{\downarrow L}) & \text{if } A = A^{\downarrow K} \times A^{\downarrow L}, \\ 0 & \text{otherwise.} \end{cases}$$

There are several generalizations of this notion of independence corresponding to conditional independence (see for example papers [YSM02b, CMW99, Kli06, She94, Stu93]). In this text we will use the generalization, which was introduced in [JV] and which differs from the notion used in [YSM02b, She94, Stu93]).

DEFINITION 10 Let m be a basic assignment on $\mathbf{X}_N$ and $K, L, M \subset N$ be disjoint, $K \neq \emptyset \neq L$. We say that groups of variables X_K and X_L are *conditionally independent given* X_M *with respect to* m (and denote it by $K \perp\!\!\!\perp L|M \ [m]$), if for all $A \subseteq \mathbf{X}_{K \cup L \cup M}$:

- if $A = A^{\downarrow K \cup M} \otimes A^{\downarrow L \cup M}$ then

$$m^{\downarrow K \cup L \cup M}(A) \cdot m^{\downarrow M}(A^{\downarrow M}) = m^{\downarrow K \cup M}(A^{\downarrow K \cup M}) \cdot m^{\downarrow L \cup M}(A^{\downarrow L \cup M}),$$

- if $A \neq A^{\downarrow K \cup M} \otimes A^{\downarrow L \cup M}$ then $m^{\downarrow K \cup L \cup M}(A) = 0$.

---

[5]Couso et al. [CMW99] call this independence *independence in random sets*, Klir [Kli06] *non-interactivity*.

Notice that for $M = \emptyset$ the concept coincides with Definition 9. In addition to this, and this is even more important, it was proven in [JV] that this notion meets all the properties required from the notion of conditional independence, so-called *semigraphoid properties* ([Lau96, Stu93]):

A1 (symmetry): $K \perp\!\!\!\perp L \,|\, M\,[m] \implies L \perp\!\!\!\perp K \,|\, M\,[m]$,

A2 (decomposition): $K \perp\!\!\!\perp L \cup M \,|\, J\,[m] \implies K \perp\!\!\!\perp M \,|\, J\,[m]$,

A3 (weak union): $K \perp\!\!\!\perp L \cup M \,|\, J\,[m] \implies K \perp\!\!\!\perp L \,|\, M \cup J\,[m]$,

A4 (contraction): $(K \perp\!\!\!\perp L \,|\, M \cup J\,[m])\ \&\ (K \perp\!\!\!\perp M \,|\, J\,[m])$
$$\implies K \perp\!\!\!\perp L \cup M \,|\, J\,[m].$$

To be honest, we have to recall that all these properties (both the semigraphoid properties and the fact that the notion is a generalization of the unconditional independence) hold true also for the concept of conditional independence used, for example, by Shenoy [She94] and Studený [Stu93] (which is the same as the *conditional non-interactivity* used by Ben Yaghlane *et al.* in [YSM02b]). In spite of the fact that their term is used also by several other authors, we do not expect that it satisfies a characterization property similar to the one proven below in Factorization lemma.

DEFINITION 11 (Simple factorization) Consider two nonempty sets $K, L \subseteq N$. We say that basic assignment m *factorizes with respect to* (K, L) if it satisfy the following two properties:

- for all $A \subseteq \mathbf{X}_{K \cup L}$, for which $A \neq A^{\downarrow K} \otimes A^{\downarrow L}$, $m(A) = 0$;

- there exist two nonnegative set functions
$$\phi : \mathcal{P}(\mathbf{X}_K) \longrightarrow [0, +\infty) \quad \text{and} \quad \psi : \mathcal{P}(\mathbf{X}_L) \longrightarrow [0, +\infty),$$

such that for each $A \subseteq \mathbf{X}_{K \cup L}$, for which $A = A^{\downarrow K} \otimes A^{\downarrow L}$, we have
$$m^{\downarrow K \cup L}(A) = \phi(A^{\downarrow K}) \cdot \psi(A^{\downarrow L})$$

LEMMA 12 (Factorization lemma) *Let* $K, L \subseteq N$ *be nonempty.* m *factorizes with respect to* (K, L) *if and only if*

$$K \setminus L \perp\!\!\!\perp L \setminus K \,|\, K \cap L\,[m].$$

Proof First notice that for $A \subset \mathbf{X}_{K \cup L}$, for which $A \neq A^{\downarrow K} \otimes A^{\downarrow L}$, $m(A) = 0$ in both situations: when m factorizes with respect to (K, L) and when $K \setminus L \perp\!\!\!\perp L \setminus K \,|\, K \cap L \, [m]$. So, to prove implication

$$K \setminus L \perp\!\!\!\perp L \setminus K \,|\, K \cap L \, [m] \quad \Longrightarrow \quad m \text{ factorizes with respect to } (K, L)$$

is trivial. It is enough to take

$$\phi = m^{\downarrow L} \qquad\qquad \psi = \frac{m^{\downarrow L}}{m^{\downarrow K \cap L}},$$

(for $m^{\downarrow K \cap L}(x^{\downarrow K \cap L}) = 0$ take $\frac{0}{0} = 0$).

To prove the opposite implication consider two functions ϕ and ψ meeting the properties required by Definition 11, and consider an arbitrary $A \subset \mathbf{X}_{K \cup L}$, for which $A = A^{\downarrow K} \otimes A^{\downarrow L}$. Before we start computing the necessary marginal basic assignments let us realize that

$$\{ B \subseteq \mathbf{X}_{K \cup L} : (B = B^{\downarrow K} \otimes B^{\downarrow L}) \& (B^{\downarrow K} = A^{\downarrow K}) \}$$
$$= \{ A^{\downarrow K} \otimes C : (C \subseteq \mathbf{X}_L) \& (C^{\downarrow K \cap L} = A^{\downarrow K \cap L}) \}.$$

When computing

$$m^{\downarrow K}(A^{\downarrow K}) = \sum_{\substack{B \subseteq \mathbf{X}_{K \cup L} \\ B^{\downarrow K} = A^{\downarrow K}}} m^{\downarrow K \cup L}(B)$$

we can summarize only over those B, for which $B = B^{\downarrow K} \otimes B^{\downarrow L}$, because if $B \neq B^{\downarrow K} \otimes B^{\downarrow L}$, as it follows from Definition 11, $m(B) = 0$. So we get

$$m^{\downarrow K}(A^{\downarrow K}) = \sum_{\substack{B \subseteq \mathbf{X}_{K \cup L} \\ B^{\downarrow K} = A^{\downarrow K}}} m^{\downarrow K \cup L}(B) = \sum_{\substack{B \subseteq \mathbf{X}_{K \cup L} \\ B^{\downarrow K} = A^{\downarrow K} \\ B = B^{\downarrow K} \otimes B^{\downarrow L}}} \phi(B^{\downarrow K}) \cdot \psi(B^{\downarrow L})$$

$$= \sum_{\substack{C \subseteq \mathbf{X}_L \\ C^{\downarrow K \cap L} = A^{\downarrow K \cap L}}} \phi(A^{\downarrow K}) \cdot \psi(C)$$

$$= \phi(A^{\downarrow K}) \cdot \sum_{\substack{C \subseteq \mathbf{X}_L \\ C^{\downarrow K \cap L} = A^{\downarrow K \cap L}}} \psi(C).$$

Computing analogously $m^{\downarrow L}(A^{\downarrow L})$ one gets

$$m^{\downarrow L}(A^{\downarrow L}) = \psi(A^{\downarrow L}) \cdot \sum_{\substack{D \subseteq \mathbf{X}_K \\ D^{\downarrow K \cap L} = A^{\downarrow K \cap L}}} \phi(D).$$

Now, we have to compute $m^{\downarrow K \cap L}(A^{\downarrow K \cap L})$. For this, realize again that

$$\{B \subseteq \mathbf{X}_{K \cup L} : (B = B^{\downarrow K} \otimes B^{\downarrow L}) \& (B^{\downarrow K \cap L} = A^{\downarrow K \cap L})\}$$
$$= \{D \otimes C : (D \subseteq \mathbf{X}_K) \& (C \subseteq \mathbf{X}_L) \& (D^{\downarrow K \cap L} = C^{\downarrow K \cap L} = A^{\downarrow K \cap L})\}.$$

Using this we get

$$m^{\downarrow K \cap L}(A^{\downarrow K \cap L}) = \sum_{\substack{B \subseteq \mathbf{X}_{K \cup L} \\ B^{\downarrow K \cap L} = A^{\downarrow K \cap L}}} m^{\downarrow K \cup L}(B)$$

$$= \sum_{\substack{B \subseteq \mathbf{X}_{K \cup L} \\ B^{\downarrow K \cap L} = A^{\downarrow K \cap L} \\ B = B^{\downarrow K} \otimes B^{\downarrow L}}} \phi(B^{\downarrow K}) \cdot \psi(B^{\downarrow L})$$

$$= \sum_{\substack{D \subseteq \mathbf{X}_K \\ D^{\downarrow K \cap L} = A^{\downarrow K \cap L}}} \sum_{\substack{C \subseteq \mathbf{X}_L \\ C^{\downarrow K \cap L} = A^{\downarrow K \cap L}}} \phi(D) \cdot \psi(C)$$

$$= \left(\sum_{\substack{D \subseteq \mathbf{X}_K \\ D^{\downarrow K \cap L} = A^{\downarrow K \cap L}}} \phi(D) \right) \cdot \left(\sum_{\substack{C \subseteq \mathbf{X}_L \\ C^{\downarrow K \cap L} = A^{\downarrow K \cap L}}} \psi(C) \right).$$

To finish the proof it is enough to substitute into the formula from Definition 10, which is in this context in the form

$$m^{\downarrow K \cup L}(A) \cdot m^{\downarrow K \cap L}(A^{\downarrow K \cap L}) = m^{\downarrow K}(A^{\downarrow K}) \cdot m^{\downarrow L}(A^{\downarrow L}),$$

the corresponding expressions computed above:

$$m^{\downarrow K \cup L}(A) \cdot m^{\downarrow K \cap L}(A^{\downarrow K \cap L})$$

$$= \phi(A^{\downarrow K}) \cdot \psi(A^{\downarrow L}) \cdot \left(\sum_{\substack{D \subseteq \mathbf{X}_K \\ D^{\downarrow K \cap L} = A^{\downarrow K \cap L}}} \phi(D) \right) \cdot \left(\sum_{\substack{C \subseteq \mathbf{X}_L \\ C^{\downarrow K \cap L} = A^{\downarrow K \cap L}}} \psi(C) \right),$$

$$m^{\downarrow K}(A^{\downarrow K}) \cdot m^{\downarrow L}(A^{\downarrow L})$$

$$= \left(\phi(A^{\downarrow K}) \cdot \sum_{\substack{C \subseteq \mathbf{X}_L \\ C^{\downarrow K \cap L} = A^{\downarrow K \cap L}}} \psi(C) \right) \cdot \left(\psi(A^{\downarrow L}) \cdot \sum_{\substack{D \subseteq \mathbf{X}_K \\ D^{\downarrow K \cap L} = A^{\downarrow K \cap L}}} \phi(D) \right).$$

□

COROLLARY 13 *Let $m_1, m_2, m_3, \ldots$ be a sequence of basic assignments each of them factorizing with respect to (K, L). If this sequence is convergent then also the limit basic assignment $\lim\limits_{j \to +\infty} m_j$ factorizes with respect to (K, L).*

Proof Since the considered frame of discernment $\mathbf{X}_N$ is finite, it is obvious that convergence of $m_1, m_2, \ldots$ implies also the convergence of all its marginals, i.e. also $\lim\limits_{j \to +\infty} m_j^{\downarrow K}$, $\lim\limits_{j \to +\infty} m_j^{\downarrow L}$ and $\lim\limits_{j \to +\infty} m_j^{\downarrow K \cap L}$. The assumption of factorization of all m_j says that, due to Lemma 12,

$$m_j^{\downarrow K \cup L} \cdot m_j^{\downarrow K \cap L} = m_j^{\downarrow K} \cdot m_j^{\downarrow L},$$

and therefore also

$$\lim_{j \to +\infty} m_j^{\downarrow K \cup L} \cdot \lim_{j \to +\infty} m_j^{\downarrow K \cap L} = \lim_{j \to +\infty} m_j^{\downarrow K} \cdot \lim_{j \to +\infty} m_j^{\downarrow L}.$$

$\square$

REMARK 14 *The notion introduced in Definition 11 is an analogy to the probabilistic simple factorization. Therefore, one can directly introduce also recursive factorization and decomposition for Dempster-Shafer theory of evidence. The problem whether analogical notions to multiple factorization and marginal factorization are meaningful remains at the moment open.*

5   Application of factorization to model construction

As said in Introduction, factorization is fully employed in probabilistic GMMs. So it is quite natural, that some authors generalized the ideas of GMMs and started considering analogous models within the framework of Dempster-Shafer theory. In this paper, we shall go different way. We shall describe a generalization of an alternative probabilistic approach, which is based on the application of the operator of composition. After describing this operator, we shall briefly illustrate its application to the construction of *compositional models* in the framework of Dempster-Shafer.

Operator of Composition

Let K and L be two subsets of N. At this moment we do not pose any restrictions on K and L; they may be but need not be disjoint, one may be a subset of the other. We even admit that one or both of them are empty but this is just a theoretical possibility without any practical impact[6]. Let m_1 and m_2 be basic assignments on $\mathbf{X}_K$ and $\mathbf{X}_L$, respectively.

---

[6]Notice that basic assignment m on $\mathbf{X}_\emptyset$ is defined $m(\emptyset) = 1$. Let us note that this is the only case when we accept $m(\emptyset) > 0$, otherwise $m(\emptyset) = 0$ according to the classical definitions of basic assignment and belief function, see [Sha76].

DEFINITION 15 For two arbitrary basic assignments m_1 on $\mathbf{X}_K$ and m_2 on $\mathbf{X}_L$ a *composition* $m_1 \triangleright m_2$ is defined for all $C \subseteq \mathbf{X}_{K \cup L}$ by one of the following expressions:

[a] if $m_2^{\downarrow K \cap L}(C^{\downarrow K \cap L}) > 0$ and $C = C^{\downarrow K} \otimes C^{\downarrow L}$ then

$$(m_1 \triangleright m_2)(C) = \frac{m_1(C^{\downarrow K}) \cdot m_2(C^{\downarrow L})}{m_2^{\downarrow K \cap L}(C^{\downarrow K \cap L})};$$

[b] if $m_2^{\downarrow K \cap L}(C^{\downarrow K \cap L}) = 0$ and $C = C^{\downarrow K} \times \mathbf{X}_{L \setminus K}$ then

$$(m_1 \triangleright m_2)(C) = m_1(C^{\downarrow K});$$

[c] in all other cases

$$(m_1 \triangleright m_2)(C) = 0.$$

REMARK 16 *Notice what this definition yields in the following trivial situations:*

- *if $K \supseteq L$ then $m_1 \triangleright m_2 = m_1$ (therefore, the operator of composition is idempotent);*

- *if $K \cap L = \emptyset$ then for each $C \subseteq \mathbf{X}_{K \cup L}$*

$$m_1 \triangleright m_2(C) = \begin{cases} m_1(C^{\downarrow K}) \cdot m_2(C^{\downarrow L}) & \text{for } C = C^{\downarrow K} \times C^{\downarrow L}, \\ 0 & \text{otherwise}, \end{cases}$$

(i.e. $m_1 \triangleright m_2$ is a basic assignment of independent groups of variables X_K and X_L).

REMARK 17 *Before presenting basic properties of this operator, we stress that*

- *operator $\triangleright$ is different from the famous Dempster's rule of combination[7] (these two rules coincide only under very special conditions, for details see [JV09]);*

- *it is neither commutative nor associative.*

---

[7]Recall that, for example, in contrast to Dempster's rule of combination, the operator of composition is idempotent.

Table 1. Basic assignments

A	$m_1(A)$	A	$m_2(A)$	A	$m_3(A)$
$\{a_1a_2\}$	0.5	$\{a_3\}$	0.5	$\{a_2a_3a_4\}$	0.5
$\mathbf{X}_1 \times \{\bar{a}_2\}$	0.5	$\{\bar{a}_3\}$	0.5	$\{\bar{a}_2\bar{a}_3\} \times \mathbf{X}_4$	0.5

The following two assertions were proven in [JVD07]

LEMMA 18 *For arbitrary two basic assignments m_1 on $\mathbf{X}_K$ and m_2 on $\mathbf{X}_L$ the following properties hold true:*

1. *$m_1 \triangleright m_2$ is a basic assignment on $\mathbf{X}_{K \cup L}$;*

2. *$(m_1 \triangleright m_2)^{\downarrow K} = m_1$;*

3. *$m_1 \triangleright m_2 = m_2 \triangleright m_1 \iff m_1^{\downarrow K \cap L} = m_2^{\downarrow K \cap L}$.*

LEMMA 19 *For arbitrary basic assignment m on $\mathbf{X}_M$ ($M = K \cup L$) the following properties hold true:*

1. *$m^{\downarrow K} \triangleright m = m$;*

2. *$m = m^{\downarrow K} \triangleright m^{\downarrow L} \iff K \setminus L \perp\!\!\!\perp L \setminus K \,|\, K \cap L\,[m]$.*

Thus, having a finite system of (low-dimensional) basic assignment m_1, $m_2, \ldots, m_p$ one can consider a (multidimensional) basic assignment

$$m_1 \triangleright m_2 \triangleright m_3 \triangleright \ldots \triangleright m_p = (\ldots ((m_1 \triangleright m_2) \triangleright m_3) \triangleright \ldots \triangleright m_p).$$

(Since the operator is not associative we have to stress that the operator is applied subsequently from right to left.) Such a basic assignment is called a *compositional model*.

Example

Let for $i = 1, 2, 3, 4$, $\mathbf{X}_i = \{a_i, \bar{a}_i\}$ and consider three basic assignments m_1, m_2 and m_3 given in Table 1. Here we present only focal elements of the respective basic assignments. This means that basic assignments equal 0 for all sets which are not presented in the tables. From the table we can also see that m_1, m_2 and m_3 are basic assignments on $\mathbf{X}_1 \times \mathbf{X}_2$, $\mathbf{X}_3$ and $\mathbf{X}_2 \times \mathbf{X}_3 \times \mathbf{X}_4$, respectively. First, notice that the given basic assignments are pairwise projective, i.e. $m_1^{\downarrow\{2\}} = m_3^{\downarrow\{2\}}$ and $m_2 = m_3^{\downarrow\{3\}}$ (m_1 and m_2 are trivially projective because they are defined on disjoint spaces). This is because both $m_1^{\downarrow\{2\}}$ and $m_3^{\downarrow\{2\}}$ acquire the same values for all (in our case only two) focal elements: $\{a_2\}$ and $\{\bar{a}_2\}$

$$m_1^{\downarrow\{2\}}(\{a_2\}) = m_3^{\downarrow\{2\}}(\{a_2\}) = 0.5,$$
$$m_1^{\downarrow\{2\}}(\{\bar{a}_2\}) = m_3^{\downarrow\{2\}}(\{\bar{a}_2\}) = 0.5,$$

and similarly

$$m_2(\{a_3\}) = m_3^{\downarrow\{3\}}(\{a_3\}) = 0.5,$$
$$m_2(\{\bar{a}_3\}) = m_3^{\downarrow\{3\}}(\{\bar{a}_3\}) = 0.5.$$

Now, we shall show that, in spite of the mentioned projectiveness (and due to the already mentioned non-associativity of the operator of composition), one can construct two different models corresponding to permutations m_1, m_2, m_3 and m_2, m_3, m_1.

First, compute $m_1 \triangleright m_2$. Since it is a composition of two basic assignments defined on disjoint spaces, the number of focal elements of $m_1 \triangleright m_2$ equals a product of numbers of focal elements of assignments m_1 and m_2: $2 \times 2 = 4$. The respective values of the composed basic assignments are computed according to case [a] of Definition 15:

$$(m_1 \triangleright m_2)(\{a_1 a_2 a_3\}) = \frac{m_1(\{a_1 a_2\}) \cdot m_2(\{a_3\})}{m_2^{\downarrow\emptyset}(\emptyset)} = \frac{0.5 \cdot 0.5}{1} = 0.25,$$

$$(m_1 \triangleright m_2)(\{a_1 a_2 \bar{a}_3\}) = \frac{m_1(\{a_1 a_2\}) \cdot m_2(\{\bar{a}_3\})}{m_2^{\downarrow\emptyset}(\emptyset)} = \frac{0.5 \cdot 0.5}{1} = 0.25,$$

$$(m_1 \triangleright m_2)(\mathbf{X}_1 \times \{\bar{a}_2 a_3\}) = \frac{m_1(\mathbf{X}_1 \times \{\bar{a}_2\}) \cdot m_2(\{a_3\})}{m_2^{\downarrow\emptyset}(\emptyset)} = \frac{0.5 \cdot 0.5}{1}$$
$$= 0.25,$$

$$(m_1 \triangleright m_2)(\mathbf{X}_1 \times \{\bar{a}_2 \bar{a}_3\}) = \frac{m_1(\mathbf{X}_1 \times \{\bar{a}_2\}) \cdot m_2(\{\bar{a}_3\})}{m_2^{\downarrow\emptyset}(\emptyset)} = \frac{0.5 \cdot 0.5}{1}$$
$$= 0.25.$$

From this one immediately sees that also its marginal basic assignment $(m_1 \triangleright m_2)^{\downarrow\{2,3\}}$ has four focal elements ($\{a_2 a_3\}, \{a_2 \bar{a}_3\}, \{\bar{a}_2 a_3\}, \{\bar{a}_2 \bar{a}_3\}$) and therefore $(m_1 \triangleright m_2)^{\downarrow\{2,3\}}$ and m_3 cannot be projective. Therefore, computation of $m_1 \triangleright m_2 \triangleright m_3$ will be a little bit more complicated. Case [a] of Definition 15 applies to two focal elements:

$$(m_1 \triangleright m_2 \triangleright m_3)(\{a_1 a_2 a_3 a_4\})$$
$$= \frac{(m_1 \triangleright m_2)(\{a_1 a_2 a_3\}) \cdot m_3(\{a_2 a_3 a_4\})}{m_3^{\downarrow\{2,3\}}(\{a_2 a_3\})} = \frac{0.25 \cdot 0.5}{0.5} = 0.25,$$

$$(m_1 \triangleright m_2 \triangleright m_3)(\mathbf{X}_1 \times \{\bar{a}_2 \bar{a}_3\} \times \mathbf{X}_4)$$
$$= \frac{(m_1 \triangleright m_2)(\mathbf{X}_1 \times \{\bar{a}_2 \bar{a}_3\}) \cdot m_3(\{\bar{a}_2 \bar{a}_3\} \times \mathbf{X}_4)}{m_3^{\downarrow\{2,3\}}(\{\bar{a}_2 \bar{a}_3\})} = \frac{0.25 \cdot 0.5}{0.5}$$
$$= 0.25.$$

Values of the remaining two focal elements are assigned according to case [b] of Definition 15:

$$(m_1 \triangleright m_2 \triangleright m_3)(\{a_1 a_2 \bar{a}_3\} \times \mathbf{X}_4) = (m_1 \triangleright m_2)(\{a_1 a_2 \bar{a}_3\}) = 0.25,$$
$$(m_1 \triangleright m_2 \triangleright m_3)(\mathbf{X}_1 \times \{\bar{a}_2 a_3\} \times \mathbf{X}_4) = (m_1 \triangleright m_2)(\mathbf{X}_1 \times \{\bar{a}_2 a_3\}) = 0.25.$$

Computation of $m_2 \triangleright m_3 \triangleright m_1$ is simple. Since m_2 is marginal to m_3 it follows from property (1) of Lemma 19 that $m_2 \triangleright m_3 = m_3$. The remaining computation of $m_3 \triangleright m_1$ consists in application of the formula from case [a] of Definition 15 just to two focal elements $\{a_1 a_2 a_3 a_4\}$ and $\mathbf{X}_1 \times \{\bar{a}_2 \bar{a}_3\} \times \mathbf{X}_4$:

$$(m_3 \triangleright m_1)(\{a_1 a_2 a_3 a_4\}) = \frac{m_3(\{a_2 a_3 a_4\}) \cdot m_1(\{a_1 a_2\})}{m_1^{\downarrow\{2\}}(\{a_2\})} = \frac{0.5 \cdot 0.5}{0.5}$$
$$= 0.5,$$

$$(m_3 \triangleright m_1)(\mathbf{X}_1 \times \{\bar{a}_2 \bar{a}_3\} \times \mathbf{X}_4) = \frac{m_3(\{\bar{a}_2 \bar{a}_3\} \times \mathbf{X}_4) \cdot m_1(\mathbf{X}_1 \times \{\bar{a}_2\})}{m_2^{\downarrow\{2\}}(\{\bar{a}_2\})}$$
$$= \frac{0.5 \cdot 0.5}{0.5} = 0.5.$$

Table 2. Composed basic assignments.

C	$(m_1 \triangleright m_2 \triangleright m_3)(C)$	$(m_2 \triangleright m_3 \triangleright m_1)(C)$
$\{a_1 a_2 a_3 a_4\}$	0.25	0.5
$\{a_1 a_2 \bar{a}_3\} \times \mathbf{X}_4$	0.25	0
$\mathbf{X}_1 \times \{\bar{a}_2 a_3\} \times \mathbf{X}_4$	0.25	0
$\mathbf{X}_1 \times \{\bar{a}_2 \bar{a}_3\} \times \mathbf{X}_4$	0.25	0.5

Both the resulting 4-dimensional basic assignments are recorded in Table 2 (recall once more that for all the other subsets of $\mathbf{X}_1 \times \mathbf{X}_2 \times \mathbf{X}_3 \times \mathbf{X}_4$, different from those included in Table 2, both the assignments equal 0).

6 Conclusions

Inspired by the (simple) factorization in probability theory, we have introduced an analogous notion in Dempster-Shafer theory of evidence. We have shown that it meets the basic property of probabilistic factorization that is anchored in the assertion widely known as *Factorization lemma*. Existence of its Dempster-Shafer version (presented here as Lemma 12) forms a new evidence supporting the definition of conditional independence introduced first in [Jir07][8] and studied in more details in [JV] (see Definition 10). The last section was included to persuade the reader that the notion of factorization is not interesting only from the theoretical point of view but that it is important also for applications, for multidimensional model construction.

Acknowledgement

The research was partially supported by Ministry of Education of the Czech Republic under grants no 1M0572 and 2C06019, and by Czech Science Foundation under the grants no. ICC/08/E010 and 201/09/1891.

BIBLIOGRAPHY

[YSM02a] B. Ben Yaghlane, P. Smets, and K. Mellouli. Belief function independence: I. The marginal case. *Int. J. Approximate Reasoning*, 29:47–70, 2002.

[YSM02b] B. Ben Yaghlane, P. Smets, and K. Mellouli. Belief function independence: II. The conditional case. *Int. J. Approximate Reasoning*, 31:31–75, 2002.

[CMW99] I. Couso, S. Moral, and P. Walley. Examples of independence for imprecise probabilities. In G. De Cooman, F. G. Cozman, S. Moral, and P. Walley, editors, *Proceedings of ISIPTA'99*, pages 121–130, 1999.

[Háj87] P. Hájek. Dempster-Shaferova teorie evidence a expertní systémy. In *Metody umělé inteligence a expertní systémy III*, pages 54–61, Prague, 1987. ČSVTS.

[Háj92] P. Hájek. Dempster-Shafer theory—what it is and how (not) to use it. In *SOFSEM '92*, pages 19–24, Brno, 1992. UVT MU.

[Háj93] P. Hájek. Deriving Dempster's rule. In B. Bouchon-Meunier, L. Valverde, and R. R. Yager, editors, *Uncertainty in Intelligent Systems*, pages 75–83, Amsterdam, 1993. North-Holland.

[Háj94] P. Hájek. Systems of conditional beliefs in Dempster-Shafer theory and expert systems. *Int. J. General Systems*, 22(2):113–124, 1994.

[HH92a] P. Hájek and D. Harmanec. An exercise in Dempster-Shafer theory. *Int. J. General Systems*, 20(2):137–142, 1992.

[HH92b] P. Hájek and D. Harmanec. On belief functions (the present state of Dempster-Shafer theory). In V. Mařík, editor, *Advanced in Artificial Intelligence*, pages 286–307, Berlin, 1992. Springer.

---

[8]In the cited paper the notion was called conditional irrelevance.

[Jir07]   R. Jiroušek. On a conditional irrelevance relation for belief functions based on the operator of composition. In C. Beierle and G. Kern-Isberner, editors, *Dynamics of Knowledge and Belief. Proceedings of the Workshop at the 30th Annual German Conference on Artificial Intelligence*, pages 28–41, Osnabrück, 2007. Fern Universität in Hagen.

[JV]   R. Jiroušek and J. Vejnarová. Compositional models and conditional independence in evidence theory. Submitted.

[JV09]   R. Jiroušek and J. Vejnarová. There are combinations and compositions in Dempster-Shafer theory of evidence. In T. Kroupa and J. Vejnarová, editors, *Proceedings of the 8th Workshop on Uncertainty Processing—WUPES'09*, pages 100–111, 2009.

[JVD07]   R. Jiroušek, J. Vejnarová, and M. Daniel. Compositional models for belief functions. In G. De Cooman, J. Vejnarová, and M. Zaffalon, editors, *Proceedings of 5th International Symposium on Imprecise Probability: Theories and Applications ISIPTA'07*, pages 243–252, Prague, 2007.

[Kli06]   G. J. Klir. *Uncertainty and Information. Foundations of Generalized Information Theory*. Wiley, Hoboken, 2006.

[Lau96]   S. L. Lauritzen. *Graphical Models*. Oxford University Press, 1996.

[Sha76]   G. Shafer. *A Mathematical Theory of Evidence*. Princeton University Press, Princeton NJ, 1976.

[She94]   P. P. Shenoy. Conditional independence in valuation-based systems. *Int. J. Approximate Reasoning*, 10:203–234, 1994.

[Stu93]   M. Studený. Formal properties of conditional independence in different calculi of AI. In K. Clarke, R. Kruse, and S. Moral, editors, *Proceedings of European Conference on Symbolic and quantitative Approaches to Reasoning and Uncertainty ECSQARU'93*, pages 341–351. Springer, 1993.

Radim Jiroušek

Faculty of Management, University of Economics

Jarošovská 1117/II, CZ377 01 Jindřichův Hradec

and

Institute of Information Theory and Automation

Academy of Sciences of the Czech Republic

Pod Vodárenskou věží 4, CZ182 08 Prague

Czech Republic

Email: radim@utia.cas.cz

Managing uncertainty in medicine

JANA ZVÁROVÁ

Dedicated to Petr Hájek
on the occasion of his 70th birthday

I have had the pleasure to follow closely the work of Petr for more than thirty years. Inference under uncertainty is a fascinating topic on the borderline of artificial intelligence and data analysis. There are several competing approaches to approximate reasoning (probabilistic, fuzzy, belief functions, many-valued logics and so on). I attended workshops and conferences where discussions reflecting Petr's presentations were very vivid. What I have always appreciated in Petr is that you could disagree with him. In universities, the ability to disagree and remain good colleagues is often not present.

Petr and his research team have been focused on fundamental research in the field of mathematical models and theories of reasoning under uncertainty (in a sense of randomness or vagueness) alternative to classical probability theory and mathematical statistics. My research team has been focused on managing uncertainty using methods of probability and mathematical statistics. The Velvet Revolution in Czechoslovakia, provoked by a students' strike on November 17, 1989, opened the door for cooperation in science and education with a number of countries worldwide. Petr invited me to participate in the proposal of a joint research project funded under the programme Copernicus. The project proposal entitled "Managing Uncertainty in Medicine" (MUM) was accepted and the project was realised in the period 1994–1996. The objectives of the project were to contribute to the emerging unifying theory of uncertainty management, including probabilistic, fuzzy, and belief-functional approaches; to incorporate these findings into the diagnostic knowledge-based systems and intelligent systems of data analysis; to gain practical experience in medical data/knowledge processing by introducing software tools into selected medical institutions; and to develop decision support systems for use in various clinical departments with the aim of providing better patient care, especially in those areas where there is a shortage of clinical expertise. Six organizations from four countries participated in the project. Bernard Richards (United Kingdom) served as a project coordinator, Petr Hájek (Czech Republic) as its scientific director, participants were from Spain (Francesco Esteva), Poland (Marek Lubitz) and the Czech Republic (Radim Jiroušek and Jana Zvárová).

One of the prime objectives of the MUM project was the production, testing, and the interchange of software packages among the MUM partners. Software packages were developed and validated in the Czech Republic (e.g. Bayesian Network Construction, Data Reduction and Constitution), in Poland (PROFILES), and in England (ESICU and ESOHS). Spanish MILORD II and EQUANT were expert system shells; ESICU and ESOHS were special purpose expert systems. The Czech program GUHA was tested for automated generation of hypotheses from incomplete data. Several data and knowledge sets were processed in parallel by the appropriate software packages. There were three large data sets analysed: Lung cancer and tuberculosis data, tumour marker data and cardiology data. Some other data sets were collected and analysed with the support of the developed software packages utilising GUHA. I refer to several papers published in cooperation with Petr which are concerned with the method GUHA [HSZ95], [RHZ$^+$95], [ZPS97]. Some results of the MUM project were also utilized in Tempus-Phare project Education in the methodology field of healthcare.

Figure 1. Researchers from the Czech Republic, Poland and Spain at the workshop in Manchester, U.K.

During the years 1994-1996 we had an opportunity to visit the project partners in Spain, Poland and United Kingdom and to take active part in evaluating meeting held in Brussels, Belgium. The workshop held in Manchester, U.K., at the University of Science and Technology is documented with the first photograph of the research group headed by Petr Hájek.

The second photograph shows Petr Hájek addressing participants of the workshop at the closing dinner organized by the host Bernard Richards and his co-workers from the University of Science and Technology in Manchester, United Kingdom.

Figure 2. Petr Hájek addresses the participants of the workshop in Manchester, U.K.

I am very pleased to have the opportunity to join in honouring my long-time good friend and highly respected colleague Petr Hájek. Clearly, Petr's contributions to managing uncertainty are of monumental importance. Besides his obvious depth of knowledge I think Petr's personal touch is particularly noteworthy. His outgoing, warm and collegial manner has done much to bring the Institute of Computer Science AS CR and the Czech school of reasoning under uncertainty to national and international awareness with a collective sense of identity and purpose. Data, information and knowledge - everything can be uncertain. We have learned from Petr about the importance of managing uncertainty and fifteen years after our joint work on the MUM project we are trying to tackle this topic again [ZVV09].

BIBLIOGRAPHY

[HSZ95]	P. Hájek, A. Sochorová, and J. Zvárová. GUHA for personal computers. *Computational Statistics and Data Analysis*, 19(2):149–153, 1995.

[RHZ⁺95]	B. Richards, P. Hájek, J. Zvárová, R. Jiroušek, F. Esteva, and M. Lubicz. MUM—managing uncertainty in medicine. In M. Laires, M. Ladeira, and J. Christensen, editors, *Health in the New Communications Age. Health Care Telematics for the 21st Century*, volume 24 of *Studies in Health Technology and Informatics*, pages 247–250. IOS Press, Amsterdam, 1995.

[ZPS97]	J. Zvárová, J. Preiss, and A. Sochorová. Analysis of data about epileptic patient using GUHA method. *Int. J. of Medical Informatics*, 45(1–2):59–64, 1997.

[ZVV09]	J. Zvárová, A. Veselý, and I. Vajda. Data, information and knowledge. In P. Berka, J. Rauch, and D. Zighed, editors, *Data Mining and Medical Knowledge Management: Cases and Applications*, Information Science Reference. Idea Group Inc., 2009.

Jana Zvárová

European Centre for Medical Informatics, Statistics and Epidemiology

Charles University in Prague and Academy of Sciences CR

Department of Medical Informatics

Institute of Computer Science AS CR

Pod Vodárenskou věží 2

182 07 Prague, Czech Republic

Email: zvarova@euromise.cz

On uncertainty and Kripke modalities in t-norm fuzzy logics

FRANCESC ESTEVA AND LLUÍS GODO

Dedicated to Professor Petr Hájek
on the occasion of his 70th anniversary

1   Introduction and some remembrances

Five years ago, in the scientific event "The Beauty of Logic" organized by the Czech Academy of Sciences (CS AR) in Prague in honour of Professor Hájek in occasion of his 65th anniversary, we already stressed that Petr Hájek is recognized as a prominent scientist not only for his outstanding theoretical contributions in many different fields but also as a researcher that has never forgotten a more applied perspective. Indeed, part of his scientific work has been devoted to develop theoretical basis of the GUHA system [HH78], a methodology to systematically generate all hypotheses (relations among properties of objects) interesting with respect to given data, to provide mathematical grounds to process uncertain information [HHJ92], and to justify from a theoretical point of view some successful applications like those based on fuzzy logic [Háj98b].

In fact, when for the first time Lluís Godo (and Carles Sierra) met Petr Hájek in Prague in June 1987 (thanks to a contact with the late prof. Jiři Bečváš), the main motivation was to talk about uncertainty inference models for expert systems. In that time, expert systems were very popular Artificial Intelligence tools with many applications in different fields, and Petr Hájek, together with some colleagues of his (among them his wife Marie Hájková, the late Tomáš Havránek, and Milan Daniel) was developing the expert system shell EQUANT-PC. It was also in that period that Petr was supervising the doctoral work of J.J. Valdés analyzing the inference models of the two pioneer expert systems like MYCIN and PROSPECTOR and relating them, for the first time as far as we know, to ordered Abelian groups. This subject was extensively developed in the doctoral dissertation of J.J. Valdés and in a chapter of the book "Uncertain Information Processing in Expert Systems" by Hájek, Havránek and Jiroušek [HHJ92].

On the other hand, in the same period, in the Department of Artificial Intelligence of the Center for Advanced Studies of Blanes (CEAB), belonging to the Spanish National Research Council (CSIC), some researchers (among them Lluis and Carles), with the supervision of R. López de Mantaras, were developing the shell for expert systems MILORD (with a fuzzy logic based inference model) together with some medical applications (see [GdMSV89]).

After that first contact, three years later the Czechoslovak Academy of Sciences (CSAS) and the Spanish CSIC signed a collaboration agreement, and the authors could make a short visit to Petr in Prague in 1991, when Petr still was at the Mathematical Institute of the Czechoslovak Academy. By that time, Petr's office was under reconstruction and we could only have some short but intensive meetings at the Mazanka residence together with a seminar at the Mathematical Institute where we talked about some initial research on formalization of fuzzy logic we were involved in. The same year, a few months later, Petr made a first visit to our institute CEAB. That visit actually was for us the beginning of a long, illuminating, fruitful and still lasting collaboration between Petr and us, which has been extended along the years to colleagues of both teams.

And, as far as we are aware, it was in that time that Petr got really interested in fuzzy logic, even more when after that visit he attended the first European Conference on Symbolic and Quantitative Approaches to Reasoning and Uncertainty ECSQARU'91 in Marseille (October 1991).

Soon after, we started our scientific collaboration and in 1994 we published our first joint paper [HHE+94], with Dagmar Harmancová and Pere Garcia, about comparative possibilistic logic. The second joint paper was in 1995 [HGE95] about probability and fuzzy logic and the third was [HGE96] where Product logic was introduced. After Łukasiewicz and Gödel-Dummett logics studied in the late fifties, Product logic was the third basic residuated fuzzy logic whose semantics are given by the three distinguished continuous t-norms and their residua. Three years after Hájek published his celebrated book [Háj98b] about Metamathematics of Fuzzy logic, where he defined the Basic Fuzzy logic BL, which is proved to be the logic of continuous t-norms [Háj98a, CEGT00]. Two years later, the authors introduced the Monoidal t-norm logic MTL [EG01], proved to be the logic of the left-continuous t-norms in [JM02]. Since then, the new discipline of *Mathematical Fuzzy logic* has become a nice and deep research area and has attracted the interest of many scholars, most of them integrating the EUSFLAT working group on Mathematical Fuzzy logic MATHFUZZLOG[1].

As a contribution to this volume as tribute to Petr Hájek we have chosen to overview a topic that we have jointly worked on all over the past years: modalities in t-norm based fuzzy logics, mainly in extensions of BL and possibly expanded with truth constants[2]. On one hand we will review the work

---

[1]See http://www.mathfuzzlog.org/.

[2]The expansion of Łukasiewicz logic with truth constants was revisited by Hájek in his

done about logics to reason about uncertainty by means of fuzzy modalities. This topic was initiated with a joint paper with Petr Hájek (see e.g. [HGE95]) where probability and other measures of uncertainty on classical propositions were modelled as modalities over Rational Pavelka logic. This approach has been extended later to modalities to model different uncertainty measures over both classical and many-valued propositions. On the other hand we will also review recent papers studying modal many-valued logics following the definition of modal operators over finite Heyting algebras in the papers by Fitting [Fit92a, Fit92b]. We will restrict ourselves to review the results on minimum modal logics over a finite BL-chain. Actually, they are in fact two different kinds of modalities, arising from different motivations and semantics and with few common features. While the former uncertainty modalities are introduced in a restricted language where e.g. nested modalities are not allowed, the latter are proper generalizations of classical modalities with Kripke semantics to a fuzzy or many-valued framework. It is worth noticing that our very first joint paper [HHE+94] was about some many-valued modalities with Kripke semantics but to model qualitative possibilistic uncertainty.

2 Uncertainty modalities: belief degrees versus truth degrees

Since the first contacts, a topic of common interest was the interpretation of certainty values associated to rules and facts in expert systems (or more in general in knowledge based systems).

Indeed, in that time, it was common in expert systems to allow rules and facts be qualified by certainty degrees, providing a measure of the strength of the statements. The way those certainty values were interpreted and the compositional way they were handled by the inference procedures varied from one system to another. However, in most of the systems, there was a mismatch between the intended semantics of the certainty degrees and the way they were used. Indeed, in some systems the certainty values were interpreted probabilistically (in some form or another), like in Prospector or Mycin, but the propagation rules were either not sound or they were making too strong conditional independence assumptions. On the other hand, other systems interpreted certainty degrees as truth degrees in a truth-functional many-valued of fuzzy logic setting. Here the problem was the misuse of partial degrees of truth as belief degrees. This kind of confusion was quite common, but Petr Hájek had already clear this distinction in mind since our first meetings in the beginning of the nineties, put forward later during the European Coper-

---

book (he called it RPL, for Rational Pavelka logic), and the general case of any logic of a continuous t-norm and its residuum has been addressed in [EGGN07]. Later, Hájek has returned to this topic in the paper [Háj06b] where he studied the complexity issues of these logics with truth constants.

nicus 10053 project MUM (*Management of Uncertainty in Medicine*) where we jointly participated, and stressed by himself in many different occasions to the fuzzy logic community.

In his excellent monograph [Háj98b] Petr Hájek deals with this problem. In the introduction he writes *Fuziness is imprecision (vagueness); a fuzzy proposition may be true in a some degree* and in the same paragraph he adds *the truth of a fuzzy proposition is a matter of degree.* He also argues a very important issue to differentiate uncertainty measures and truth values in a logical setting. He states that *most many-valued logics are truth functional*[3] but this is not true for uncertainty measures that, as it is well known, are not truth functional.

However, as Petr Hájek and Dagmar Harmancová noticed in [HH95], one can safely interpret a probability degree on a Boolean proposition φ as a truth degree, but not of φ itself but of another proposition $P\varphi$, read as "φ is *probable*". The point is that "being probable" is actually a fuzzy predicate, which can be more or less true, depending on how much probable is φ. Hence, it is meaningful to take the truth-degree of $P\varphi$ as the probability-degree of φ. The second important observation is the fact that the standard Lukasiewicz logic connectives provide a proper modelling of the Kolmogorov axioms of finitely additive probabilities. Indeed, these were the key issues that are behind the probability logic first defined in our joint UAI'95 paper [HGE95] as a theory over Rational Pavelka logic, and later described with an improved presentation in Hájek's monograph [Háj98b, Chapter 8] where P is introduced as a modality.

In the following subsections we recall the definition of the basic probability logic and then succinctly describe some other fuzzy probability logics we have jointly studied extending the original approach.

2.1   The basic probability logic FP(CPL, Ł)

The basic probability logic, as defined in [Háj98b], takes formulas of classical propositional logic φ and defines atomic probability formulas to be of the form $P\varphi$, and compound probability formulas are then defined from the atomic ones by means of Lukasiewicz logic connectives. So, the language is not a full modal language, actually it contains two kind of formulas, Boolean formulas of classical propositional logic (CPL) and probability formulas. Notice that nesting of the P operator is not allowed as well as formulas combining Boolean and probability subformulas. The axioms of FP(CPL, RPL), standing for fuzzy probability logic to reason about the probability of CPL formulas using Lukasiewicz logic, are the following: Axioms and rules of CPL for Boolean

---

[3]This is also true for the combination of membership degrees in fuzzy set theory since from Zadeh most of fuzzy set systems assume functional combination of membership degrees.

formulas, axioms and rules of Łukasiewicz logic for probability formulas, plus the following three probabilistic axioms:

(FP1) $(P\varphi \,\&\, P(\varphi \to \psi)) \to_{\mathrm{L}} P\psi$

(FP2) $P(\neg\varphi) \equiv_{\mathrm{L}} \neg_{\mathrm{L}} P\varphi$

(FP3) $P(\varphi \lor \psi) \equiv_{\mathrm{L}} (P\varphi \to_{\mathrm{L}} P(\varphi \land \psi)) \to_{\mathrm{L}} P\psi.$

and the necessitation rule for P: if φ is a theorem of CPL, then derive $P\varphi$.

The crucial point is that axiom (FP3) exactly captures finite additivity. In fact, an easy checking shows that a Łukasiewicz truth evaluation e satisfying (FP3) is such that $e(P(\varphi\lor\psi)) = e(P\varphi) + e(P\psi) - e(P(\varphi\land\psi))$. Moreover, any e satisfying axioms (FP1), (FP2) and (FP3) and respecting the necessitation rule (i.e. making $e(P\varphi) = 1$ for each CPL theorem φ) induces a probability μ_e on Boolean formulas, namely, if we define $\mu_e(\varphi) = e(P\varphi)$, then $\mu_e : \mathcal{L} \to [0,1]$ is a finitely additive probability preserving classical logical equivalence on the set of Boolean formulas $\mathcal{L}$. And conversely, given such a probability on formulas μ, it induces a Łukasiewicz truth-evaluation of atomic probability formulas e_μ, by defining $e_\mu(P\varphi) = \mu(\varphi)$, which is a model of the above axioms.

These considerations, combined together with Łukasiewicz logic finite standard completeness gives the following probabilistic completeness of FP(CPL, Ł): any finite theory T of probability formulas proves in FP(CPL, Ł) a probability formula Φ iff for any probability on formulas μ such that e_μ is a model of all formulas of T, e_μ is also a model of Φ.

If one wants to reason with explicit probability degrees one can switch from Łukasiewicz logic to Rational Pavelka logic by introducing rational truth-constants into the language of probability formulas, and adding the corresponding book-keeping axioms to the FP(CPL, Ł) logic. In this way we obtain the logic FP(CPL, RPL), which inherits the Pavelka-style completeness of RPL, stating that the truth degree of a probability formula Φ over an arbitrary theory T of probability formulas equals its provability degree over FP(CPL, RPL).

Let us mention that the same approach can be used to reason about other uncertainty measures, like possibility and necessity measures [Háj98b] or upper and lower probability measures [Mar08]. The case of belief functions is recalled in the next subsection.

On the other hand, if one is interested in reasoning about conditional probabilities, then one definitely needs to replace Ł, or RPL, by the more expressive logic $\mathrm{L\Pi}\frac{1}{2}$ [EGM01], putting together Łukasiewicz and Product logic connectives. If $\to_\Pi$ denotes Product logic implication, then one can mimic the usual definition of conditional probability in terms of the 1-place probability by defining $P(\varphi \mid \psi)$ as a shorthand for

$$P\psi \to_\Pi P(\varphi \wedge \psi).$$

The corresponding logic FP(CPL, LΠ$\frac{1}{2}$) was defined and studied in a joint paper with Petr Hájek [GEH00]. Another alternative is to start with conditional probability as a primitive notion, so to introduce a binary probability operator $P(\cdot \mid \cdot)$, redefine accordingly the set of probability formulas and to replace the above axioms (FP1)-(FP3) by a suitable set of axioms for conditional probability. This approach was taken in [GM06] and the corresponding logic was denoted FCP(CPL, LΠ$\frac{1}{2}$). An alternative approach to reason about conditional probability, based on non-standard probabilities, has been developed by Flaminio and Montagna in [FM05], where they define the logic FP(SLΠ), and further studied in [Fla07]. In this framework of fuzzy modal logics of conditional measures, let us mention as well the paper by Marchioni [Mar06] where he defines a logic to reason about conditional possibilities.

2.2 FP(S5, RPL): a logic for belief functions

In [GHE03] we generalized the approach to deal with belief functions, i.e. we took the following identity

belief degree of φ = truth degree of $B\varphi$,

where now belief degrees are related to Dempster-Shafer belief functions, and $B\varphi$ stands for the fuzzy proposition "φ is believed". For getting a complete axiomatization we used one of possible definitions of Dempster-Shafer belief functions, namely as probability of knowledge, in the epistemic logic sense (see for instance [Rus87, HKR94, Háj96]. This enabled us to combine the above approach to probabilistic reasoning with the modal logic S5, the most common logic for knowledge, i.e. to define our belief formulas $B\varphi$ to be defined as $P\square\varphi$, where $\square$ is a S5 modality and φ is a propositional modality-free formula.

We recall that a belief function on Boolean formulas is a mapping into $[0,1]$ preserving classical logical equivalence and satisfying the following properties:

(B1) $bel(\top) = 1$,

(B2) $bel(\bot) = 0$,

(B3) $bel(\varphi_1 \vee \ldots \vee \varphi_n) \geq \displaystyle\sum_{\emptyset \neq I \subseteq \{1,\ldots,n\}} (-1)^{|I|+1} bel(\bigwedge_{i \in I} \varphi_i)$, for each n.

As it was proved in [HKR94, FHM90] for the case of finitely-many propositional variables and generalized in [GHE03] for the countable case, all belief functions on formulas are given by probabilities on S5-formulas. More precisely, for each belief function on formulas *bel* there is a $\Box$-probabilistic Kripke model $K = (W, e, R, \mu)$, where (W, e, R) is a S5 Kripke model and μ is a probability measure on a subalgebra of 2^W such that $bel(\varphi) = \mu(\{w \in W \mid e(w, \Box\varphi) = 1\})$.

In the logic FP(S5, RPL) we distinguish again two kinds of *formulas*:

(1) *Boolean formulas* (or S5-formulas) are built from propositional variables $p_0, p_1, \ldots$ using connectives $\rightarrow, \vee, \wedge, \neg$ (say) and the modality $\Box$ as described above. *Modality-free* formulas do not contain $\Box$; a formula is *closed* if each propositional variable is in the scope of a modality. Closed formulas without nested modalities are called *normal;* they are just built from formulas of the form $\Box\varphi$, where φ if modality-free, using the connectives above.

(2) *P-formulas* (or many-valued formulas). Atomic P-formulas have the form $P\varphi$, where φ is a normal S5-formula. P-formulas are built from atomic P-formulas using the connectives of Lukasiewicz logic, i.e. $\rightarrow_L, \&$, and truth constants $\bar{r}$ for each rational $r \in [0, 1]$. As mentioned above, $B\varphi$ is an abbreviation for $P(\Box\varphi)$, where φ is a modality-free formula; Formulas built from these using L-connectives and truth-constants are called *belief formulas* or B-formulas.

The deduction rule of FP(S5, RPL) is modus ponens for Lukasiewicz implication $\rightarrow_L$ and its axioms are:

(FB1) φ, for each S5-formula φ provable in S5
(FB2) $P\varphi$, for each normal S5-formula φ provable in S5
(FB3)   the schemes (FP1)–(FP3) above for *P*-formulas
(FB4)   axioms of RPL for *P*-formulas.

The main completeness result for FP(S5, RPL) in [GHE03] reads as follows. Let T be a finite set of belief formulas and Φ a belief formula. Then, T proves Φ in FP(S5, RPL) iff $\|\Phi\|_{bel} = 1$ for each belief function *bel* which is a model of T.

2.3   Logics combining belief and fuzziness

An extension of the notion of probability to the framework of fuzzy sets was early defined by Zadeh in order to represent and reasoning with sentences like *the probability that the traffic in Rome will be chaotic tomorrow is 0.7*. Clearly, the modeling of this kind of knowledge cannot be done using the classical approach to probability since, given the unsharp nature of events like *chaotic traffic*, the structure of such fuzzy events cannot be considered to be a Boolean algebra any longer. The study of finitely-additive measures in the context of MV-algebras was started by Mundici in [Mun95] and has been further developed in the last years, see e.g. [RM02, Kro06]. Similar studies of probabilities on Gödel algebras have been carried out by Aguzzoli, Gerla and Marra [AGM08].

Therefore, a fuzzy logical approach to reason about the probability of fuzzy events is a natural generalization of the previous works. In logical terms, this can be approached by assuming that the logic of events is a (suitable) many-valued logic and by defining and axiomatizing appropriate probability-like measures on top of the many-valued propositions.

In [Háj98b] Hájek himself already proposed a logic built up over the Łukasiewicz predicate calculus LꓖA allowing a treatment of (simple) probability of fuzzy events. To model this kind of probability, Hájek introduced in LꓖA a generalized fuzzy quantifier standing for *most* together with a set of characteristic axioms and denoted his logic by Lꓖ$\forall\int$. In his monograph Hájek also proposed two (Kripke-style) probabilistic semantics for this logic, called *weak* and *strong*, that can be roughly described as follows:

- A weak probabilistic model for Lꓖ$\forall\int$ evaluates a modal formula $P(\varphi)$ by means of a finitely additive measure (or *state*) defined over the MV-algebra of provably equivalent Łukasiewicz formulas.

- A strong probabilistic model for Lꓖ$\forall\int$ consists on a probability distribution σ over the set of all the evaluations of the events (remember that an event is now a formula of the Łukasiewicz calculus). Then the truth value of a modal formula $P(\varphi)$ is defined as the *integral* of the fuzzy-set of all the evaluations of φ under the measure σ.

Hájek showed that his logic is Pavelka-style complete w.r.t. weak probabilistic models, but the issue of completeness w.r.t. to the strong semantics remains as an open problem. More recently, Flaminio and one of the authors have defined in [FG07] a logic to reason about the probability of many-valued events of finitely-valued Łukasiewicz logic, following the approach described in the previous subsections and called it $FP(Ł_n,Ł)$. Here the basic difference with the previous logic $FP(CPL, Ł)$ is of course the inner logic of events. Indeed, the axioms $FP(Ł_n,Ł)$ are those of $Ł_n$ for non-modal formulas, axioms of $Ł$ for modal formulas, the following version of the above probabilistic axioms

$$(FP1) \quad P(\varphi \to_Ł \psi) \to_Ł (P\varphi \to_Ł P\psi)$$

$$(FP2) \quad P(\neg_Ł\varphi) \equiv_Ł \neg_Ł P\varphi$$

$$(FP3) \quad P(\varphi \oplus \psi) \equiv_Ł [(P\varphi \to_Ł P(\varphi \,\&\, \psi)) \to_Ł P\psi]$$

together with the rule of modus ponens and the rule of necessitation for P: if φ is a theorem of $Ł_n$ then derive $P\varphi$. This logic was shown to be complete for finite theories of modal formulas with respect to both weak and strong probabilistic models as defined above, i.e. defined by finitely additive measures on $Ł_n$-formulas (and hence respecting $Ł_n$ logical equivalence) and by probability distributions on sets of worlds (which may be identified with L_n-evaluations). Completeness of the analogous logic over events in the infinitely-valued Łukasiewiz logic Ł remains as an open problem.

It is worth pointing out that Petr Hájek has studied in [Háj07] the complexity of a variety of fuzzy probability logics, extending first results of Hájek and Tulipani for FP(CPL, Ł) and FP(CPL, ŁΠ) [HT01] showing that the corresponding sets of 1-satisfiable formulas are NP-complete and in PSPACE respectively. Indeed, he considers arbitrary fuzzy probability logics FP(L_1, L_2), where L_1, the logic of events, can be either Boolean logic (CPL) or a fuzzy logic, and L_2 is a fuzzy logic. The fuzzy probability logic FP(L_1, L_2) is then a fuzzy modal logic whose non-modal formulas are those of L_1, atomic modal formulas have the form $P\varphi$ and other modal formulas are formed from atomic ones by means of connectives of the logic L_2. Probabilistic Kripke models have the form $M = (W, e, \sigma)$ where W is a non-empty (at most countable) set of possible worlds, e evaluates in each possible world all atomic formulas by truth values from the standard set of truth values of L_1 and $\sigma : W \to [0, 1]$ with $\Sigma_{w \in W} \ \sigma(w) = 1$. For each L_1-formula φ, the evaluation of $P\varphi$ in the model is the probability of φ defined as $\|P\varphi\|_M = \Sigma_{w \in W} \ \sigma(w) \cdot e_{L_1}(\varphi, w)$. Hájek then shows, among others, complexity results for different sets of FP(L_1, L_2) satisfiable formulas. In particular, he shows that the sets of all satisfiable formulas of FP($Ł_n$, Ł) and FP(G_n, Ł), where G_n denotes the n-valued Gödel logic, are NP-complete. Also, the set of satisfiable formulas of FP(G, L_2), for L_2 arbitrary suitable[4], and the set of 1-tautologies of FP(L_1, Ł) for L_1 arbitrary suitable, are in PSPACE.

To close this section, let us mention that for the case of *possibilistic* uncertainty, some logics have been also defined to reason about the necessity of fuzzy events. Indeed, in our very first joint paper with Petr Hájek and Dagmar Harmancová [HHE+94], we dealt with logics of necessity of events of $Ł_n$, a very similar topic of the more recent paper [FGM], while in [DGM09] several necessity logics are defined over events of Gödel logic.

3   Kripke modalities and fuzzy logic

In this section we review some results about modal many-valued logics understood as logics defined by Kripke frames (including the possibility of many-valued accessibility relations) where every world follows the rules of a many-valued logic, this many-valued logic being the same for every world.

One can find in the literature some attempts and approaches to generalize modal logic formalisms to the many-valued setting. Roughly speaking, we can classify the approaches in three groups depending how the corresponding

---

[4]Hájek defines a fuzzy logic to be suitable when its standard set of truth values is the real unit interval [0, 1] and the truth functions of its (finitely many) connectives are definable by open formulas in the (ordered) field of reals.

Kripke frames look like, in the sense of how many-valuedness affects the worlds and the accesibility relations. Next we describe these three groups and comment about our work in each of them.

A first group (see e.g. [Gob70, CF92, EGGR97, Suz97]) is formed by those logical systems whose class of Kripke frames are such that their *worlds are classical* (i.e. they follow the rules of classical logic) but their *accessibility relations are many-valued*, with values in some suitable lattice A. In such a case, the usual approach to capture the many-valuedness of an accessibility relation $R : W \times W \to A$ is by considering the induced set of classical accessibility relations $\{R_a \mid a \in A\}$ defined by the different level-cuts of R, i.e. $\langle w, w' \rangle \in R_a$ iff $R(w, w') \geq a$. At the syntactical level, one then introduces as many (classical) necessity operators $\Box_a$ (or possibility operators $\Diamond_a$) as elements a of the lattice A, interpreted by (classical) relations R_a. Therefore, in this kind of approach, one is led to a multi-modal language but where (both modal and non modal) formulas are Boolean in each world.

In this setting the work of our team is related to models of similarity-based reasoning developed in [EGCG94, DEG+95, DPE+97]. The starting point is the paper by Ruspini [Rus91] about a possible semantics for fuzzy set theory. He develops the idea that we could represent a fuzzy concept by its set of prototypical elements (which will have fully membership to the corresponding fuzzy set) together with a similarity relation giving the degree of similarity of each element of the universe to the closest prototype. This degree is taken then as the membership to the fuzzy set. From this basic idea, we have developed and characterized some forms of graded entailments that can be represented in multi-modal systems with frames where the (graded) accessibility is given by a similarity relation on pairs of worlds, and for which we have proved completeness in several cases [EGGR98].

A second group of approaches are the ones whose corresponding Kripke frames have *many-valued worlds*, evaluating propositional variables in a suitable lattice of truth-values A, but with *classical accessibility relations* (see e.g. [HH96, Háj98b, HT06, Háj]). In this case, we have languages with only one necessity and/or possibility operator ($\Box, \Diamond$), but whose truth-evaluation rules in the worlds is many-valued, so modal (and non-modal formulas) are many-valued.

Main Hájek's contributions to many-valued modal logic in the sense reviewed in this section fall into this group, see e.g. [HH96, Háj98b, Háj]. We have also worked in this setting but only as a particular case of the general modal many-valued setting that we will review after introducing the third group.

Finally, a third group of approaches are *fully many-valued*, in the sense that in their Kripke frames, both worlds and accessibility relations are many-valued, again over a suitable lattice A. In that case, some approaches (like

[Fit92a, Fit92b, KNP02, CR]) have a language with a single necessity/ possibility operator ($\square, \diamond$), and some (like [Mir05]) consider a multi-modal language with a family of indexed operators $\square_a$ and $\diamond_a$ for each $a \in A$, interpreted in the Kripke models via the level-cuts R_a of a many-valued accessibility relations R. Actually, these two kinds of approaches are not always equivalent, in the sense that the operator $\square$ and the set of operators $\{\square_a \mid a \in A\}$ are not always interdefinable (or analogously with the possibility operators).

In what follows we review the main results contained in the recent papers [BEG08, BEGR09] falling in this third group. To do so we first introduce the logic of a finite BL-chain and then we focus on the minimum modal and multi-modal systems over these logics.

3.1 The logic of a finite BL-chain

Let $\mathbf{A} = \langle A, \wedge, \vee, \odot, \rightarrow, 1, 0 \rangle$ be a finite BL-chain and define the corresponding propositional logic, denoted $\Lambda(\mathbf{A})$, as the logic such that for all (finite) set of formulas Γ and a formula φ,

$$\Gamma \models \varphi \text{ if and only if } e(\varphi) = 1 \text{ for any } A\text{-evaluation } e \text{ that is a model of } \Gamma$$

It is known that $t\Lambda(\mathbf{A})$ has a finite axiomatization, see e.g. [AM03]. Now consider the logic $\Lambda(\mathbf{A}^c)$ obtained from $\Lambda(\mathbf{A})$ by adding a set of truth constants in the language (one $\overline{a}$ for each $a \in \mathbf{A}$) and the following axioms and rules of inference:

- *Book-keeping axioms*

$$(\overline{a} * \overline{b}) \leftrightarrow \overline{a * b} \quad \text{where } a, b \in A \text{ and } * \in \{\odot, \rightarrow\}.$$

- *Witnessing axiom*

$$\bigvee\nolimits_{a \in A}(\varphi \leftrightarrow \overline{a})$$

- And the rule of inference

$$\overline{r} \vee \varphi \vdash \varphi$$

where $r \in \mathbf{A}$ is the coatom of the finite chain $\mathbf{A}$.

In this setting one can prove that the logic $\Lambda(\mathbf{A}^c)$ is finite strong canonical complete, i.e. complete with respect to evaluations over the chain $\mathbf{A}^c$ and interpreting each constant by its corresponding element of the chain.

3.2 The minimum modal many-valued logic

The main goal of [BEGR09] was to define and study $\square$-modal systems over the logic $\Lambda(\mathbf{A}^c)$, i.e., starting from an axiomatization of the logic $\Lambda(\mathbf{A}^c)$, to obtain an axiomatization of the modal logics over it. The basic definition of the minimum modal logic over the logic of a finite BL-chain is the following:

Modal language. The modal language is the expansion of the language of either $\Lambda(\mathbf{A}^c)$ by a new unary connective: the *necessity operator* $\Box$.

Kripke frames. A Kripke frame is a pair $F = \langle W, R \rangle$ where W is a non empty set (the *worlds*) and R is a binary relation valued in A (i.e., $R : W \times W \longrightarrow A$) called *accessibility relation*. The Kripke frame is said to be *crisp* if the range of R is included in $\{0,1\}$ and *idempotent* if included in the set of idempotent elements. The classes of Kripke frames, idempotent frames and crisp Kripke frames are denoted, respectively, by $Fr(\mathbf{A})$, $IFr(\mathbf{A})$ and $CFr(\mathbf{A})$ (or simply Fr, IFr and CFr if there is no ambiguity).

Kripke models. A Kripke model is a 3-tuple $M = \langle W, R, V \rangle$ where $\langle W, R \rangle$ is an $\mathbf{A}$-valued Kripke frame and V is a map, called *valuation*, assigning to each variable in Var and each world in W an element of A (i.e., $V : Var \times W \longrightarrow A$). The map V can be uniquely extended to arbitrary formulas by interpreting connectives and truth-constants as usual by the corresponding operations in the chain $\mathbf{A}$ and stipulating

- $V(\Box\varphi, w) = \bigwedge\{R(w, w') \to V(\varphi, w') : w' \in W\}.$

Let us denote by $\Lambda(Fr, B)$, $\Lambda(IFr, B)$ and $\Lambda(CFr, B)$ the set of valid formulas in the class of frames Fr, IFr and CFr respectively over B, where B is either $\mathbf{A}$ or $\mathbf{A}^c$.

Some available results that can be found in the literature about many-valued systems are the following:

- Fitting in [Fit92a, Fit92b] studied $\Lambda(Fr, A^c)$ when $\mathbf{A}$ is a finite Heyting Algebra.

- Caicedo and Rodríguez in [CR] study $\Lambda(Fr, [0, 1]_G)$ and proved its equivalence with $\Lambda(CFr, [0, 1]_G)$

- Metcalfe and Olivetti in [MO09] give a proof theory for $\Lambda(Fr, [0, 1]_G)$

But none of them deals with the general case, studying the minimum modal many-valued logic in a systematic way. Among the difficulties to find such an axiomatization is that while the meet-distributivity axiom

(MD) $(\Box\varphi \wedge \Box\psi) \leftrightarrow \Box(\varphi \wedge \psi)$

is valid (as in the classical modal case), in general this is not the case in the many-valued modal setting for the normality axiom

(K) $\quad \Box(\varphi \to \psi) \to (\Box\varphi \to \Box\psi)$

In fact the normality axiom K is only valid if either we restrict to idempotent or crisp Kripke frames (*IFr* or *CFr* respectively) or when **A** is a Heyting algebra (when the interpretation of the strong conjunction coincides with the meet).

In [BEGR09], using the canonical model method, we have proved completeness of the axiomatization of $\Lambda(Fr, A^c)$, the minimum modal logic over a finite BL-chain **A**, given below.

Axiomatization of the set $\Lambda(Fr, A^c)$ when **A** is a finite BL-chain

- the set of axioms is the smallest set closed under substitutions containing

  - an axiomatic basis for $\Lambda(\mathbf{A})$,
  - the witnessing axiom $\bigvee_{a\in A}(\varphi \leftrightarrow \overline{a})$
  - the book-keeping axioms $(\overline{a_1} * \overline{a_2}) \leftrightarrow \overline{a_1 * a_2}$ (for every $a_1, a_2 \in A$ and every $* \in \{\wedge, \vee, \odot, \rightarrow\}$),
  - $\Box\overline{1}$, $(\Box\varphi \wedge \Box\psi) \rightarrow \Box(\varphi \wedge \psi)$ and $\Box(\overline{a} \rightarrow \varphi) \leftrightarrow (\overline{a} \rightarrow \Box\varphi)$ (for every $a \in A$),

- Modus Ponens plus the rule $\overline{k} \vee \varphi \vdash \varphi$ (where k is the coatom of **A**) and the Monotonicity rule $\varphi \rightarrow \psi \vdash \Box\varphi \rightarrow \Box\psi$.

Notice that this axiomatization is over $\Lambda(\mathbf{A}^c)$, i.e., having truth constants in the language. The problem of axiomatizing the minimum modal logic over a finite residuated chain without truth constants in the language is still an open problem except for the case of the minimum modal logic over finite Łukasiewicz logic that was also axiomatized in [BEGR09] but in a rather complicated way.

The case of idempotent frames was easily solved because they are characterized as the general frames where the axiom K is valid. From that the proof that the logic $\Lambda(IFr, \mathbf{A}^c)$ is axiomatized by the axioms and rules of $\Lambda(Fr, \mathbf{A}^c)$ adding the axiom K is easy and proved also in [BEGR09].

If we restrict ourselves to *Crisp* Kripke models then the axiom K is valid even though it does not characterize them. The found axiomatization of $\Lambda(CFr, A^c)$ is given below. Notice that, due to axiom K, in this case one can replace the monotonicity rule for the more usual and simple necessitation rule.

Axiomatization of the set $\Lambda(CFr, \mathbf{A}^c)$ when $\mathbf{A}$ is a finite BL-chain

- the set of axioms is the smallest set closed under substitutions containing

  - an axiomatic basis for $\Lambda(\mathbf{A})$,
  - the witnessing axiom $\bigvee_{a \in A}(\varphi \leftrightarrow \bar{a})$
  - the book-keeping axioms $\overline{(\bar{a_1} * \bar{a_2})} \leftrightarrow \overline{a_1 * a_2}$ (for every $a_1, a_2 \in A$ and every $* \in \{\wedge, \vee, \odot, \to\}$),
  - $\Box\bar{1}$, $(\Box\varphi \wedge \Box\psi) \to \Box(\varphi \wedge \psi)$ and $\Box(\varphi \to \psi) \to (\Box\varphi \to \Box\psi)$
  - $\Box(\bar{a} \to \varphi) \leftrightarrow (\bar{a} \to \Box\varphi)$ (for every $a \in A$) and $\Box(\bar{k}\vee\varphi) \to (\bar{k}\vee\Box\varphi)$ (where k is the coatom of $\mathbf{A}$),

- Modus Ponens plus the rule $\bar{k} \vee \varphi \vdash \varphi$ (where k is the coatom of $\mathbf{A}$) and the Necessity rule $\varphi \vdash \Box\varphi$.

3.3   Multi-modal many-valued versus Modal many-valued

As in the first group of approaches mentioned above, one way to capture the many-valuedness of the accessibility relation R is by considering the induced set of classical accessibility relations defined by the different level-cuts of R. Namely it is well-known the correspondence between a fuzzy binary relation R over A, i.e. a function $R : W \times W \longrightarrow A$, and an $A \setminus \{0\}$-indexed and decreasing family of crisp binary relations, i.e. a family $\{R_a : a \in A, a \neq 0\}$ such that if $a \leq b$ then $R_b \subseteq R_a \subseteq W \times W$. This correspondence is given by the following identities:

(1) $R_a = \{(w_1, w_2) \in W \times W : R(w_1, w_2) \geq a\}$

(2) $R(w_1, w_2) = \max\{a \in A : (w_1, w_2) \in R_a\}$

In the same way, an $\mathbf{A}$-valued Kripke frame $\mathcal{F} = (W, R)$ can be transformed into a family of crisp Kripke frames

$$F_a = (W, R_a))_{a \in A \setminus \{0\}}.$$

At the syntactical level, one then introduces as many necessity operators $\Box_a$ as elements a of the lattice A interpreted by the (classical) relations R_a. The advantage is that each modal operator $\Box_a$ satisfies the axiom K. Therefore, in this kind of approach, one is led to a multi-modal language which is built from the set of propositional variables and connectives adding the truth constants $\{\bar{a} \mid a \in A\}$ and the family of modal operators $\{\Box_a \mid a \in A\}$. The Kripke frames $\mathcal{M} = (W, R, V)$ are defined as in the precedent case and the valuation V is extended to modal formulas by the following condition:

$$V(\Box_a \varphi, w) = \bigwedge_{w' \in W} R_a(w, w') \to V(\varphi, w') = \bigwedge_{w' \in W} \{V(\varphi, w') : R(w, w') \geq a\}.$$

The resulting multi-modal system, denoted as $M\Lambda(Fr, \mathbf{A}^c)$, has been axiomatized in [BEGR09] by the following axioms and rules of inference:

Axiomatization of the set $M\Lambda(Fr, \mathbf{A}^c)$ when $\mathbf{A}$ is a finite BL-chain

- the set of axioms is the smallest set closed under substitutions containing

  - Axiomatic basis for $\Lambda(A^c)$,
  - $\Box_b(\varphi \to \psi) \to (\Box_b \varphi \to \Box_b \psi)$, for every $b \in A \setminus \{0\}$).
  - $\Box_b(\overline{a} \to \varphi) \leftrightarrow (\overline{a} \to \Box_b \varphi)$, for every $a \in A$ and $b \in A \setminus \{0\}$),
  - $\Box_b(\overline{k} \vee \varphi) \to (\overline{k} \vee \Box_b \varphi)$, for every $b \in A \setminus \{0\}$, where k is the coatom of $\mathbf{A}$),
  - $\Box_{b_1} \varphi \to \Box_{b_2} \varphi$ for every $b_1, b_2 \in A \setminus \{0\}$ such that $b_1 \leq b_2$),

- Modus Ponens plus the rule $\overline{k} \vee \varphi \vdash \varphi$ (where k is the coatom of $\mathbf{A}$) and the Necessity rules: $\varphi \vdash \Box_b \varphi$ for every $b \in A \setminus \{0\}$.

An interesting problem is to study and compare the expressive power of many-valued modal and multi-modal systems. To this purpose it is useful to introduce the following concepts. Given two pointed[5] Kripke models $\langle M, w \rangle$ and $\langle M', w' \rangle$, we say that they are *modally equivalent* in the case that $V(\varphi, w) = V'(\varphi, w')$ for every modal formula φ. Analogously we will talk about *multimodally equivalent* and *full modally equivalent* in the case we focus, respectively, on multi-modal formulas or full modal formulas.

A first remark comparing the expressive power is that in every finite chain (and even in every finite residuated lattice) $\mathbf{A}$ it holds that

(3) $\Box \varphi \equiv \bigwedge \{\overline{a} \to \Box_a \varphi : a \in A \setminus \{0\}\}$

Therefore, the modality $\Box$ is explicitly definable using the modalities $\Box_a$'s (and these last ones have the advantage that satisfy the normality axiom).

---

[5]By a *pointed Kripke model* we mean a Kripke model together with a distinguished point or world.

In other words, the expressive power of the modal language is smaller than the one of the multi-modal one. Thus, $\Lambda(Fr, A^c)$ can be seen as a fragment of $M\Lambda(Fr, A^c)$. It is obvious that if two pointed Kripke models are multimodally equivalent then they are also modally equivalent.

But the inverse sense is not valid in general. As a matter of fact it is necessary to distinguish between having an empty set Var of propositional variables or not:

- If $Var = \emptyset$ and two pointed Kripke models are modally equivalent, then they are also multimodally equivalent.

- If $Var \neq \emptyset$, let $\mathbf{A}$ be the ordinal sum $\mathbf{A}_1 \oplus \mathbf{A}_2$ of two finite BL chains such that $\mathbf{A}_1$ and $\mathbf{A}_2$ are non trivial (i.e., $\min\{|A_1|, |A_2|\} \geq 2$). Then, there are two pointed Kripke models that are modally equivalent but not multimodally equivalent (in [BEGR09] a counterexample proving this fact is shown).

However, the expressive power of both logics become the same in case $\mathbf{A}$ is a finite MV-chain since in such a case we have the following inter-definability result: let $\mathbf{A}$ be the finite Łukasiewicz chain of cardinal n; then, for every $a \in A \setminus \{0\}$ it holds that

$$\Box_a \varphi \equiv \bigwedge \left\{ (\overline{a} \to \neg\Box\neg((\varphi \leftrightarrow \overline{b})^{n-1}))^{n-1} \to \overline{b} : b \in A \right\}$$

The proof is based on the fact that φ^{n-1} only takes crisp values (i.e., in $\{0, 1\}$). Indeed, φ^{n-1} takes value 1 when φ takes value 1, and φ^{n-1} takes value 0 elsewhere. Hence, φ^{n-1} takes the same value than $\Delta\varphi$, where Δ is the well-known Baaz projection (see [Háj98b]).

This leads to the following general result, for $\mathbf{A}$ being a finite BL-chain, stating that the following statements are equivalent:

1. $\mathbf{A}$ is a finite MV-chain (i.e., the only idempotent elements of A are 0 and 1),

2. The modalities $\Box_a$'s are explicitly definable in the modal language.

3. Two pointed Kripke models are modally equivalent iff they are multi-modally equivalent

The problem of axiomatizing the modal and multi-modal systems obtained by adding either a possibility operator $\Diamond$ or a family of possibility operators $\Diamond_a$ for $a \in \mathbf{A}$ are open problems. Of course, it is solved in the case of $\mathbf{A}$ being a finite MV chain since possibility modal operators are definable as in the classical case as $\Diamond \equiv \neg\Box\neg$ and the same for $\Diamond_a$ and $\Box_a$. Moreover in [CR] the axiomatization of the modal system obtained adding either □ or ◇ over Gödel Logic are given. But the problem of the axiomatization of a system obtained by adding both modal operators over Gödel Logic is still open.

4 And the collaboration is going on

The relation initiated by Petr Hájek (with Dagmar Darmancová and Milan Daniel) with the authors almost twenty years ago is not only very alive but also has improved with the incorporation of young colleagues of both groups, that have a very good and fruitful collaboration both at the scientific and personal level. In this contribution we have overviewed only a common topic in our research, leaving aside many other topics that are of our common interest. Nevertheless one of the first common topics of interest, the difference between degrees of truth and uncertainty measures has been deeply studied and the first part of this paper is devoted to this subject. The second part (that deals with many-valued modal logics) is very related to Fuzzy Description logics, a topic that has been studied by our teams in the last period. Following the first ideas of Hájek in [Háj05, Háj06a] we have developed the logical basis for Fuzzy Description languages in [GCAE]. It is well known that description logics have a nice translation to modal logic in the classical case and the same remains valid for the fuzzy case. In this sense the study of many-valued modal logics we think will be very useful for the theoretical study of fuzzy description logis.

We believe that the collaboration between the two teams has been really fruitful, we have learnt a lot from Petr Hájek, and during all this time we have enjoyed his personality, ideas and friendship. We dedicate this paper to him in the occasion of his seventieth anniversary. Thank you Petr for being a reference, for your stimulating scientific ideas and for your friendship, in summary, for being as you are.

BIBLIOGRAPHY

[AGM08]   S. Aguzzoli, B. Gerla, and V. Marra. De Finetti's no-Dutch-book criterion for Gödel logic. *Studia Logica*, 90(1):25–41, 2008.

[AM03]   P. Aglianò and F. Montagna. Varieties of BL-algebras I: General properties. *J. Pure Appl. Algebra*, 181:105–129, 2003.

[BEG08]   F. Bou, F. Esteva, and L. Godo. Exploring a syntactic notion of modal many-valued logics. *Mathware & Soft Computing*, 15(2):175–188, 2008.

[BEGR09] F. Bou, F. Esteva, L. Godo, and R. Rodríguez. On the minimum many-valued modal logic over a finite residuated lattice. To appear in *J. Logic Comput.*, 2009.

[CEGT00] R. Cignoli, F. Esteva, L. Godo, and A. Torrens. Basic fuzzy logic is the logic of continuous t-norms and their residua. *Soft Computing*, 4:106–112, 2000.

[CF92]   P. Chatalic and C. Froidevaux. Lattice-based graded logic: a multimodal approach. In *Proceedings of the eighth conference on Uncertainty in Artificial Intelligence*, pages 33–40, San Francisco CA, 1992. Morgan Kaufmann Publishers Inc.

[CR]   X. Caicedo and R. O. Rodríguez. Standard Gödel modal logics. To appear
              in *Studia Logica*.
[DEG⁺95]   D. Dubois, F. Esteva, P. Garcia, L. Godo, and H. Prade. Similarity-based
              consequence relations. In *Symbolic and Quantitative Approaches to Reason-
              ing and Uncertainty, Proc. of ECSQARU 95 (Fribourg 1995)*, volume 946
              of *LNCS*, pages 171–179. Springer, Berlin, 1995.
[DGM09]   P. Dellunde, L. Godo, and E. Marchioni. Exploring extensions of possibilistic
              logic over Gödel logic. In *Proc. of the 10th European Conference on Symbolic
              and Quantitative Approaches to Reasoning with Uncertainty*, volume 5590
              of *LNAI*, pages 923–934. Springer, 2009.
[DPE⁺97]   D. Dubois, H. Prade, F. Esteva, P. Garcia, and L. Godo. A logical approach
              to interpolation based on similarity relations. *Int. J. Approximate Reasoning*,
              17(1):1–36, 1997.
[EG01]   F. Esteva and L. Godo. Monoidal t-norm based logic: Towards a logic for
              left-continuous t-norms. *Fuzzy Sets Syst.*, 124(3):271–288, 2001.
[EGCG94]   F. Esteva, P. Garcia-Calvés, and L. Godo. Relating and extending seman-
              tical approaches to possibilistic reasoning. *Int. J. Approximate Reasoning*,
              10(4):311–344, 1994.
[EGGN07]   F. Esteva, J. Gispert, L. Godo, and C. Noguera. Adding truth-constants
              to logics of continuous t-norms: Axiomatization and completeness results.
              Fuzzy Sets Syst., 158(6):597–618, 2007.
[EGGR97]   F. Esteva, P. Garcia, L. Godo, and R. Rodríguez. A modal account of
              similarity-based reasoning. *Int. J. Approximate Reasoning*, 16(3-4):235–260,
              1997.
[EGGR98]   F. Esteva, P. Garcia, L. Godo, and R. O. Rodriguez. Fuzzy approximation
              relations, modal structures and possibilistic logic. *Mathware & Soft Com-
              puting*, 5(2-3):151–166, 1998.
[EGM01]   F. Esteva, L. Godo, and F. Montagna. The ŁΠ and ŁΠ$\frac{1}{2}$ logics: Two com-
              plete fuzzy systems joining Łukasiewicz and product logics. *Arch. Math.
              Logic*, 40(1):39–67, 2001.
[FG07]   T. Flaminio and L. Godo. A logic for reasoning about the probability of
              fuzzy events. *Fuzzy Sets Syst.*, 158:625–638, 2007.
[FGM]   T. Flaminio, L. Godo, and E. Marchioni. On the logical formalization of
              possibilistic counterparts of states over n-valued Łukasiewicz events. To
              appear in *J. Logic Comput.*
[FHM90]   R. Fagin, J. Y. Halpern, and N. Megiddo. A logic for reasoning about prob-
              abilities. *Information and Computation*, 87(1-2):78–128, 1990.
[Fit92a]   M. Fitting. Many-valued modal logics. *Fund. Informaticae*, 15:235–254,
              1992.
[Fit92b]   M. Fitting. Many-valued modal logics, II. *Fund. Informaticae*, 17:55–73,
              1992.
[Fla07]   T. Flaminio. NP-containment for the coherence test of assessments of condi-
              tional probability: a fuzzy logical approach. *Arch. Math. Logic*, 46:301–319,
              2007.
[FM05]   T. Flaminio and F. Montagna. A logical and algebraic treatment of condi-
              tional probability. *Ann. Math. Logic*, 44(2):245–262, 2005.
[GCAE]   A. Garcia-Cerdaña, E. Armengol, and F. Esteva. Fuzzy description logics and
              t-norm based fuzzy logics. To appear in *Int. J. of Approximate Reasoning*.
[GdMSV89]   L. Godo, R. L. de Mántaras, C. Sierra, and A. Verdaguer. Milord: The
              architecture and the management of linguistically expressed uncertainty. *Int.
              J. Intell. Syst.*, 4:471–501, 1989.
[GEH00]   L. Godo, F. Esteva, and P. Hájek. Reasoning about probability using fuzzy
              logic. *Neural Network World*, 2:811–824, 2000.
[GHE03]   L. Godo, P. Hájek, and F. Esteva. A fuzzy modal logic for belief functions.
              Fund. Informaticae, 57(2-4):127–146, 2003.

[GM06]   L. Godo and E. Marchioni. Coherent conditional probability in a fuzzy logic setting. *Logic J. of the IGPL*, 14(3):457–481, 2006.

[Gob70]   L. F. Goble. Grades of modality. *Logique et Analyse (N.S.)*, 13:323–334, 1970.

[Háj]   P. Hájek. On fuzzy modal logics S5(C). To appear in *Fuzzy Sets Syst.*

[Háj96]   P. Hájek. Getting belief functions from Kripke models. *Int. J. General Systems*, 24(3):325–327, 1996.

[Háj98a]   P. Hájek. Basic fuzzy logic and BL-algebras. *Soft Computing*, 2:124–128, 1998.

[Háj98b]   P. Hájek. *Metamathematics of Fuzzy Logic*, volume 4 of *Trends in Logic*. Kluwer, Dordrecht, 1998.

[Háj05]   P. Hájek. Making fuzzy description logic more general. *Fuzzy Sets Syst.*, 154(1):1–15, 2005.

[Háj06a]   P. Hájek. What does mathematical fuzzy logic offer to description logic? In E. Sanchez, editor, *Fuzzy Logic and the Semantic Web*, pages 91–100. Elsevier, Amsterdam, 2006.

[Háj06b]   P. Hájek. Computational complexity of t-norm based propositional fuzzy logics with rational truth constants. *Fuzzy Sets Syst.*, 157(5):677–682, 2006.

[Háj07]   P. Hájek. Complexity of fuzzy probability logics II. *Fuzzy Sets Syst.*, 158(23):2605–2611, 2007.

[HGE95]   P. Hájek, L. Godo, and F. Esteva. Fuzzy logic and probability. In P. Besnard and S. Hanks, editors, *Uncertainty in Artificial Intelligence. Proceedings of the 11th Conference*, pages 237–244, San Francisco, 1995. Morgan Kaufmann.

[HGE96]   P. Hájek, L. Godo, and F. Esteva. A complete many-valued logic with product-conjunction. *Arch. Math. Logic*, 35:191–208, 1996.

[HH78]   P. Hájek and T. Havránek. *Mechanizing Hypothesis Formation— Mathematical Foundations of a General Theory*. Springer, Berlin, 1978.

[HH95]   P. Hájek and D. Harmancová. Medical fuzzy expert systems and reasoning about beliefs. In P. Barahona, M. Stefanelli, and J. Wyatt, editors, *Artificial Intelligence in Medicine*, pages 403–404, Berlin, 1995. Springer.

[HH96]   P. Hájek and D. Harmancová. A many-valued modal logic. In *Proceedings of IPMU'96*, pages 1021–1024, Granada, 1996. Universidad de Granada.

[HHE+94]   P. Hájek, D. Harmancová, F. Esteva, P. García, and L. Godo. On modal logics of qualitative possibility in a fuzzy setting. In L. de Mantaras and D. Poole, editors, *Uncertainty in Artificial Intelligence. Proceedings of the 10th Conference*, pages 278–285, San Francisco, 1994. Morgan Kaufmann.

[HHJ92]   P. Hájek, T. Havránek, and R. Jiroušek. *Uncertain Information Processing in Expert Systems*. CRC Press, Boca Raton, 1992.

[HKR94]   D. Harmanec, G. J. Klir, and G. Resconi. On modal logic interpretation of Dempster-Shafer theory of evidence. *Int. J. of Uncertain. Fuzziness Knowledge-Based Syst.*, 9:941–951, 1994.

[HT01]   P. Hájek and S. Tulipani. Complexity of fuzzy probability logics. *Fund. Informaticae*, 45(3):207–213, 2001.

[HT06]   G. Hansoul and B. Teheux. Completeness results for many-valued Łukasiewicz modal systems and relational semantics, 2006. Available at http://arxiv.org/abs/math/0612542.

[JM02]   S. Jenei and F. Montagna. A proof of standard completeness for Esteva and Godo's logic MTL. *Studia Logica*, 70(2):183–192, 2002.

[KNP02]   C. D. Koutras, C. Nomikos, and P. Peppas. Canonicity and completeness results for many-valued modal logics. *J. Appl. Non-Classical Logics*, 12(1):7–41, 2002.

[Kro06]   T. Kroupa. Representation and extension of states on mv-algebras. *Arch. Math. Logic*, 45:381–392, 2006.

[Mar06] E. Marchioni. Possibilistic conditioning framed in fuzzy logics. *Int. J. Approximate Reasoning*, 43:133–165, 2006.

[Mar08] E. Marchioni. Representing upper probability measures over rational Łukasiewicz logic. *Mathware & Soft Computing*, 15(2):159–173, 2008.

[Mir05] A. M. Mironov. Fuzzy modal logics. *J. Math. Sci.*, 128(6):3641–3483, 2005.

[MO09] G. Metcalfe and N. Olivetti. Proof systems for a Gödel modal logic. In M. Giese and A. Waaler, editors, *Proceedings of TABLEAUX 2009*, volume 5607 of *LNAI*. Springer, 2009.

[Mun95] D. Mundici. Averaging the truth-value in Łukasiewicz logic. *Studia Logica*, 55(1):113–127, 1995.

[RM02] B. Riečan and D. Mundici. Probability on MV-algebras. In E. Pap, editor, *Handbook of Measure Theory, Vol. II*, pages 869–909. Elsevier, Amsterdam, 2002.

[Rus87] E. H. Ruspini. Epistemic logics, probability and the calculus of evidence. In *Proc. Tenth International Joint Conference on Artificial Intelligence, IJCAI'87*, pages 924–931, 1987.

[Rus91] E. H. Ruspini. On the semantics of fuzzy logic. *Int. J. Approximate Reasoning*, 5(1):45–88, 1991.

[Suz97] N. Y. Suzuki. Kripke frame with graded accessibility and fuzzy possible world semantics. *Studia Logica*, 59(2):249–269, 1997.

Francesc Esteva
Artificial Intelligence Research Institute (IIIA), CSIC
Campus de la Universitat Autònoma de Barcelona s/n
08193 Bellaterra, Spain
Email: esteva@iiia.csic.es

Lluís Godo
Artificial Intelligence Research Institute (IIIA), CSIC
Campus de la Universitat Autònoma de Barcelona s/n
08193 Bellaterra, Spain
Email: godo@iiia.csic.es

Shaping the logic of fuzzy set theory

Siegfried Gottwald

Dedicated to Petr Hájek
on his 70th anniversary

ABSTRACT. Actually it is a well accepted fact that fuzzy set theory, as a mathematical theory of a sort of generalized sets, has a natural relationship to particular kinds of non-classical logics of comparable truth degrees, called mathematical fuzzy logics.

The paper discusses aspects of the historical development of the relationship of fuzzy sets with formal logics suitable for a natural presentation of fuzzy set theory.

It also sketches the core mathematical ideas behind mathematical fuzzy logics, explains the key rôle played by Petr Hájek in this development, presents some of the basic results for these logics, and offers some indications for ongoing developments in the field.

1 Introduction

The notion of fuzzy set is a technical tool to mathematically grasp the use and the effect of vague notions in a manner completely different from the way classical mathematics is treating them, if unavoidable, i.e. by "precisifying" them into crisp notions. Formally, each fuzzy set A is a fuzzy subset of a given universe of discourse U, characterized by its membership function $\mu_A : U \to [0, 1]$. The value $\mu_A(x)$ is the membership degree of x with respect to the fuzzy set A.

As Lotfi A. Zadeh introduced fuzzy sets in his seminal paper [Zad65] he essentially did not relate them, or at least their suitable treatment, to non-classical logics. There was, however, a minor exception: in discussing the meaning of the membership degrees he mentioned (in a "comment" on pages 341–342; and with reference to the monograph [Kle52] and Kleene's three valued logic) with respect to two thresholds $0 < \beta < \alpha < 1$ that one may interpret the case $\mu_A(x) \geq \alpha$ as saying that x belongs to the fuzzy set A, that one may interpret $\mu_A(x) \leq \beta$ as saying that x does not belong to the fuzzy set A, and leaving the case $\beta < \mu_A(x) < \alpha$ as an indeterminate status for the membership of x in A.

Nevertheless, the overwhelming majority of fuzzy set papers that followed [Zad65] and the other early Zadeh papers on fuzzy sets treated fuzzy sets

in the standard mathematical context, i.e. with an implicit reference to a
naively understood classical logic as argumentation structure.

Even philosophically oriented predecessors of Zadeh in the discussion of
vague notions, like Max Black [Bla37] and Carl Hempel [Hem39], did refer
only to classical logic, even in those parts of these papers in which they
discuss the problem of some incompatibilities of the naively correct use of
vague notions and principles of classical logic, e.g., concerning the treatment
of negation-like statements.

Formally, however, it was important that Zadeh not only proposed to
define union $A \cup B$ and intersection $A \cap B$ of fuzzy sets A, B by the well
known formulas

$$(1) \qquad \mu_{A \cup B}(x) \;=\; \max\{\mu_A(x), \mu_B(x)\},$$
$$(2) \qquad \mu_{A \cap B}(x) \;=\; \min\{\mu_A(x), \mu_B(x)\},$$

but that he also introduced in [Zad65] other operations for fuzzy sets, called
"algebraic" by him, as, e.g., an algebraic product AB and an algebraic sum
$A + B$ defined via the equations

$$(3) \qquad \mu_{AB}(x) \;=\; \mu_A(x) \cdot \mu_B(x),$$
$$(4) \qquad \mu_{A+B}(x) \;=\; \min\{\mu_A(x) + \mu_B(x), 1\}.$$

The core point here is that it is mathematically more or less obvious that
these two additional operations are particular cases of further generalized intersection and union operations for fuzzy sets besides the "standard" versions
(2) and (1).

2 Early approaches involving non-classical logics

2.1 The "many-valued sets" of Klaua

As early as 1965, and hence independent of the approach by Zadeh, the
German mathematician Dieter Klaua presented two versions for a cumulative
hierarchy of *many-valued* sets.[1] In both cases he considered as membership
degrees the real unit interval $W_\infty = [0, 1]$ or a finite, m-element set $W_m =
\{\frac{k}{m-1} \mid 0 \leq k < m\}$ of equidistant points of $[0, 1]$. And he understood these
membership degrees as the truth degrees of the corresponding Łukasiewicz
systems L_∞ or L_m, respectively. Additionally, he started these hierarchies
with a given set U of urelements.

The first one of these hierarchies, presented in [Kla65, Kla67] in 1965,
offered an interesting simultaneous definition of a graded membership and a
graded equality predicate, but did not work well and was almost immediately

---

[1]The stimulus for these investigations came from discussions following a colloquium
talk which Karl Menger had given in Berlin (East) in the first half of the 1960s.

abandoned. The main reason for this failure, cf. [Got], was that the class of objects that was intended to act as many-valued sets was not well chosen.

The second one of these hierarchies, presented in 1966 in [Kla66b, Kla66a], had as its objects A functions into the truth degree set W, the values $A(x)$ being the membership degrees of the object x in the generalized set A.

The first level of this second hierarchy was formed just of all W-valued fuzzy subsets of the set U. Thus, quite naturally, the set of urelements in the second Klaua approach played just the role of the universe of discourse in the Zadeh approach. And each first-level many-valued set of this hierarchy is nothing but a particular fuzzy subset of U.

So it is reasonable to identify the many-valued sets of Klaua with the fuzzy sets of Zadeh, as shall be done further on in this paper.

Therefore the 1966 approach by Klaua offered immediately the Łukasiewicz systems of many-valued logic as the suitable logics to develop fuzzy set theory within their realm.

And indeed, the majority of results in [Kla66b, Kla66a] were presented using the language of these Łukasiewicz systems. Some examples are:

$$\models \quad A \subseteq B \,\&\, B \subseteq C \rightarrow_{\mathsf{L}} A \subseteq C \,,$$
$$\models \quad a\,\varepsilon\,B \,\&\, B \subseteq C \rightarrow_{\mathsf{L}} a\,\varepsilon\,C \,,$$
$$\models \quad A \equiv B \,\&\, B \subseteq C \rightarrow_{\mathsf{L}} A \subseteq C \,.$$

Here $\rightarrow_{\mathsf{L}}$ is the Łukasiewicz implication, $\&$ the strong (or: arithmetical) conjunction with truth degree function $(u, v) \mapsto \max\{0, u + v - 1\}$, ε the graded membership predicate, and $\models \varphi$ means that the formula φ of the language of Łukasiewicz logic is logically valid, i.e. assumes always truth degree 1. A graded inclusion relation $\subseteq$ is defined (for fuzzy sets of the same level in the hierarchy) as

(5) $A \subseteq B =_{\text{def}} \forall x (x\,\varepsilon\,A \rightarrow_{\mathsf{L}} x\,\varepsilon\,B)\,,$

and a graded equality $\equiv$ for fuzzy sets is defined as

(6) $A \equiv B =_{\text{def}} A \subseteq B \wedge B \subseteq A\,.$

This line of approach was continued in the early 1970s, e.g., in this author's papers [Got74, Got76]. The topic of [Got74] is the formulation of (crisp) properties of fuzzy relations – the natural continuation, to consider graded properties of fuzzy relations, followed only in the 1991 paper [Got91]. The topic of [Got76] was the formulation of generalized versions of the standard ZF axioms valid in a modified version of Klaua's second hierarchy of fuzzy sets.

2.2   Relating the Zadeh approach to non-classical logics

It was Joseph A. Goguen who, starting only from Zadeh's approach, was the
first to emphasize an intimate relationship to non-classical logics. In his 1969
paper [Gog69], he considers membership degrees as generalized truth values,
i.e. as truth degrees. Additionally he sketches a "solution" of the sorites
paradox, i.e. the heap paradox, using – but only implicitly – the ordinary
product $*$ in $[0, 1]$ as a generalized conjunction operation. Based upon these
ideas, and having in mind suitable analogies to the situation for intuitionistic
logic, he proposes completely distributive lattice ordered monoids, called *closg*
by him, enriched with a (right) residuation operation $\rightarrow$ characterized by the
well known adjointness condition

(7) $a * b \leq c \;\Leftrightarrow\; b \leq a \rightarrow c,$

and with the "implies falsum"-negation, as suitable structures for the membership degrees of fuzzy sets. He introduces in this context the notion of
tautology, with the neutral element of the monoid as the only designated
truth degree. He defines a graded notion of inclusion in the same natural
way as Klaua (5) did, of course with the residual implication $\rightarrow$ instead of
the implication $\rightarrow_L$ of the Łukasiewicz systems. But he does not mention any
results for this graded implication.

   Additionally, because of an inadequate understanding of logical calculi, he
does not see a possibility to develop a suitable formalized logic of closg's, as
may be seen from his statement:

> Tautologies have the advantage of independence of truth set, but
> no list of tautologies can encompass the entire system because we
> want to perform calculations with degrees of validity between 0
> and 1. In this sense the logic of inexact concepts does not have
> a *purely* syntactic form. Semantics, in the form of specific truth
> values of certain assertions, is sometimes required.

Another author to point out a strong relationship between fuzzy sets and
many-valued logic is Robin Giles. Starting in 1975, he proposed in a series
of papers [Gil75, Gil76, Gil79], and again in [Gil88], a general treatment of
reasoning with vague predicates by means of a formal system based upon a
convenient dialogue interpretation. This dialogue interpretation he had already used in other papers, like [Gil74], dealing with subjective belief and the
foundations of physics. The main idea is to let "a sentence represent a belief
by expressing it tangibly in the form of a bet". In this setting then a "sentence ψ is considered to follow from sentences $\varphi_1, \ldots, \varphi_n$ just when he who
accepts the bets $\varphi_1, \ldots, \varphi_n$ can at the same time bet ψ without fear of loss".

   The (formal) language obtained in this way is closely related to Łukasiewicz's infinite-valued logic L_∞: in fact the two systems coincide if one assigns

to a sentence φ the truth value $1 - \langle \varphi \rangle$, with $\langle \varphi \rangle$ for the risk value of asserting φ. And he even adds the remark "that, with this dialogue interpretation, Łukasiewicz logic is exactly appropriate for the formulation of the 'fuzzy set theory' first described by L. A. Zadeh [Zad65]; indeed, it is not too much to claim that L_∞ is related to fuzzy set theory exactly as classical logic is related to ordinary set theory".

Almost from the very beginning it was, however, clear from the mathematical point of view that set-algebraic operations for fuzzy sets can be reduced, in a many-valued setting, to generalized connectives in essentially the same way as standard set-algebraic operations for crisp, i.e. classical sets can be reduced to connectives of classical logic.

So, the question was what structural consequences the acceptance of definitions like (3) and (4) would have for generalized intersections and unions. A particular, somehow "reverse" question was which structural conditions, besides (2) and (1), could eliminate such generalizations. An answer to this "reverse" question was given in [BG73], cf. also [Gai76]: some rather natural "boundary conditions" together with the inclusion maximality of the standard intersection w.r.t. each other generalized intersection, with the inclusion minimality of the standard union w.r.t. each other generalized union, with commutativity and associativity, and with the mutual distributivity of the generalized union and intersection force a restriction to the "standard" case (2) and (1).

However, the set of all these structural restrictions from [BG73] seems to be very restrictive, and hence it did not really look convincing. Therefore the restriction to the "standard" operations (2) and (1) was never accepted by the majority of the mathematically oriented people of the fuzzy community.

As a consequence, a group of authors, a lot of them from the Spanish fuzzy community, discussed what might be suitable choices of such "fuzzy" connectives which might be used to define unions and intersections for fuzzy sets different from (2) and (1). One of the leading ideas in their considerations was to look at the types of restrictive conditions discussed in [BG73] as functional equations or functional inequalities, to reduce this set of functional conditions, to look also at other conditions, and to discuss the solutions of suitable sets of such functional conditions. The paper [ATV83] is a typical example, its focus is on pairs of generalized conjunctions and disjunctions. Other papers, with emphasis on generalized implication operations are, e.g., [TV85] and [BK80].

3   Invoking t-norms

It the beginning 1980s it became common use in the mathematical fuzzy community to consider t-norms as suitable candidates for connectives upon which generalized intersection operations for fuzzy sets should be based, see

[ATV80, Dub80, Pra80] or a bit later [Kle82, Web83]. These t-norms, a shorthand for "triangular norms", first became important in discussions of the triangle inequality within probabilistic metric spaces, see [SS83, KMP00]. They are binary operations in the real unit interval which make this interval into an ordered abelian monoid with 1 as unit element of the monoid.

The most basic examples of t-norms for the present context are the Łukasiewicz t-norm T_L, the Gödel t-norm T_G, and the product t-norm T_P defined by the equations

$$\begin{aligned}
T_L(u,v) &= \max\{u+v-1,0\}, \\
T_G(u,v) &= \min\{u,v\}, \\
T_P(u,v) &= u \cdot v.
\end{aligned}$$

The general opinion in the context of fuzzy connectives is that t-norms form a suitable class of generalized conjunction operators.

For logical considerations the class of left-continuous t-norms is of particular interest. Here left-continuity for a t-norm $T : [0,1]^2 \to [0,1]$ means that for each $a \in [0,1]$ the unary function $T_a(x) = T(a,x)$ is left-continuous. The core result, which motivates the interest in left-continuous t-norms, is the fact that just for left-continuous t-norms $*$ a suitable implication function, usually called R-implication, is uniquely determined via the adjointness condition (7). Suitability of an implication function here means that it allows for a corresponding sound detachment, or *modus ponens* rule: if one infers a formula ψ from formulas $\varphi \to \psi$ and φ then the logical validity

(8) $\quad \models \varphi \,\&\, (\varphi \to \psi) \to \psi$

yields the inequality $[\![\varphi]\!] * [\![\varphi \to \psi]\!] \leq [\![\psi]\!]$ for the truth degrees.

It was almost immediately clear that a propositional language with connectives $\wedge, \vee$ for the truth degree functions $\min, \max$, and with connectives $\&, \to$ for a left-continuous t-norm T and its residuation operation offered a suitable framework to do fuzzy set theory within – at least as long as the complementation of fuzzy sets remains out of scope.

With this limitation, i.e. disregarding complementation, this framework offers a suitable extension of Zadeh's standard set-algebraic operations.

Additionally, this framework, with the "implies falsum" construction, yields a natural way to define a negation, i.e. to introduce a t-norm related complementation operation for fuzzy sets, via the definition $-_T\varphi =_{\text{def}} \varphi \to \overline{0}$ using a truth degree constant $\overline{0}$ for the truth degree 0. However, this particular complementation operation does not always become the standard complementation of Zadeh's approach.

This t-norm based construction gives the infinite-valued Łukasiewicz system L_∞ if one starts from the t-norm T_L, and thus the right negation for

Zadeh's complementation. This construction gives the infinite-valued Gödel system G_∞ if one starts with the t-norm T_G, and it gives the product logic [HGE96] if one starts with the t-norm T_Π. The "implies falsum" negations of the latter two systems coincide, but are different from the negation operation of the Łukasiewicz system L_∞. So these two cases do not offer Zadeh's complementation. But this can be reached if one adds the Łukasiewicz negation to these systems, as done in [EGHN00].

It was essentially a routine matter to develop this type of t-norm based logic to some suitable extent, as was done 1984 in this author's paper [Got84]. Also the development of fuzzy set theory on this basis did not offer problems, and it was done in [Got86], including essential parts of fuzzy set algebra, some fuzzy relation theory up to a fuzzified version of the Szpilrajn order extension theorem, and some solvability considerations for systems of fuzzy relation equations. (All these considerations have later been included into the monograph [Got93].)[2]

There is, however, also another way to develop t-norm based logics for fuzzy set theory. This way avoids the introduction of the R-implications via the residuation operation—and so it does not need the restriction to left-continuous t-norms. Instead it uses additionally negation functions, i.e. unary functions $N : [0,1] \rightarrow [0,1]$ which are at least order reversing and satisfy $N(0) = 1$ as well as $N(1) = 0$. The strategy to introduce an implication function $I_{T,N}$ in this setting is to define

(9) $I_{T,N}(u,v) = N(T(u, N(v)))$.

The implication connectives defined in this way usually are called S-implications. A prominent paper which studies this type of approach is [BKZ95].

But the fact that S-implications do not necessarily satisfy (8) means that the corresponding rule of detachment is not always correct. And this seems to be the main reason that this type of approach never became popular among logicians interested in fuzzy set matters.

4   Logics of t-norms

What was missing in all the previously mentioned approaches toward a suitable logic for fuzzy set theory, as long as this logic should be different from

---

[2] *Personal reminiscence*: In these papers appears the notion of a φ-operator of a t-norm (so in [Got84, Got86]; in [Got93] called Φ-operator instead). I learned this notion from Witold Pedrycz. He used it in his PhD work on fuzzy relation equations. In [Ped85] it is called Ψ-operator, and in [Ped83] a particular case appears as τ-operator.

Clearly this was a suitable implication operation, and it is just the R-implication for the given t-norm. But in that time I was unaware of the equivalent characterizability of the φ-operator by the adjointness condition (7).

the infinite-valued Lukasiewicz system L_∞, that was an adequate axiomatization of such a logic. All these approaches offered interesting semantics, but did not provide suitable logical calculi – neither for the propositional nor for the first-order level.

The first proposal to fill in this gap was made by Ulrich Höhle [Höh94, Höh95, Höh96] who offered his *monoidal logic*. This common generalization of the Lukasiewicz logic L_∞, the intuitionistic logic, and Girard's integral, commutative linear logic [Gir87] was determined by an algebraic semantics, viz. the class M-alg of all integral residuated abelian lattice-ordered monoids with the unit element of the monoid, i.e. the universal upper bound of the lattice, as the only designated element. So this monoidal logic was determined by a particular subclass of Goguen's closg's, indeed by a variety of algebras. And adequate axiomatizations for the propositional as well as for the first-order version of this logic were given in [Höh94, Höh96].

Of course, this monoidal logic had the whole matter of the relationship of fuzzy set theory and the t-norm basedness of their set-algebraic operations in the background. But it was not really strongly tied with this background.

Guided by the idea that it would be a suitable restriction for the context of fuzzy set theory to confine the considerations to the case of continuous t-norms, instead of allowing also non-continuous but left-continuous t-norms, it was the idea of Petr Hájek to ask for the common part of all those t-norm based logics which refer to a continuous t-norm: in short, to ask for the common logic of all continuous t-norms.

This logic was called *basic logic* by Hájek, later also *basic fuzzy logic* or *basic t-norm logic*,[3] usually denoted BL, and based upon an algebraic semantics.

There are two crucial observations which pave the way to the original, and still mainly used algebraic semantics for BL. The first one is that for any t-norm $*$ and their residuation operation $\to$ one has

$$(10) \quad (u \to v) \vee (v \to u) = 1\,,$$

with $\vee$ to denote the lattice join here, i.e. the max-operation for a linearly ordered carrier. This *prelinearity* condition (10) is a first restriction on the variety M-alg which determines the monoidal logic, and it yields the variety MTL-alg of all MTL-algebras – now with $*$ denoting the semigroup operation.

Moreover, by the way, if this condition is imposed upon the Heyting algebras, which form an adequate algebraic semantics for intuitionistic logic, the resulting class of preliniear Heyting algebras is an adequate algebraic semantics for the infinite-valued Gödel logic.

---

[3]These latter names are preferable because the terminus "basic logic" is also in use in a completely different sense: as some weakening of the standard system of intuitionistic logic, e.g. in [AR98, Rui98].

The second observation is that the continuity condition can be given in algebraic terms: for any t-norm $*$ and its residuum $\to$ one has that the *divisibility* condition

$$(11) \quad u*(u \to v) = u \wedge v$$

is satisfied if and only if $*$ is a continuous t-norm, see [Höh95]. Condition (11), again with $*$ denoting the semigroup operation and $\wedge$ the lattice meet, is the second restriction here. The subclass of all those algebras from MTL-alg which satisfy this divisibility condition (11) is the subvariety BL-alg of all BL-algebras.

Hájek characterized his basic fuzzy logic by this class BL-alg as algebraic semantics – again with the universal upper bound of the lattice as the only designated element. And he gave adequate axiomatizations for the propositional version BL as well as for the first-order version BL∀ of this basic fuzzy logic in his highly influential monograph [Háj98b].

Despite the fact – caused by the properties of the R-implications – that the set $\mathcal{E}$ of equations, which characterizes the algebras of the variety BL-alg as the model class of $\mathcal{E}$, could routinely be rewritten as a set $\mathcal{E}^*$ of implications such that BL-alg is also the model class of $\mathcal{E}^*$, Petr Hájek offered in [Háj98b] a much shorter and considerably more compact axiomatic basis for the propositional system BL:

$(\text{Ax}_{\text{BL}}1) \quad (\varphi \to \psi) \to ((\psi \to \chi) \to (\varphi \to \chi))$,

$(\text{Ax}_{\text{BL}}2) \quad \varphi \& \psi \to \varphi$,

$(\text{Ax}_{\text{BL}}3) \quad \varphi \& \psi \to \psi \& \varphi$,

$(\text{Ax}_{\text{BL}}4) \quad (\varphi \to (\psi \to \chi)) \to (\varphi \& \psi \to \chi)$,

$(\text{Ax}_{\text{BL}}5) \quad (\varphi \& \psi \to \chi) \to (\varphi \to (\psi \to \chi))$,

$(\text{Ax}_{\text{BL}}6) \quad \varphi \& (\varphi \to \psi) \to \psi \& (\psi \to \varphi)$,

$(\text{Ax}_{\text{BL}}7) \quad ((\varphi \to \psi) \to \chi) \to (((\psi \to \varphi) \to \chi) \to \chi)$,

$(\text{Ax}_{\text{BL}}8) \quad \bar{0} \to \varphi$,

with the rule of detachment as its (only) inference rule.

Routine calculations show that the axioms $\text{Ax}_{\text{BL}}4$ and $\text{Ax}_{\text{BL}}5$ essentially code the adjointess condition (7). Also by elementary calculations one can show that $\text{Ax}_{\text{BL}}7$ codes the prelinearity condition (10). This was one of the interesting reformulations Hájek gave to the standard algebraic properties. Another one was that he recognized that the weak disjunction, i.e. the connective which corresponds to the lattice join operation in the truth degree structures, could be defined as

$$(12) \quad \varphi \vee \psi =_{\text{def}} ((\varphi \to \psi) \to \psi) \wedge ((\psi \to \varphi) \to \varphi).$$

Here $\wedge$ is the weak conjunction with the lattice meet as truth degree function which can, according to the divisibility condition, be defined as

(13) $\varphi \wedge \psi \ =_{\text{def}} \ \varphi \,\&\, (\varphi \to \psi)\,.$

A feeling for the compactness of this system may come from the hint that Höhle's axiom system for the monoidal logic consisted of 14 axioms, and did not have to state the prelinearity and the divisibility conditions.

But Hájek's presentation of the basic fuzzy logic BL was only a partial realization of the plan to give the logic of all continuous t-norms. The most natural, somehow standard algebraic semantics for such a logic of all continuous t-norms would be the subclass T-alg of BL-alg consisting of all BL-algebras with carrier $[0, 1]$, i.e. the subclass of all t-algebras.[4]

It was the guess of Petr Hájek that this standard semantics, determined by the class T-alg of all t-algebras, should be an adequate semantics for the fuzzy logic BL too. He was able to reduce the problem to the BL-provability of two particular formulas [Háj98a], but the final result was proved by Roberto Cignoli et al. in [CEGT00].

And yet another fundamental property of BL could be proved by Francesc Esteva, Lluís Godo, and Franco Montagna [EGM04]: all the t-norm based residuated many-valued logics with *one* continuous t-norm algebra as their standard semantics can be adequately axiomatized as finite extensions of BL. The proof comes by algebraic methods, viz. through a study of the variety of all BL-algebras and their subvarieties which are generated by continuous t-norm algebras: for each one of these subvarieties a finite system of defining equations is algorithmically determined.

Only a short time after Hájek's axiomatization of the logic of continuous t-norms also the logic of all left-continuous t-norms was adequately axiomatized. It was the guess of [EG01] that the class MTL-alg should give an adequate semantics for this logic. First they offered an adequate axiomatization of the logic MTL, a shorthand for *monoidal t-norm logic*, which is determined by the class MTL-alg. And later on Sandor Jenei and Franco Montagna [JM02] proved that MTL is really the logic of all left-continuous t-norms: the logical calculus MTL has an adequate algebraic semantics formed by the subclass of MTL-alg consisting of all MTL-algebras with carrier $[0, 1]$.[5]

The extensions of these propositional logics to first-order ones follows the standard lines of approach: one has to start from a first-order language[6] $\mathcal{L}$

---

[4]If a BL-algebra has carrier $[0, 1]$ then its semigroup operation is automatically a continuous t-norm.

[5]If an MTL-algebra has carrier $[0, 1]$ then its semigroup operation is automatically a left-continuous t-norm.

[6]With the two standard quantifiers $\forall, \exists$, but without function symbols for the present considerations.

and a suitable residuated lattice $\mathbf{A}$, and has to define $\mathbf{A}$-interpretations $\mathbf{M}$ by fixing a nonempty domain $M = |\mathbf{M}|$ and by assigning to each predicate symbol of $\mathcal{L}$ an $\mathbf{A}$-valued relation in M (of suitable arity) and to each constant an element from (the carrier of) $\mathbf{A}$.

The satisfaction relation is also defined in the standard way. The quantifiers $\forall$ and $\exists$ are interpreted as taking the infimum or supremum, respectively, of all the values of the relevant instances.

In order to show that this approach worked well one had either to suppose that the underlying lattices of the interpretations are complete lattices, or at least that all the necessary infima and suprema do exist in these lattices. Interpretations over lattices which satisfy this last condition are called *safe* by Hájek [Háj98b].

For the logic BL of continuous t-norms, Hájek [Háj98b] added the axioms

($\forall$1) $(\forall x)\varphi(x) \to \varphi(t)$, where t is substitutable for x in φ,

($\exists$1) $\varphi(t) \to (\exists x)\varphi(x)$, where t is substitutable for x in φ,

($\forall$2) $(\forall x)(\chi \to \varphi) \to (\chi \to (\forall x)\varphi)$, where x is not free in χ,

($\exists$2) $(\forall x)(\varphi \to \chi) \to ((\exists x)\varphi \to \chi)$, where x is not free in χ,

($\forall$3) $(\forall x)(\chi \vee \varphi) \to \chi \vee (\forall x)\varphi$, where x is not free in χ,

and the rule of generalization to the propositional system BL yielding the system BL$\forall$.

Then he was able to prove the following general *chain completeness theorem*: A first-order formula φ is BL$\forall$-provable iff it is valid in all safe interpretations over BL-chains.

This result can be extended to finite theories as well as to a lot of other first-order fuzzy logics, e.g. to MTL$\forall$.

We will not discuss further completeness results here but refer to the extended survey [CH10]. But it should be mentioned that, as suprema are not always maxima and infima not always minima, the truth degree of an existentially/universally quantified formula may not be the maximum/minimum of the truth degrees of the instances. It is, however, interesting to have conditions which characterize models in which the truth degree of each existentially/universally quantified formula is witnessed as the truth degree of an instance. Petr Cintula and Petr Hájek [HC06] study this problem.

5   Basing fuzzy set theory on t-norm logics

With the previously discussed t-norm based fuzzy logics a toolbox is given to develop fuzzy set theory. Of course, there are still quite different ways to approach this problem depending, e.g., on whether one is interested to

204   Siegfried Gottwald

have some more model-based approach, or whether one prefers a primarily axiomatic one.

In both respects Petr Hájek has offered ideas how to attach this problem. They are special cases in a much wider spectrum of approaches as explained in [Got06a, Got06b]. Nevertheless they deserve to be mentioned here.

5.1   ZF-style approaches

Two (slightly different) model-based approaches of a ZF-like fuzzy set theory have been presented by Petr Hájek and Zuzana Haniková in [HH01] and in [HH03]. They are based upon the first-order *basic t-norm logic* BL∀△, enriched with the △-operator of Matthias Baaz [Baa96].[7]

In a language with primitive predicates $\in, \subseteq, =$ the axioms chosen in [HH01] are suitable versions of (i) extensionality, (ii) pairing, (iii) union, (iv) powerset, (v) $\in$-induction (i.e. foundation), (vi) separation, (vii) collection, i.e. comprehension, (viii) infinity, together with (ix) an axiom stating the existence of the *support* of each fuzzy set. The axiom set of [HH03] offers additionally an axiom of $\in$-induction, i.e. of foundation.

A kind of "standard" model for these theories is formed w.r.t. some complete BL-chain $\mathbf{L} = \langle L, \wedge, \vee, *, \rightarrow, 0, 1 \rangle$ and designed in the style of the Boolean-valued models for standard ZF set theory, see e.g. [Bel85]. This model is based upon the hierarchy

$$V_0^{\mathbf{L}} = \emptyset, \qquad V_{\alpha+1}^{\mathbf{L}} = \left\{ f \in {}^{\mathrm{dom}(u)}L \mid \mathrm{dom}(u) \subseteq V_\alpha^{\mathbf{L}} \right\}$$

with unions at limit stages. In [HH01], he primitive predicates $\in, \subseteq, =$ are interpreted as

$$[\![x \in y]\!] = \bigcup_{u \in \mathrm{dom}(y)} ([\![u = x]\!] * y(u)),$$

$$[\![x \subseteq y]\!] = \bigcap_{u \in \mathrm{dom}(x)} (x(u) \Rightarrow [\![u \in y]\!]),$$

$$[\![x = y]\!] = \triangle [\![x \subseteq y]\!] * \triangle [\![y \subseteq x]\!].$$

The last condition forces the equality to be crisp, and makes the authors' standard form of the axiom of extensionality trivially true in the model. In [HH03], however, these primitive predicates are determined in a simpler way:

$$[\![x \in y]\!] = y(x),$$

together with

$$[\![x = y]\!] = \begin{cases} 1, & \text{if } x = y, \\ 0 & \text{otherwise}. \end{cases}$$

---

[7]This logic is in detail explained, e.g., in [Háj98b].

The main results are that the structure $V^{\mathbf{L}} = \bigcup_{\alpha \in On} V_{\alpha}^{\mathbf{L}}$ together with the different interpretations of the primitive predicates gives in both cases a model of all the (respective) axioms chosen by the authors.

It is interesting to see that the modification in the interpretations of the primitive predicates which distinguishes [HH01] and [HH03] essentially mirrors a similar difference between [Kla65] and [Kla66b].

5.2 A Cantor-style approach

Another, primarily axiomatic approach by Hájek [Háj05] toward a fuzzy set theory, in the sense of a set theory based upon a many-valued logic, is going back to an older approach and has the form of a *Cantorian set theory* over L_{∞}.

That older approach toward a consistency proof of naive set theory, i.e. set theory with the axioms of *comprehension* and *extensionality* only, in the realm of Łukasiewicz logic was initiated by Thoralf Skolem [Sko57] and resulted—after a series of intermediate steps mentioned, e.g., in [Got01]—in a proof theoretic proof (in the realm of L_{∞}) of the consistency of naive set theory with *comprehension only* by Richard B. White [Whi79].

Two equality predicates come into consideration here—Leibniz equality $=_l$ and extensional equality $=_e$ with definitions

$$x =_l y \quad =_{def} \quad \forall z(x \in z \leftrightarrow y \in z),$$
$$x =_e y \quad =_{def} \quad \forall z(z \in x \leftrightarrow z \in y).$$

Leibniz equality is shown to be a *crisp* predicate, but extensional equality is *not*. The whole system becomes *inconsistent* by the coincidence assumption

$$x =_l y \quad \leftrightarrow \quad x =_e y.$$

A set of natural numbers can be added. This yields an essentially undecidable and essentially incomplete system, see [Háj].

6 Conclusion

Actually, approximately four decades after the introduction of fuzzy sets into knowledge engineering and mathematics, the scientific community owes to Petr Hájek's work, particularly to his system BL and its extensions and generalizations, convincing systems of logics for fuzzy sets. These t-norm based systems seem to offer a family of "canonical" logics for fuzzy sets – at least as long as the choice of the class of t-norms as suitable candidates for non-idempotent, i.e. "interactive" versions of intersections for fuzzy sets remains favored in the fuzzy sets community.

But independent of this situation, the class of t-norm based fuzzy logics, pioneered by the inception of the basic fuzzy logic BL, has become an interesting research area for logicians. And this topic of mathematical fuzzy logics is not only related to fuzzy set theory, as was the main focus of this paper, it has its independent interest as a field of logic in which one studies logics of comparable truth degrees. And additionally these logics can be understood as particular cases of substructural logics, see [KO10].

A series of recent surveys, as well as the current research activities in the field, indicate that Hájek's monograph [Háj98b] opened a kind of gold mine for investigations in the wider field of non-classical logics.

BIBLIOGRAPHY

[AR98]   M. Ardeshir and W. Ruitenburg. Basic propositional calculus I. *Math. Logic Quarterly*, 44(3):317–343, 1998.

[ATV80]   C. Alsina, E. Trillas, and L. Valverde. On non-distributive logical connectives for fuzzy sets theory. *Bull. for Studies and Exchanges on Fuzziness and Its Appl.*, 3:18–29, 1980.

[ATV83]   C. Alsina, E. Trillas, and L. Valverde. On some logical connectives for fuzzy sets theory. *J. Math. Anal. Appl.*, 93:15–26, 1983.

[Baa96]   M. Baaz. Infinite-valued Gödel logics with 0-1 projections and relativizations. In P. Hájek, editor, *Gödel '96*, volume 6 of *Lecture Notes in Logic*, pages 23–33, New York, 1996. Springer.

[Bel85]   J. L. Bell. *Boolean-valued Models and Independence Proofs in Set Theory.* Number 12 in Oxford Logic Guides. Oxford University Press, Oxford, 2nd edition, 1985.

[BG73]   R. Bellman and M. Giertz. On the analytic formalism of the theory of fuzzy sets. *Information Sci.*, 5:149–156, 1973.

[BK80]   W. Bandler and L. Kohout. Fuzzy power sets and fuzzy implication operators. *Fuzzy Sets Syst.*, 4:183–190, 1980.

[BKZ95]   D. Butnariu, E. P. Klement, and S. Zafrany. On triangular norm-based propositional fuzzy logics. *Fuzzy Sets Syst.*, 69:241–255, 1995.

[Bla37]   M. Black. Vagueness. An exercise in logical analysis. *Philosophy of Science*, 4:427–455, 1937.

[CEGT00]   R. Cignoli, F. Esteva, L. Godo, and A. Torrens. Basic fuzzy logic is the logic of continuous t-norms and their residua. *Soft Computing*, 4:106–112, 2000.

[CH10]   P. Cintula and P. Hájek. Triangular norm based predicate fuzzy logics. *Fuzzy Sets Syst.*, 161:311–346, 2010.

[Dub80]   D. Dubois. Triangular norms for fuzzy sets. In E. P. Klement, editor, *Proc. 2nd Internat. Seminar Fuzzy Set Theory*, pages 39–68, Linz, 1980. Johannes Kepler University.

[EG01]   F. Esteva and L. Godo. Monoidal t-norm based logic: Towards a logic for left-continuous t-norms. *Fuzzy Sets Syst.*, 124(3):271–288, 2001.

[EGHN00]   F. Esteva, L. Godo, P. Hájek, and M. Navara. Residuated fuzzy logic with an involutive negation. *Arch. Math. Logic*, 39:103–124, 2000.

[EGM04]   F. Esteva, L. Godo, and F. Montagna. Equational characterization of the subvarieties of BL generated by t-norm algebras. *Studia Logica*, 76:161–200, 2004.

[Gai76]   B. R. Gaines. Foundations of fuzzy reasoning. *Int. J. Man-Machine Stud.*, 8(6):623–668, 1976.

[Gil74]   R. Giles. A non-classical logic for physics. *Studia Logica*, 33:397–415, 1974.

[Gil75]   R. Giles. Łukasiewicz logic and fuzzy set theory. In *Proceedings 1975 Internat. Symposium Multiple-Valued Logic (Indiana Univ., Bloomington/IN)*, pages 197–211, Long Beach CA, 1975. IEEE Computer Society Press.

[Gil76]   R. Giles. Łukasiewicz logic and fuzzy set theory. *Int. J. Man-Machine Stud.*, 8:313–327, 1976.

[Gil79]   R. Giles. A formal system for fuzzy reasoning. *Fuzzy Sets Syst.*, 2:233–257, 1979.

[Gil88]   R. Giles. The concept of grade of membership. *Fuzzy Sets Syst.*, 25:297–323, 1988.

[Gir87]   J.-Y. Girard. Linear logic. *Theoretical Comp. Sci.*, 50:1–101, 1987.

[Gog69]   J. A. Goguen. The logic of inexact concepts. *Synthese*, 19:325–373, 1968–69.

[Got]   S. Gottwald. An early approach toward graded identity and graded membership in set theory. To appear in *Fuzzy Sets Syst.*

[Got74]   S. Gottwald. Mehrwertige Anordnungsrelationen in klassischen Mengen. *Math. Nachr.*, 63:205–212, 1974.

[Got76]   S. Gottwald. A cumulative system of fuzzy sets. In A. Zarach, editor, *Set Theory Hierarchy Theory, Mem. Tribute A. Mostowski, Bierutowice 1975*, Lect. Notes Math. 537, pages 109–119. Springer, Berlin, 1976.

[Got84]   S. Gottwald. *T*-Normen und φ-Operatoren als Wahrheitswertfunktionen mehrwertiger Junktoren. In G. Wechsung, editor, *Frege Conference 1984 (Schwerin, 1984)*, Math. Research 20, pages 121–128. Akademie Verlag, Berlin, 1984.

[Got86]   S. Gottwald. Fuzzy set theory with *t*-norms and φ-operators. In A. Di Nola and A. G. S. Ventre, editors, *The Mathematics of Fuzzy Systems*, volume 88 of *Interdisciplinary Systems Res.*, pages 143–195. TÜV Rheinland, Cologne, 1986.

[Got91]   S. Gottwald. Fuzzified fuzzy relations. In R. Lowen and M. Roubens, editors, *Proc. IFSA'91*, volume Mathematics, pages 82–86, Brussels, 1991.

[Got93]   S. Gottwald. *Fuzzy Sets and Fuzzy Logic*. Artificial Intelligence. Verlag Vieweg, Wiesbaden, Tecnea, Toulouse, 1993.

[Got01]   S. Gottwald. *A Treatise on Many-Valued Logics*, volume 9 of *Studies in Logic and Computation*. Research Studies Press, Baldock, 2001.

[Got06a]   S. Gottwald. Universes of fuzzy sets and axiomatizations of fuzzy set theory. Part I: Model-based and axiomatic approaches. *Studia Logica*, 82:211–244, 2006.

[Got06b]   S. Gottwald. Universes of fuzzy sets and axiomatizations of fuzzy set theory. Part II: Category theoretic approaches. *Studia Logica*, 84:23–50, 2006.

[Háj]   P. Hájek. On equality and natural numbers in Cantor-Łukasiewicz set theory. Submitted.

[Háj98a]   P. Hájek. Basic fuzzy logic and BL-algebras. *Soft Computing*, 2:124–128, 1998.

[Háj98b]   P. Hájek. *Metamathematics of Fuzzy Logic*, volume 4 of *Trends in Logic*. Kluwer, Dordrecht, 1998.

[Háj05]   P. Hájek. On arithmetic in the Cantor-Łukasiewicz fuzzy set theory. *Arch. Math. Logic*, 44:763–782, 2005.

[HC06]   P. Hájek and P. Cintula. On theories and models in fuzzy predicate logics. *J. Symb. Logic*, 71:863–880, 2006.

[Hem39]   C. G. Hempel. Vagueness and logic. *Philosophy of Science*, 6:163–180, 1939.

[HGE96]   P. Hájek, L. Godo, and F. Esteva. A complete many-valued logic with product-conjunction. *Arch. Math. Logic*, 35:191–208, 1996.

[HH01]   P. Hájek and Z. Haniková. A set theory within fuzzy logic. In *31st IEEE Internat. Symp. Multiple-Valued Logic, Warsaw, 2001*, pages 319–323. IEEE Computer Society Press, Los Alamitos CA, 2001.

[HH03]   P. Hájek and Z. Haniková. A development of set theory in fuzzy logic. In M. Fitting and E. Orlowska, editors, *Beyond Two: Theory and Applications of Multiple-Valued Logic*, Studies in Soft Computing, pages 273–285. Physica Verlag, Heidelberg, 2003.

[Höh94]   U. Höhle. Monoidal logic. In R. Kruse, J. Gebhardt, and R. Palm, editors, *Fuzzy Systems in Computer Science*, Artificial Intelligence, pages 233–243. Verlag Vieweg, Wiesbaden, 1994.

[Höh95]   U. Höhle. Commutative, residuated l-monoids. In U. Höhle and E. P. Klement, editors, *Non-Classical Logics and Their Applications to Fuzzy Subsets*, volume 32 of *Theory Decis. Libr., Ser. B.*, pages 53–106. Kluwer, Dordrecht, 1995.

[Höh96]   U. Höhle. On the fundamentals of fuzzy set theory. *J. Math. Anal. Appl.*, 201:786–826, 1996.

[JM02]   S. Jenei and F. Montagna. A proof of standard completeness for Esteva and Godo's logic MTL. *Studia Logica*, 70(2):183–192, 2002.

[Kla65]   D. Klaua. Über einen Ansatz zur mehrwertigen Mengenlehre. *Monatsber. Deutsch. Akad. Wiss. Berlin*, 7:859–867, 1965.

[Kla66a]   D. Klaua. Grundbegriffe einer mehrwertigen Mengenlehre. *Monatsber. Deutsch. Akad. Wiss. Berlin*, 8:782–802, 1966.

[Kla66b]   D. Klaua. Über einen zweiten Ansatz zur mehrwertigen Mengenlehre. *Monatsber. Deutsch. Akad. Wiss. Berlin*, 8:161–177, 1966.

[Kla67]   D. Klaua. Ein Ansatz zur mehrwertigen Mengenlehre. *Math. Nachr.*, 33:273–296, 1967.

[Kle52]   S. C. Kleene. *Introduction to Metamathematics*. North-Holland / van Nostrand, Amsterdam, New York, 1952.

[Kle82]   E. P. Klement. Construction of fuzzy σ-algebras using triangular norms. *J. Math. Anal. Appl.*, 85:543–565, 1982.

[KMP00]   E. P. Klement, R. Mesiar, and E. Pap. *Triangular Norms*, volume 8 of *Trends in Logic*. Kluwer, Dordrecht, 2000.

[KO10]   T. Kowalski and H. Ono. Fuzzy logics from substructural perspective. *Fuzzy Sets Syst.*, 161:301–310, 2010.

[Ped83]   W. Pedrycz. Fuzzy relational equations with generalized connectives and their applications. *Fuzzy Sets Syst.*, 10:185–201, 1983.

[Ped85]   W. Pedrycz. On generalized fuzzy relational equations and their applications. *J. Math. Anal. Appl.*, 107:520–536, 1985.

[Pra80]   H. Prade. Unions et intersections d'ensembles flous. *Bull. for Studies and Exchanges on Fuzziness and Its Appl.*, 3:58–62, 1980.

[Rui98]   W. Ruitenburg. Basic predicate calculus. *Notre Dame J. Formal Logic*, 39:18–46, 1998.

[Sko57]   T. Skolem. Bemerkungen zum Komprehensionsaxiom. *Zeitschr. Math. Logik Grundlagen Math.*, 3:1–17, 1957.

[SS83]   B. Schweizer and A. Sklar. *Probabilistic Metric Spaces*. North-Holland, Amsterdam, 1983.

[TV85]   E. Trillas and L. Valverde. On mode and implication in approximate reasoning. In *Approximate Reasoning in Expert Systems*, pages 157–166. North-Holland, Amsterdam, 1985.

[Web83]   S. Weber. A general concept of fuzzy connectives, negations and implications based on *t*-norms and *t*-conorms. *Fuzzy Sets Syst.*, 11:115–134, 1983.

[Whi79]   R. B. White. The consistency of the axiom of comprehension in the infinite-valued predicate logic of Łukasiewicz. *J. Phil. Logic*, 8:509–534, 1979.

[Zad65]   L. A. Zadeh. Fuzzy sets. *Information and Control*, 8:338–353, 1965.

Siegfried Gottwald
Abteilung Logik am Institut fuer Philosophie
Leipzig University
D-04109 Leipzig, Germany
E-mail: gottwald@uni-leipzig.de

A brief history of fuzzy logic in the Czech Republic and the significance of P. Hájek for its development

VILÉM NOVÁK[1]

ABSTRACT. In this paper, we will briefly look at the history of mathematical fuzzy logic in Czechoslovakia (and the Czech Republic) starting from the 1970s and extending until 2009. The role of P. Hájek in the development of fuzzy logic is especially emphasized.

1 Introduction

This paper presents a brief look at the history of mathematical fuzzy logic in Czechoslovakia (and the Czech Republic) from the 1970s to 2009. We will especially emphasize the crucial role of P. Hájek in its development. Let us recall that in 1993, Czechoslovakia split into two countries: the Czech Republic and the Slovak Republic. Therefore, we will refer mostly to mathematical fuzzy logic in the Czech Republic.

This paper is founded on personal remembrances, and claims neither completeness nor objectivity. The list of publications presented accompanies our description and is far from being complete.

We focus on mathematical fuzzy logic and some closely related topics. Therefore, various other branches intensively studied in fuzzy set theory as well as in logic are not included.

It should be noted that some areas that we suppose to be less known are discussed in more detail. Finally, let us remark that occasionally we accompany the description of our memories with mathematical concepts or formulas to provide a more clear or specific explanation but that, because of different focus of this paper, we cannot be too detailed or exact.

2 The Seventies

There was no significant movement in life in Czechoslovakia in the seventies. Those years were negatively influenced by communist normalization after the

---

[1]The research was supported by project MSM 6198898701.

year 1968, and so those who stayed in Czechoslovakia[1] were subject to various kinds of restrictions motivated by the communist party's attempt to have a voice in everything. This pressure was especially serious in the humanities. Luckily, in mathematics, the pressure was not as strong because orthodox enthusiastic communists were not scientists and so they could not understand abstract sciences such as mathematics. On the other hand, access to western literature was quite limited, mainly for economical reasons rather than political ones; Czechoslovakia simply had very limited sources of western currency, and thus, research was partly closed off from western science. Luckily, a lot of good western books were available in Russian because the Soviet Union at that time did not respect copyright laws and freely published most good English books in Russian.

With regard to fuzzy logic, two great ideas were raised in these gray days. The first idea was the concept of *Alternative Set Theory* (AST), and the second idea was establishing a precise formal system of fuzzy logic $Ev_Ł$ that is now known as *fuzzy logic with evaluated syntax*.

The basic concept in AST, with a motivation similar to that of the fuzzy set, is that of *semiset*. This is a proper class (not a set) X as a subclass of some set a, i.e.

(1) $X \subseteq a.$

As has been shown by P. Vopěnka and P. Hájek in the book [VH72], semisets can be consistently introduced inside classical Gödel-Bernays axiomatic set theory by slightly weakening its axioms. Note that semisets are by no means as strange as they might seem at first glance. For example, let us consider a class Sm of *small natural numbers*. Such numbers are 0, 1, ... but this sequence is not finished—there is no last small number and, therefore, also no first large number. On the other hand, there is surely a number n that is large (i.e., surely not small); for example, $n = 10^{10}$. Thus, we have $Sm \subseteq n$. Let us stress that Sm cannot be a set because then it would have to be finite (because n is finite) and so the last small number would have to exist. The class Sm is a typical semiset.

These considerations led P. Vopěnka to develop the AST theory. In about 1974, a group of logicians[2] had gathered around him and commenced developing a new set theory that it was thought might become an alternative (hence the name) to the classical Cantorian one (ST). Several important principles of AST depart from ST. The most important among them is a new model of infinity. In classical set theory, an infinite set is obtained via its never-ending

---

[1]Many thousands of mostly well educated people emigrated from Czechoslovakia in 1968–69.

[2]Among other active people in this group, let us mention that K. Čuda, A. Sochor and J. Mlček.

extension by new elements, while in AST, the infinity emerges when one is dealing with sufficiently large but otherwise classically finite sets. We arrive at the concept of the so-called *natural infinity*, which, unlike ST, is directed inside: a set is infinite if it contains a proper subclass. In other words, a part of an infinite set is not a set. One can verify that a in (1), as well as the number 10^{10} in our example, is infinite. The crucial concept in AST is that of a semiset $\mathbb{FN}$ of all *finite natural numbers* that is a proper subclass of a class $\mathbb{N}$ of all natural numbers. This means that any number $\alpha \in \mathbb{N} - \mathbb{FN}$ is infinite. But recall again that such a number α is *classically finite* (though very large in the given context). We encountered this infinity phenomenon emerging inside very large sets. The semiset $\mathbb{FN}$ is *infinitely countable*, while $\mathbb{N}$ is *uncountable*.

In addition to the many papers published, mostly in *Commentationes Mathematicae Universitatis Carolinae*, the small book [Vop79], written by P. Vopěnka, was published. It presents the essential results for AST up to 1979. Ten years later, a much more extensive book [Vop89] presenting a wide array of well established philosophical background material for AST appeared. Unfortunately, it has never been translated into English. The most recent book on AST focusing especially on its philosophy is [Vop09].

The reader has surely noticed that the elements of the above-introduced semiset Sm represent numbers of grains that do not form a heap in the known *sorites paradox*:

One grain does not form a heap. Adding one grain to what is not yet a heap does not make a heap. Consequently, there are no heaps.

This paradox is resolved in AST. Moreover, semisets represent vaguely defined classes: i.e., the objects that are in fuzzy logic modeled by *fuzzy sets*.

The *fuzzy logic with evaluated syntax* Ev_L was developed by J. Pavelka in his PhD thesis defended in 1976 and published as a sequence of three papers in [Pav79]. Let us emphasize that the main reviewer of the thesis was P. Hájek. This was probably his first encounter with mathematical fuzzy logic.

Pavelka's papers contain full descriptions of the formal system of propositional Ev_L including the algebraic analysis of possible structures of its truth values and its metatheory. The third of the papers [Pav79] contains a complicated algebraic proof of the syntactico-semantical completeness of Ev_L.

The second of the papers [Pav79] contains a deep algebraic analysis of the possible structures of truth values and refers especially to the concept of residuated lattice, thus anticipating the theory of fuzzy logic developed later by P. Hájek. The basic algebra considered there is (what is now called) *standard Łukasiewicz MV-algebra*

(2) $\langle [0,1], \vee, \wedge, \otimes, \rightarrow, 0, 1 \rangle$

where $\otimes$ is the Łukasiewicz conjunction (also called the Łukasiewicz or bold product) defined by $a \otimes b = \max(0, a + b - 1)$ and $\rightarrow$ is the Łukasiewicz implication defined by $a \rightarrow b = \min(1, 1 - a + b)$, $a, b \in [0,1]$. Another structure considered is the finite Łukasiewicz chain (J. Pavelka proved that the infinitely countable chain is not acceptable).

The logic developed by J. Pavelka is consistent with the principal idea of fuzziness, namely not only by assuming truth values between 0 and 1 but also allowing axioms not to be fully convincing. This means that a formula A can be an axiom only in some (arbitrary) degree[3]. This assumption implies the concept of proof accompanied by a value that expresses the degree to which the proof is convincing. As a result, we arrive at the concept of *provability degree* (of a formula A). The prevailing principle in $\mathrm{Ev_L}$ is the *maximality principle* stating that if there are more proofs of a formula A with various degrees then the final provability degree is equal to the maximum (supremum) of all of the former ones. This also implies that all these degrees are equally important; essentially, if we cannot reach 1, then any other degree is satisfactory, provided that we know that it cannot be increased.

A very important consequence of metamathematical considerations about $\mathrm{Ev_L}$ is the theorem stating that, if we consider a residuated lattice on $[0,1]$ and the implication (residuation) operation is not continuous, then it is *impossible to form a syntactico-semantically complete fuzzy logic with evaluated syntax*. This theorem was originally proved by Pavelka in [Pav79]; see also [NPM00], Theorem 4.2. Consequently, the only plausible structure of truth values on $[0,1]$ for this logic is the standard Łukasiewicz MV-algebra and its isomorphs.

3   The Eighties

These years were characterized by increasing interest in the fuzzy set theory and fuzzy logic. A notable paper was that of A. Pultr [Pul84] emphasizing the role of fuzzy equality in fuzzy set theory where the essential transitivity axiom is stated as follows:

$$(a \sim b) \otimes (b \sim c) \leq a \sim c, \qquad a, b, c \in L.$$

By $x \sim y$, we denote the degree of equality between x and y, and $\otimes$ is the multiplication operation in a considered residuated lattice of truth values. This concept has been extensively developed in many papers and books in fuzzy set theory since then. Recall, among other ideas, the important concept of the *real semiset* in AST; i.e. a semisets that is fully characterized by some indiscernibility equivalence relation $\equiv$. The latter is obtained as the

---

[3]In fact, all formulas are then axioms, but most of them only in the degree 0.

intersection of a decreasing countable sequence of set-definable relations[4]

$$(3) \quad \equiv \quad = \quad \bigcap \{R_n \mid n \in \mathbb{FN}\}.$$

This definition is motivated as follows: two objects are indiscernible if all the requirements for still higher and higher precision fail to be met. Hence, fuzzy sets can be seen as reasonable approximations of semisets when one is replacing the indiscernibility relation with the corresponding fuzzy equality. This idea is formally developed in [Nov84].

It seems that the main impulse for the growing interest in fuzzy logic in Czechoslovakia came from the book [Nov86] written by the author of this paper. This book was quite successful and thus was published as a revised English edition in [Nov89] and then again in Czech [Nov90a]. It is important to emphasize that the fuzzy set theory is explained in this book in a unified way from the point of view of Ev_L (which is described in detail in Chapter 3 of the volume in question) as the metatheory at play. As a result, one can see that the concept of formal fuzzy mathematics as proposed much later (see below) is anticipated in this book.

At the same time, the author of this paper followed the ideas of J. Pavelka and developed a predicate version of Ev_L in his PhD thesis defended in 1988. Let us emphasize that the main reviewer of the thesis was again P. Hájek. The 1980s also represented years of intensive discussions between the author of this paper and P. Hájek about fuzzy logic and its meaning.

One part of the thesis was published in two papers [Nov90b]. It contains a proof (based on Pavelka's technique) of a generalization of the Gödel-Henkin completeness theorem:

$$T \vdash_a A \quad \text{iff} \quad T \models_a A$$

for all fuzzy theories T and formulas A where by $T \vdash_a A$ we mean that a formula A is *provable in a degree a* in T and by $T \models_a A$, that it is *true in a degree a*. Another crucial property proven there is a generalized deduction theorem:

$$(4) \quad T \cup \{1/A\} \vdash_a B \quad \text{iff} \quad \text{there is } n > 0 \text{ such that } T \vdash_a A^n \Rightarrow B.$$

Note that a slightly weaker form of this theorem is now known as the *local deduction theorem* in fuzzy logic (see below).

It is worth mentioning that a working group for *Fuzzy Sets and Their Applications* was established within the Society of Czechoslovak Mathematicians and Physicists in 1985. This group has had annual meetings since then. The most important of its activities has been the *International Symposium on*

---

[4]Many technical details are omitted here. The interested reader should refer to [Vop79].

Fuzzy Approach to Reasoning and Decision-Making (FARDM'90), which took place in a small town, Bechyně, in June 1990. The main organizers of this event were V. Novák, M. Mareš, J. Ramík, J. Nekola and M. Černý. The symposium had a key influence on the further development of fuzzy logic in Czechoslovakia. On November 17, 1989, the Czechoslovak communistic regime broke down. The country became open, people were allowed to travel freely and thus, closer contact with western scientists could be established. Because the freshly liberated communist country was interesting to visitors, many people took part in FARDM'90, coming both from the other communist countries and from western ones (the number of participants was slightly over 100). For the first time, we had an opportunity to discuss freely and to meet with the people whose papers we rarely had had the opportunity to read before. A large group of Russian scientists came (actually 27); furthermore, there were several people from Hungary, Poland, and East Germany, as well as D. Dubois from France, M. Roubens from Belgium, G.J. Klir, L. Kohout and J. Bezdek from the USA, among others . The impact of this symposium on fuzzy logic in the Czech Republic can still be felt today. Selected papers from this symposium have been published in the edited volume [NRM+92].

4   The Nineties

4.1   Beginning of the decade

This decade was probably the most active and important for the development of fuzzy logic not only in the Czech Republic but also on an international scale. After long and fruitless quarrels between Czech and Slovak representatives that delayed the economical and political progress of Czechoslovakia, it was decided that Czechoslovakia would be split up into two countries: the *Czech Republic* and *Slovakia* (the Slovak Republic). Though many people were regretful, this decision was very useful from the point of view of the development of both countries. At the same time, the existing close relations between scientists in the two nations were not harmed. Concerning fuzzy set theory, while Slovakia was more oriented towards algebraic and analytical aspects of fuzzy set theory, Czech Republic was more oriented towards fuzzy logic and logical foundations of fuzzy mathematics.

The PhD thesis by the author of this paper contained another result: namely a model of natural language semantics using fuzzy logic. The idea of developing such a model was suggested by P. Sgall (head of the Prague linguistic circle) and finally resulted in the book [Nov92][5] that emerged as an unprecedented attempt at applying AST (hence the title) to the theory of linguistic semantics. The reason for choosing AST was the greater philosophical depth and the more sound and better justified tools in comparison

---

[5]The book was published especially due to the kind support of G.J. Klir.

with fuzzy set theory. While semisets seem to capture the core of the vagueness phenomenon, fuzzy sets in a certain sense seem to quantify it as if from the outside. The book contains a model of the meaning of the most important classes of words and linguistic syntagms. The meaning is represented by special real semisets. Because they are characterized by some indiscernibility relation, they can be technically approximated using fuzzy sets in the sense of [Nov84]. Unfortunately, this book remained almost unnoticed for two reasons in different contexts: it was not mathematical enough and too non-standard for mathematicians, and it was too mathematical for linguists.

This book remained a permanent source of inspiration for the author in using formal fuzzy logic for various kinds of applications, especially when linguistic semantics is at play. The first direction pursued was that of *linguistically oriented control* [Nov95a], whose main idea was to develop a technique using which the computer could act as if it understood the expert's linguistic instructions regarding how a given plant should be controlled. The essential difference between this and (classical) fuzzy control is that the expert specifies his description of control directly using words *without thinking about fuzzy sets hidden inside*. This technique has been implemented and applied to control several real processes. The most successful was the control of an aluminium melting furnace in 1996 (see [NK00])[6].

Another direction was establishing the program of *fuzzy logic in broader sense* in [Nov95b]. The main idea was to develop a formal model of natural human reasoning with the emphasis on employing vague semantics of natural language and solving various related practical problems.

4.2   Entering of P. Hájek into fuzzy logic

For a long time, P. Hájek had been quite hesitant about fuzzy logic, arguing against it in many discussions with the author of this paper over the course of about 10 years. During one of these discussions in 1990 at the Conference of the Czech Mathematicians and Physicists in Zvíkovské Podhradı, he let himself be persuaded that fuzzy logic is not probability theory in disguise. Another point in favor of fuzzy logic came from G. Takeuti, namely from his paper [TT92] and from his visit to Prague.

After that time, P. Hájek's name was seen more and more often in connection with fuzzy logic (see, e.g. [HH93, HH94, HHE+94]. In the paper [Háj95b], he for the first time clearly formulated his view of fuzzy logic as a special branch of mathematical logic.

The event of crucial importance for the further development of mathematical fuzzy logic was the monograph [Háj98]. In it, P. Hájek established mathe-

---

[6]It should be noted that this application turned out to be very successful and was extended to four more furnaces in subsequent years. The application has already been continuously working for more than 13 years.

matical fuzzy logic as a generalization of classical logic. This logic has slightly
modified syntax (through the addition of new connectives and different set of
axioms) and non-classical semantics. He actually introduced a class of formal
logical systems now called (mathematical) *fuzzy logic in narrow sense* (FLn)
with traditional syntax. In addition to G. Takeuti, he was inspired mainly by
the papers of J. Pavelka, V. Novák, U. Höhle [Höh95, Höh94], and D. Mundici
on MV-algebras (e.g. [Mun93]), as well as S. Gottwald's book on many-valued
logic [Got88], and also by the people developing the mathematical theory of
t-norms, namely R. Mesiar, E.P. Klement, E. Pap, D. Butnariu, M. Navara,
and others (an essential monograph about t-norms that was published at the
turn of the 21st century is [KMP00]).

Fuzzy logic in P. Hájek's terms stems from the assumption that the truth
values form a special residuated lattice that then yields rules for the specification of its syntax and semantics. Furthermore, the attempt to coin the
interval of reals $[0,1]$ as a set of truth values led him to establish his *basic
fuzzy logic* (BL-fuzzy logic, or simply BL). Based on the Moster-Shields theorem for t-norms stating that each continuous t-norm is either isomorphic to
minimum, product, or Łukasiewicz conjunction (which is also a t-norm) or
otherwise an ordinal sum of them (see [KMP00] for details), he could confine
to four kinds of main fuzzy logic: *Gödel logic*, which was initiated by K. Gödel
and further developed by G. Takeuti in [TT92]; *Łukasiewicz logic*, which has
been studied already for many years; *product logic*, which was established as
a new logic by P. Hájek; and BL-fuzzy logic, comprised of all three previous
ones. He also recognized the essential role of the prelinearity axiom

$$(a \to b) \vee (b \to a) = 1$$

which then enabled him to prove the following *completeness theorem* for all of
the above-discussed logics: For every formula, A the following is equivalent:

(a) A is provable.

(b) A is a 1-tautology w.r.t. the corresponding algebras of truth values.

(c) A is a 1-tautology w.r.t. the corresponding *linearly ordered* algebras of
truth values.

This theorem supports the idea that the truth values of fuzzy logics should
always be linearly ordered. The famous *standard completeness theorem* stating that semantics of BL-fuzzy logic can be confined only to BL-algebras on
$[0,1]$ in item (iii) (cf. [CEGT00]) is a successful culmination of P. Hájek's
ideas.

As for Ev_L, P. Hájek never accepted the idea of evaluated syntax; thus, this
logic was not for him a "genuine" logic. He reformulated it inside Łukasiewicz
logic as a side logic among other fuzzy ones and calls it *Rational Pavelka*

Logic. In a more or less artificial way, it is possible to introduce a special degree called the *provability degree*. The author of this paper, however, does not think that this term is convenient because it occludes the essential idea that axioms need not be fully convincing and thus that evaluated syntax is obtained as a nice generalization of the classical one. Namely, it can be demonstrated (cf. [NPM00]) that the concept of provability degree can be seen as a natural generalization of classical provability and, consequently, that traditional syntax is a special case of the evaluated one. From this point of view, J. Pavelka left the realm of classical logic in a most radical way, thus introducing a new kind of logic. On the other hand, there is only one Ev_L (cf. our discussion on Ev_L above), thus it seems to be too narrow for many mathematicians.

4.3   Ending of the decade

P. Hájek's monograph is the first of a series of books providing, in our opinion, a full picture of the state of the art in fuzzy logic in the nineties. The second book is [NPM00], written by V. Novák, I. Perfilieva and J. Močkoř. It contains a complete presentation of Ev_L. In addition, the book also contains detailed elaboration of the approximation properties of fuzzy models, the model theory of fuzzy logics in categories, and a more detailed elaboration of FLb (at that time, developed on the basis of Ev_L only). A notable result by I. Perfilieva is an original constructive proof of the McNaughton theorem for Łukasiewicz logic, which states that each formula of the latter can be uniquely represented by a piecewise linear function on $[0,1]$.

The third book of this series is the volume [NP00] edited by V. Novák and I. Perfilieva. It contains contributions by several authors[7] and is focused on various further aspects of fuzzy logic not covered by the previous two books. A full picture of the state of the art of mathematical fuzzy logic in 20th century is completed by the book [KMP00] on t-norms written by E.P. Klement, R. Mesiar and E. Pap.

The successful development of fuzzy logic in the Czech Republic also opened the door to an institutional establishment. Namely, in 1996, the Institute for Research and Applications of Fuzzy Modeling was established in the University of Ostrava in Ostrava (IRAFM OU). Its founders were V. Novák, J. Močkoř, A. Dvořák, R. Bělohlávek, R. Smolíková and D. Jedelský. It became a renowned place of work that successfully combines theoretical and applied research in one place.

---

[7]The contributors were: L.A. Zadeh, G.J. Klir, E.P. Klement, R. Mesiar, E. Pap, J. Paris, A. Di Nola, G. Georgescu, I. Leuştean, E. Trillas, A.R. de Soto, S. Cubillo, P. Hájek, G. Gerla, S. Lehmke, E. Turunen, V. Novák, I. Perfilieva, H. Thiele, D. Mundici, F. Esteva, P. Garcia, L. Godo, S. Gottwald, F. Klawonn, R. Bělohlávek, K. Broda, A. Russo, D. Gabbay

In 1999, IRAFM OU was significantly strengthened by Russian mathematician I. Perfilieva, who naturalized in the Czech Republic. She introduced many new impulses and, in addition to her work in fuzzy logic, she introduced an interesting and related area: the *fuzzy approximation* of functions.

At the turn of the century, a group of young people gathered around P. Hájek, mostly at the Institute of Computer Science of the Academy of Sciences of the Czech Republic in Prague (ICS AV ČR). Among them, very active people include P. Cintula and L. Běhounek, R. Horčík, Z. Haniková, D. Coufal and T. Kroupa (who works at the Inst. of Information Theory and Automation of the ASCR).

Both places of work became leading centers of research in fuzzy logic, not only in the Czech Republic, but also internationally Their researchers meet regularly several times per year and organize various meetings and conferences. The most important event in the nineties was the Seventh IFSA'97 (International Fuzzy Systems Association) World Congress, which took place in June 25–29, 1997 and was co-organized by the University of Ostrava, the School of Economics in Prague, the Institute of Computer Science ASCR, the Institute of Theory of Information and Automation ASCR in Prague, Czech the Technical University in Prague, Slovak Technical University in Bratislava, and Silesian University in Karviná. The general chair was V. Novák, and the other organizers were J. Ivánek, M. Mareš, R. Mesiar, P. Hájek, J. Ramík, P. Vysoký, P. Přibyl, and M. Zeithamlová. The IFSA'97 World Congress in Prague was evaluated by many of its participants as the best of all IFSA congresses. The proceedings was published by the famous publishing house Academia in Prague [NMRI97].

5   21st Century

5.1   P. Hájek and his group in fuzzy logic

Thanks to P. Hájek's monograph (and also to the other mentioned books), the beginning of 21st was devoted to the intensive study of mathematical fuzzy logic not only in the Czech Republic but also in several other European countries, especially Italy and Spain. The results penetrated to a great depth both in the corresponding algebraic structures of truth values and in the properties of the corresponding formal logical systems. Various other logics have been studied, for example, ΠMTL by R. Horčík in [HC04, Hor05]. The work led to the combination of various axioms and was concluded in the so-called LΠ fuzzy logic initiated by Spanish and Italian colleagues in [EGM01]. Their logic has been extended to a predicate first-order version and elaborated on in detail by P. Cintula in [Cin01, Cin03].

One of the problems intensively studied by P. Hájek since the 1990s is complexity of fuzzy logics. This study was begun in [Háj95a] and then continued in a sequence of papers [Háj97, Háj01, Háj04, Háj05, Háj09]. The most recent one is [CH09]. The complexity properties of fuzzy logic are very intricate and depend on the chosen class of logics and their semantics. The following sets of formulas have been studied: 1-tautologies, 1-satisfiable formulas (i.e., those that are 1-true for at least one model), positive tautologies (true in a degree > 0) and positively satisfiable formulas. Their complexity ranges, in general, from Σ_1 to Π_2, but in some cases, it is even non-arithmetical (usually if the given logic contains the product connective: e.g., BL). For example, Łukasiewicz predicate logic with respect to the standard algebra (2) has Π_2-complete set of 1-tautologies, Π_1-complete set of 1-satisfiable formulas, Σ_1-complete set of positive tautologies and Σ_1 complete set of positively satisfiable formulas. If we consider general tautologies (i.e., tautologies with respect to linearly ordered algebras from the given variety) then the given fuzzy logic is only Σ_1-complete. Moreover, for each set $\mathcal{K}$ of continuous t-norms such that propositional $\mathcal{K}$-tautologies are axiomatizable, general tautologies with respect to $\mathcal{K}$ are Σ_1-complete (cf. [Háj05]). It is not surprising that complexity of fuzzy logics is sometimes fairly high given that fuzzy logics should be used in modeling various complex phenomena, namely if vagueness is at play and one cannot expect miracles. Thus, to apply it in practice, we must seek special local solutions.

Various kinds of fuzzy logics and their structure have been extensively studied. The most comprehensive study is the paper [CHH07] written by P. Hájek, P. Cintula and R. Horčík, who studied 60 various systems of fuzzy logic simultaneously. A similar comprehensive paper on algebraic semantics of fuzzy logic is [CEG$^+$09].

The fact that there are so many systems of fuzzy logic has inspired discussion about how fuzzy logic can be defined. In [Cin06], P. Cintula proposes to delineate fuzzy logic mathematically, namely as a logic of the comparative notion of truth. A special issue entitled "What is Fuzzy Logic" from the journal Fuzzy Sets and Systems (Volume 157, Issue 5, Pages 595–718, March 2006) also contains the opinion of P. Hájek's group in the papers [BC06b, Háj06]. This discussion led them to the concept of *core fuzzy logic* (see [HC06]): it is a logic that is an expansion of MTL logic fulfilling

$$A \Leftrightarrow B \vdash C(A) \Leftrightarrow C(B),$$

(for all formulas A, B, C) and holding the *local deduction theorem* (cf. 4):

$$T \cup \{A\} \vdash B \quad \text{iff} \quad \text{there is } n > 0 \text{ such that } T \vdash A^n \Rightarrow B$$

(for all formulas A, B). One of the problems coming out of a formal definition is the following: one can argue that not all logics fulfilling it are indeed capable of providing a model of the vagueness phenomenon (e.g., Gödel logic). This leads to an informal, taxative definition—fuzzy logic is one of several explicitly named logics chosen from the point of view of their capability to provide a model of essential features of the vagueness phenomenon and the human way of reasoning. A contribution to this direction is presented in [Nov05a, Nov06].

A special direction that is very important for the development of fuzzy set theory is the program of *formal fuzzy mathematics* introduced by P. Cintula and L. Běhounek in [BC06a]. Their idea is to develop various areas of fuzzy mathematics formally using mathematical fuzzy logic in the style introduced by P. Hájek as the metamathematical frame. The mathematical background, which is based on the class of fuzzy logics developed before, is the *fuzzy class theory* [BC05]. In a sense, this is a somewhat restricted higher-order fuzzy logic that might be equivalent with the fuzzy type theory mentioned below. Recall that the program of formal fuzzy mathematics is a precise formulation of the program anticipated already in the book [Nov89].

P. Cintula and L. Běhounek not only introduced the program of fuzzy mathematics but also established a group of people (T. Kroupa, R. Horčík, M. Daňková, C. Noguera and few other ones) who began to realize it step by step. In any case, this is a significant attempt to put fuzzy set theory on very well and strongly established grounds.

5.2   Other activities in fuzzy logic in the Czech Republic

Several PhD theses have been defended dealing with various aspects of fuzzy logic, like the fundamental issues (P. Cintula), algebraic aspects (R. Horčík, Z. Haniková), generalized quantifiers (M. Holčapek), normal forms of approximation theory (M. Daňková) and applications (A. Dvořák, M. Štěpnička).

The work of IRAFM OU in fuzzy logic has focused on several aspects. In 2005, higher-order fuzzy logic was introduced by V. Novák in [Nov05b]. In accordance with the classical type theory[8] it is called the *fuzzy type theory*. In principle, the λ-calculus has been extended and many-valued semantics has been introduced for it. The main motivation was to introduce a sound formal means for the development of FLb, which would be more convenient than $\mathrm{Ev_L}$ as applied so far. Let us remark that everything formulated in formal fuzzy mathematics using the fuzzy class theory can also easily be formulated in the fuzzy type theory (using the λ-calculus).

Among the topics which were elaborated more deeply, let us mention the concept of *fuzzy logic normal forms* in predicate fuzzy logic and their use in the fuzzy approximation: i.e., a formal theory of approximation of functions on the basis of imprecise information. This concept was initiated by

---

[8]The type theory was introduced already by B. Russel and A. Whitehead in their Principia Mathematica; it was later on developed more formally by A. Church in [Chu40] and elaborated in detail by L. Henkin in [Hen50, Hen63] and P. Andrews in [And02].

I. Perfilieva already in [Per99] and elaborated on further by M. Daňková (cf. [Daň02, Per01, Per02, Per04]).

In 2006, the Russian edition of the book by V. Novák, I. Perfilieva and J. Močkoř was published in [NPM06]. The book has been quite successful in this vast market. Notable results have been obtained also in the theory of generalized (fuzzy) quantifiers by M. Holčapek and A. Dvořák [Hol08, DH09], who introduced the results of the "classical theory" of generalized quantifiers (cf. [PW06]) into mathematical fuzzy logic, and V. Novák [Nov08], who proposed a formal model of so-called intermediate quantifiers (e.g., "most, many, a lot of, several, a few", etc.).

Let us also mention the attempt at leaving the realm of residuated lattice-based fuzzy logic by introducing a new special algebraic structure called EQ-algebra, in which the fundamental operation is that of *fuzzy equality* $\sim$. The implication in it is a derived operation (cf. [Nov, NdB09, ND09]) defined by

$$(5) \quad a \to b = (a \wedge b) \sim a.$$

This direction in some sense follows the ideas of G.W. Lebniz, who claimed that "a fully satisfactory logical calculus must be an equational one". As a matter of fact, formula (5) also follows the Leibnitz idea.[9]

Another small group led by R. Bělohlávek (he left IRAFM OU in 2004) was established in Olomouc. Let us refer especially to two books [Běl02] and [BV05] where the fuzzy equational logic has been established. It has been shown in this book that when confining only to equalities, generalized completeness in the sense of Ev_L can be obtained even without the assumption that the implication operation is continuous on $[0, 1]$.

Several scientific meetings have taken place in the Czech Republic: in 2005, the conference "The Logic of Soft Computing V" was organized in Ostrava by IRAFM OU. The conference was the penultimate of the series of this kind of conferences on formal fuzzy logic[10]. A special issue of some selected papers has been published in Fuzzy Sets and Systems vol. 158, 2007.

An attempt at organizing a meeting of philosophers and mathematicians in one place was "Conference on Uncertainty & Vagueness" which took place in Prague, September 5–8, 2006. The main goal of this conference was to make philosophers familiar with mathematical fuzzy logic and to explain to them that their criticism of the latter is unjustified (cf. [Kee00]). Unfortunately, it seems that the philosophers (though hopefully not all of them) are to a great extent closed towards ideas coming from outside and thus, from this point of view, the conference was not fully successful.

---

[9]Let $A \equiv B$ be an equality of concepts and AB their intersection. Then Leibnitz interprets the implication "if A then B" as $AB \equiv A$.

[10]The first one was organized in Capri in 2001 and the last one in Malaga in 2006.

The most important meeting in the first decade of the 21st century was the 5th EUSFLAT Conference (European Society for Fuzzy Logic and Technology), which took place in Ostrava (Czech Republic) during September 11–14, 2007. The organizer of this conference was IRAFM OU. The conference was successful and the proceedings were published in [ŠNB07]. In this conference, the Working Group on Mathematical Fuzzy Logic (MathFuzzLog) of EUSFLAT was established. It aims to conduct and promote research in Mathematical Fuzzy Logic.

It is difficult to write a complete presentation of the results related to fuzzy logic in the Czech Republic in 21st century because many activities of various kinds have been accomplished and so, much more space than we have would be needed. We refer the readers to the webpages http://irafm.osu.cz and http://www.mathfuzzlog.org where more information, including full references to many publications or to their preprints, can be found.

6 Conclusion

This paper aims especially to pay tribute to P. Hájek and his work in fuzzy logic on the occasion of 70th birthday celebration. We have briefly overviewed some of the main points in the development of fuzzy logic in Czechoslovakia and, since 1993, in the Czech Republic. Our presentation is far from being complete because we focused mainly on fuzzy logic and because we omitted a lot of results obtained in other areas of fuzzy mathematics such as fuzzy mathematical programming and other ones.

Our goal was to show that the contribution of P. Hájek to fuzzy logic is crucial. Without him, the subject would be incomparably less developed. He deserves many thanks for his work. We wish him good health and many inspiring ideas in the following years.

BIBLIOGRAPHY

[And02]   P. Andrews. *An Introduction to Mathematical Logic and Type Theory: To Truth Through Proof.* Kluwer, Dordrecht, 2002.

[BC05]   L. Běhounek and P. Cintula. Fuzzy class theory. *Fuzzy Sets Syst.*, 154(1):34–55, 2005.

[BC06a]   L. Běhounek and P. Cintula. From fuzzy logic to fuzzy mathematics: A methodological manifesto. *Fuzzy Sets Syst.*, 157(5):642–646, 2006.

[BC06b]   L. Běhounek and P. Cintula. Fuzzy logics as the logics of chains. *Fuzzy Sets Syst.*, 157(5):604–610, 2006.

[Běl02]   R. Bělohlávek. *Fuzzy Relational Systems: Foundations and Principles*, volume 20 of *IFSR International Series on Systems Science and Engineering*. Kluwer and Plenum Press, New York, 2002.

[BV05]   R. Bělohlávek and V. Vychodil. *Fuzzy Equational Logic*, volume 186 of *Stud. Fuzziness Soft Comput.* Springer, Berlin, Heidelberg, 2005.

[CEG+09] P. Cintula, F. Esteva, J. Gispert, L. Godo, F. Montagna, and C. Noguera. Distinguished algebraic semantics for t-norm based fuzzy logics: Methods and algebraic equivalencies. *Ann. Pure Appl. Logic*, 160(1):53–81, 2009.

[CEGT00]   R. Cignoli, F. Esteva, L. Godo, and A. Torrens. Basic fuzzy logic is the logic
           of continuous t-norms and their residua. *Soft Computing*, 4:106–112, 2000.
[CH09]   P. Cintula and P. Hájek. Complexity issues in axiomatic extensions of
           Łukasiewicz logic. *J. Logic Comput.*, 19(2):245–260, 2009.
[CHH07]   P. Cintula, P. Hájek, and R. Horčík. Formal systems of fuzzy logic and their
           fragments. *Ann. Pure Appl. Logic*, 150(1–3):40–65, 2007.
[Chu40]   A. Church. A formulation of the simple theory of types. *J. Symb. Logic*,
           5:56–68, 1940.
[Cin01]   P. Cintula. The ŁΠ and ŁΠ$\frac{1}{2}$ propositional and predicate logics. *Fuzzy Sets
           Syst.*, 124(3):289–302, 2001.
[Cin03]   P. Cintula. Advances in the ŁΠ and ŁΠ$\frac{1}{2}$ logics. *Arch. Math. Logic*, 42(5):449–
           468, 2003.
[Cin06]   P. Cintula. Weakly implicative (fuzzy) logics I: Basic properties. *Arch. Math.
           Logic*, 45(6):673–704, 2006.
[Daň02]   M. Daňková. Representation of logic formulas by normal forms. *Kybernetika*,
           38:711–728, 2002.
[DH09]   A. Dvořák and M. Holčapek. L-fuzzy quantifiers of the type ⟨1⟩ determined
           by measures. To appear in *Fuzzy Sets Syst.*, 2009.
[EGM01]   F. Esteva, L. Godo, and F. Montagna. The ŁΠ and ŁΠ$\frac{1}{2}$ logics: Two complete
           fuzzy systems joining Łukasiewicz and product logics. *Arch. Math. Logic*,
           40(1):39–67, 2001.
[Got88]   S. Gottwald. *Mehrwertige Logik*. Akedemie Verlag, Berlin, 1988.
[Háj95a]   P. Hájek. Fuzzy logic and arithmetical hierarchy. *Fuzzy Sets Syst.*, 73(3):359–
           363, 1995.
[Háj95b]   P. Hájek. Fuzzy logic as logic. In G. Coletti, D. Dubois, and R. Scozzafava,
           editors, *Mathematical Models of Handling Partial Knowledge in Artificial
           Intelligence*, pages 21–30, New York, 1995. Plenum Press.
[Háj97]   P. Hájek. Fuzzy logic and arithmetical hierarchy II. *Studia Logica*, 58(1):129–
           141, 1997.
[Háj98]   P. Hájek. *Metamathematics of Fuzzy Logic*, volume 4 of *Trends in Logic*.
           Kluwer, Dordrecht, 1998.
[Háj01]   P. Hájek. Fuzzy logic and arithmetical hierarchy III. *Studia Logica*, 68(1):129–
           142, 2001.
[Háj04]   P. Hájek. Fuzzy logic and arithmetical hierarchy IV. In V. Hendricks,
           F. Neuhaus, S. A. Pedersen, U. Scheffler, and H. Wansing, editors, *First-
           Order Logic Revised*, pages 107–115. Logos Verlag, Berlin, 2004.
[Háj05]   P. Hájek. Arithmetical complexity of fuzzy predicate logics—a survey. *Soft
           Computing*, 9(12):935–941, 2005.
[Háj06]   P. Hájek. What is mathematical fuzzy logic. *Fuzzy Sets Syst.*, 157(5):597–603,
           2006.
[Háj09]   P. Hájek. Arithmetical complexity of fuzzy predicate logics—a survey II. *Ann.
           Pure Appl. Logic*, 161(2):212–219, 2009.
[HC04]   R. Horčík and P. Cintula. Product Łukasiewicz logic. *Arch. Math. Logic*,
           43(4):477–503, 2004.
[HC06]   P. Hájek and P. Cintula. On theories and models in fuzzy predicate logics.
           J. Symb. Logic, 71:863–880, 2006.
[Hen50]   L. Henkin. Completeness in the theory of types. *J. Symb. Logic*, 15:81–91,
           1950.
[Hen63]   L. Henkin. A theory of propositional types. *Fund. Mathematicae*, 52:323–344,
           1963.
[HH93]   P. Hájek and D. Harmancová. A comparative fuzzy modal logic. In E. P.
           Klement, editor, *Fuzzy Logic in Artificial Intelligence FLAI'93*, volume 695
           of *LNAI*, pages 27–34, Berlin, 1993. Springer.
[HH94]   D. Harmanec and P. Hájek. A qualitative belief logic. *Int. J. of Uncertain.
           Fuzziness Knowledge-Based Syst.*, 2(2):227–236, 1994.

[HHE⁺94]   P. Hájek, D. Harmancová, F. Esteva, P. García, and L. Godo. On modal
            logics of qualitative possibility in a fuzzy setting. In L. de Mantaras and
            D. Poole, editors, *Uncertainty in Artificial Intelligence. Proceedings of the
            10th Conference*, pages 278–285, San Francisco, 1994. Morgan Kaufmann.

[Höh94]   U. Höhle. Monoidal logic. In R. Kruse, J. Gebhardt, and R. Palm, editors,
            Fuzzy Systems in Computer Science, Artificial Intelligence, pages 233–243.
            Verlag Vieweg, Wiesbaden, 1994.

[Höh95]   U. Höhle. Commutative, residuated l-monoids. In U. Höhle and E. P. Kle-
            ment, editors, *Non-Classical Logics and Their Applications to Fuzzy Subsets*,
            volume 32 of *Theory Decis. Libr., Ser. B.*, pages 53–106. Kluwer, Dordrecht,
            1995.

[Hol08]   M. Holčapek. Monadic L-fuzzy quantifiers of the type $\langle 1^n, 1 \rangle$. *Fuzzy Sets
            Syst.*, 159:1811–1835, 2008.

[Hor05]   R. Horčík. Standard completeness theorem for ΠMTL. *Arch. Math. Logic*,
            44(4):413–424, 2005.

[Kee00]   R. Keefe. *Theories of Vagueness*. Cambridge University Press, Cambridge,
            2000.

[KMP00]   E. P. Klement, R. Mesiar, and E. Pap. *Triangular Norms*, volume 8 of *Trends
            in Logic*. Kluwer, Dordrecht, 2000.

[Mun93]   D. Mundici. Ulam games, Łukasiewicz logic and AFC*-algebras. *Fund. In-
            formaticae*, 18:151–161, 1993.

[ND09]   V. Novák and M. Dyba. Non-commutative EQ-logics and their extensions. In
            Proc. World Congress IFSA-EUSFLAT 2009, Lisbon, 2009.

[NdB09]   V. Novák and B. de Baets. EQ-algebras. *Fuzzy Sets Syst.*, 160:2956–2978,
            2009.

[NK00]   V. Novák and J. Kovář. Linguistic IF-THEN rules in large scale applica-
            tion of fuzzy control. In D. Ruan and E. E. Kerre, editors, *Fuzzy If-Then
            Rules in Computational Intelligence: Theory and Applications*, pages 223–
            241. Kluwer, Boston, 2000.

[NMRI97]   V. Novák, M. Mareš, J. Ramík, and J. Ivánek, editors. *IFSA'97 Prague.
            Proceedings*. Academia, Prauge, 1997.

[Nov]   V. Novák. EQ-algebra-based fuzzy type theory and its extensions. To appear
            in *Logic J. of the IGPL*.

[Nov84]   V. Novák. Fuzzy sets—the approximation of semisets. *Fuzzy Sets Syst.*,
            14:259–272, 1984.

[Nov86]   V. Novák. *Fuzzy Sets and Their Applications*. SNTL, Prague, 1986. In Czech.

[Nov89]   V. Novák. *Fuzzy Sets and Their Applications*. Adam Hilger, Bristol, 1989.

[Nov90a]   V. Novák. *Fuzzy Sets and Their Applications*. SNTL, Prague, 2nd edition,
            1990. In Czech.

[Nov90b]   V. Novák. On the syntactico-semantical completeness of first-order fuzzy logic
            I, II. *Kybernetika*, 26:47–66, 134–154, 1990.

[Nov92]   V. Novák. *The Alternative Mathematical Model of Linguistic Semantics and
            Pragmatics*. Plenum Press, New York, 1992.

[Nov95a]   V. Novák. Linguistically oriented fuzzy logic controller and its design. *Int. J.
            Approximate Reasoning*, 12:263–277, 1995.

[Nov95b]   V. Novák. Towards formalized integrated theory of fuzzy logic. In Z. Bien and
            K. Min, editors, *Fuzzy Logic and Its Applications to Engineering, Information
            Sciences, and Intelligent Systems*, pages 353–363. Kluwer, Dordrecht, 1995.

[Nov05a]   V. Novák. Are fuzzy sets a reasonable tool for modeling vague phenomena?
            Fuzzy Sets Syst., 156:341–348, 2005.

[Nov05b]   V. Novák. On fuzzy type theory. *Fuzzy Sets Syst.*, 149:235–273, 2005.

[Nov06]   V. Novák. Which logic is the real fuzzy logic? *Fuzzy Sets Syst.*, 157:635–641,
            2006.

[Nov08]   V. Novák. A formal theory of intermediate quantifiers. *Fuzzy Sets Syst.*,
            159(10):1229–1246, 2008. DOI 10.1016/j.fss.2007.12.008.

[NP00]   V. Novák and I. Perfilieva, editors. *Discovering the World With Fuzzy Logic*, volume 57 of *Stud. Fuzziness Soft Comput.* Springer, Heidelberg, 2000.
[NPM00]   V. Novák, I. Perfilieva, and J. Močkoř. *Mathematical Principles of Fuzzy Logic.* Kluwer, Dordrecht, 2000.
[NPM06]   V. Novák, I. Perfilieva, and J. Močkoř. *Mathematical Principles of Fuzzy Logic.* Fizmatlit, Moscow, 2006.
[NRM+92]   V. Novák, J. Ramík, M. Mareš, M. Černý, and J. Nekola, editors. *Fuzzy Approach to Reasoning and Decision-Making.* Academia and Kluwer, Prague, Dordrecht, 1992.
[Pav79]   J. Pavelka. On fuzzy logic I, II, III. *Zeitschr. Math. Logik Grundlagen Math.*, 25:45–52, 119–134, 447–464, 1979.
[Per99]   I. Perfilieva. Fuzzy logic normal forms for control law representation. In V. H., H. J. Zimmermann, and B. R., editors, *Fuzzy Algorithms for Control*, pages 111–125. Kluwer, Boston, 1999.
[Per01]   I. Perfilieva. Normal forms for fuzzy logic functions and their approximation ability. *Fuzzy Sets Syst.*, 124:371–384, 2001.
[Per02]   I. Perfilieva. Logical approximation. *Soft Computing*, 7(2):73–78, 2002.
[Per04]   I. Perfilieva. Normal forms in BL-algebra of functions and their contribution to universal approximation. *Fuzzy Sets Syst.*, 143:111–127, 2004.
[Pul84]   A. Pultr. Fuzziness and fuzzy equality. In H. J. Skala, S. Termini, and E. Trillas, editors, *Aspects of Vaguenes*, pages 119–135. D. Reidel Publishing Company, Dordrecht, 1984.
[PW06]   S. Peters and D. Westerståhl. *Quantifiers in Language and Logic.* Claredon Press, Oxford, 2006.
[ŠNB07]   M. Štěpnička, V. Novák, and U. Bodenhofer, editors. *New Dimensions in Fuzzy Logic and Related Technologies. Volume I, II. Proceedings of the 5th EUSFLAT Conference.* University of Ostrava, Ostrava, 2007.
[TT92]   G. Takeuti and S. Titani. Fuzzy logic and fuzzy set theory. *Arch. Math. Logic*, 32:1–32, 1992.
[VH72]   P. Vopěnka and P. Hájek. *The Theory of Semisets.* Academia, Prague, 1972.
[Vop79]   P. Vopěnka. *Mathematics in the Alternative Set Theory.* Teubner, Leipzig, 1979.
[Vop89]   P. Vopěnka. *Úvod do Matematiky v Alternatívnej Teórii Množín (An Introduction to Mathematics in the Alternative Set Theory).* Alfa, Bratislava, 1989. In Slovak.
[Vop09]   P. Vopěnka. *Pojednání o jevech povstávajících na množstvích (A Treatise on Phenomena that Arise on Multitudes).* OPS, Plzeň, Nymburk, 2009. In Czech.

Vilém Novák
Institute for Research and Applications of Fuzzy Modeling
University of Ostrava
30. dubna 22, 701 03 Ostrava 1
Czech Republic
Email: Vilem.Novak@osu.cz

What is primary: negation or implication?

MIRKO NAVARA

In honour of Petr Hájek
on the occasion of his 70th birthday

ABSTRACT. We compare R-fuzzy logics (based on an implication interpreted by the residuum of a triangular norm) and S-fuzzy logics
(based on an involutive negation and using a different interpretation
of implication). We show that the former approach (promoted by Petr
Hájek) leads to much richer logical results. Then we discuss extensions
of R-fuzzy logics by an involutive negation as a new connective. These
combine the features of both preceding approaches. We summarize
recent results of this research (which was also initiated by Petr Hájek).

1 Introduction

Fuzzy logic was suggested as a tool extending the classical (Boolean) logic
to manipulation with information which is vague (partial truth is allowed).
Two approaches which were equivalent in the classical logic lead to different
types of fuzzy logics.

The first approach starts from the interpretation of basic connectives,
negation $\neg$, conjunction &, and disjunction $\underline{\vee}$. The negation is interpreted
by a fuzzy negation which is usually involutive, i.e., a decreasing bijection
$n\colon [0,1] \to [0,1]$. The conjunction is interpreted by a continuous *triangular
norm* (shortly, a *t-norm*), i.e., a continuous function $*\colon [0,1]^2 \to [0,1]$ which
is commutative, associative, nondecreasing in both arguments, and which
satisfies $x * 1 = x$ for every $x \in [0,1]$. The disjunction is interpreted by a continuous *triangular conorm* (shortly, a *t-conorm*), i.e., a continuous function
$\oplus\colon [0,1]^2 \to [0,1]$ which is commutative, associative, nondecreasing in both
arguments, and which satisfies $x \oplus 0 = x$ for every $x \in [0,1]$.

REMARK 1 *We distinguish here the symbols for logical connectives and the
corresponding algebraic operations. We mostly follow the notation of [Háj98]
and [EGHN00], although it is quite unusual when used in the context of
Boolean logic.*

We describe other fuzzy operations (including the implication) using only
this set of basic operations. This approach leads to so-called *S-fuzzy logics*.

As an alternative, promoted by Petr Hájek in [Háj98], we may start from a logic based on deduction, and the implication plays a key role in the syntax. Other operations are mostly derived from it. This approach results in so-called *R-fuzzy logics*.

In the sequel, we compare these two approaches. Trying to combine the advantages of both, *fuzzy logics with involutive negations* (called also *RS-fuzzy logics*) were suggested in [EGHN00] and studied in many subsequent papers, e.g. [FM06, Han03, HS08]. We add some recent results in this line. We restrict attention to *propositional* fuzzy logics.

2   Fuzzy algebras as a basis of many-valued hardware

For electrical engineers (including me), fuzzy logic is a tool for building hardware based on fuzzy logical circuits. They open a possibility to describe operations with more than two truth values, possibly also a continuum (analog circuits). Such circuits could be faster because binary logic requires too many digits which have to be processed sequentially in basic mathematical operations. In some particular applications, many-valued hardware could be simpler than its two-valued equivalent. *(E.g., some control tasks can be performed on the analog level with sufficient performance, without the need of extensive hardware used in any current processor.)*

REMARK 2 *In fact, there is one reason for the preference of the binary logic—its reliability. Distinguishing only two truth values allows to build up electronic circuits which are very complex and reliable. Surprisingly, their components are very imprecise, a tolerance of transistor parameters of tens of percent is usual. This contrasts with the situation in mechanical engineering, where much higher precision of components is required and still the reliability of very complex systems is low.*

Whatever hardware for a many-valued circuit is used, we need many-valued (=fuzzy) logic for its description and design. The choice of the basic elements—*gates*—can be inspired by the development of two-valued circuits. These were originally based on gates performing conjunction (**and**, $*$), disjunction (**or**, $\oplus$) and negation (**not**, **neg**, n). These operations form a *sufficient set of operations* which allows to express any Boolean formula. (Moreover, either the conjunction or the disjunction can be omitted.) Later on, the unification of gates has led to the use of universal gates—one type of gates is sufficient for the implementation of any combinational circuit. Such a connective itself forms a sufficient set of operations. Among these operations, negated conjunction (**nand**, |) prevailed in applications. Negated disjunction (**nor**) was equally good from the point of view of design, only hardware preferences decided in favour of **nand**. Also implication ($\Rightarrow$) together with the constant **false** (0) forms a sufficient set of connectives. *(Using a constant is*

no restriction in real hardware where constant values are always available.)
Thus it was a candidate for a universal gate. However, it was not good
from the point of view of implementation. The inputs (arguments) have a
non-symmetrical role which causes, among other problems, different delays
in transfer of signals from input to output. This may cause dynamic hazards.
Thus the implication gate was never used as a basic component for the implementation of binary logical circuits. Instead, implication was expressed
using the negation (implemented by a **nand** gate with a single input or with
two identical inputs, $n(x) = x \mid x$),

$$(x \Rightarrow y) = n(x) \oplus y = n(x * n(y)) = x \mid n(y) = x \mid (y \mid y).$$

REMARK 3 *In fact, although **nand** gates were chosen as the basic implementation tool, they were not the only elements. To simplify the resulting
circuits, other gates, e.g., **and–nor** were used for functions whose implementation using **nand** gates was too complex and slow.*

The possible application of fuzzy logic in many-valued hardware has led
to the study of fuzzy logic based on a conjunction (interpreted by a triangular norm, t-norm $*$), a disjunction (interpreted by a triangular conorm,
t-conorm $\oplus$), and a fuzzy negation n. The implication defined by

$$(x \overset{S}{\Rightarrow} y) = n(x) \oplus y$$

is called the *S-implication* induced by the t-conorm $\oplus$. This approach leads
to so-called *S-fuzzy logics*.

The design of fuzzy logical circuits based on S-fuzzy logics is developed,
e.g., in [LKG09, Pet07, Pet08], where fuzzy generalizations of flip-flops and
Quine-McCluskey method were found.

Here we restrict attention to *continuous* operations satisfying De Morgan
laws, i.e., to a De Morgan triple $(*, \oplus, n)$. Thus the fuzzy negation n has to
be *involutive (=strong)*.

3 Fuzzy algebras based on implication

The approach of Petr Hájek was different. From his point of view, logic is
mainly a tool of reasoning and deduction is one of the basic methods. In order
to describe deduction, implication plays the key role, while other connectives
can be often expressed as secondary notions or omitted in the language of
the logic. The rule *modus ponens* is formulated using an implication.

Boolean algebra allows to express all formulas using only the implication
($\Rightarrow$) and the logical constant **false** (0), thus we do not need other operations.
In particular, the (involutive) negation (n) is considered a derived operation

$$n(x) = (x \Rightarrow 0)$$

and the conjunction ($*$) is expressed as

(1) $x * y = n(x \Rightarrow n(y)) = ((x \Rightarrow (y \Rightarrow 0)) \Rightarrow 0)$.

From this point of view, the implication is essential, other operations (except for the constant 0) are unnecessary.

In the context of fuzzy logics, the conjunction ($*$) is introduced explicitly as a monoidal operation. It is assumed that the conjunction and the implication form an *adjoint pair*, i.e.,

$$x * y \leq z \qquad \text{iff} \qquad x \leq y \Rightarrow z .$$

We call $\Rightarrow$ the *residuum (residuated implication, R-implication)* induced by the t-norm $*$. Thus the interpretation of the implication is uniquely determined by the interpretation of the conjunction (and vice versa), but not as simply as in (1) and in general not by any algebraical formula. As in [Háj98], when speaking of *standard* models (on the interval $[0, 1]$), we restrict attention to implications derived from *continuous* t-norms. To distinguish this approach from that of the preceding section, we speak of *R-fuzzy logics*.

One of the key notions is that of a BL-algebra, introduced by Petr Hájek [Háj98]. A *BL-algebra* is a *residuated lattice* $(L, \cap, \cup, *, \Rightarrow, 0, 1)$ satisfying

$$x \cap y = x * (x \Rightarrow y),$$

$$(x \Rightarrow y) \cup (y \Rightarrow x) = 1 .$$

The latter condition is called *prelinearity*. It is a consequence of the fact that, in a *linearly ordered* BL-algebra,

$$(x \Rightarrow y) = 1 \qquad \text{iff} \qquad x \leq y .$$

This allows to express the ordering by an equation. In S-fuzzy logics, this is obtained only for some t-norms, e.g., Łukasiewicz t-norm.

4   Łukasiewicz logic and MV-algebras

Łukasiewicz logic (and MV-algebras [CDM99] as its algebraic counterpart) play an exceptional role. As in Boolean logic, here it is also possible to express all formulas using only the implication ($\Rightarrow_L$) and the logical constant **false** (0). *Łukasiewicz implication*

$$x \Rightarrow_L y = \begin{cases} 1 - x + y & \text{if } x > y, \\ 1 & \text{otherwise} \end{cases}$$

induces the *standard fuzzy negation*, $n_S(x) = (x \Rightarrow_L 0) = 1 - x$. Conversely, *Łukasiewicz t-conorm*

$$x \oplus_L y = \begin{cases} x + y & \text{if } x + y < 1, \\ 1 & \text{otherwise} \end{cases}$$

and the standard negation allow to express Łukasiewicz implication by the usual formula $(x \Rightarrow_L y) = n_S(x) \oplus_L y$. *Łukasiewicz t-norm*

$$x *_L y = \begin{cases} x + y - 1 & \text{if } x + y \geq 1, \\ 0 & \text{otherwise,} \end{cases}$$

together with Łukasiewicz t-conorm and the standard negation form a De Morgan triple, i.e., $x \oplus_L y = n_S(n_S(x) *_L n_S(y))$. Łukasiewicz implication is obtained also as the residuum (R-implication) of Łukasiewicz t-norm. In other words, Łukasiewicz R-implication and S-implication are the same.

In this particular situation, the approaches of the two preceding sections (R- and S-fuzzy logics) coincide. We may start from the t-norm (or t-conorm) and either the implication or the standard negation; in both cases we obtain the same operations and the same results. Analogous principle applies to some other t-norms, except that the involutive negation is not standard.

5   T-norms without zero divisors

A continuous t-norm T is called

strict if $\forall x \in \,]0, 1[\, : 0 < T(x, x) < x$,

nilpotent if $\forall x \in \,]0, 1[\, : T(x, x) < x$ and $\exists y \in \,]0, 1[\, : T(y, y) = 0$.

If a continuous t-norm is not nilpotent, the corresponding R-implication is not continuous. Thus it cannot be expressed by any formula using only the t-norm and an involutive negation.

REMARK 4 *The analogue of the notion of* sufficient set of operations *can hardly be found in fuzzy logics. In contrast to Boolean logic, fuzzy logics admit uncountably many logical functions and there is no hope to express them by finite expressions using a* finite *set of operations. Satisfactory results are obtained only if we admit (countable)* infinitary *operations, e.g., application of a t-norm to sequences [BKMN05].*

Any continuous t-norm can be expressed as an ordinal sum whose summands are either nilpotent or strict (see, e.g., [Háj98, KMP00]). If no ordinal summands are used, we obtain *Gödel t-norm*, $x *_G y = \min(x, y)$. Reducing our attention to t-norms which cannot be expressed as non-trivial ordinal sums, we deal with three cases—the t-norm is Gödel, strict, or nilpotent. Nilpotent t-norms do not cause problems, as shown in the latter section; here we concentrate on Gödel and strict t-norms. Their common feature is that they have no *zero divisors*, i.e., $x * y = 0$ implies $x = 0$ or $y = 0$. Thus the corresponding R-implication is discontinuous (at least at $(0,0)$) and so is the induced negation, called *Gödel negation* n_G,

$$n_G(x) = (x \Rightarrow 0) = \begin{cases} 1 & \text{if } x = 0, \\ 0 & \text{otherwise.} \end{cases}$$

For the design of fuzzy hardware, discontinuities are undesirable; they do not admit a reliable implementation. Even if they are implemented, they may cause undesirable sensitivity of output to small noise on input. Another disadvantage of Gödel negation is that it attains only the two Boolean values and that all arguments strictly between 0 and 1 are handled as 1. Thus it looses a natural fuzzy logical interpretation. However, Petr Hájek [Háj98] has presented *logical* arguments defending this approach and showing its advantages which we discuss in the sequel.

6   Fuzzy logic as a basis of many-valued deduction

In [Háj98], Petr Hájek has shown that fuzzy logical operations represent the semantics (discussed in the next section) of logics which have reasonable axiomatic systems and the notion of consequence. Thanks to his contribution, fuzzy logic is not only a marginal part of mathematics, but a fully recognized branch of logic. "Fuzzy logic is neither a poor man's logic nor poor man's probability. ... Fuzzy logic in the narrow sense is a beautiful logic, but is also important for applications" [Háj98]. The frequently discussed truth-functionality of fuzzy logic is a natural feature of reasoning with uncertainty which is not probabilistic, but originates from vagueness of our knowledge.

The unifying basis of several other fuzzy logics is the *basic logic* (BL), introduced by Petr Hájek in [Háj98]. Its language is built in the usual way from a countable set of propositional variables, a conjunction &, an implication $\rightarrow$, and the truth constant $\overline{0}$. Further connectives are defined as follows:

$\varphi \wedge \psi$ is $\varphi \,\&\, (\varphi \rightarrow \psi)$,
$\varphi \vee \psi$ is $((\varphi \rightarrow \psi) \rightarrow \psi) \wedge ((\psi \rightarrow \varphi) \rightarrow \varphi)$,
$\neg\varphi$ is $\varphi \rightarrow 0$,
$\varphi \equiv \psi$ is $(\varphi \rightarrow \psi) \,\&\, (\psi \rightarrow \varphi)$.

The axioms of BL are

(A1) $(\varphi \rightarrow \psi) \rightarrow ((\psi \rightarrow \chi) \rightarrow (\varphi \rightarrow \chi))$,
(A2) $(\varphi \,\&\, \psi) \rightarrow \varphi$,
(A3) $(\varphi \,\&\, \psi) \rightarrow (\psi \,\&\, \varphi)$,
(A4) $(\varphi \,\&\, (\varphi \rightarrow \psi)) \rightarrow (\psi \,\&\, (\psi \rightarrow \varphi))$,
(A5a) $(\varphi \rightarrow (\psi \rightarrow \chi)) \rightarrow ((\varphi \,\&\, \psi) \rightarrow \chi)$,
(A5b) $((\varphi \,\&\, \psi) \rightarrow \chi) \rightarrow (\varphi \rightarrow (\psi \rightarrow \chi))$,
(A6) $((\varphi \rightarrow \psi) \rightarrow \chi) \rightarrow (((\psi \rightarrow \varphi) \rightarrow \chi) \rightarrow \chi)$,
(A7) $\overline{0} \rightarrow \varphi$.

The deduction rule of BL is *modus ponens*: from φ and $\varphi \to \psi$ infer ψ.[1]

Further fuzzy logics are obtained as extensions of the basic logic by specific axioms (see [Háj98] for details), in particular the following:

Łukasiewicz logic $\neg\neg\varphi \to \varphi$,

product logic $\qquad \neg\neg\varphi \to ((\varphi \to \varphi \,\&\, \psi) \to (\neg\neg\psi \,\&\, \psi))$,

Gödel logic $\qquad \varphi \to (\varphi \,\&\, \varphi)$.

(Łukasiewicz logic has also an equivalent axiomatization which does not use the conjunction &.)

All these fuzzy logics admit a generalization of the deduction theorem from the classical logic. Its general form is the following:

THEOREM 5 *[Háj98] Let T be a theory (i.e., a set of formulas) and let φ, ψ be formulas. Then $T \cup \{\varphi\} \vdash \psi$ iff there is an $n \in \mathbb{N}$ such that $T \vdash \varphi^n \to \psi$, where $\varphi^n = \underbrace{\varphi \,\&\, \ldots \,\&\, \varphi}_{n \ times}$.*

As we shall discuss in the next section, the corresponding semantics is based on R-implications and hence R-fuzzy logics. In contrast to this, S-fuzzy logics do not admit an analogue. The attempt by Butnariu, Klement, and Zafrany [BKZ95] ended with the return to the syntax of the classical logic. This works, but does not bring anything new resulting from the fuzzification. *(Only Łukasiewicz S-fuzzy logic admits another deductive system, identical to the approach based on the R-implication.)* Here the contribution by Petr Hájek opened a new field which not only shows the richness of fuzzy logics, but contributed also to general theory of logics.

7 Tautologies and completeness

The notion of tautology is essential in semantics of a logic. In fuzzy logics, the abundance of truth values in general BL-algebras gives us a rich spectrum of possible definitions of tautologies. E.g., one could study formulas φ satisfying $e(\varphi) > 0$ for all evaluations e. Alternatively, we could study the inequalities $e(\varphi) \geq c$ for a fixed c, $0 < c \leq 1$. Following Petr Hájek [Háj98], we shall restrict attention to the latter case with $c = 1$. Thus we look for *1-tautologies*, i.e., formulas φ satisfying $e(\varphi) = 1$ for all evaluations e.

In general, the semantics of the basic logic is based on BL-algebras (which were introduced for this purpose by Petr Hájek [Háj98]). Analogously, the

---

[1]This original set of axioms is not independent. P. Cintula [Cin05] proved that (A3) depends on the other axioms. By the use of extensive machine syntactic proofs, S. Lehmke [Leh05] got the same result and also the dependence of (A2) if (A3) is assumed. Recently, K. Chvalovský [Chv08] proved that both axioms (A2) and (A3) can be omitted (simultaneuously) and we still obtain an equivalent axiomatic system of the basic logic.

semantics of its extensions, Gödel, Łukasiewicz, and product logic, is based on the respective algebras, namely *G-algebras* [Háj98], *MV-algebras* [CDM99], and *product algebras* [Háj98]. These are specific types of BL-algebras which satisfy additional equations corresponding to the added logical axioms (see [Háj98]). In all these logics, the semantics is based on residuated lattices, thus on R-implications, and contributes to the study of R-fuzzy logics.

BL-algebras can be much more general than the real interval $[0, 1]$ with a continuous t-norm and its residuum. This particular case is called the *standard semantics* (standard model) of the fuzzy logic in question. In general, BL-algebras can be only partially ordered. However, each BL-algebra can be decomposed as a subdirect product of linearly ordered BL-algebras.

It has been shown in [Háj98] that R-fuzzy logics (in particular, the basic logic and its extensions, Gödel, Łukasiewicz, and product logic) have rich sets of 1-tautologies. The axiomatic schemata were chosen so that these logics are complete, i.e., 1-tautologies coincide with provable formulas. Completeness has been proved for Łukasiewicz [RR58], Gödel [Dum59], product [HGE96], and basic [CEGT00] logics. (See [Háj98] for all these results.) Moreover, *standard* completeness was obtained, i.e., provable formulas are exactly those which are 1-tautologies with respect to all *standard* models (on $[0, 1]$).

Again, S-fuzzy logics do not lead to any new contribution. For Łukasiewicz t-norm, we obtain the situation known from Łukasiewicz (R-fuzzy) logic. For S-fuzzy logics based on a t-norms without zero divisors (in particular, Gödel or product t-norm), we do not obtain consistency with respect to a reasonably rich set of axioms. In [BKZ95], the axioms of Boolean logic are suggested; then provable formulas are those which are theorems of Boolean logic. On the other hand, there are very few 1-tautologies of an S-fuzzy logic based on a t-norm without zero divisors because

$$
\begin{aligned}
n(x) &= 1 \quad &\text{iff} \quad & x = 0, \\
x * y &= 1 \quad &\text{iff} \quad & x = 1 \ \text{ and } \ y = 1, \\
x \oplus y &= 1 \quad &\text{iff} \quad & x = 1 \ \text{ or } \ y = 1, \\
(x \Rightarrow y) &= 1 \quad &\text{iff} \quad & x = 0 \ \text{ or } \ y = 1.
\end{aligned}
$$

In particular if $x, y \in \,]0, 1[$, so are the results of the above operations. Among the axioms of the basic logic, only $(A7)$ is a 1-tautology of an S-fuzzy logic based on a t-norm without zero divisors.

Thus the study of tautologies of S-fuzzy logics did not bring anything new, while the R-fuzzy logic approach promoted by Petr Hájek resulted in rich and deep results on completeness.

On the other hand, compactness has been naturally generalized to S-fuzzy logics [BKZ95]. In R-fuzzy logics different from Łukasiewicz logic, it holds only in special cases [CKMN10, CN04].

8 Probability of fuzzy events

It is a natural requirement to introduce probability on systems of events which are only partially satisfied, therefore described in terms of fuzzy sets. These attempts required a highly non-trivial generalization of classical probability spaces—we have to specify collections of fuzzy sets (generalizing classical σ-algebras) on which a probability can be defined. Then also the axioms of probability measures have to be modified because their usual formulation has the required meaning only in Lukasiewicz logic.

As the domain of probability, *tribes* of fuzzy sets have been suggested by Butnariu and Klement [BK93]. These are non-empty collections of fuzzy sets closed under the formation of (standard) complements and countable fuzzy intersections (performed by a pairwise application of a given continuous t-norm $*$).

REMARK 6 *Minor modifications of the original definition of a tribe have been suggested later on, e.g., in [BNW03, Nav05].*

In order to define a measure, m, on a tribe, consider the ordinary additivity condition

if $A * B = \mathbf{0}$, then $m(A \oplus B) = m(A) + m(B)$,

where the operations $*, \oplus$ are extended from truth values to fuzzy sets pointwise and $\mathbf{0}$ denotes the empty set (whose membership function is the constant 0). This condition has the desired meaning for Lukasiewicz operations, but it is very week if $*$ is a t-norm without zero divisors. Therefore D. Butnariu and E. P. Klement replaced it with the valuation equation

$$m(A \oplus B) + m(A * B) = m(A) + m(B),$$

which led to meaningful notions. They proved deep representation theorems for this type of measures [BK93]. Further generalizations were obtained, e.g., in [BNW03, Nav99, Nav05].

The approach follows the idea of S-fuzzy logics. It uses only the standard fuzzy negation, a continuous t-norm, and (infinitary, countable) operations derived from them; no implication plays an essential role here. Results concerning measures on BL-algebras (and their non-commutative generalizations) appeared only recently, e.g., in [DH06]. So far, studies of measures on collections of fuzzy sets dealt mainly with systems based on S-fuzzy logics with an involutive (usually standard) negation rather than on R-fuzzy logics and a (residuated) implication.

9 Logics with involutive negation

In the basic logic, we have a derived connective $\vee$ whose standard interpretation is Gödel t-conorm (=maximum). However, this operation is dual to & only in Gödel logic. The lack of a disjunction dual to the conjunction has led to an expansion of the basic logic. Another negation $\sim$ has been introduced as an additional unary connective (possibly different from $\neg\varphi = \varphi \to 0$ and interpreted by an involutive fuzzy negation n). Petr Hájek was again one of the authors of this innovation [EGHN00][2]. The resulting fuzzy logics could be adequately called *RS-fuzzy logics*. In these logics, further connectives can be defined, in particular the projection connective Δ (*Baaz delta*):

$$\Delta\varphi \text{ is } \neg\sim\varphi.$$

It will be used to simplify the axiomatization. Its interpretation is

$$\triangle x = \begin{cases} 1 & \text{if } x = 1, \\ 0 & \text{otherwise.} \end{cases}$$

A strong disjunction $\varphi \underline{\vee} \psi$ is definable as $\sim(\sim\varphi \,\&\sim\psi)$; it is interpreted by the t-conorm $\oplus$ dual to $*$, $x \oplus y = n(n(x) * n(y))$. A contrapositive implication $\varphi \hookrightarrow \psi$ is definable as $\sim\varphi \underline{\vee} \psi$; it is interpreted by the S-implication $\overset{c}{\Rightarrow}$ defined as $(x \overset{c}{\Rightarrow} y) = n(x) \oplus y$.

We obtain a new (RS-)fuzzy logic, the *strict basic involutive logic*, SBL$_\sim$ [EGHN00]. Its axioms of are those of BL plus

(SBL) $\neg\neg\varphi \vee \neg\varphi$
($\sim$1) $\sim\sim\varphi \equiv \varphi$
($\sim$2) $\neg\varphi \to \sim\varphi$
($\sim$3) $\Delta(\varphi \to \psi) \to \Delta(\sim\psi \to \sim\varphi)$

(Originally, another set of axioms has been introduced in [EGHN00]. Here we use a simplified axiomatization due to [Cin01].) Deduction rules of SBL$_\sim$ are modus ponens and *necessitation* for Δ: from φ derive $\Delta\varphi$.

Logic SBL$_\sim$ proves, among other consequences, De Morgan laws and a derived inference rule

(DM1) $\sim(\varphi \wedge \psi) \equiv (\sim\varphi \vee \sim\psi)$,
(DM2) $\sim(\varphi \vee \psi) \equiv (\sim\varphi \wedge \sim\psi)$,
(CP)   from $\varphi \to \psi$ derive $\sim\psi \to \sim\varphi$.

---

[2]Incidentally, I read the preliminary version of this paper and contributed by a counterexample to the *standard* completeness of the product logic with an involutive negation. It was a great honour for me being included in this distinguished team and having the chance to co-operate as a coauthor on the final version of the paper. I regret to say that (so far) this has been my only joint paper with Petr Hájek.

In SBL$_\sim$, the classical deduction theorem fails, we have a weaker formulation (cf. [Háj98, 2.4.14]):

THEOREM 7 *[EGHN00] Let T be a theory over SBL$_\sim$. Then $T \cup \{\varphi\} \vdash \psi$ iff $T \vdash \Delta\varphi \to \psi$.*

The semantics of SBL$_\sim$ is based on the following notion:

DEFINITION 8 *An* SBL$_\sim$*-algebra $(L, \cap, \cup, *, \Rightarrow, n, 0, 1)$ is a BL-algebra expanded with a unary operation n satisfying the following conditions:*

  1) $(x \Rightarrow 0) \cap x = 0$,
  2) $n(n(x)) = x$,
  3) $\Delta(x \Rightarrow y) = \Delta(n(y) \Rightarrow n(x))$,

where $\Delta x = (n(x) \Rightarrow 0)$.

A crucial example of an SBL$_\sim$-algebra is the algebra $([0, 1], \max, \min, *, \Rightarrow, n_S, 0, 1)$ of the unit interval of the real line with a t-norm $*$ without zero divisors, its corresponding R-implication $\Rightarrow$, and with the standard negation n_S.

SBL$_\sim$ is sound with respect to the class of SBL$_\sim$-algebras.

In a linearly ordered SBL$_\sim$-algebra, $\Delta x = 0$ for all $x \neq 1$.

THEOREM 9 *[EGHN00] Any* SBL$_\sim$*-algebra is a subdirect product of linearly ordered* SBL$_\sim$*-algebras.*

THEOREM 10 *[EGHN00] Logic SBL$_\sim$ is complete w.r.t. the class of SBL$_\sim$-algebras. In more details, for each formula φ, the following are equivalent:*

  (i) SBL$_\sim$ $\vdash \varphi$,

  (ii) φ is a 1-tautology in each SBL$_\sim$-algebra,

  (iii) φ is a 1-tautology in each linearly ordered SBL$_\sim$-algebra.

The axiom (SBL) excludes Łukasiewicz logic and is satisfied in Gödel and product logic. The extension of SBL$_\sim$ by the corresponding axioms of Gödel or product logic leads to logics G$_\sim$ or Π$_\sim$, respectively. (See [EGHN00] for more details.) Adding the corresponding equations to the definition of an SBL$_\sim$-algebra, we obtain G$_\sim$-algebras or Π$_\sim$-algebras, respectively. Both G$_\sim$ and Π$_\sim$ are complete with respect to the corresponding algebras. (Moreover, it suffices to consider linearly ordered algebras.) The logic G$_\sim$ satisfies also *standard* completeness, i.e., provable formulas are exactly those which are 1-tautologies of the *standard* G$_\sim$-algebra (with domain $[0, 1]$ and $*$ equal to the Gödel t-norm) [EGHN00]. (The choice of the involutive negation is unimportant in this case.) In contrast to this, *standard* completeness is not obtained in Π$_\sim$. We get only *semistandard* completeness:

THEOREM 11 *[EGHN00]* *A formula in* $\Pi_\sim$ *is provable iff it is a 1-tautology in all* semistandard $\Pi_\sim$*-algebras, i.e.,* $\Pi_\sim$*-algebras on* $[0,1]$ *in which* $*$ *is the product t-norm,* $\Rightarrow$ *its corresponding (Goguen) R-implication, and* n *is any involutive negation.*

REMARK 12 *In Theorem 11, we can equivalently describe semistandard* $\Pi_\sim$*-algebras on* $[0,1]$ *as those in which* $*$ *is any* strict t-norm, $\Rightarrow$ *its corresponding R-implication, and* n *is the* standard *negation.*

10   Varieties of algebras with involutive negation

The class of all BL-algebras can be defined by equations, thus it is a variety. Other classes are obtained as its subvarieties, in particular G-algebras, MV-algebras, and product algebras. Also the algebras corresponding to expansions of the basic logic, namely SBL$_\sim$-algebras, G$_\sim$-algebras, and $\Pi_\sim$-algebras, form varieties.

In the context of $\Pi_\sim$-algebras, one important difference seems not to have been paid sufficient attention: In the standard model of the product logic, the conjunction is interpreted by any strict t-norm, but without loss of generality we may restrict attention to the product t-norm only; all other standard models are isomorphic to this one (see [CP89, Lin65, MS57, PC54, Tom87] or overviews in [Als92, AFS06, KMP00]). In algebraic terms, we may say that there is only one standard product algebra. Moreover, the only non-trivial subvariety of product algebras is the variety of Boolean algebras [CT00]. The situation is completely different if we add an involutive negation and consider $\Pi_\sim$-algebras.

The class of $\Pi_\sim$-algebras is a variety. We shall show that, in contrast to product algebras, it has many non-trivial subvarieties. For $k \in \mathbb{N}$ and $x \in [0,1]$, we use the notations

$$x^k = \underbrace{x * \cdots * x}_{k \text{ times}},$$
$$k\,x = \underbrace{x \oplus \cdots \oplus x}_{k \text{ times}},$$

and define equations

(C$_k$) $k\,x^k = x$.

Any set of these equations defines a subvariety of $\Pi_\sim$-algebras. It is shown in [CKMN10] that infinitely many of these varieties are distinct. In particular, there is a non-Boolean semistandard $\Pi_\sim$-algebra which belongs to all these varieties. It is isomorphic (in the sense of Remark 12) to the semistandard

$\Pi_\sim$-algebra in which the negation is standard and the t-norm is the *Hamacher product* $*_H$, defined by

$$x *_H y = \frac{xy}{x + y - xy}$$

for all $(x, y) \in \,]0, 1]^2$. This is based on the observation that Hamacher product and its dual t-conorm satisfy (C_k) for all $k \in \mathbb{N}$. Up to an isomorphism (preserving the standard negation!), Hamacher product is the only t-norm with this property [CKMN10]. The corresponding $\Pi_\sim$-algebra generates a non-trivial subvariety of $\Pi_\sim$-algebras.

Thus the algebraic product is only one of infinitely many possible interpretations of the conjunction in semistandard models of the logic $\Pi_\sim$ (provided that the negation is taken standard). Multiplication plays neither universal nor a privileged role in the semantics of $\Pi_\sim$. Therefore another terminology has been suggested in [CKMN06, CKMN10]: instead of $\Pi_\sim$, the name $CBL_\sim$ (*cancellative basic logic with involutive negation*) is proposed for this logic. The corresponding algebras are then called $CBL_\sim$-algebras.

Alternatively, we may restrict attention to the logic with only one model on $[0, 1]$, where $*$ is the algebraic product and n is the standard negation. Then we obtain the logic $L\Pi$ (see [EGM01]). It combines three basic connectives: the Łukasiewicz implication $\to_L$, the product implication $\to_P$ and the product conjunction $\&_P$.

We have seen that the introduction of an involutive negation makes a distinction between various strict t-norms and gives rise to an abundance of varieties of algebras offering standard interpretations of the respective RS-fuzzy logics.

11   Conclusion

We compared two approaches to fuzzy logics, one based on an involutive negation (and considering an implication a derived operation), the other based on implication (and deriving the negation from it). Both approaches have their advantages and disadvantages in different fields of study; the latter (promoted by Petr Hájek) was particularly successful in logical foundations.

A recent approach tries to combine both attitudes, starting from both an implication and an involutive negation. (These two operations are chosen independently and cannot be derived from each other.) This research enriched our knowledge and proposed perspective models of fuzzy logic. On the other hand, it stated new questions which will be subject to further study.

Acknowledgements. This research was supported by the Czech Ministry of Education under project MSM 6840770038. The author thanks to Petr Cintula for improvements of the manuscript.

BIBLIOGRAPHY

[AFS06]   C. Alsina, M. J. Frank, and B. Schweizer. *Associative Functions: Triangular Norms and Copulas*. World Scientific, Singapore, 2006.

[Als92]   C. Alsina. On a method of Pi-Calleja for describing additive generators of associative functions. *Aequationes Mathematicae*, 43:14–20, 1992.

[BK93]   D. Butnariu and E. P. Klement. *Triangular Norm-Based Measures and Games with Fuzzy Coalitions*. Kluwer, Dordrecht, 1993.

[BKMN05]   D. Butnariu, E. P. Klement, R. Mesiar, and M. Navara. Sufficient triangular norms in many-valued logics with standard negation. *Arch. Math. Logic*, 44:829–849, 2005.

[BKZ95]   D. Butnariu, E. P. Klement, and S. Zafrany. On triangular norm-based propositional fuzzy logics. *Fuzzy Sets Syst.*, 69:241–255, 1995.

[BNW03]   G. Barbieri, M. Navara, and H. Weber. Characterization of T-measures. *Soft Computing*, 8:44–50, 2003.

[CDM99]   R. Cignoli, I. M. L. D'Ottaviano, and D. Mundici. *Algebraic Foundations of Many-Valued Reasoning*, volume 7 of *Trends in Logic*. Kluwer, Dordrecht, 1999.

[CEGT00]   R. Cignoli, F. Esteva, L. Godo, and A. Torrens. Basic fuzzy logic is the logic of continuous t-norms and their residua. *Soft Computing*, 4:106–112, 2000.

[Chv08]   K. Chvalovský. On the independence of axioms in BL and MTL. In *Doktorandské dny 2008 Ústavu informatiky AV ČR*, pages 28–36. ICS AS, Prague, 2008.

[Cin01]   P. Cintula. An alternative approach to the ŁΠ logic. *Neural Network World*, 11:561–572, 2001.

[Cin05]   P. Cintula. Short note: on the redundancy of axiom (A3) in BL and MTL. *Soft Computing*, 9(12):942–942, 2005.

[CKMN06]   P. Cintula, E. P. Klement, R. Mesiar, and M. Navara. Residuated logics based on strict t-norms with an involutive negation. *Math. Logic Quarterly*, 52(3):269–282, 2006.

[CKMN10]   P. Cintula, E. P. Klement, R. Mesiar, and M. Navara. Fuzzy logics with an additional involutive negation. *Fuzzy Sets Syst.*, 161(3):390–411, 2010.

[CN04]   P. Cintula and M. Navara. Compactness of fuzzy logics. *Fuzzy Sets Syst.*, 143(1):59–73, 2004.

[CP89]   R. Craigen and Z. Páles. The associativity equation revisited. *Aequationes Mathematicae*, 37:306–312, 1989.

[CT00]   R. Cignoli and A. Torrens. An algebraic analysis of product logic. *Multiple-Valued Logic*, 5:45–65, 2000.

[DH06]   A. Dvurečenskij and M. Hyčko. Subinterval algebras of BL-algebras, pseudo BL-algebras and bounded residuated ℓ-monoids. *Math. Slovaca*, 56(2):125–144, 2006.

[Dum59]   M. Dummett. A propositional calculus with denumerable matrix. *J. Symb. Logic*, 27:97–106, 1959.

[EGHN00]   F. Esteva, L. Godo, P. Hájek, and M. Navara. Residuated fuzzy logic with an involutive negation. *Arch. Math. Logic*, 39:103–124, 2000.

[EGM01]   F. Esteva, L. Godo, and F. Montagna. The ŁΠ and ŁΠ$\frac{1}{2}$ logics: Two complete fuzzy systems joining Łukasiewicz and product logics. *Arch. Math. Logic*, 40(1):39–67, 2001.

[FM06]   T. Flaminio and E. Marchioni. T-norm based logics with an independent involutive negation. *Fuzzy Sets Syst.*, 157(4):3125–3144, 2006.

[Háj98]   P. Hájek. *Metamathematics of Fuzzy Logic*, volume 4 of *Trends in Logic*. Kluwer, Dordrecht, 1998.

[Han03]   Z. Haniková. On the complexity of propositional logic with an involutive negation. In M. Wagenknecht and R. Hampel, editors, *Proceedings of EUSFLAT Conference*, pages 636–639, Zittau, 2003.

[HGE96]   P. Hájek, L. Godo, and F. Esteva. A complete many-valued logic with product-conjunction. *Arch. Math. Logic*, 35:191–208, 1996.

[HS08]   Z. Haniková and P. Savický. Distinguishing standard SBL-algebras with involutive negations by propositional formulas. *Math. Logic Quarterly*, 54:579–596, 2008.

[KMP00]   E. P. Klement, R. Mesiar, and E. Pap. *Triangular Norms*, volume 8 of *Trends in Logic*. Kluwer, Dordrecht, 2000.

[Leh05]   S. Lehmke. Mechanical proof of the theory of P. Hájek's basic many-valued propositional logic. Technical report, TU Dortmund, Germany, 2005.

[Lin65]   C. H. Ling. Representation of associative functions. *Publicationes Mathematicae Debrecen*, 12:189–212, 1965.

[LKG09]   R. Lovassy, L. T. Kóczy, and L. Gál. Multilayer perceptrons constructed of fuzzy flip-flops. In *4th Int. Symp. Computational Intelligence and Intelligent Informatics*, pages 9–14, Egypt, 2009.

[MS57]   P. S. Mostert and A. L. Shields. On the structure of semigroups on a compact manifold with boundary. *Annals of Mathematics, Second Series*, 65:117–143, 1957.

[Nav99]   M. Navara. Characterization of measures based on strict triangular norms. *J. Math. Anal. Appl.*, 236:370–383, 1999.

[Nav05]   M. Navara. Triangular norms and measures of fuzzy sets. In E. P. Klement and R. Mesiar, editors, *Logical, Algebraic, Analytic, and Probabilistic Aspects of Triangular Norms*, pages 345–390. Elsevier, 2005.

[PC54]   P. Pi-Calleja. Las ecuacionas funcionales de la teoría de magnitudes. In *Segundo Symposium de Matemática, Villavicencio, Mendoza*, pages 199–280, Buenos Aires, 1954.

[Pet07]   M. Petrík. Quine-McCluskey method for many-valued logical functions. *Soft Computing*, 12:393–402, 2007.

[Pet08]   M. Petrík. Many-valued R-S memory circuits. *Int. J. of Uncertain. Fuzziness Knowledge-Based Syst.*, 16:495–518, 2008.

[RR58]   A. Rose and J. B. Rosser. Fragments of many-valued statement calculi. *Trans. Amer. Math. Soc.*, 87:1–53, 1958.

[Tom87]   M. S. Tomás. Sobre algunas medias de funciones asociativas. *Stochastica*, 11(1):25–34, 1987.

Mirko Navara
Center for Machine Perception, Department of Cybernetics
Faculty of Electrical Engineering
Czech Technical University in Prague
Technická 2, CZ-166 27 Prague 6
Czech Republic
Email: navara@cmp.felk.cvut.cz

Why some fuzzifications are easier than others

Aleš Pultr[1]

In honour of Petr Hájek
on the occasion of his 70th birthday

ABSTRACT. When building a fuzzy variant of a theory of Gabriel–Ulmer type (such as a variety of algebras, a relational system, classical automata, etc.) it does not matter whether one models the theory directly in the universe of fuzzy sets or takes the corresponding crisp theory and fuzzifies it ex post.

Introduction

The concept of fuzzy set naturally led to fuzzy variants of crisp theories (fuzzy algebras, fuzzy automata, fuzzy spaces, etc.). Some of such efforts brought deep results, some others, however, appeared to be rather straightforward. In this paper we present a partial explanation of this phenomenon. Namely, it turns out that if the fuzzification is based on a value lattice without an additional structure (such as an additional operation, say residuation), and if the theory in question is locally presentable (varieties of algebras, relational systems, automata, etc., see Section 2) it does not matter whether one treats the theory in the fuzzy context, or takes the ready crisp theory and fuzzifies it afterwards (4.4 below). Thus, the message is that one can expect more interesting results either with enriched value systems, or when discussing theories that are not locally presentable (like for instance spaces).

I use the technique of category theory, but I wish the text to be easily readable for everyone who just knows the basic concepts (category, functor, transformation) – or is willing to look for them in the first pages of a textbook. Therefore, although the requirements are not very demanding, I am rather explicit in presenting the necessary (known) facts.

A reader wishing for more about categories can consult, e.g., [ML71].

[1]The author would like to express his thanks for the support by the project LN 1M0021620808 of the Ministry of Education of the Czech Republic.

1   Preliminaries

1.1. If $\mathfrak{C}$ is a category and a, b some objects of $\mathfrak{C}$ we write

$$(*) \qquad\qquad\qquad \mathfrak{C}(a, b)$$

for the set of all morphisms $\phi\colon a \to b$ in $\mathfrak{C}$. It will be indeed always assumed a set, while the system $|\mathfrak{C}|$ of all the objects of $\mathfrak{C}$ may be a proper class. If $|\mathfrak{C}|$ is a set we speak of a *small category*. As a rule, small categories will be denoted by roman capitals.

The identical morphism of an object a will be denoted by 1_a.

1.2. The opposite (dual) category of $\mathfrak{C}$, that is, the category obtained by keeping $|\mathfrak{C}|$ and formally reversing the directions of morphisms, will be denoted by $\mathfrak{C}^{\mathrm{op}}$. The category of all sets and mappings will be denoted by **Set**.

1.2.1. With a category $\mathfrak{C}$ we associate a functor

$$\mathfrak{C}(-, -)\colon \mathfrak{C}^{\mathrm{op}} \times \mathfrak{C} \to \mathbf{Set}$$

(the cartesian product $\mathfrak{A} \times \mathfrak{B}$ is structured by setting $1_{(a,b)} = (1_a, 1_b)$ and $(\alpha, \beta)(\alpha', \beta') = (\alpha\alpha', \beta\beta')$, of course) defined by

$$\mathfrak{C}(\alpha, \beta) = (\phi \mapsto \beta\phi\alpha)\colon \mathfrak{C}(a, b) \to \mathfrak{C}(a', b') \quad \text{for} \quad \alpha\colon a' \to a, \beta\colon b \to b'$$

($\mathfrak{C}(a, b)$ as in $(*)$ above). This functor will be often restricted to just one of the variables, say $\mathfrak{C}(a, -)\colon \mathfrak{C} \to \mathbf{Set}$. Then we write $\mathfrak{C}(a, \beta)$ for $\mathfrak{C}(1_a, \beta)$ which will hopefully create no confusion. Similarly, $\mathfrak{C}(-, b)$.

1.3. A *diagram* in $\mathfrak{C}$ is a functor $D\colon K \to \mathfrak{C}$ with K small. A *lower bound* of such D is a collection of morphisms $(\phi_k\colon x \to D(k))_{k \in |K|}$ such that for each $\kappa\colon k \to l$ in K, $D(\kappa)\phi_k = \phi_l$. A lower bound $(\lambda_k\colon a \to D(k))_{k \in |K|}$ is a *limit* of D if for each lower bound $(\phi_k\colon x \to D(k))_{k \in |K|}$ there is precisely one $\phi\colon x \to a$ such that $\lambda_k\phi = \phi_k$ for all $k \in |K|$.

A category $\mathfrak{C}$ is said to be *complete* if each diagram in $\mathfrak{C}$ has a limit.

1.4. Fact *Each $\mathfrak{C}(c, -)\colon \mathfrak{C} \to \mathbf{Set}$ preserves all the existing limits.*

Proof   Indeed, let $\lambda = (\lambda_k\colon a \to D(k))_k$ be a limit of $D\colon K \to \mathfrak{C}$ and let $(f_k\colon X \to \mathfrak{C}(c, D(k))_k$ be a lower bound of $\mathfrak{C}(c, D(-))$ Then for a $\kappa\colon k \to l$ in K and $x \in X$, $D(\kappa) \cdot f_k(x) = (\mathfrak{C}(c, \kappa)f_k)(x)$ and since λ is a limit there is exactly one $f(x)\colon c \to a$ such that $\lambda_k \cdot f(x) = f_k(x)$ for all k. This yields a mapping $f\colon X \to \mathfrak{C}(c, a)$ satisfying $\mathfrak{C}(c, \lambda_k) \cdot f = f_k$. $\square$

1.5. If A is a small category and $\mathfrak{C}$ any category we denote by $\mathfrak{C}^A$ the category of all functors $F\colon A \to \mathfrak{C}$ (as objects) and all transformations $\tau\colon F \to G$ between them as morphisms.

1.5.1. Proposition *Let $\mathfrak{C}$ be a complete category. Then each $\mathfrak{C}^A$ is a complete category with the limits given componentwise, that is, a limit $(\lambda^r\colon L \to D(r))_r$ is obtained from the limits $(\lambda_a^r\colon L(a) \to D(r)(a))_r$ taken individually for all $a \in |A|$.*

Proof For a diagram $D\colon R \to \mathfrak{C}^A$ consider the $D(-)(a)\colon R \to \mathfrak{C}$ and for each of them choose a limit

(∗) $(\lambda_a^r\colon L(a) \to D(r)(a))_r.$

For a morphism $\alpha\colon a \to b$ in A we have $D(\rho)_b \cdot D(r)(\alpha) \cdot \lambda_a^r = D(s)(\alpha) \cdot D(\rho)_a \cdot \lambda_a^r = D(s)(\alpha) \cdot \lambda_a^s$; hence, applying (∗) for $L(b)$, we obtain a unique $L(\alpha)\colon L(a) \to L(b)$ such that

$$\lambda_b^r \cdot L(\alpha) = D(r)(\alpha) \cdot \lambda_a^r.$$

This (with the unicity) makes L a functor, and all the $\lambda^r = (\lambda_a^r\colon L \to D(r))_a$ natural transformations.

Now let $\tau^r\colon F \to D(r))_r$ be a lower bound of D. Then for each $a \in |A|$, $(\tau_a^r\colon F(a) \to D(r)(a))_r$ is a lower bound of $D(-)(a)$ and hence there is precisely one $\theta_a\colon F(a) \to L(a)$ such that $\lambda_a^r \cdot \theta_a = \tau_a^r$ for all r. Now $\lambda_b^r \cdot L(\alpha) \cdot \theta_a = D(r)(\alpha) \cdot \lambda_a^r \cdot \theta_a = D(r)(\alpha) \cdot \tau_a^r = \tau_b^r \cdot F(\alpha) = \lambda_b^r \cdot \theta_b \cdot F(\alpha)$, and since $(\lambda_b^r)_r$ is a limit we obtain $L(\alpha) \cdot \theta_a = \theta_b \cdot F(\alpha)$; thus, $\theta\colon F \to L$ is a transformation, unique such that $\lambda_-^r \cdot \theta = \tau^r$ for all r.   □

1.5.2. Observation *For the product (as in 1.2.1) of small categories there is a natural isomorphism of categories*

$$\mathfrak{C}^{A \times B} \cong (\mathfrak{C}^A)^B.$$

(Namely, for a functor $F\colon A \times B \to \mathfrak{C}$ define $\tilde{F}\colon A \to \mathfrak{C}^B$ by setting $\tilde{F}(a)(b) = F(a,b)$, $\tilde{F}(a)(\beta) = F(1_a, \beta)$ and $\tilde{F}(\alpha)_b = F(\alpha, 1_b)$; for $G\colon A \to \mathfrak{C}^B$ and $\alpha\colon a \to a', \beta\colon b \to b'$ set $\hat{G}(a,b) = G(a)(b)$, $\hat{G}(\alpha, \beta) = G(\alpha)_{a'} \cdot G(a)(\beta) = G(a')(\beta) \cdot G(\alpha)_b$.)

1.6. A partially ordered set $L = (X, \leq)$ can be viewed as a category with $|L| = X$ and morphisms

$$(x \leq y)\colon x \to y$$

(thus, for any x, y there is at most one morphism $x \to y$).

The category L^{op} as in 1.2 then coincides with thus interpreted reversely ordered poset. In L^{op} we will often write

$$(x \geq y)\colon x \to y$$

using the original order of L.

Note that the limits in $L = (X, \leq)$ are the infima, and hence L is complete as a category iff it is a complete lattice.

1.7. The Yoneda embedding Recall 1.2.1. For a small category A define a functor

$$\mathsf{Y}\colon A^{\mathrm{op}} \to \mathbf{Set}^A$$

by setting $\mathsf{Y}(a) = A(a, -)$ and $\mathsf{Y}(\alpha) = (\mathsf{Y}(\alpha)_c = A(\alpha, 1_c))_c \colon \mathsf{Y}(a) \to \mathsf{Y}(b)$ for $\alpha\colon b \to a$ (hence $\mathsf{Y}(\alpha)_c(\phi) = \phi\alpha$). $\mathsf{Y}(\alpha)$ is obviously a natural transformation since $A(\alpha, 1_c)A(1_a, \phi) = A(\alpha, \phi) = A(1_b, \phi)A(\alpha, 1_d)$.

1.7.1. Fact Y *is an embedding of* A^{op} *into* $\mathbf{Set}^A$ *as a full subcategory.*

Proof Trivially $\mathsf{Y}(a) \neq \mathsf{Y}(b)$ if $a \neq b$ and if $\alpha \neq \beta$, $\alpha, \beta\colon b \to a$ then $\mathsf{Y}(\alpha)_b(1_b) = \alpha \neq \beta = \mathsf{Y}(\beta)_b(1_b)$; hence Y is one-one. Now let $\tau\colon \mathsf{Y}(a) \to \mathsf{Y}(b)$ be a natural transformation. Set $\alpha = \tau_a(1_a) \in \mathsf{Y}(b)(a) = A(b, a)$. Then since for any $\phi\colon a \to c$ the diagram

$$
\begin{array}{ccc}
\mathsf{Y}(a)(c) & \xrightarrow{\ \tau_c\ } & \mathsf{Y}(b)(c) \\[4pt]
{\scriptstyle A(1_a, \phi)}\big\uparrow & & \big\uparrow{\scriptstyle A(1_b, \phi)} \\[4pt]
\mathsf{Y}(a)(a) & \xrightarrow{\ \tau_a\ } & \mathsf{Y}(b, a)
\end{array}
$$

commutes we have $\tau_c(\phi) = \tau_c(A(1_a, \phi)(1_a)) = A(1_a, \phi)(\tau_a(1_a)) = \phi\alpha = \mathsf{Y}(\alpha)_c(\phi)$, and the embedding is full. $\qquad\square$

2 Gabriel–Ulmer theories

2.1. Examples: Multigraphs and graphs A multigraph $X = (A, V, b, e)$ consists of a set of arrows A, a set of vertices V, and two maps $b, e\colon A \to V$ (the beginning and the end of an arrow). A multigraph homomorphism consists of two maps $f_a\colon A_1 \to A_2$ and $f_v\colon V_1 \to V_2$ respecting the beginnings and ends, that is,

(2.1.1) $b_2(f_a(x)) = f_v(b_1(x))$ and $e_2(f_a(x)) = f_v(e_1(x))$.

Thus we see that we can view a multigraph as a functor $X\colon M \to \mathbf{Set}$ where M has two objects a and v and just two nontrivial morphisms $\beta, \epsilon\colon a \to v$, as indicated in the following picture:

The conditions (2.1.1) say nothing else but that a multigraph homomorphism $f\colon X \to Y$ is a natural transformation between the functors X, Y: we have the commuting diagram

$$
\begin{array}{ccc}
X(a) & \xrightarrow{\ f_a\ } & Y(a) \\
{\scriptstyle X(\beta)}\Big\downarrow{\scriptstyle X(\epsilon)} & & {\scriptstyle X(\beta)}\Big\downarrow{\scriptstyle X(\epsilon)} \\
X(v) & \xrightarrow{\ f_v\ } & Y(v)\ .
\end{array}
$$

Thus we have represented the category of multigraphs as

$$\mathbf{Set}^M.$$

Now if we would wish to represent the category of (directed) graphs instead, we can restrict ourselves to the functors X in which the pair $X(\beta), X(\epsilon)$ is collectionwise monomorphic, that is, if $X(\beta)g = X(\beta)h$ and $X(\epsilon)g = X(\epsilon)h$ then $g = h$. This can be expressed as a limit-preserving condition. Note that (m_1, m_2) is a collectionwise monomorphic pair iff in the diagram

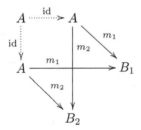

the dotted arrows constitute a limit of the rest.

2.2. Example: (Mealy) automata Consider the category A described by the diagram

$$
p \overset{\sigma}{\underset{\pi_s}{\rightrightarrows}} s
$$
$$
{\scriptstyle \pi_a}\Big\downarrow
$$
$$
a\ .
$$

The category of Mealy automata can be represented as a full subcategory of $\mathbf{Set}^A$ determined by the functors $X\colon A \to \mathbf{Set}$ that send

(2.2.1)
$$
p \xrightarrow{\ \pi_s\ } s
$$
$$
{\scriptstyle \pi_a}\Big\downarrow
$$
$$
a
$$

to a product diagram. For such functors one sees easily that in a transformation $\tau\colon X \to Y$ one has $\tau_p = \tau_s \times \tau_a$ and the commutativity of the
diagram

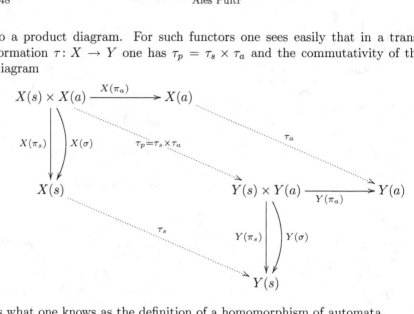

is what one knows as the definition of a homomorphism of automata.

2.3.   Lawvere's representation of varieties of algebras Construct a
category A as follows. First, take the sets $n = \{0, 1, \ldots, n-1\}$ and all the
maps between them, obtaining a category B. Note that in B, n is a coproduct
of n-times 1, with the coproduct injections $\pi_j = (0 \mapsto j)$. Now

1. add new formal morphisms taking care that n remains the coproduct
   $^n 1$ as before (which creates further additional morphism with each new
   one – we may need new ϕ's such that $\phi\iota_i = \phi_i$ if $(\iota_i)_i$ is a system of
   coproduct injections and some of the ϕ_i have been added), obtaining a
   category C, and finally

2. define A as the dual of C; thus, in A we have products

$$(2.3.1) \qquad\qquad \pi_j\colon n = 1^n \to 1.$$

Now consider the subcategory of $\mathbf{Set}^A$ determined by all the functors that
preserve all the products (2.3.1).

What it amounts to: First, the values $X(n)$ are determined by that of
$X(1)$, namely as $X(1)^n$, with $X(\pi_j)$ the standard projections. Second, we
have new mappings $X(\omega)\colon X(1)^n \to X(1)$ obtained from the extras $\omega\colon n \to 1$,
the n-ary operations in the theory, making X to an algebraic structure. Finally, the compositions in A give rise to equalities of an algebraic theory.
Now the fact of life is that each variety of algebras of a finite type can be
represented this way.

2.4. Gabriel–Ulmer theories and locally presentable categories A *Gabriel–Ulmer theory* $\mathbb{A} = (A, \mathcal{A})$ (further often just *theory*) consists of

- a small category A, and

- a system $\mathcal{A}$ of diagrams endowed with fixed lower bounds.

Let $\mathfrak{C}$ be a complete category. A model of a theory $\mathbb{A} = (A, \mathcal{A})$ in $\mathfrak{C}$ is a functor $X \colon A \to \mathfrak{C}$ such that it sends all the $D \in \mathcal{A}$ into limits in $\mathfrak{C}$. The category of models of $\mathbb{A}$ in $\mathfrak{C}$ will be denoted by $\mathfrak{C}^{\mathbb{A}}$. The categories $\mathbf{Set}^{\mathbb{A}}$ are often referred to as *locally presentable categories*.

2.4.1. Notes A concept of the theory of the type presented above appeared, first, in an unpublished paper by Gabriel in the sixties. About the same time, Lawvere dealt with varieties of algebras in this vein ([Law63]). An early treatment can be found in [GU71] (where the term "locally presentable" occurs). For a reader wishing for more on the subject we can recommend the modern monograph [AR97].

It should be noted that important relevant ideas also appeared in Isbell's paper [Isb72].

2.5. Notes 1. Thus, we have seen in 2.1 the theory of multigraphs as $(M, \emptyset)$ and that of (directed) graphs as $(M, \mathcal{A})$ where $\mathcal{A}$ consists of the single lower-bound diagram

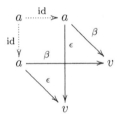

In 2.2 we have the $\mathcal{A}$ consisting of the lower-bound diagram (2.2.1), and in 2.3 the $\mathcal{A}$ consists of all the coproduct diagrams (2.3.1).

2. A simpler fact that will be used in the sequel is that $\alpha \colon A \to b$ is a monomorphism iff

$$
\begin{array}{ccc}
a & \overset{1_a}{\dashrightarrow} & a \\
{\scriptstyle 1_a}\downarrow & & \downarrow{\scriptstyle \alpha} \\
a & \underset{\alpha}{\longrightarrow} & b
\end{array}
$$

is a limit (if it is and $\alpha\beta = \alpha\gamma$ then there is exactly one δ such that $1_a\delta = \beta$ and $1_a\delta = \gamma$; if α is a monomorphism and $\alpha\beta = \alpha\gamma$ then $\beta = \gamma$ and $1_a\beta = \beta$ and $1_a\beta = \gamma$).

3. With the exception of the automata in 2.2, the systems of lower-bound diagrams were themselves limit diagrams in the small categories in question. In fact, as a rule it is not hard to extend the category A in such a way that we have only limit diagrams in $\mathcal{A}$ (one can then speak of the models as the functors that preserve certain specified limits; of course such an extension can be somewhat messy).

$\mathcal{A}$ consisting of limits has its advantages. The Yoneda functor Y (recall 1.7) then embeds the dual category A^{op} into $\mathbf{Set}^A$ as a full subcategory and the objects $Y(a)$ are typically rather important (for instance in the Lawvere representation the $Y(n)$ is the free algebra with n generators). Generally, the copy $Y[A^{\mathrm{op}}]$ of A^{op} in $\mathbf{Set}^A$ is something like a basis from which all the objects of $\mathbf{Set}^A$ can be constructed in a canonical way (see e.g. [Ulm68], [Pul70]).

4. All our examples concerned models in the category of sets. To give another example, consider the category $\mathbf{Top}$ of topological spaces and continuous mappings. Then for a Lawvere theory $\mathbb{A}$, $\mathbf{Top}^A$ represents the corresponding category of topological (continuous) algebras; if, say, $\mathbb{A}$ is the theory of groups, $\mathbf{Top}^A$ is the category of topological groups. Similarly, topological graphs, ordered semigroups, etc.

2.6. Proposition $\mathfrak{C}^{\mathbb{A}}$ *is closed under limits in* $\mathfrak{C}^A$.

Proof Let $S\colon R \to \mathfrak{C}^A$ be a diagram and let $(\theta_r\colon L \to S_r)_{r\in|R|}$ be its limit in $\mathfrak{C}^A$ (we write S_r instead of $S(r)$ to simplfy the notation). Hence (recall 1.5.1), for all $x \in |A|$

$(*)$ $((\theta_r)_x\colon L(x) \to S_r(x))_r$ is a limit in $\mathfrak{C}$.

Let all the S_r be in $\mathfrak{C}^{\mathbb{A}}$ end let $(\alpha_k\colon a \to D(k))_{k\in|K|}$ be in $\mathcal{A}$. Then for all $r \in |R|$

$(**)$ $(S_r(\alpha_k)\colon S_r(a) \to S_r(D(k)))_{k\in|K|}$ is a limit in $\mathfrak{C}$.

Let $\phi_k\colon X \to L(D(k)))_k$ be a lower bound of LD, that is, for each $\kappa\colon k \to l$ in K, $LD(\kappa)\phi_k = \phi_l$. Consider the system

$$((\theta_k)_{D(k)} \cdot \phi_k\colon X \to S_r(D(k)), \quad k \in |K|.$$

It is a lower bound of $S_r \cdot D$: indeed, for $\kappa\colon k \to l$,

$$S_r(D(\kappa)) \cdot (\theta_r)_{D(k)} \cdot \phi_k = (\theta_r)_{D(l)} \cdot L(D(\kappa)) \cdot \phi_k = (\theta_r)_{D(k)} \cdot \phi_l.$$

Thus, there is a $\psi_r\colon X \to S_r(a)$ such that $S_r(\alpha_k)\cdot\psi_r = (\theta_r)_{D(l)}\cdot\phi_k$. We show that $(\psi_r\colon X \to S_r(a))_r$ is a lower bound of the functor $(S_r(a), S(\rho)_a)_{r\in|R|,\rho\in R}$. Indeed, let $\rho\colon r \to s$ be a morphism in S. Then since $S(\rho) \cdot \theta_r = \theta_s$ we have

$$S_r(\alpha_k) \cdot S(\rho)_a \cdot \psi_r = S(\rho)_{D(k)} \cdot S_r(\alpha_k) \cdot \psi_S = S(\rho)_{D(k)} \cdot (\theta_r)_{D(k)} \cdot \phi_k =$$
$$= (\theta_s)_{D(k)} \cdot \phi_k = S_s(\alpha_k)_{D(k)} \cdot \psi_s$$

and since $(**)$ is a limit, $S(\rho)_a \cdot \psi_r = \psi_s$.

Thus, using the limit $(\theta_k)_a$, $r \in |S|$, from $(*)$ we obtain a $\phi\colon X \to L(a)$ such that $(\theta_r)_a \cdot \phi = \psi_r$ for all r. Now

$$(\theta_r)_{D(k)} \cdot L(\alpha_k) \cdot \phi = S_r(\alpha_k) \cdot (\theta_r)_a \cdot \phi = S_r(\alpha_k) \cdot \psi_r = (\theta_r)_{D(k)} \cdot \phi_k$$

and since $((\theta_r)_{D(k)}\colon L(D(k)) \to S_r(D(k))_r$ is a limit, finally $L(\alpha_k) \cdot \phi = \phi_k$.

The unicity of such ϕ is easily proved by backtracking the equalities and can be left to the reader. $\qquad\qquad\qquad\qquad\qquad\qquad\qquad\qquad\qquad\qquad\square$

3 Fuzzification as a GU-theory

3.1. Let L be a lattice (in the original papers on fuzziness, e.g. [Zad65], L was the unit interval, here we will have in 3.1 and 3.3 a general lattice, in 3.2 and 3.4 a complete one).

In (perhaps) the simplest form, a *fuzzy set* (more precisely an *L-fuzzy set*) X is a set with fuzzy membership of elements $(\dot{X}, (\in_a)_{a \in L})$ satisfying

$$a \leq b \ \& \ x \in_b X \ \Rightarrow \ x \in_a X.$$

Thus, if we set

$$X(a) = \{x \mid x \in_a X\},$$

the fuzzy structure X appears as a stratification $(X(a))_{a \in L}$ of the set $\dot{X}$ such that

$$a \leq b \quad \Rightarrow \quad X(b) \subseteq X(a).$$

A morphism $f\colon (\dot{X}, (\in_a^X)_a) \to (\dot{Y}, (\in_a^Y)_a)$ between fuzzy sets is an $f\colon \dot{X} \to \dot{Y}$ such that

$$x \in_a^X X \ \Rightarrow \ f(x) \in_a^Y Y \qquad (\text{in other words, } f[X(a)] \subseteq Y(a))$$

for all $a \in L$.

The resulting category will be denoted by

$$L\mathbf{Fuzz}_w.$$

3.2. The definition above is rather weak and, in particular, does not produce a satisfactory system of fuzzy subsets.

Take the definition of a fuzzy map. One can think of a fuzzy subset $(\dot{Y}, (\epsilon_a^Y)_a)$ of $(\dot{X}, (\epsilon_a^X)_a)$ if $\dot{Y} \subseteq \dot{X}$ and if the embedding map is a morphism, that is, $x \in_a^Y Y \ \Rightarrow \ x \in_a^X X$. Now consider an $x \in \dot{X}$ with $x \in_a X$ and the fuzzy set $\overline{x}_a$ carried by $\{x\}$ to which x belongs to degree a but not more. If a union $u = \bigcup\{\overline{x}_a \mid x \in_a X\}$ should exist it can be expected to be carried by $\{x\}$ again, with the degree of membership $\geq a$ for all a with $x \in_a X$; on the other hand u should be a fuzzy subset of our $X = (\dot{X}, (\in_a)_a)$. Thus,

- for every x there is a maximum $\alpha(x)$ such that $x \in_{\alpha(x)} X$,

and the full information on the fuzziness is given by the map $\alpha \colon \dot{X} \to L$ (that is, $x \in_a X$ iff $a \leq \alpha(x)$). The morphisms $f \colon (\dot{X}, \alpha) \to (\dot{Y}, \beta)$ are then characterized by the property $\beta(f(x)) \geq \alpha(x)$.

The category of thus represented (more special) fuzzy sets will be denoted

$$LFuzz.$$

3.2.1. Remark $LFuzz$ is what one usually takes for the plain category of fuzzy sets (that is, without more structure – an extra operation, for instance – on L), see e.g. [Gog67],[Pul76]. It should be noted that here one already has a satisfactory structure of fuzzy subsets: we can take the $(\dot{X}, \beta) \subseteq (\dot{X}, \alpha)$ with $\beta \leq \alpha$ (the $(\dot{Y}, (\epsilon_a^Y)_a) \subseteq (\dot{X}, (\epsilon_a^X)_a)$ with $\dot{Y} \subseteq \dot{X}$ above are more handily represented by extending β by 0 on $\dot{X} \setminus \dot{Y}$). We then have, of course, $\bigcup_{i \in J}(\dot{X}, \alpha_i) = (\dot{X}, \sup_{i \in J} \alpha_i)$ and $\bigcap_{i \in J}(\dot{X}, \alpha_i) = (\dot{X}, \inf_{i \in J} \alpha_i)$.

3.2.2. Lemma *Let L be a complete lattice. Then the following statements on a fuzzy structure $(\in_a)_a$ on $\dot{X}$ (resulting in the stratification $(X(a))_a$) are equivalent.*

1. *There is a mapping $\alpha \colon \dot{X} \to L$ such that*

$$x \in_a X \quad (\text{that is, } x \in X(a)) \quad \text{iff} \quad a \leq \alpha(x).$$

2. *For every $M \subseteq L$,*

$$\bigcap_{m \in M} X(m) = X(\sup M).$$

3. $\bigcap\{X(a) \mid x \in_a X\} = X(\sup\{a \mid x \in_a X\})$.

4. *For every $x \in X$ there is a maximum $s \in L$ such that $x \in_s X$.*

Proof $(1) \Rightarrow (2)$: $x \in \bigcap_{m \in M} X(m)$ iff $\forall m \in M$, $m \leq \alpha(x)$ iff $\sup M \leq \alpha(x)$ iff $x \in X(\sup M)$.

$(2) \Rightarrow (3)$ is trivial.

$(3) \Rightarrow (1)$: Set $\alpha(x) = \sup\{a \mid x \in X(a)\} = \sup\{a \mid x \in_a X\}$. Then trivially $x \in_a X$ implies $a \leq \alpha(x)$. On the other hand, by definition $x \in X(a)$ for $x \in_a X$ and hence $x \in \bigcap\{X(a) \mid x \in_a XX(a)\} = X(\alpha(x))$, and hence if $a \leq \alpha(x)$ then also $x \in X(a)$.

$(3) \equiv (4)$: the statements are just reformulations of each other. Of course, $s = \sup\{a \mid x \in_a X\}$. $\square$

3.3. Let L^{op} be the lattice L with the inverse order, taken as a small category (see 1.6). Note that all of its morphisms are monomorphisms. Set

$$\mathbb{F}w = (L^{\mathrm{op}}, \mathcal{A}')$$

where $\mathcal{A}'$ is the system of all the lower bounds

(amounting to the condition that in the functors $X \in \mathfrak{C}^{\mathbb{F}w}$ all the $X((a \geq b))$ are monomorphic – recall 2.5.2).

In particular, for $\mathfrak{C} = \mathbf{Set}$, we have all the $X((a \geq b))\colon X(a) \to X(b)$ one-one.

3.3.1. For an $X \in \mathbf{Set}^{\mathbb{F}w}$ define $X' = X'(0) = X(0)$ and for general $a \in L$,

$$X'(a) = X((a \geq 0))[X(a)].$$

Let $a \geq b$ in L. We have $(a \geq 0) = (b \geq 0)(a \geq b)$ in L^{op} and hence

$$X'(a) = X((b \geq 0))[X((a \geq b))[X(a)]] \subseteq X'(b)$$

so that

$$a \geq b \quad \Rightarrow \quad X'(a) \subseteq X'(b),$$

and we have a fuzzy set as in 3.1. Further, defining $X'((a \geq b))$ as the inclusion maps we obtain another functor $X' \in \mathbf{Set}^{\mathbb{F}w}$.

3.3.2. Observation In $\mathbf{Set}^{\mathbb{F}w}$, X' is isomorphic to X.

(Indeed, define $\tau_a\colon X(a) \to X'(a)$ by $\tau_a(x) = X((a \geq a))(x)$. Obviously each τ_a is one-one onto, and it is easy to check that $\tau = (\tau_a)_a$ is a natural transformation.)

3.3.3. Corollary $\mathbf{Fuzz}_w$ *is equivalent to* $\mathbf{Set}^{\mathbb{F}w}$. *More precisely, although* $\mathbf{Set}^{\mathbb{F}w}$ *has more objects, each of them is isomorphic to an object of* $\mathbf{Fuzz}_w$.

3.4. To represent the more satisfactory $L\mathbf{Fuzz}$ (now, L is complete), consider

$$\mathbb{F} = (L^{\mathrm{op}}, \mathcal{A})$$

where $\mathcal{A}$ is $\mathcal{A}'$ extended by all the infimum diagrams in L^{op} (that is, all the supremum diagrams in L). By 3.2.2(2) and 3.3,

$L\mathbf{Fuzz}$ *is, up to isomorphisms, represented as* $\mathbf{Set}^{\mathbb{F}}$.

4   Product of GU-theories. Application

4.1. Let A, B be small categories. For an object b in B and a lower bound $\mathfrak{a} = (\alpha_k \colon a \to D(k))_{k \in |K|}$ define a lower bound $\mathfrak{a} * b$ in $A \times B$ by setting

$$\mathfrak{a} * b = (\alpha_k \times 1_b \colon (a, b) \to (D(k), b))_{k \in |K|}.$$

Similarly, for $a \in |A|$ and lower bounds $\mathfrak{b}$ in B, we define

$$a * \mathfrak{b} = (1_a \times \beta_k \colon (a, b) \to (a, D(k)))_{k \in |K|}.$$

Note To avoid too many symbols we keep denoting the diagram in a lower bound by D and its domain by K. Of course they typically differ for distinct elements of $\mathcal{A}$ resp. $\mathcal{B}$.

4.1.2. Define a *product of theories* $\mathbb{A} \times \mathbb{B}$ as $(A \times B, \mathcal{A} \oplus \mathcal{B})$ by setting

$$\mathcal{A} \oplus \mathcal{B} = \{\mathfrak{a} * b \mid \mathfrak{a} \in \mathcal{A}, \ b \in |B|\} \cup \{a * \mathfrak{b} \mid \mathfrak{b} \in \mathcal{B}, \ a \in |A|\}.$$

(The word "product" is here used simply for an operation producing a new object from two others, not for a categorical product. We do not introduce a category of theories.)

4.2. Theorem *The equivalence of categories* $\mathfrak{C}^{A \times B} \cong (\mathfrak{C}^B)^A$ *from 1.5.2 restricts to an equivalence*

$$\mathfrak{C}^{\mathbb{A} \times \mathbb{B}} \cong (\mathfrak{C}^{\mathbb{B}})^{\mathbb{A}}.$$

Proof Recall functor $\widehat{F} \colon A \times B \to \mathfrak{C}$ from 1.5.2 associated with $F \colon A \to \mathfrak{C}^B$. We have to prove that the following claims are equivalent:

(1) $\widehat{F}$ sends all the lower bounds from $\mathcal{A} \otimes \mathcal{B}$ to limits in $\mathfrak{C}$

(2) each $F(a)$ sends all the lower bounds from $\mathcal{B}$ to limits in $\mathfrak{C}$, and F sends the lower bounds from $\mathcal{A}$ to limits in $\mathfrak{C}^B$ (that is, by 2.6, limits in $\mathfrak{C}^{\mathbb{B}}$).

(1)$\Rightarrow$(2): Consider a $(\beta_k \colon b \to D(k))_{k \in |K|}$ from $\mathcal{B}$. Now we have $(F(a)(\beta_k) \colon F(a)(b) \to F(a)(D(k))_k = (\widehat{F}(1_a, \beta_k) \colon \widehat{F}(a, b) \to \widehat{F}(1_a, D(k)))_k$ is a limit in $\mathfrak{C}$ since $((1_a, \beta_k) \colon (a, b) \to (1_a, D(k)))_k$ is in $\mathcal{A} \otimes \mathcal{B}$. Thus, $F(a)$ is in $\mathfrak{C}^{\mathbb{B}}$. Now let $(\alpha_k \colon a \to D(k))_k$ be in $\mathcal{A}$. Then the system $F(\alpha_k) \colon F(a) \to F(D(k)))_k$ is a limit in $\mathfrak{C}^B$ iff each $F(\alpha_k)(1_b) \colon F(a)(b) \to F(D(k))(b))_k$ is a limit in $\mathfrak{C}$. But it is $(\widehat{F}(\alpha_k, 1_b) \colon \widehat{F}(a, b) \to \widehat{F}(D(k), b))_k$, and $(\alpha_k, 1_b) \colon (a, b) \to (D(k), b))_k$ is in $\mathcal{A} \otimes \mathcal{B}$.

(2)$\Rightarrow$(1): Consider an $(\alpha_k \colon a \to D(k))_k$ from $\mathcal{A}$ and any of the corresponding $(\alpha_k, 1_b) \colon (a, b) \to (D(k), b))_k$ in $\mathcal{A} \otimes \mathcal{B}$. Now we have $(\widehat{F}(\alpha_k, 1_b) \colon \widehat{F}(a, b) \to \widehat{F}(D(k), b))_k = (F(\alpha_k)(b) \colon F(a)(b) \to F(D(k))(b))_k$ limits by 1.5.1 since $(F(\alpha_k) \colon F(a) \to F(D(k))_k$ is a limit. If $\mathfrak{b} = (\beta_k \colon b \to D(k))_k$ is in $\mathcal{B}$, $(\widehat{F}(1_a, \beta_k) \colon \widehat{F}(a, b) \to \widehat{F}(a, D(k))_k$ is a limit since $F(a)$ sends $\mathcal{B}$ to a limit. $\qquad\square$

4.3. Corollary $(\mathfrak{C}^{\mathbb{B}})^{\mathbb{A}} \cong (\mathfrak{C}^{\mathbb{A}})^{\mathbb{B}}$.

(Indeed, we have $(\mathfrak{C}^{\mathbb{B}})^{\mathbb{A}} \cong \mathfrak{C}^{\mathbb{B} \times \mathbb{A}} \cong \mathfrak{C}^{\mathbb{A} \times \mathbb{B}} \cong (\mathfrak{C}^{\mathbb{A}})^{\mathbb{B}}$ where the middle isomorphism follows from the obvious correspondence $F'(\alpha, \beta) = F(\beta, \alpha)$.)

4.3.1. A simple example Ordered semigroups can be studied as semigroups in the category of posets (that is, assuming the operation monotone), or as an order in the category of semigroups (that is, the relations $\leq$ are subsemigroups of the semigroups $S \times S$).

4.4. Using the representations from Section 3 we obtain

4.4.1. Corollary For any Gabriel–Ulmer theory $\mathbb{A}$,

$$LFuzz_w^{\mathbb{A}} \cong (\mathbf{Set}^{\mathbb{A}})^{\mathbb{F}w} \quad \text{and} \quad LFuzz^{\mathbb{A}} \cong (\mathbf{Set}^{\mathbb{A}})^{\mathbb{F}}.$$

(Roughly speaking, treating a theory $\mathbb{A}$ in the context of fuzzy sets is the same as taking the standard crisp models of the theory and fuzzifying them afterwards.)

BIBLIOGRAPHY

[AR97]   J. Adámek and J. Rosický. *Locally Presentable and Accessible Categories*. Cambridge University Press, 1997.

[Gog67] J. A. Goguen. *L*-fuzzy sets. *J. Math. Anal. Appl.*, 18:145–174, 1967.

[GU71]   P. Gabriel and F. Ulmer. *Local Präsentierbare Kategorien*, volume 221 of *Lecture Notes in Mathematics*. Springer, Berlin, 1971.

[Isb72] J. R. Isbell. General functorial semantics I. *Amer. J. Math.*, 94:535–596, 1972.

[Law63] F. W. Lawvere. *Functorial Semantics of Algebraic Theories*. PhD thesis, Columbia University, 1963.

[ML71]   S. Mac Lane. *Categories for the Working Mathematician*. Number 5 in Graduate Texts in Mathematics. Springer, 1971.

[Pul70] A. Pultr. The right adjoints into the categories of relational systems. In *Reports of the Midwest Category Seminar IV*, volume 137 of *Lecture Notes in Mathematics*, pages 100–113. Springer, Berlin, 1970.

[Pul76] A. Pultr. Fuzzy mappings and fuzzy sets. *Comm. Math. Univ. Carolinae*, 17(3):441–459, 1976.

[Ulm68] F. Ulmer. Properties of dense and relative adjoint functors. *J. Algebra*, 8:77–95, 1968.

[Zad65] L. A. Zadeh. Fuzzy sets. *Information and Control*, 8:338–353, 1965.

Aleš Pultr
Department of Applied Mathematics and ITI
Faculty of Mathematics and Physics, Charles University in Prague
CZ 11800 Praha 1, Malostranské nám. 25
Email: pultr@kam.ms.mff.cuni.cz

Petr Hájek, BL logic and BL-algebras

FRANCO MONTAGNA

Dedicated to Petr Hájek
on the occasion of his 70th birthday

1   Foreword

If you look for Petr Hájek in Wikipedia, you find the following:

> Petr Hájek, born 6 February 1940 in Prague (CZ) is a Czech scientist in the area of mathematical logic and a professor of mathematics. He works at the Institute of Computer Science at the Academy of Sciences of the Czech Republic and worked as a lecturer at the Faculty of Mathematics and Physics at the Charles University in Prague and at the Faculty of Nuclear Sciences and Physical Engineering of the Czech Technical University in Prague, now just at the last one.

> Petr Hájek studied at the Faculty of Mathematics and Physics of the Charles University in Prague and afterwards at the Academy of Performing Arts in Prague as well. He specialized in set theory, arithmetic, later also in logic and artificial intelligence. He contributed to establishing the mathematical fundamentals of fuzzy logic. During the communist totality in Czechoslovakia he refused to cooperate with the State Security. Following the velvet revolution he was appointed a senior lecturer (1993), and a professor (1997). From 1992 to 2000 he held the position of chairman of the Institute of Computer Science at the Academy of Sciences of the Czech Republic. From 1996 to 2003 he was also president of the Kurt Gödel Society. In 2006, he was awarded the Medal for Merits in the area of sciences by President of the Czech Republic.

This information tells you that Petr Hájek is a great scientist, a very active man and a democratic person. This is absolutely true, but I would like to add that he is also a good heartened man, and that if you behave well, you may easily become a friend of him. The best comment I ever heard about his personality is due to Matthias Baaz, who, in occasion of Petr's 65th birthday, said the following seemingly trivial, but deep sentence: *Petr Hájek is always Petr Hájek*. Let me illustrate this sentence with a few examples, taken from my experience with him. During the communist regime, Petr Hájek was rather isolated, not rich and not powerful at all. Nevertheless, he could manage to invite me in Czech Republic, to find a house for my family and even to invite us to his home. He was always generous, in spite of his problems, but never slavish with me.

Then, in the nineties, after the velvet revolution, I visited him again in Prague. Things had changed very much, one could breath a free air, and Petr had become an important person (for instance, he was the chairman of his Department). However, his behavior with me didn't change: he remained a good friend, and he was always willing to invite me at his home (and to let me sit in the chair which years ago was occupied by Alfred Tarski, a great honor for me).

Petr was also always willing to help me in case of need. I remember that one day I broke my pants (don't tell this to my students!), and that Petr took me to his home and his wife Marie patiently could sew them in such a way that they became as new.

But there is another important thing that I wish to point out: after the velvet revolution Petr changed his official position, but not his (rather modest) style of life. It is not fair to investigate the reasons for this choice, which belong to his private life, but in my feeling, this only means that he did not use his power for himself, but rather for young researchers and for promotion science. In other words, *Petr Hájek remained Petr Hájek.*

2   My scientific collaboration with Petr Hájek

I am not in a position to describe completely the scientific activity of Petr Hájek, because the scope of his interests was (and still is) very wide. But let me just mention some of them, in which I was involved, directly or indirectly. When I was young, I became interested in Set Theory: I was charmed by the undecidable problems like the Continuum Hypothesis or the Souslin Hypothesis. At that time, there was a big wave of independence results, using Cohen's forcing, but the method was already well consolidated, and it was hard to tell something new in this field. Looking for new ideas, I found in the library of my Department a book by Hájek and Vopěnka, *The Theory of Semisets*, which contained a new and revolutionary idea (namely, any property determines a class, but a subclass of a set need not be a set), and which constituted a very promising approach to Non-Standard Analysis. But unfortunately I was not able to develop this very interesting idea.

Then, due both to my interest in Gödel theorems and to the law *publish or perish*, I changed my topic, and I moved to the metamathematics of arithmetic. I published several papers on the logic of provability, but I was also attracted by interpretability, a very powerful tool to prove the undecidability of axiomatic theories.

Somewhat surprisingly, when looking at the Mathematical Reviews, I found several papers by Hájek on this field, for instance, *On Interpretability in Theories Containing Arithmetic*, with Marie Hájková. Then, I decided to invite him to Siena, and our collaboration began. Already at that time, Petr was also studying Fuzzy Logic. I remember that when he visited Siena he was invited to Naples by Gianni Gerla and Antonio Di Nola, who, together with Daniele Mundici, were the Italian pioneers of Fuzzy Logic. During Petr's stay in Siena, we wrote two joint research papers on interpretability and Π_1 conservativity, thus generalizing an important result by Alessandro Berarducci.

In the early nineties, Petr Hájek, together with Pavel Pudlák, finished his monography *Metamathematics of First Order Arithmetic*, a really deep book ranging from Peano Arithmetic to weak fragments.

In the subsequent years, we still met each other, but our research fields diverged for a while. Nevertheless, although I was not yet interested in Fuzzy Logic, I was attracted by his paper [HPS00] in collaboration with Jeff Paris and John Sheperdson, about the problem of adding a truth predicate to Peano Arithmetic built in Fuzzy Logic instead of in Classical Logic.

Then, in 1998, our scientific interests met again. In spring 1998, during a meeting in Pontignano, Petr presented his Basic Fuzzy Logic, and I became so interested in this subject that I decided to start doing research in Fuzzy Logic. Then, in summer 1998 Petr presented his Basic Logic at the Logic Colloquium in Prague, and I delivered my first talk in Fuzzy Logic, together with Francesc Esteva. The rest of the story of my collaboration with Petr is made of scientific results and will be presented in the next sections.

3 BL-algebras

One of the most important contributions of Petr Hájek is constituted by the Basic Fuzzy Logic BL and its algebraic models, the BL-algebras. The logic BL is a common fragment of the three main Fuzzy Logics existing at that time, namely, Łukasiewicz logic, Gödel logic and product logic.

DEFINITION 1 *The logic BL is formulated in a propositional language with the binary connectives & and $\rightarrow$ and with a propositional constant 0. BL is axiomatized as follows:*

(BL1) $(\varphi \rightarrow \psi) \rightarrow ((\psi \rightarrow \gamma) \rightarrow (\varphi \rightarrow \gamma))$
(BL2) $\varphi \,\&\, \psi \rightarrow \varphi$
(BL3) $\varphi \,\&\, \psi \rightarrow \psi \,\&\, \varphi$
(BL4) $\varphi \,\&\, (\varphi \rightarrow \psi) \rightarrow \psi \,\&\, (\psi \rightarrow \varphi)$
(BL5) $(\varphi \rightarrow (\psi \rightarrow \gamma)) \rightarrow (\varphi \,\&\, \psi \rightarrow \gamma)$
(BL6) $(\varphi \,\&\, \psi \rightarrow \gamma) \rightarrow (\varphi \rightarrow (\psi \rightarrow \gamma))$
(BL7) $((\varphi \rightarrow \psi) \rightarrow \gamma) \rightarrow (((\psi \rightarrow \varphi) \rightarrow \gamma) \rightarrow \gamma)$
(BL8) $0 \rightarrow \varphi$

The only rule of BL is Modus Ponens:

$$(MP) \quad \frac{\varphi \quad \varphi \rightarrow \psi}{\psi}.$$

We will write $\neg\varphi$ for $\varphi \to 0$, $\varphi \leftrightarrow \psi$ for $(\varphi \to \psi) \,\&\, (\psi \to \varphi)$, $\varphi \wedge \psi$ for $\varphi \,\&\, (\varphi \to \psi)$, $\varphi \vee \psi$ for $((\varphi \to \psi) \to \psi) \wedge ((\psi \to \varphi) \to \varphi)$ and 1 for $0 \to 0$.

The equivalent algebraic semantics for Basic Logic is constituted by the BL-algebras.

DEFINITION 2 *A* BL-algebra *is an algebra* $(A, \&, \to, \vee, \wedge, 0, 1)$ *such that:*

- $(A, \&, 1)$ *is a commutative monoid.*

- $(A, \vee, \wedge, 0, 1)$ *is a bounded lattice with minimum* 0 *and maximum* 1.

- $\to$ *is a binary operation where the following* residuation *property holds with respect to the order* $\leq$ *induced by the lattice operations:*

  (RP) $x \,\&\, y \leq z$ iff $x \leq y \to z$

- *The following divisibility property holds:*

  (DIV) $x \,\&\, (x \to y) = x \wedge y.$

- *The following prelinearity property holds:*

  (PREL) $(x \to y) \vee (y \to x) = 1$

An MV-algebra is a BL-algebra satisfying the equation $\neg\neg x = x$, *where* $\neg x = x \to 0$. *A Gödel algebra is a BL-algebra satisfying the equation* $x \,\&\, x = x$. *Finally, a product algebra is a BL-algebra satisfying the equation* $\neg x \vee ((x \to (x \,\&\, y)) \to y) = 1$.

BL-algebras constitute a complete algebraic semantics for the logic BL. This means the following: let $\mathbf{A}$ be any BL-algebra. By the name *valuation* into $\mathbf{A}$ we mean a homomorphism from the algebra of formulas into $\mathbf{A}$. Then, the following *strong completeness theorem* holds:

PROPOSITION 3 *Let* Γ *be a set of BL-formulas and* φ *be a BL-formula. The following are equivalent:*

(1) $\Gamma \vdash_{BL} \varphi$, *that is, there is a derivation of* φ *from* Γ *in BL.*

(2) $\Gamma \models_{BL} \varphi$, *that is for every BL-algebra* $\mathbf{A}$ *and for every valuation* v *into* $\mathbf{A}$, *if* $v(\psi) = 1$ *for every* $\psi \in \Gamma$, *then* $v(\varphi) = 1$.

(3) $\Gamma \models_{BL_{lin}} \varphi$, *that is for every totally ordered BL-algebra* $\mathbf{A}$ *and for every valuation* v *into* $\mathbf{A}$, *if* $v(\psi) = 1$ *for every* $\psi \in \Gamma$, *then* $v(\varphi) = 1$.

An important class of BL-algebras is constituted by the *standard* BL-algebras. In order to introduce them, we start from the following definition.

DEFINITION 4 *A* continuous t-norm *is a continuous binary operation* $*$ *on* $[0,1]$ *such that* $([0,1], *, 1, \leq)$ *is an ordered commutative monoid, that is, a commutative monoid such that if* $x \leq z$ *and* $y \leq u$, *then* $x * y \leq z * u$. *Given a continuous t-norm* $*$, *define*

$$x \rightarrow_* y = \sup \{z : z * x \leq y\}.$$

In [Háj98b] it is proved that $([0,1], *, \rightarrow_*, \max, \min, 0, 1)$ is a BL-algebra iff $*$ is a continuous t-norm. Such a BL-algebra will be called a *standard BL-algebra* or a *t-algebra*. The class of all t-algebras will be denoted by $\mathcal{STBL}$.

There is another form of completeness for BL, namely the *standard completeness*. For this theorem we need to assume that the set Γ is finite.

PROPOSITION 5 (Standard completeness theorem [CEGT00]) *Let* Γ *be a finite set of BL-formulas and* φ *be a BL formula. The following are equivalent:*

(1) $\Gamma \vdash_{BL} \varphi$, *that is, there is a derivation of* φ *from* Γ *in BL.*

(2) $\Gamma \models_{\mathcal{STBL}} \varphi$, *that is for every standard BL-algebra* **A** *and for every valuation* v *into* **A**, *if* $v(\psi) = 1$ *for every* $\psi \in \Gamma$, *then* $v(\varphi) = 1$.

The standard completeness was proved by Cignoli, Esteva, Godo and Torrens in [CEGT00]. However, Hájek in [Háj98a] was rather close to prove the result. Let us say that a triple x, y, z in a BL-chain is *pathological* iff $x < y < z$, $x \,\&\, z = x$, $x \,\&\, y < x$ and $y \,\&\, z < y$. Then Hájek proved the following:

PROPOSITION 6

(1) *In any standard BL-algebra there is no pathological triple.*

(2) *There is a finitely axiomatizable subvariety of BL-algebras whose totally ordered models are precisely the BL-chains without pathological triples.*

(3) *There is a finitely axiomatizable extension BL* of BL, obtained by the adding of axioms which forbid pathological triples, which is sound and complete with respect to the class of standard BL-algebras.*

Cignoli, Esteva, Godo and Torrens showed that in any BL-chain pathological triples do not exist, and hence the extension BL* introduced by Hájek is just BL. It follows that BL is sound and complete with respect to the class $\mathcal{STBL}$ of standard BL-algebras. We may comment on this saying that among many very important things, Petr Hájek also invented something which does not exist, namely, the pathological triples.

When trying to prove the standard completeness of BL, Hájek made an important step towards a decomposition theorem for BL-chains along the lines of a famous result, due to Mostert and Shields [MS57]. The latter result roughly says that a continuous t-norm can be decomposed into a number of components such that each component is isomorphic either to the Lukasiewicz t-norm $x *_L y = \max\{x + y - 1, 0\}$ or to the product t-norm $x *_\Pi y = xy$. Moreover, outside these components the t-norm behaves as the Gödel t-norm $x *_G y = \min\{x, y\}$. More precisely:

PROPOSITION 7 *Given a continuous t-norm* $*$, *there is a family of mutually disjoint open subintervals* $(a_i, b_i) : i \in I$ *of* $[0, 1]$ *such that:*

(1) *In the closure* $[a_i, b_i]$ *of each interval, the t-norm* $*$ *is isomorphic to* $*_L$ *or to* $*_\Pi$. *(This means: there is an increasing bijection h from* $[0, 1]$ *onto* $[a_i, b_i]$ *such that either for all* $x, y \in [a_i, b_i]$, $x * y = h(h^{-1}(x) *_L h^{-1}(y))$ *or for all* $x, y \in [a_i, b_i]$, $x * y = h(h^{-1}(x) *_\Pi h^{-1}(y))$).

(2) *If there is no interval* $[a_i, b_i]$ *such that* $x, y \in [a_i, b_i]$, *then* $x * y = \min\{x, y\}$, *that is,* $*$ *behaves as the Gödel t-norm.*

Note that the implication $\rightarrow_*$ associated to $*$ is as follows:

(a) If $x \leq y$, then $x \rightarrow y = 1$.

(b) If $x > y$ and x, y are in the same interval $[a_i, b_i]$, then:

(b1) If in $[a_i, b_i]$ the t-norm $*$ is isomorphic to $*_L$ via the isomorphism h, then $x \rightarrow_* y = h(h^{-1}(x) \rightarrow_L h^{-1}(y))$, where $\rightarrow_L$ is the Lukasiewicz implication defined by $x \rightarrow_L y = \min\{1 - x + y, 1\}$.

(b2) If in $[a_i, b_i]$ the t-norm $*$ is isomorphic to the product t-norm $*_\Pi$ via the isomorphism h, then $x \rightarrow_* y = h(h^{-1}(x) \rightarrow_\Pi h^{-1}(y))$, where $\rightarrow_\Pi$ is the product implication defined as follows: if $x > y$, then $x \rightarrow_\Pi y = \frac{y}{x}$ and if $x \leq y$, then $x \rightarrow_\Pi y = 1$.

(c) If $x > y$ and there is no interval $[a_i, b_i]$ such that $x, y \in [a_i, b_i]$, then $x \rightarrow_* y = y$.

REMARK 8 *It can be shown (cf. [Háj98b]) that the variety of BL-algebras generated by* $([0, 1], *_L, \rightarrow_L, \max, \min, 0, 1)$, $(([0, 1], *_G, \rightarrow_G, \max, \min, 0, 1)$, $([0, 1], *_\Pi, \rightarrow_\Pi, \max, \min, 0, 1)$ *respectively) is the class of MV-algebras (Gödel algebras, product algebras respectively).*

The Mostert and Shields theorem is very important, because it allows to decompose arbitrary continuous t-norm into simpler ones. Hence, one may wonder if it is possible to prove a similar result for arbitrary BL-chains. In his paper [Háj98a], Hájek does an important step in this direction.

Let us say that a *cut* of a BL-chain $\mathbf{A}$ is a pair (X, Y) of non-empty subsets of $\mathbf{A}$ such that $\mathbf{A} = X \cup Y$, Y is closed under $\&$ and for all $x \in X$ and $y \in Y$, $x \& y = x$. Note that if (X, Y) is a cut of $\mathbf{A}$, then for all $x \in X$ and $y \in Y$, we have $x = x \& y \leq 1 \& y = y$. Hence, X is downward closed and hence it is also closed under $\&$.

We say that a BL-chain $\mathbf{A}$ is *saturated* iff for every cut (X, Y) of $\mathbf{A}$ there is $d \in \mathbf{A}$ such that for all $x \in X$ and $y \in Y$ we have $x \leq d \leq y$.

Suppose that (X, Y) is a cut of a saturated BL-chain $\mathbf{A}$ and that X and Y have at least two elements. Let d be an idempotent such that, for all $x \in X$ and $y \in Y$, $x \leq d \leq y$. Then $X \cup \{d\}$ and $Y \cup \{d\}$ are the domains of two BL-chains $\mathbf{X}$ and $\mathbf{Y}$ defined as follows:

(a) the minimum of $\mathbf{X}$ is 0 and the maximum of $\mathbf{X}$ is d;

(b) the minimum of $\mathbf{Y}$ is d and its maximum is 1;

(c) the monoid and lattice operations $\&_X, \vee_X, \wedge_X$ and $\&_Y, \vee_Y, \wedge_Y$ on $\mathbf{X}$ and $\mathbf{Y}$ respectively are the restrictions of the monoid operation $\&$ and of the lattice operations $\vee$ and $\wedge$ on $\mathbf{A}$.

(d) The implication $\rightarrow_Y$ on $\mathbf{Y}$ is the restriction of the implication $\rightarrow$ on $\mathbf{A}$, and the implication $\rightarrow_X$ on $\mathbf{X}$ is defined by $x \rightarrow_X y = d$ if $x \leq y$ and $x \rightarrow_X y = x \rightarrow y$ otherwise.

We then say that $\mathbf{A}$ is the ordinal sum of $\mathbf{X}$ and $\mathbf{Y}$ and we write $\mathbf{A} = \mathbf{X} \oplus \mathbf{Y}$. Conversely, given two BL-chains $\mathbf{A}_1$ and $\mathbf{A}_2$ such that $\mathbf{A}_1 \cap \mathbf{A}_2$ consists of a single element which is the maximum of $\mathbf{A}_1$ and the minimum of $\mathbf{A}_2$, the *ordinal sum* $\mathbf{A}_1 \oplus \mathbf{A}_2$ is the algebra $\mathbf{A}$ defined as follows:

(1) The domain of $\mathbf{A}$ is the union of the domains of $\mathbf{A}_1$ and $\mathbf{A}_2$.

(2) The order of $\mathbf{A}$ restricted to $\mathbf{A}_1$ ($\mathbf{A}_2$ respectively) coincides with the order of $\mathbf{A}_1$ ($\mathbf{A}_2$ respectively). Moreover every element of $\mathbf{A}_1$ precedes every element of $\mathbf{A}_2$. In particular, the minimum of $\mathbf{A}$ is the minimum of $\mathbf{A}_1$ and the maximum of $\mathbf{A}$ is the maximum of $\mathbf{A}_2$.

(3) The lattice operations are uniquely determined by the (total) order $\mathbf{A}$ and the restrictions of the monoid operation $\&$ to $\mathbf{A}_1$ ($\mathbf{A}_2$ respectively) coincide with the monoid operation of $\mathbf{A}_1$ ($\mathbf{A}_2$ respectively). Moreover, if $a_1 \in \mathbf{A}_1$ and $a_2 \in \mathbf{A}_2$, then $a_1 \& a_2 = a_2 \& a_1 = a_1$.

(4) If $a \leq b$, then $a \rightarrow b = 1$. If $a > b$ and $a, b \in \mathbf{A}_1$ ($\mathbf{A}_2$ respectively), then $a \rightarrow b = a \rightarrow_1 b$ ($a \rightarrow b = a \rightarrow_2 b$ respectively), where $\rightarrow_1$ and $\rightarrow_2$ denote the implications in $\mathbf{A}_1$ and in $\mathbf{A}_2$ respectively. If $a > b$ and $a \in \mathbf{A}_2$ and $b \in \mathbf{A}_1$, then $a \rightarrow b = b$.

In a similar way, one can define the ordinal sum $\bigoplus_{i \in I} \mathbf{A}_i$ of a (possibly infinite) family $\{\mathbf{A}_i : i \in I\}$ of BL-chains. Some caution is needed here: first of all, the index set I has to be totally ordered. Moreover, if I has a minimum i_0, then the minimum of the ordinal sum $\bigoplus_{i \in I} \mathbf{A}_i$ is the minimum of $\mathbf{A}_{i_0}$, otherwise we have to add a new minimum 0 which does not belong to any $\mathbf{A}_i$, and we must stipulate that $0 \ \& \ x = x \ \& \ 0 = 0$, $0 \to x = 1$ for every x, and if $x > 0$, then $x \to 0 = 0$.

Now let us say that a BL-chain is *irreducible* iff either it is a Gödel chain, i.e., it entirely consists of idempotent elements, or there is no cut (X, Y) such that both X and Y have at least two elements. (Warning: I have changed Petr's definition here, because otherwise I was not able to reconstruct his proof). Then Hájek in [Háj98a] proved the following:

THEOREM 9 *Every saturated BL chain is the ordinal sum of an ordered family of irreducible saturated BL-chains.*

In order to derive an analogue of the Mostert and Shields theorem, one should prove that every irreducible BL-chain is either an MV-chain, or a Gödel chain or a product chain. Moreover, it would be nice to find a general formulation in which it is not assumed that the algebra is saturated. To this purpose, it is convenient to change a little bit the definition of ordinal sum in order to adapt it to the case where the BL-chain is not saturated. The main differences are:

(1) In the new definition, the components, except from the lowest one, need not have a minimum, and hence they are only assumed to be subreducts (subalgebras of reducts) of BL-algebras in the language with $\&, \to, 1$ and without 0 (note that lattice operations are definable in terms of $\&$ and $\to$ and hence it is unnecessary to assume that they are in the language). Such subreducts are usually called *BL-hoops*.

(2) In the new definition, we assume that any two different components in the ordinal sum have in common only the top element 1.

In this way, we arrive to the following definition.

DEFINITION 10 *Let I be a totally ordered set and let for $i \in I$, $\mathbf{A}_i$ be a totally ordered BL-hoop. Suppose that if $i, j \in I$ and $i \neq j$, then $\mathbf{A}_i \cap \mathbf{A}_j = \{1\}$. The ordinal sum (in the new sense) $\bigoplus_{i \in I} \mathbf{A}_i$ of the ordered family $\{\mathbf{A}_i : i \in I\}$ is the algebra $\mathbf{A}$ defined as follows:*

(a) The domain of $\mathbf{A}$ is the union of the domains of all $\mathbf{A}_i : i \in I$.

(b) The order $\leq$ of $\mathbf{A}$ is as follows: the top element is 1. Moreover if $i < j$, then every element of $\mathbf{A}_i \setminus \{1\}$ precedes every element of $\mathbf{A}_j$. Finally, the restriction of $\leq$ to a component $\mathbf{A}_i$ coincides with the order in $\mathbf{A}_i$.

(c) *The restriction of the monoid operation* & *to* $\mathbf{A}_i$ *is the monoid operation* $\&_i$ *of* $\mathbf{A}_i$. *Moreover, if* $x \in \mathbf{A}_i \backslash \{1\}$ *and* $y \in \mathbf{A}_j$ *with* $i < j$, *then* $x \& y = y \& x = x$.

(d) *The restriction of the implication* $\rightarrow$ *to* $\mathbf{A}_i$ *is the implication* $\rightarrow_i$ *of* $\mathbf{A}_i$. *Moreover, if* $x \leq y$, *then* $x \rightarrow y = 1$ *and if* $x \in \mathbf{A}_j$ *and* $y \in \mathbf{A}_i$ *with* $i < j$, *then* $x \rightarrow y = y$.

The ordinal sum $\bigoplus_{i \in I} \mathbf{A}_i$ (in the new sense) of an ordered family of totally ordered BL-hoops is a totally ordered BL-hoop. However, if I has a minimum i_0 and $\mathbf{A}_{i_0}$ is a BL-algebra, then $\bigoplus_{i \in I} \mathbf{A}_i$ becomes a BL-algebra if we interpret 0 as the minimum of $\mathbf{A}_{i_0}$.

A BL-hoop (BL-algebra respectively) is said to be *sum-irreducible* if it cannot be decomposed as an ordinal sum (in the new sense) of two BL-hoops (a BL-algebra and a BL-hoop respectively) with at least two elements. In [AM03] we proved the following:

THEOREM 11   *(1) Every totally ordered BL-hoop or BL-algebra can be decomposed into an ordinal sum of sum-irreducible BL-hoops (in the case of BL-chains, the lowest hoop is a BL-algebra).*

(2) *Every totally ordered sum-irreducible BL-hoop is either the reduct of an MV-chain or a product chain deprived of its minimum.*

The main steps toward the proof of (2) are:

(a) If $\mathbf{A}$ is a totally ordered BL-hoop or BL-algebra, and there are elements $a, b \in \mathbf{A} \backslash \{1\}$ such that $a \rightarrow b = b$, then $\mathbf{A}$ is not sum-irreducible.

(b) If $\mathbf{A}$ is a totally ordered BL-hoop or BL-algebra, and there are no $a, b \in \mathbf{A} \backslash \{1\}$ such that $a \rightarrow b = b$, then $\mathbf{A}$ is either an MV-chain, or the reduct of an MV-chain or a product chain deprived of its minimum.

From Theorem 11, we can derive the following strong generalization of the Mostert and Shields theorem.

THEOREM 12 ([AM03])   *(1) Every totally ordered BL-hoop* $\mathbf{A}$ *is the ordinal sum (in the new sense) of an ordered family of BL-hoops* $\mathbf{A}_i : i \in I$ *which are either the reduct of an MV-chain or a product chain deprived of its minimum. Such BL-hoops* $\mathbf{A}_i$ *are called the* components *of* $\mathbf{A}$.

(2) *Every BL-chain is an ordinal sum of a family of components as in (1), with the difference that* I *has minimum* i_0 *and* $\mathbf{A}_{i_0}$ *is an MV-algebra.*

REMARK 13 *In this setting, Gödel components disappear. A Gödel chain with more than two elements is the ordinal sum of an ordered family of copies of (reducts of) the two element MV-chain. Moreover a product chain is not sum-irreducible: indeed, it is the ordinal sum of the two element MV-algebra and the product chain itself deprived of its minimum.*

4   Other common research topics

BL-algebras constitute my most important area of research and of course the influence of Petr Hájek was fundamental for them. But even if we restrict to Fuzzy Logic, there are other important joint research subjects. An interesting one is constituted by first-order fuzzy logics. Although we do not give a complete treatment, in this section we will present some scattered results which in our opinion are of considerable interest. For simplicity, we will restrict ourselves to predicate logics over extensions of BL.

DEFINITION 14 *Let $\mathbf{L} = (L, \&, \rightarrow, \max, \min, 0, 1)$ be a BL-chain. A first order structure over $\mathbf{L}$ or simply an $\mathbf{L}$-structure is a pair $\mathbf{M} = (M, {}^M)$ where M is a non-empty set, called* the domain of the interpretation, *and ^{M} is a map which associates to each constant c an element c^M of M, to each n-ary function symbol f a function f^M from M^n into M, and to every predicate P of arity n a map P^M from M^n into $\mathbf{L}$.*
An assignment *is a map v from the set of individual variables into M. The interpretation of a term $t^{v,M}$ in $\mathbf{M}$ under the assignment v is defined as usual in the classical models.*

DEFINITION 15 *Given a formula φ, the* truth value *$||\varphi||^{\mathbf{L}}_{\mathbf{M},v}$ of φ in the $\mathbf{L}$-structure $\mathbf{M}$ under the assignment v is defined by induction as follows:*

(1) *If $\varphi = P(t_1, ..., t_n)$ is atomic, then $||\varphi||^{\mathbf{L}}_{\mathbf{M},v} = P^M(t_1^{v,M}, ..., t_n^{v,M})$.*

(2) *$||...||^{\mathbf{L}}_{\mathbf{M},v}$ commutes with $\&$ and with $\rightarrow$.*

(3) *Let $A_{v,x}$ be the set of all assignments which coincide with v on all variables different from x. Then $||\exists x\psi||^{\mathbf{L}}_{\mathbf{M},v} = \sup_{w \in A_{v,x}} ||\psi||^{\mathbf{L}}_{\mathbf{M},w}$ if such a supremum exists in $\mathbf{L}$, and is undefined otherwise.*

(4) *$||\forall x\psi||^{\mathbf{L}}_{\mathbf{M},v} = \inf_{w \in A_{v,x}} ||\psi||^{\mathbf{L}}_{\mathbf{M},w}$ if such an infimum exists in $\mathbf{L}$, and is undefined otherwise.*

DEFINITION 16 *Let $\mathcal{K}$ be a class of BL-chains. A $\mathcal{K}$-interpretation is a triple $(\mathbf{L}, \mathbf{M}, v)$ where $\mathbf{L} \in \mathcal{K}$, $\mathbf{M}$ is an $\mathbf{L}$-structure and v is an assignment in $\mathbf{M}$. A $\mathcal{K}$-interpretation $(\mathbf{L}, \mathbf{M}, v)$ is* safe *iff $||\varphi||^{\mathbf{L}}_{\mathbf{M},v}$ is defined for every formula φ.*

A formula φ is $\mathcal{K}$-1-satisfiable if there is a safe $\mathcal{K}$-interpretation $(\mathbf{L}, \mathbf{M}, v)$ *such that* $\|\varphi\|_{\mathbf{M},v}^{\mathbf{L}} = 1$. *A formula φ is a $\mathcal{K}$-1-tautology iff for every safe $\mathcal{K}$-interpretation* $(\mathbf{L}, \mathbf{M}, v)$ *we have* $\|\varphi\|_{\mathbf{M},v}^{\mathbf{L}} = 1$.

Finally, given a set Γ of formulas and a formula φ, we say that φ is a semantic consequence *of Γ in $\mathcal{K}$, abbreviated as $\Gamma \models_{\mathcal{K}} \varphi$ iff for every safe $\mathcal{K}$-interpretation* $(\mathbf{L}, \mathbf{M}, v)$, *if* $\|\psi\|_{\mathbf{M},v}^{\mathbf{L}} = 1$ *for all $\psi \in \Gamma$, then* $\|\varphi\|_{\mathbf{M},v}^{\mathbf{L}} = 1$.

DEFINITION 17 *Let L be a schematic extension of BL. The logic $L\forall$ is defined as follows:*

(1) *The language of L is the language of first order predicate logic with connectives & and $\rightarrow$ and with the propositional constant 0.*

(2) *The axioms of $L\forall$ are:*

    (1) *All axiom schemas of L, extended to all first-order formulas.*

    (2) *The schema $\forall x\psi \rightarrow \psi(x/t)$, t substitutable for x in ψ.*

    (3) *The schema $\psi(x/t) \rightarrow \exists x\psi$, t substitutable for x in ψ.*

    (4) *The schema $\forall x(\psi \rightarrow \varphi) \rightarrow (\psi \rightarrow \forall x\varphi)$ where x is not free in ψ.*

    (5) *The schema $\forall x(\psi \rightarrow \varphi) \rightarrow (\exists x\psi \rightarrow \varphi)$, where x is not free in φ.*

    (6) *The schema $\forall x(\psi \vee \varphi) \rightarrow ((\forall x\psi) \vee \varphi)$, where x is not free in φ.*

(3) *The rules of $L\forall$ are Modus Ponens:* $\dfrac{\varphi \quad \varphi \rightarrow \psi}{\psi}$ *and Gen:* $\dfrac{\varphi}{\forall x\varphi}$.

In [Háj98b], the following is shown:

THEOREM 18 *Let $\mathcal{K}$ be the class of all BL-chains such that every theorem of L is a $\mathcal{K}$-1-tautology. Then, for every set Γ of formulas and for every formula φ we have $\Gamma \vdash_{L\forall} \varphi$ iff $\Gamma \models_{\mathcal{K}} \varphi$.*

During a visit to Siena in 2000, Petr showed me a trick to prove that the set of first order formulas which are 1-satisfiable in the standard product algebra $([0,1], *_{\Pi}, \rightarrow_{\Pi}, \max, \min, 0)$ is not arithmetical. The idea is that given any first-order formula φ of arithmetic we can obtain a BL$\forall$ formula which is a combination of the axioms of a finitely axiomatizable theory of arithmetic and another formula which implies that for every n, the truth value $\|P(n+1)\|_{\mathbf{M},v}^{\mathbf{L}}$ of $P(n+1)$ (P a unary predicate) is the square of $\|\forall x \leq nP(x)\|_{\mathbf{M},v}^{\mathbf{L}}$. Then if $0 < \|P(0)\|_{\mathbf{M},v}^{\mathbf{L}} < 1$ we have that the standard natural numbers are precisely those elements a for which $\|P(a)\|_{\mathbf{M},v}^{\mathbf{L}}$ is strictly positive. This allows us to obtain a BL$\forall$ formula φ^* such that φ^* is satisfiable in the standard product algebra iff φ is true in the standard model of natural numbers.

Elaborating this idea, I was able to prove the non-arithmeticity of the 1-tautologies in the standard product algebra and of 1-tautologies in the class of all standard BL-algebras [Mon01].

A few years later, in 2004, I met Petr in Vienna. There, he told me a nice counterexample by Felix Bou, in which it was shown that in a complete BL-chain the monoid operation need not distribute over infinite infima, that is, it is not true that $x \,\&\, (\bigwedge_{i \in I} x_i) = \bigwedge_{i \in I}(x \,\&\, x_i)$. The counterexample is as follows: take the ordinal sum (in the new sense) of the three-element MV-chain and the negative cone of the integers. If I is the set of natural numbers, if $x_i = -i$ and x is the coatom of the three element of the MV-chain, then $x \,\&\, (\bigwedge_{i \in I} x_i) = x \,\&\, x = 0$, and $\bigwedge_{i \in I}(x \,\&\, x_i) = x > 0$.

From this fact it follows that the formula $\forall x(\varphi(x) \,\&\, \psi) \to (\forall x \varphi(x) \,\&\, \psi)$, where x has no free occurrence in ψ, is a not 1-tautology with respect to the class of all complete BL-chains, although it is a 1-tautology in the class of all standard BL-algebras.

Just after hearing this example, I realized that I had done an error in my paper [MS03] with Lorenzo Sacchetti, in which we claimed that 1-tautologies for the class of all complete BL-chains and 1-tautologies for the class of all standard BL-algebra coincide. Just after I returned from Vienna, we immediately wrote to the journal trying to withdraw the paper, but it was too late, and hence we were forced to write a corrigendum [MS04].

A good scientist should learn from his errors, but I didn't. In a preliminary version of a joint paper with Cintula, Esteva, Godo, Gispert and Noguera, we claimed that the 1-tautologies with respect to the class of all BL-chains and the 1-tautologies with respect to all BL-chains over the rational interval $[0,1]$ coincide. Unfortunately, this is wrong: the formula $\forall x(\varphi(x) \,\&\, \psi) \to (\forall x \varphi(x) \,\&\, \psi)$ is not a 1-tautology with respect to the class of all complete BL-chains, but, as shown in Hájek's book [Háj98b], it is a 1-tautology with respect to the class of densely ordered BL-chains and hence for the class of BL-chains over the rational interval $[0,1]$. Fortunately, we discovered the error (George Metcalfe has pointed out to us the result from [Háj98b]).

Not happy with this, Petr pointed out to me my error in my paper with Sacchetti and proposed to me to investigate this topic further. As a result, Petr and I wrote a joint paper in which we proved the following theorem, which characterizes the complete BL-chains which are models of all standard 1-tautologies.

THEOREM 19 *The following are equivalent for any complete BL-chain* **B**:

(1) *For every family $\{x_i : i \in I\}$ of elements of* **B** *we have:* $(\bigwedge_{i \in I} x_i) \,\&\, (\bigwedge_{i \in I} x_i) = \bigwedge_{i \in I}(x_i \,\&\, x_i)$.

(2) *The formula $\forall x(\varphi(x) \,\&\, \varphi(x)) \to \forall x \varphi(x) \,\&\, \forall x \varphi(x)$ is a 1-tautology in* **B**.

(3) *Every 1-tautology for the class of standard BL-chains is a 1-tautology in* **B**.

Petr's work on first-order Fuzzy Logics is very wide. For instance, together with Petr Cintula he suggested a new kind of semantics for first-order Fuzzy Logics, namely, the semantics of witnessed models in which suprema and infima in the interpretation of quantifiers are actually reached, cf. [HC06]. Moreover, he treated not only 1-tautologies, but also positive tautologies and (1-/positively) satisfiable formulas, cf. [Háj05] for a survey.

Yet another research subject which was suggested to me and to Matteo Bianchi by ideas of Petr Hájek is the problem of supersoundness. Let L be a schematic extension of BL and let $\mathcal{K}$ be the class of all chains in which every theorem of L is valid. We say that L$\forall$ is *supersound* if every theorem φ of L$\forall$ has truth value 1 in every (possibly not safe) $\mathcal{K}$-interpretation $(\mathbf{L}, \mathbf{M}, v)$ such that $||\varphi||_{\mathbf{M},v}^{\mathbf{L}}$ is defined. Hájek and Sheperdson [HS01] proved the following:

THEOREM 20 *First-order Łukasiewicz logic, first-order product logic and first-order BL are not supersound. Moreover, Gödel logic is the only supersound logic which is complete with respect to its standard semantics on $[0,1]$.*

The proof contains an ingenious trick which has been extended by Matteo Bianchi and myself to a wider class of logics. Although the obtained results hold in a more general context of extensions of MTL$\forall$, we will present the result for extensions of BL$\forall$. First of all, note that every variety $\mathcal{V}$ of BL-algebras satisfies the following condition:

(*) for every chain $\mathbf{A} \in \mathcal{V}$ and for every $a, b, c \in \mathbf{A}$ with $a \leq b \leq c$, if $b \,\&\, c = b$, then $a \,\&\, c = a$.

Then Matteo Bianchi and I ([BM]) proved:

THEOREM 21 *Let $\mathcal{V}$ be any variety of BL-algebras. Let L be the logic corresponding to $\mathcal{V}$, that is, the logic whose theorems are precisely the formulas valid in $\mathcal{V}$, and let L$\forall$ be its first order extension. Let P be a unary predicate and C be a propositional constant. Let φ and ψ be the formulas $\forall x \exists y (P(x) \to (P(y) \,\&\, C))$ and $\exists y (P(y) \to (P(y) \,\&\, C))$ respectively. Then, $\varphi \vdash_L \psi$, and hence, for some n, $\vdash_L \varphi^n \to \psi$.*

Proof Let $(\mathbf{L}, \mathbf{M}, v)$ be a safe interpretation. Let us set $c = ||C||_{\mathbf{M},v}^{\mathbf{L}}$ and let $b = ||\exists x P(x)||_{\mathbf{M},v}^{\mathbf{L}}$ and assume $||\forall x \exists y (P(x) \to (P(y) \,\&\, C))||_{\mathbf{M},v}^{\mathbf{L}} = 1$. We claim that $b \leq c$ and that $b \,\&\, c = b$.

For all $m \in \mathbf{M}$, $\bigvee_{m' \in \mathbf{M}} ||P(m) \to (P(m') \,\&\, C)||_{\mathbf{M},v}^{\mathbf{L}} = 1$, and hence $||P(m)||_{\mathbf{M},v}^{\mathbf{L}} \to (b \,\&\, c) = 1$.

It follows that $||P(m)||_{\mathbf{M},v}^{\mathbf{L}} \leq b \,\&\, c$. Therefore, taking the supremum of all $||P(m)||_{\mathbf{M},v}^{\mathbf{L}}$ we get $b \leq b \,\&\, c$ and finally $b = b \,\&\, c$. By condition (*), for every $m \in \mathbf{M}$, $c \,\&\, ||P(m)||_{\mathbf{M},v}^{\mathbf{L}} = ||P(m)||_{\mathbf{M},v}^{\mathbf{L}}$, and $||\exists y (P(y) \to (P(y) \,\&\, C))||_{\mathbf{M},v}^{\mathbf{L}} = \bigvee_{m \in M} (||P(m)||_{\mathbf{M},v}^{\mathbf{L}} \to (c \,\&\, ||P(m)||_{\mathbf{M},v}^{\mathbf{L}})) = 1$. $\qquad\square$

THEOREM 22 *Suppose that a variety $\mathcal{V}$ of BL-algebras contains either an infinite product chain or the Chang MV-algebra. Then the consequence relation $\varphi \vdash \psi$ in Theorem 21 can be invalidated in a (non safe) interpretation in which the truth values of both formulas φ and ψ are defined.*

Proof Suppose that Chang's algebra $\mathbf{L}$ is in $\mathcal{V}$. Let ε be the atom of $\mathbf{L}$. Take M to be the set of natural numbers, and let for $m \in M$, $||P(m)||_{\mathbf{M},v}^{\mathbf{L}} = (m+1)\varepsilon$ and $||C||_{\mathbf{M},v}^{\mathbf{L}} = 1 - \varepsilon$. Then $||P(m+1)\,\&\,C||_{\mathbf{M},v}^{\mathbf{L}} = (m+1)\varepsilon$ and $||P(m) \to (P(m+1)\,\&\,C)||_{\mathbf{M},v}^{\mathbf{L}} = 1$. Moreover, $||P(m) \to (P(m)\,\&\,C)||_{\mathbf{M},v}^{\mathbf{L}} = 1 - \varepsilon$. Therefore, $||\exists y(P(y) \to (P(y)\,\&\,C))||_{\mathbf{M},v}^{\mathbf{L}} = 1 - \varepsilon < 1$.

In the case of a product algebra, any infinite product chain generates the whole variety of product algebras. Thus $\mathcal{V}$ contains a non-standard extension, $[0,1]_{\Pi}^*$, of the standard product algebra on $[0,1]_{\Pi}$. Let M be a set of natural numbers, ε a positive infinitesimal in $[0,1]^*$, and for $m \in M$, let $||P(m)||_{\mathbf{M},v}^{\mathbf{L}} = \frac{1}{2^{\frac{1}{m+1}}}$ and $||C||_{\mathbf{M},v}^{\mathbf{L}} = 1 - \varepsilon$. Then $||P(m)||_{\mathbf{M},v}^{\mathbf{L}} = \frac{1}{2^{\frac{1}{m+1}}} \leq \frac{1}{2^{\frac{1}{m+2}}}(1-\varepsilon) = ||P(m+1)\,\&\,C||_{\mathbf{M},v}^{\mathbf{L}}$ and $||P(m) \to (P(m+1)\,\&\,C)||_{\mathbf{M},v}^{\mathbf{L}} = 1$. On the other hand, for every m we have $||P(m) \to (P(m)\,\&\,C)||_{\mathbf{M},v}^{\mathbf{L}} = 1 - \varepsilon < 1$. Therefore, $||\exists y(P(y) \to (P(y)\,\&\,C))||_{\mathbf{M},v}^{\mathbf{L}} = 1 - \varepsilon < 1$. This ends the proof. $\square$

As a consequence, Matteo Bianchi and I [BM] proved a characterization of supersound extensions of BL. In order to illustrate it, we start from the following definition.

DEFINITION 23 *An axiomatic extension L of BL is n-contractive iff it satisfies the axiom $\varphi^n \to \varphi^{n+1}$. A variety of BL-algebras is n-contractive iff its members satisfy the identity $x^n = x^{n+1}$.*

Let L be a schematic extension of BL and let $\mathcal{V}$ be its corresponding variety of BL-algebras. If for every n L is not n-contractive, then $\mathcal{V}$ contains either Chang's algebra or an infinite product chain, and hence L$\forall$ is not supersound. The converse also holds, that is, if L$\forall$ is not supersound, then for every n, L is not n-contractive. Therefore:

THEOREM 24 *Let L be a schematic extension of BL. Then, L$\forall$ is supersound iff there is a natural number n such that L is n-contractive.*

5   Conclusions

I hope that in this paper I was able to show the importance of Hájek's contribution to the scientific community and to the Fuzzy Logic community in particular. But I would like to add that, even if we restrict ourselves to Fuzzy Logic Petr's contribution is not limited to propositional or to first-order

Fuzzy Logics, but it also involves other important topics like probability and generalized quantifiers. In my opinion, many of these topics, like the fuzzy quantifier *many* are not sufficiently developed and deserve further investigation. Hence, the moral of the whole story is the following: Petr Hájek planted a lot of fruit trees in the field of Fuzzy Logic. Even though he did a great job, much work still remains to be done, and we hope he will continue to help us for long. But now it is also our turn to take care of the plants and to let then grow up and give good fruits.

BIBLIOGRAPHY

[AM03]   P. Aglianò and F. Montagna. Varieties of BL-algebras I: General properties. *J. Pure Appl. Algebra*, 181:105–129, 2003.
[BM]   M. Bianchi and F. Montagna. Supersound many-valued logics and Dedekind-MacNeille completions. To appear in *Arch. Math. Logic*.
[CDM99]   R. Cignoli, I. M. L. D'Ottaviano, and D. Mundici. *Algebraic Foundations of Many-Valued Reasoning*, volume 7 of *Trends in Logic*. Kluwer, Dordrecht, 1999.
[CEGT00]   R. Cignoli, F. Esteva, L. Godo, and A. Torrens. Basic fuzzy logic is the logic of continuous t-norms and their residua. *Soft Computing*, 4:106–112, 2000.
[CH09]   P. Cintula and P. Hájek. Complexity issues in axiomatic extensions of Łukasiewicz logic. *J. Logic Comput.*, 19(2):245–260, 2009.
[Háj98a]   P. Hájek. Basic fuzzy logic and BL-algebras. *Soft Computing*, 2:124–128, 1998.
[Háj98b]   P. Hájek. *Metamathematics of Fuzzy Logic*, volume 4 of *Trends in Logic*. Kluwer, Dordrecht, 1998.
[Háj01]   P. Hájek. Fuzzy logic and arithmetical hierarchy III. *Studia Logica*, 68(1):129–142, 2001.
[Háj05]   P. Hájek. Arithmetical complexity of fuzzy predicate logics—a survey. *Soft Computing*, 9(12):935–941, 2005.
[HC06]   P. Hájek and P. Cintula. On theories and models in fuzzy predicate logics. *J. Symb. Logic*, 71:863–880, 2006.
[HPS00]   P. Hájek, J. Paris, and J. C. Shepherdson. The liar paradox and fuzzy logic. *J. Symb. Logic*, 65(1):339–346, 2000.
[HS01]   P. Hájek and J. C. Shepherdson. A note on the notion of truth in fuzzy logic. *Ann. Pure Appl. Logic*, 109:65–69, 2001.
[Mon01]   F. Montagna. Three complexity problems in quantified fuzzy logic. *Studia Logica*, 68(1):143–152, 2001.
[MS57]   P. S. Mostert and A. L. Shields. On the structure of semigroups on a compact manifold with boundary. *Annals of Mathematics, Second Series*, 65:117–143, 1957.
[MS03]   F. Montagna and L. Sacchetti. Kripke-style semantics for many-valued logics. *Math. Logic Quarterly*, 49(6):629–641, 2003.
[MS04]   F. Montagna and L. Sacchetti. Corrigendum to 'Kripke-style semantics for many-valued logics'. *Math. Logic Quarterly*, 50(1):104–107, 2004.

Franco Montagna
Department of Mathematics and Computer Science,
Pian dei Mantellini 44,
53100 Siena, Italy
Email: montagna@unisi.it

From BL to MV

DANIELE MUNDICI

Dedicated to Petr Hájek
on the occasion of his 70th anniversary

ABSTRACT. BL-algebras were invented by Petr Hájek to prove the completeness of (propositional) Basic Logic, the logic of all continuous t-norms and their residual implications. MV-algebras were invented earlier by C.C. Chang to prove the completeness theorem of Łukasiewicz (propositional) logic L_∞, the logic of a very special continuous t-norm. Upon writing $A = \{x \in B \mid \neg\neg x = x\}$, and letting B range over all BL-algebras we recover every MV-algebra A. Hájek showed that BL-algebras are a fundamental tool to reason about uncertain information. In this note we will discuss two such applications, dealing with many-valued probability and coding-theoretic semantics. We will provide all necessary background results, with appropriate references for readers interested in the proofs.

Antefact: Wendisch-Rietz (Berlin), April 1988

Organized by the Logic Group of the Humboldt University of Berlin, the Sixth Easter Conference on Model Theory was held in Wendisch-Rietz, near Berlin, during April 4–9, 1988. Petr Hájek gave a talk on "Problems on uncertainty processing", and I discussed "Sylvester's law of unit determinants, Handelman's problem on p.o. groups, and Chang's MV-algebras". While I'm not sure this was the first time Petr heard of MV-algebras, from his talk I learned—from the most authoritative possible source—that many-valued logic has an important role in the processing of uncertain information.

In the last twenty years Hájek has developed a monumental logic-algebraic apparatus centered on Basic Logic and BL-algebras, for the formal treatment of imprecise information. During the same period, also the literature on Łukasiewicz logic and their algebras (the MV-algebras of C.C. Chang) has expanded vigorously. Of the many points of contact between these two theories, only a few will be touched upon in this note. I have allowed my personal tastes to decide what to include, and I apologize in advance if the content looks too eccentric. The reader is assumed to be familiar with BL-algebras and MV-algebras. For background on all undefined notions Hájek's

monograph [Háj98] is recommended, together with [CDM99]. For the sake of
brevity, most results will be stated in a terse algebraic language.

1 Events and possible worlds

This rather technical section can be either skimmed or omitted on a first
reading. Its aim is to give "events" and "possible worlds" a sufficiently
general definition using the standard C^*-algebraic formalization of classical
physical systems—which is the commutative case of the general formulation
in [Emc84, pp. 362, 378].

Let $\mathcal{B}$ be the commutative C^*-algebra of a classical physical system $\mathcal{P}$,
with the set $\mathcal{A} \subseteq \mathcal{B}$ of self-adjoint elements (=the observables of $\mathcal{P}$) and
the set $\mathcal{S}^*$ of real-valued normalized positive linear functionals on $\mathcal{A}$. As explained in [Con85, VIII, 2.1], $\mathcal{B}$ can be identified with the C^*-algebra $C(\Sigma)$
of all complex-valued continuous functions over the compact Hausdorff space
Σ of maximal ideals of $\mathcal{B}$, [Con85, pp. 224–225]. Σ represents the space of all
possible "phases" of $\mathcal{P}$, and $\mathcal{A}$ is the set of all real-valued continuous functions on Σ. In many important concrete cases, $\mathcal{A}$ includes such observables
as "energy", momentum", "angular momentum". $\mathcal{S}^*$ represents the set of
modes of preparation of $\mathcal{P}$ which, by Riesz representation theorem, is in one-
one correspondence with the set of regular Borel probability measures on Σ.
For any $\rho \in \mathcal{S}^*$ and $A \in \mathcal{A}$ the real number $\rho(A)$ is the expectation value of
the observable A when $\mathcal{P}$ is prepared in mode ρ. Let $E = \{X_1, \ldots, X_m\}$ be a
set of nonzero positive self-adjoint elements of $\mathcal{B}$. Thus, each X_i is a positive
continuous function over Σ. For each X_i let $\sup X_i$ denote the sup norm
of X_i. Then by elementary C^*-algebra theory (which is the special case of
[Emc84, pp. 363–369] for commutative $\mathcal{B}$), each preparation mode $\rho \in \mathcal{S}^*$ determines a map $w_\rho \colon E \to [0,1]$ by the stipulation $w_\rho(X_i) = \rho(X_i)/(\sup X_i)$.
The set $W = \{w_\rho \mid \rho \in \mathcal{S}^*\}$ is closed in the m-cube $[0,1]^E$. Intuitively, X_i is
the "event" "the value of the observable X_i will be high," and the "possible
world" $w_\rho \in W$ gives a precise "truth-value" to this event.

2 From coherent probability assessments to
many-valued logic

As we have just seen, "events" and "possible worlds" live symbiotically: a
yes-no event X will or will not occur, in a possible world w. The quantity
$w(X) \in \{yes, no\} = \{1, 0\}$ is the "truth-value" of X in the possible world w.
Events with continuous spectrum are natural generalizations of yes-no events.
For any such event X and possible world w, the truth-value $w(X) \in [0,1]$
can be conveniently thought of as the result of the measurement of a suitably
normalized observable, also denoted X. Our expectation "X has a large
value" is made precise by the result of the measurement of X.

Stripping all references to physical systems and their associated C^*-algebras, given a finite but otherwise arbitrary set $E = \{X_1, \ldots, X_n\}$, whose elements we call *events*, and a set $W \subseteq [0,1]^E$ of *possible worlds*, we say that a map $\beta \colon E \to [0,1]$ is *coherent*, (or, W-coherent, for greater definiteness) if for no $\sigma_1, \ldots, \sigma_m \in \mathbb{R}$ one has

$$\sum_{i=1}^m \sigma_i(\beta(X_i) - v(X_i)) < 0 \quad \text{for all } v \in W.$$

This is a generalization of De Finetti's well known coherence criterion (see [dF31, pp. 311–312] and [dF74, pp. 85–90]).

To fully understand this definition, let us imagine two players, Ada (the bookmaker) and Blaise (the bettor). For each event $X_i \in E$, Ada proclaims her "betting odd" $\beta(X_i)$, and Blaise chooses a "stake" $\sigma_i \in \mathbb{R}$. If $\sigma_i > 0$, he pays Ada $\sigma_i \beta(X_i)$ euro, with the proviso that, in the world $w \in W$, Ada will pay back $\sigma_i w(X_i)$ euro. If $\sigma_i < 0$, the same amounts of money are exchanged, but in the opposite directions. Supposing money transfers are oriented so that "positive" means "Blaise-to-Ada", then Ada's book β is coherent in the precise sense of the above definition iff Blaise cannot choose stakes ensuring him a net profit (equivalently, a profit of at least one zillion euro) in every possible world $v \in W$.

Łukasiewicz logic $\mathrm{Ł}_\infty$ ([Luk20, LT30, LT56, Háj98, CDM99]) has a universal role in the interpretation of events, possible worlds, and coherent books:

THEOREM 1 ([Mun09, Theorem 1.2]) *For any set $E = \{X_1, \ldots, X_m\}$ and closed nonempty set $W \subseteq [0,1]^E = [0,1]^m$, there is a set Θ of formulas in the variables $\{X_1, \ldots, X_m\}$ such that $W = \{v{\restriction}E \mid v \text{ satisfies } \Theta\}$, where v ranges over valuations of m-variable formulas in Łukasiewicz logic $\mathrm{Ł}_\infty$. Further, a map $\beta \colon E \to [0,1]$ is W-coherent iff it can be extended to a convex combination of valuations of m-variable formulas satisfying Θ in $\mathrm{Ł}_\infty$.*

The proof relies on the continuity of all connectives of $\mathrm{Ł}_\infty$. One might then hope to be able to prove the analog of Theorem 1 for other $[0,1]$-valued logics with continuous connectives. The following results restrict our choice:

THEOREM 2 ([Pav79, pp. 121–122]) *Let $(L, \vee, \wedge, 0, 1)$ be a complete lattice with smallest element 0 and greatest element 1. Suppose L is enriched with a binary operation $*$ such that*

(i) $*$ *is monotone in both variables;*

(ii) *for each subset I of L, $(\bigvee_{i \in I} i) * j = \bigvee_{i \in I}(i * j)$.*

Then there exists a unique operation $\to$ over L such that for all $i, j, k \in L$: $i \leq j \to k$ iff $i * j \leq k$. This operation is given by $j \to k = \bigvee\{i \mid i * j \leq k\}$.

THEOREM 3 ([Pav79, pp. 121–122]) *Let* $(L, \vee, \wedge, 0, 1)$ *be as in Theorem 2. Suppose* L *is enriched with a binary operation* $\Rightarrow$ *such that*

(iii) $\Rightarrow$ *is anti-monotone in the first variable and monotone in the second;*

(iv) for each subset K *of* L, $j \Rightarrow (\bigwedge_{k \in K} k) = \bigwedge_{k \in K} (j \Rightarrow k)$.

Then there exists a unique operation $\cdot$ *over* L *such that for all* $i, j, k \in L$: $i \leq j \Rightarrow k$ *iff* $i \cdot j \leq k$. *This operation is given by* $i \cdot j = \bigwedge \{k \mid i \leq j \Rightarrow k\}$.

Following [WD39], a *residuated lattice is a structure* $(L, \vee, \wedge, *, \Rightarrow, 0, 1)$ such that: $(L, \vee, \wedge, 0, 1)$ is a lattice with bottom 0 and top 1; $(L, *, 1)$ is a commutative monoid; for any $i, j, k \in L$, we have $i \leq j \Rightarrow k$ iff $i * j \leq k$.

THEOREM 4 *In complete residuated lattices, conditions (i)–(iv) above hold.*

THEOREM 5 ([MP76]) *Let* $([0,1], \vee, \wedge, *, \Rightarrow, 0, 1)$ *be a residuated lattice in which* $\vee$ *and* $\wedge$ *are the* max *and* min *operations. Suppose that the map* $\Rightarrow \colon [0,1]^2 \to [0,1]$ *is continuous. Then* $*$ *and* $\Rightarrow$ *are the Łukasiewicz conjunction* $x * y = \max(0, x+y-1)$ *and implication* $x \Rightarrow y = \min(1, 1-x+y)$.

In the language of *t-norms* (i.e., commutative associative monotone binary operations on $[0,1]$ having 1 as their neutral element, [KMP00]) the above results show that, among all continuous t-norms, Łukasiewicz conjunction is the only one yielding a logic with a continuous implication connective.

3   Representation theorems

In Theorem 6 below it is shown that Hájek's Basic Logic is the logic of *all* continuous t-norms and their residual implication connectives. In the next section we will see that "truth-values" in Basic Logic need no longer lie in the real unit interval $[0,1]$, and equivalence classes of formulas are no longer represented by continuous $[0,1]$-valued functions. Yet, the rich algebraic and order-theoretic structure of Basic Logic affords an interesting geometric visualization, where the functional representation of free MV-algebras has a relevant role, [AB]. By a *BL-algebra* we mean a residuated lattice satisfying the two additional equations

$$x \wedge y = x * (x \Rightarrow y) \qquad \text{and} \qquad (x \Rightarrow y) \vee (y \Rightarrow x) = 1.$$

When the lattice is totally ordered we have a *BL-chain*. BL-algebras stand to Basic Logic as Boolean algebras stand to Boolean logic. More precisely, in [Háj98, CEGT00] one finds a proof of the following completeness theorem:

THEOREM 6 *A formula is provable in Basic Logic iff it is satisfied by every BL-algebra iff it is satisfied by every BL-chain iff it holds whenever it is interpreted by every continuous t-norm.*

The MV-algebraic counterpart of Theorem 6 was proved by Chang [Cha58, Cha59] (see [RR58] for a different proof of the completeness of Ł$_\infty$):

THEOREM 7 *A formula is provable in Łukasiewicz logic* Ł$_\infty$ *iff it holds in the MV-algebra* $[0, 1]$ *iff it holds in every MV-chain iff it holds in every MV-algebra.*

As will be apparent from the rest of this section, the representation theory of BL-algebras is tightly related to that of MV-algebras. We recall here only the results used in [CM07] for the proof of Theorem 15 below.

Following [CDM99], let $A = (A, 0, \neg, \oplus)$ be an MV-algebra. Upon setting $x \Rightarrow y = \neg x \oplus y$ and $x \odot y = \neg(\neg x \oplus \neg y)$, the structure A' obtained from A replacing $\neg$ and $\oplus$ by $\odot$ and $\Rightarrow$ is called a *Wajsberg algebra*.

THEOREM 8 *A* Wajsberg hoop *can be defined as a subalgebra of a reduct of a Wajsberg algebra obtained by deleting 0 from the language, [BF00, p. 241]. Every Wajsberg hoop inherits the natural MV-algebraic order, and we have* $x \leq y$ *iff* $x \Rightarrow y = 1$, *[CDM99, Lemma 1.1.2]. A Wajsberg hoop with an additional constant interpreted as the bottom element is said to be* bounded. *Thus bounded Wajsberg hoops coincide with Wajsberg algebras.*

Let $(I, \leq)$ be a totally ordered set with smallest element o. For all $i \in I$ let H_i be a Wajsberg hoop such that for $i \neq j$, $H_i \cap H_j = \{1\}$. Assume H_o has a bottom element 0. Then the *ordinal sum* $\bigoplus_{i \in I} H_i$ is the structure whose universe is given by $\bigcup_{i \in I} H_i$, and whose operations (following [BF00, page 247]) are defined by

$$x \Rightarrow y = \begin{cases} x \stackrel{H_i}{\Rightarrow} y & \text{if } x, y \in H_i \\ y & \text{if } \exists j < i (x \in H_i, y \in H_j) \\ 1 & \text{if } \exists i < j (1 \neq x \in H_i, y \in H_j) \end{cases}$$

$$x \odot y = \begin{cases} x \odot^{H_i} y & \text{if } x, y \in H_i \\ y & \text{if } \exists j < i (x \in H_i, 1 \neq y \in H_j) \\ x & \text{if } \exists i < j (1 \neq x \in H_i, y \in H_j) \end{cases}$$

and whose bottom element is 0.

THEOREM 9 ([AM03, LS02]) *Up to isomorphism, BL-chains coincide with ordinal sums of Wajsberg hoops whose first component is a Wajsberg algebra.*

For any totally ordered BL-algebra A, the Wajsberg hoops W_i such that $A \cong \bigoplus W_i$ will be called the *components* of A. By definition, an algebra (BL, MV, or a hoop) is *trivial* if it is a singleton.

THEOREM 10 ([AM03]) *Suppose the BL-chain A can be represented as $A = \bigoplus_{i=0}^{n} W_i$ with components $W_0, \ldots, W_n$. Then its nontrivial subalgebras are precisely the ordinal sums $B = \bigoplus_{i=0}^{n} U_i$, where U_0 is a nontrivial subalgebra of W_0, and for each $i = 1, \ldots, n$, U_i is a (possibly trivial) subalgebra of W_i.*

THEOREM 11 ([BF00]) *Every finite partial subalgebra of a Wajsberg hoop can be embedded into a finite Wajsberg hoop.*

As an immediate consequence of the definition of ordinal sum, for any BL-chain A and elements $a_1, \ldots, a_n \in A$, all in the same component W of A, the subalgebra generated by $\{a_1, \ldots, a_n\}$ is contained in W. We then have

THEOREM 12 *Every finitely generated BL-chain A is the ordinal sum of finitely many Wajsberg hoops, where every component contains at least one generator of A.*

THEOREM 13 ([AFM07]) *The variety of BL-algebras is generated by its finite totally ordered members.*

4   Multichannel error-correcting codes, BL- and MV-algebras

In this section Basic Logic will be endowed with a game-theoretic semantics. We will tacitly use the representation theorems of the foregoing section, referring the reader to [CM07] for full details. We assume familiarity with the Rényi and Ulam game of Twenty Questions with errors/lies in the answers, [CDM99, § 5], [Pel02, Rén76, Ula76]. This is a chapter of Berlekamp's fault-tolerant communication theory with feedback, [Ber68]. The two players of the game are called Questioner and Responder. They initially fix a finite nonempty set $\mathcal{S}$ of numbers, called the *search space*. Responder chooses a number $x_{secret} \in \mathcal{S}$, which Questioner must find by asking (as few as possible) yes-no questions. While questions are sent through a noiseless feedback channel, there are m channels $1, 2, 3, \ldots, m$, for the answers. The reliability degree of each channel i is given by a label $e_i \in \{0, 1, 2, \ldots\}$, where e_i is the maximum number of lies/errors/distortions $b \mapsto 1 - b$ that may affect the bits b carrying Responder's answers during a game. We also assume that all e_i's are known to both players, and $0 \leq e_1 < \ldots < e_m$.

A *question* Q is a subset of $\mathcal{S}$, together with an integer $j \in \{1, 2, \ldots, m\}$ obliging Responder to reply on channel j. An *answer* to Q is a pair $A = (j, T)$ where $T \in \{Q, \mathcal{S} \setminus Q\}$. We naturally say that an element $z \in \mathcal{S}$ *satisfies* A iff $z \in T$. Otherwise, z *falsifies* A. Since equal answers to the same repeated question carry more information than single answers, all that Questioner knows about x_{secret} is given by the finite *multiset* $\mathfrak{A}$ of answers: each answer

$A \in \mathfrak{A}$ carries a multiplicity, telling how many times the same answer A has been given in a game.

We set $\mathfrak{A}_l = \{A \in \mathfrak{A} \mid A = (l, U) \text{ for some } U \subseteq S\}$ for any channel $l = 1, \ldots, m$. For every element $z \in S$ and multiset $\mathfrak{A}$, the *truth value* $\mathfrak{A}^{\#}(z)$ of z in $\mathfrak{A}$ is the pair (j, r) where $(j, r) = (m, 1)$ if z satisfies all answers in $\mathfrak{A}$, and otherwise

$$
\begin{aligned}
j &= \text{ first channel } l \in \{1, \ldots, m\} \text{ such that for some } A \in \mathfrak{A}_l, z \text{ falsifies } A \\
r &= \max(0,\ 1 - (\text{number of answers in } \mathfrak{A}_l \text{falsified by } z)/(e_l + 1)).
\end{aligned}
$$

Intuitively, the quantity $\mathfrak{A}^{\#}(z)$ measures (in units of $e_l + 1$) the distance of z from the condition of being excluded from the set of possible candidates for x_{secret} *as an effect of the information sent on channel* l. While only the first channel can permanently ban an element x from the set of possible candidates for x_{secret}, yet the information supplied by the other channels has a decisive role in reducing the size of the search space. Truth-values form a totally ordered set, by stipulating

$$(j, r) < (j', r') \text{ iff (either } j < j' \text{ or } (j = j' \text{ and } r < r')).$$

Intuitively: the information provided by cheap noisy channels is superseded by low-noise expensive channels.

Two multisets $\mathfrak{A}$ and $\mathfrak{B}$ are said to be *equivalent* iff they assign the same truth-value to each $z \in S$. We respectively denote by $a, b, c, \ldots$ the equivalence class of $\mathfrak{A}, \mathfrak{B}, \mathfrak{C}, \ldots$. A *state of knowledge* in an m-channel Rényi-Ulam game over search space S with $\mathbf{e} = (e_1 < \ldots < e_m)$ lies, is an equivalence class of multisets. The *initial* state of knowledge $\mathbf{1}$ is the equivalence class of the empty multiset of answers. $\mathbf{1}$ assigns truth value $(m, 1)$ to each $z \in S$. We let $\mathcal{K}_{S, \mathbf{e}}$ denote the set of states of knowledge.

Direct inspection shows:

THEOREM 14 *Given states of knowledge a and b, let us write $a \leq b$ (read: "a is more restrictive than b") if and only if $\mathfrak{A}^{\#}(z) \leq \mathfrak{B}^{\#}(z)$ for all $z \in S$. We then have:*

1. *The relation $\leq$ equips $\mathcal{K}_{S, \mathbf{e}}$ with a partial order.*

2. *Let us define the* juxtaposition *$\mathfrak{A} \odot \mathfrak{B}$ of two multisets $\mathfrak{A}$ and $\mathfrak{B}$ as the multiset obtained by giving each answer A a multiplicity equal to the sum of its multiplicity in $\mathfrak{A}$ plus its multiplicity in $\mathfrak{B}$. Then $\odot$ equips the set $\mathcal{K}_{S, \mathbf{e}}$ with an operation, also denoted $\odot$, which is defined by $a \odot b =$ the equivalence class of $\mathfrak{A} \odot \mathfrak{B}$. This operation is commutative, associative, and has the initial state of knowledge $\mathbf{1}$ as its neutral element.*

3. *For any two states of knowledge a and b, among all $t \in \mathcal{K}_{\mathcal{S},\mathbf{e}}$ such that $a \odot t \leq b$ there is a least restrictive state of knowledge, denoted $a \Rightarrow b$. The set $\mathcal{K}_{\mathcal{S},\mathbf{e}}$ equipped with the $\odot$ and $\Rightarrow$ operations and with the distinguished constant $\mathbf{1}$, is a BL-algebra. The partial order of $\mathcal{K}_{\mathcal{S},\mathbf{e}}$ is definable in $(\mathcal{K}_{\mathcal{S},\mathbf{e}}, \odot, \Rightarrow, \mathbf{1})$ via the stipulation $a \leq b$ iff $a \Rightarrow b = \mathbf{1}$.*

In view of the above result, one is naturally led to study the equational properties of the BL-algebras $\mathcal{K}_{\mathcal{S},\mathbf{e}}$. The proof of the following result uses Theorems 12–13 of the foregoing section:

THEOREM 15 ([CM07, 2.4]) *Given BL-terms $\sigma = \sigma(x_1,\ldots,x_n)$ and $\tau = \tau(x_1,\ldots,x_n)$, the following conditions are equivalent for the equation $\sigma = \tau$:*

(i) *For every finite set $\mathcal{S} \neq \emptyset$ and m-tuple $\mathbf{e} = (e_1 < \ldots < e_m)$ of integers ≥ 0, the equation is valid in the BL-algebra $\mathcal{K}_{\mathcal{S},\mathbf{e}}$ in the m-channel Rényi-Ulam game over search space $\mathcal{S}$ with $\mathbf{e}$ errors.*

(ii) *The equation holds in every BL-algebra.*

The following result relies on Chang completeness theorem (Theorem 7):

THEOREM 16 *BL-algebras stand to multi-channel Rényi-Ulam games with errors/lies as MV-algebras stand to one-channel games, as Boolean algebras stand to Twenty Questions games without lies.*

See [CDM99, 5.3.1] for a proof.

5 Closing a circle of ideas: From many-valued logic to probability

As the reader will recall, a *state* of an MV-algebra A (whence in particular, a state of a Boolean algebra) is a map $s\colon A \to [0,1]$ such that $s(1) = 1$ and $s(x \oplus y) = s(x) + s(y)$ whenever $x \odot y = 0$. We let $S(A)$ denote the set of states of A, as a compact convex subset of the Tychonov cube $[0,1]^A$. A state s of A is *extreme* if $s \notin \text{conv}(S(A) \setminus \{s\})$.

Among the results showing the connections between $[0,1]$-valued logic and probability theory, the following has a particularly terse form:

THEOREM 17 ([Mun09, Proof of 4.1 and references therein]) *The extremal states of any MV-algebra B are exactly the homomorphisms of B into $[0,1]$. The states of B are the limits in $[0,1]^B$ of convex combinations of homomorphisms of B into $[0,1]$. Once B is thought of as the Lindenbaum algebra of some set Φ of formulas of Łukasiewicz logic, these homomorphisms are precisely the truth-valuations satisfying Φ.*

A well known classical corollary is

THEOREM 18 *The states of any Boolean algebra B are the limits in $\{0,1\}^B$ of convex combinations of homomorphisms of B into $\{0,1\}$. The homomorphisms of B into $\{0,1\}$ are exactly the extremal states of B.*

The next result shows that (finitely additive) states are just the algebraic counterpart of (countably-additive) measures on maximal spectral spaces:

THEOREM 19 (Kroupa-Panti theorem [Kro06, Pan08]) *In any MV-algebra B the integral induces a one-one correspondence between states of B and regular Borel probability measures over the maximal spectral space of B equipped with the hull kernel topology.*

We are now in a position to extend the universality result (Theorem 1), showing that De Finetti's notion of coherence actually provides a foundation of Kolmogorov probability theory:

THEOREM 20 (Continuation of Theorem 1, [Mun09, 4.1])
Let $E = \{X_1,\ldots,X_m\}$ and $W \subseteq [0,1]^E = [0,1]^m$ be a nonempty closed subset of the m-cube. Let $\mathcal{M}(W)$ be the MV-algebra of restrictions to W of the McNaughton functions on $[0,1]^m$. Then for any map $\beta\colon E \to [0,1]$ the following conditions are equivalent:

(i) *β is W-coherent.*

(ii) *For some state s of $\mathcal{M}(W)$ we have $\beta(X_i) = s(x_i \restriction W)$ for all $i = 1,\ldots,m$.*

(iii) *For some (automatically regular) Borel probability measure μ on W we have*

$$\beta(X_i) = \int_W x_i \, d\mu, \quad for \quad all \; i = 1,\ldots,m,$$

with x_i denoting the ith coordinate function.

For the case of yes-no events, for any set $E = \{X_1,\ldots,X_m\}$ and set $W \subseteq \{0,1\}^E$ of possible worlds, De Finetti showed that W-coherent maps on E are the same as restrictions to E of probability measures on W, (see [dF31, pp. 311–312] and [dF74, pp. 85–90]). De Finetti's theorem was generalized to modal logics in [Par01], to Gödel logic in [AGM08], to all finite-valued logics in [KM07] and, more generally, to all $[0,1]$-valued logics with continuous connectives.

The proof of the following result will appear elsewhere:

THEOREM 21 L_∞ *has a Rényi conditional probability, i.e., a map* P: $(\phi, \theta) \mapsto \mathsf{P}(\phi \mid \theta)$, *defined whenever* θ *satisfiable, and having the following properties for all elements* $\alpha, \beta, \phi, \psi, \theta$ *of the set* FORM_n *of formulas in the variables* $X_1, \ldots, X_n$:

(I) $\mathsf{P}(\phi \mid \theta) \geq 0$ *and* $\mathsf{P}(\theta \mid \theta) = 1$.

(II) *If* $\neg(\phi \odot \psi)$ *is a tautology,* $\mathsf{P}(\phi \oplus \psi \mid \theta) = \mathsf{P}(\phi \mid \theta) + \mathsf{P}(\psi \mid \theta)$.

(III) *If* $\psi \odot \theta$ *is satisfiable, then* $\mathsf{P}((\phi \odot \psi)^\infty \mid \theta) = \mathsf{P}(\phi^\infty \mid \psi \odot \theta) \cdot \mathsf{P}(\psi^\infty \mid \theta)$, *where* $\mathsf{P}(\alpha^\infty \mid \beta) = \lim_{t \to \infty} \mathsf{P}(\underbrace{\alpha \odot \ldots \odot \alpha}_{t \text{ times}} \mid \beta)$.

Further, P *has the following properties:*

(a) *(Effectiveness):* $\mathsf{P}(\phi \mid \theta)$ *is a Turing computable rational.*

(b) *(Invariance under isomorphisms): If* $\phi', \theta' \in \mathsf{FORM}_{n'}$, *and* ϕ *and* ϕ' *correspond via an isomorphism of the Lindenbaum algebras of* θ *and* θ', *then* $\mathsf{P}(\phi \mid \theta) = \mathsf{P}(\phi' \mid \theta')$.

(c) *(Invariance under substitutions): Let* $var(\theta)$ *denote the set of variables in* θ. *If* $X \notin var(\theta) \cup var(\phi)$ *then* $\mathsf{P}(\phi \mid \theta) = \mathsf{P}(X \mid \theta \odot (\phi \leftrightarrow X))$.

(d) *(Independence): If* $var(\phi) \cap var(\theta) = \emptyset$ *then* ϕ *is independent of* θ, *i.e.,* $\mathsf{P}(\phi \mid \theta) = \mathsf{P}(\phi \mid \phi \leftrightarrow \phi) = \int \hat\phi(x) dx$.

Properties (I)–(III) are a many-valued generalization of Rényi's original axioms (I)–(III) [Rén55, p. 289] for random yes-no events and their corresponding Boolean algebras of sets given by Stone duality. According to Rényi, a conditional probability is a set of ordinary probability spaces which are connected with each other by Axiom (III), in conformity with the usual definition of conditional probability. For any $\alpha, \beta, \theta \in \mathsf{FORM}$ let us write $\alpha \equiv_\theta \beta$ if $\theta \vdash \alpha \leftrightarrow \beta$. Then the map

$$\theta \mapsto \mathsf{P}_\theta = \mathsf{P}\left(\frac{\cdot}{\equiv_\theta} \mid \theta\right)$$

sends each formula θ into a state of the Lindenbaum algebra of θ. In his Axiom (II), Rényi also assumed countable additivity, whereas (II) above amounts to the usual finite additivity of states. However, the Kroupa-Panti theorem 19 yields a map $\theta \mapsto \mathsf{P}_\theta \mapsto \mu_\theta$ where μ_θ is a (countably additive, regular) Borel probability measure on the maximal spectral space of the Lindenbaum algebra of θ.

As an alternative to (the BL-algebraic generalization of) the above results, in [Háj98, 8.4] one finds a different approach to probability in BL-algebras: for every logic given by a continuous t-norm $*$, the author introduces a new unary connective P whose intended meaning is "probably". For the particular case when $*$ is Łukasiewicz conjunction, axioms are given and a completeness theorem is proved. Hájek's internalized approach to probability is further developed in [FM09]–[FM]. Given the essential role played by the geometric representation of free MV-algebras in the study of MV-algebraic probability, there should be little doubt that the geometric representation of free BL-algebras in [AB] will provide a key tool in the further developments of BL-algebraic probability theory.

6 Conclusion

Petr Hájek has given fundamental contributions to understand "fuzzy logic", not only as an important artifact originating from engineering problems, but also as an interesting logic in se, with its rules of inference, semantics, algorithms, completeness phenomena.

The apparatus of symbols and rules that one finds here, rather than tying one's legs and wings, aims to fix a rigorous framework where deductions can be carried over as an algorithmic procedure acting on sentences rather than on numbers, generalizing classical deductions. Any progress in the understanding of many-valued inference has a positive effect in our concrete way of coping with uncertain information, as is commonly done in probability, data analysis, fuzzy control, error-correcting codes.

Prior and after [Háj98], Hájek's papers on the logical treatment of uncertain information have covered an impressive area, ranging from automatic hypothesis formation, [HHC66, HH78, HH03] to belief, possibilistic, modal, experimental, intuitionistic, interpretability, dynamic, three-valued, probability and other logics, [HBR71, Háj77, Háj83, Háj86, HH94, Háj95b, Háj95c, HH96, Háj96, DGH$^+$05].

His constant interest in concrete applications (prior and after his talk in Wendisch Rietz) ranges from expert systems [HBR71, Háj88, HV94, HH95, Háj04, HHR10], and their implementations on personal computers [HSZ95], to complexity-theoretic issues, [HT01, BHMV02, Háj05, Háj06, CH09].

His analysis of many formal systems, with their axioms, rules, semantics and decision problems, has set a standard of clarity and rigor for all writers of fuzzy logical papers.

BIBLIOGRAPHY

[AB]   S. Aguzzoli and S. Bova. The free n-generated BL-algebra. To appear in *Ann. Pure Appl. Logic*.

[AFM07]   P. Aglianò, I. M. A. Ferreirim, and F. Montagna. Basic hoops: An algebraic study of continuous t-norms. *Studia Logica*, 87(1):73–98, 2007.

[AGM08]   S. Aguzzoli, B. Gerla, and V. Marra. De Finetti's no-Dutch-book criterion for Gödel logic. *Studia Logica*, 90(1):25–41, 2008.

[AM03]   P. Aglianò and F. Montagna. Varieties of BL-algebras I: General properties. *J. Pure Appl. Algebra*, 181:105–129, 2003.

[Ber68]   E. R. Berlekamp. Block coding for the binary symmetric channel with noise-less, delayless feedback. In H. B. Mann, editor, *Error-correcting Codes*, pages 330–335. Wiley, New York, 1968.

[BF00]   W. J. Blok and I. M. A. Ferreirim. On the structure of hoops. *Algebra Universalis*, 43(2–3):233–257, 2000.

[BHMV02]   M. Baaz, P. Hájek, F. Montagna, and H. Veith. Complexity of t-tautologies. *Ann. Pure Appl. Logic*, 113:3–11, 2002.

[CDM99]   R. Cignoli, I. M. L. D'Ottaviano, and D. Mundici. *Algebraic Foundations of Many-Valued Reasoning*, volume 7 of *Trends in Logic*. Kluwer, Dordrecht, 1999.

[CEGT00]   R. Cignoli, F. Esteva, L. Godo, and A. Torrens. Basic fuzzy logic is the logic of continuous t-norms and their residua. *Soft Computing*, 4:106–112, 2000.

[CH09]   P. Cintula and P. Hájek. Complexity issues in axiomatic extensions of Łukasiewicz logic. *J. Logic Comput.*, 19(2):245–260, 2009.

[Cha58]   C. C. Chang. Algebraic analysis of many-valued logics. *Trans. Amer. Math. Soc.*, 88:456–490, 1958.

[Cha59]   C. C. Chang. A new proof of the completeness of Łukasiewicz axioms. *Trans. Amer. Math. Soc.*, 93:74–80, 1959.

[CM07]   F. Cicalese and D. Mundici. Recent developments of feedback coding and its relations with many-valued logic. In A. Gupta, R. Parikh, and J. van Benthem, editors, *Logic at the Crossroads: an Interdisciplinary View, Proceedings of the First Indian Congress on Logic and Applications*, pages 222–240, New Delhi, 2007. Allied Publishers Pvt.Ltd.

[Con85]   J. B. Conway. *A Course in Functional Analysis*, volume 96 of *Graduate Texts in Mathematics*. Springer, New York, 1985.

[dF31]   B. de Finetti. Sul significato soggettivo della probabilità. *Fund. Mathematicae*, 17:298–329, 1931. Translated into English as: On the subjective meaning of probability. In P. Monari & D. Cocchi (Eds.), *Probabilità e Induzione*, pages 291–321. Clueb, Bologna, 1993.

[dF74]   B. de Finetti. *Theory of Probability. Vol. 1*. John Wiley & Sons, Chichester, 1974.

[DGH+05]   D. Dubois, S. Gottwald, P. Hájek, J. Kacprzyk, and H. Prade. Terminological difficulties in fuzzy set theory—the case of "intuitionistic fuzzy sets". *Fuzzy Sets Syst.*, 156(3):485–491, 2005.

[Emc84]   G. G. Emch. *Mathematical and Conceptual Foundations of 20th Century Physics*, volume 100 of *Notas de Matemática*. North-Holland, Amsterdam, 1984.

[FM]   T. Flaminio and F. Montagna. Models for many-valued probabilistic reasoning. To appear in *J. Logic Comput.* DOI 10.1093/logcom/exp013.

[FM09]   T. Flaminio and F. Montagna. MV-algebras with internal states and probabilistic fuzzy logics. *Int. J. Approximate Reasoning*, 50:138–152, 2009.

[Háj77]   P. Hájek. Experimental logics and Π_3^0 theories. *J. Symb. Logic*, 42(4):515–522, 1977.

[Háj83]   P. Hájek. Arithmetical interpretations of dynamic logic. *J. Symb. Logic*, 48(3):704–713, 1983.

[Háj86]   P. Hájek. A simple dynamic logic. *Theoretical Comp. Sci.*, 46(2–3):239–259, 1986.

[Háj88]   P. Hájek. Towards a probabilistic analysis of MYCIN-like expert systems. In D. Edwards and N. E. Raun, editors, *COMPSTAT '88*, pages 117–121, Heidelberg, 1988. Physica Verlag.

[Háj95a]   P. Hájek. Fuzzy logic as logic. In G. Coletti, D. Dubois, and R. Scozzafava, editors, *Mathematical Models of Handling Partial Knowledge in Artificial Intelligence*, pages 21–30, New York, 1995. Plenum Press.

[Háj95b]   P. Hájek. On logics of approximate reasoning II. In G. della Riccia, R. Kruse, and R. Viertl, editors, *Proceedings of the ISSEK94 Workshop on Mathematical and Statistical Methods in Artificial Intelligence*, volume 363 of *Courses and Lectures*, pages 147–156, Wien, 1995. Springer.

[Háj95c]   P. Hájek. Possibilistic logic as interpretability logic. In B. Bouchon-Meunier, R. R. Yager, and L. A. Zadeh, editors, *Advances in Intelligent Computing— IPMU'94*, volume 945 of *LNCS*, pages 243–280, Berlin, 1995. Springer.

[Háj96]   P. Hájek. Getting belief functions from Kripke models. *Int. J. General Systems*, 24(3):325–327, 1996.

[Háj98]   P. Hájek. *Metamathematics of Fuzzy Logic*, volume 4 of *Trends in Logic*. Kluwer, Dordrecht, 1998.

[Háj04]   P. Hájek. Relations in GUHA style data mining II. In R. Berghammer and B. Möller, editors, *Relational and Kleene-Algebraic Methods in Computer Science*, volume 3051 of *LNCS*, pages 163–170, Berlin, 2004. Springer.

[Háj05]   P. Hájek. Arithmetical complexity of fuzzy predicate logics—a survey. *Soft Computing*, 9(12):935–941, 2005.

[Háj06]   P. Hájek. Computational complexity of t-norm based propositional fuzzy logics with rational truth constants. *Fuzzy Sets Syst.*, 157(5):677–682, 2006.

[Háj07]   P. Hájek. Complexity of fuzzy probability logics II. *Fuzzy Sets Syst.*, 158(23):2605–2611, 2007.

[Háj09]   P. Hájek. Arithmetical complexity of fuzzy predicate logics—a survey II. *Ann. Pure Appl. Logic*, 161(2):212–219, 2009.

[HBR71]   P. Hájek, K. Bendová, and Z. Renc. The GUHA method and the three valued logic. *Kybernetika*, 7(6):421–435, 1971.

[HH78]   P. Hájek and T. Havránek. *Mechanizing Hypothesis Formation— Mathematical Foundations of a General Theory*. Springer, Berlin, 1978.

[HH94]   D. Harmanec and P. Hájek. A qualitative belief logic. *Int. J. of Uncertain. Fuzziness Knowledge-Based Syst.*, 2(2):227–236, 1994.

[HH95]   P. Hájek and D. Harmancová. Medical fuzzy expert systems and reasoning about beliefs. In P. Barahona, M. Stefanelli, and J. Wyatt, editors, *Artificial Intelligence in Medicine*, pages 403–404, Berlin, 1995. Springer.

[HH96]   P. Hájek and D. Harmancová. A many-valued modal logic. In *Proceedings of IPMU'96*, pages 1021–1024, Granada, 1996. Universidad de Granada.

[HH03]   P. Hájek and M. Holeňa. Formal logics of discovery and hypothesis formation by machine. *Theoretical Comp. Sci.*, 292(2):345–357, 2003.

[HHC66]   P. Hájek, I. Havel, and M. Chytil. The GUHA-method of automatic hypotheses determination. *Computing*, 1(4):293–308, 1966.

[HHJ92]   P. Hájek, T. Havránek, and R. Jiroušek. *Uncertain Information Processing in Expert Systems*. CRC Press, Boca Raton, 1992.

[HHR10]   P. Hájek, M. Holeňa, and J. Rauch. The GUHA method and its meaning for data mining. *J. of Computer and System Sci.*, 76(1):34–48, 2010.

[HHV95]   P. Hájek, D. Harmancová, and R. Verbrugge. A qualitative fuzzy possibilistic logic. *Int. J. Approximate Reasoning*, 12(1):1–19, 1995.

[HSZ95]   P. Hájek, A. Sochorová, and J. Zvárová. GUHA for personal computers. *Computational Statistics and Data Analysis*, 19(2):149–153, 1995.

[HT01]   P. Hájek and S. Tulipani. Complexity of fuzzy probability logics. *Fund. Informaticae*, 45(3):207–213, 2001.

[HV90]   P. Hájek and J. J. Valdés. Algebraic foundations of uncertainty processing in rule-based expert systems (group-theoretic approach). *Computers and Artificial Intelligence*, 9(4):325–344, 1990.

[HV91]   P. Hájek and J. J. Valdés. A generalized algebraic approach to uncertainty processing in rule-based expert systems (dempsteroids). *Computers and Artificial Intelligence*, 10(1):29–42, 1991.

[HV94]   P. Hájek and J. J. Valdés. An analysis of MYCIN-like expert systems. *Mathware & Soft Computing*, 1(1):45–68, 1994.

[KM07]   J. Kühr and D. Mundici. De Finetti theorem and Borel states in [0,1]-valued algebraic logic. *Int. J. Approximate Reasoning*, 46:605–616, 2007.

[KMP00]   E. P. Klement, R. Mesiar, and E. Pap. *Triangular Norms*, volume 8 of *Trends in Logic*. Kluwer, Dordrecht, 2000.

[Kro06]   T. Kroupa. Every state on semisimple MV-algebra is integral. *Fuzzy Sets Syst.*, 157(20):2771–2782, 2006.

[LS02]   M. C. Laskowski and Y. V. Shashoua. A classification of BL-algebras. *Fuzzy Sets Syst.*, 131(3):271–282, 2002.

[LT30]   J. Łukasiewicz and A. Tarski. Untersuchungen über den Aussagenkalkül. *Comptes Rendus des Séances de la Société des Sciences et des Lettres de Varsovie*, 23(iii):30–50, 1930.

[LT56]   J. Łukasiewicz and A. Tarski. Investigations into the sentential calculus. In *Logic, Semantics, Metamathematics*, pages 38–59. Oxford University Press, 1956. Reprinted by Hackett Publishing Company, Indianapolis, (1983).

[Łuk20]   J. Łukasiewicz. O logice trójwartościowej (On three-valued logic). *Ruch filozoficzny*, 5:170–171, 1920.

[MP76]   J. Menu and J. Pavelka. A note on tensor products on the unit interval. *Comm. Math. Univ. Carolinae*, 17:71–83, 1976.

[Mun09]   D. Mundici. Interpretation of de Finetti coherence criterion in Łukasiewicz logic. *Ann. Pure Appl. Logic*, 161(2):235–245, 2009.

[Pan08]   G. Panti. Invariant measures in free MV-algebras. *Communications in Algebra*, 36:2849–2861, 2008.

[Par01]   J. B. Paris. A note on the Dutch Book method. In G. De Cooman, T. Fine, and T. Seidenfeld, editors, *Proc. of the 2nd Int. Symp. on Imprecise Probabilities and their Applications ISIPTA 2001*, pages 301–306, Ithaca NY, 2001. Shaker Publishing Company. Available at http://www.maths.man.ac.uk/DeptWeb/Homepages/jbp/.

[Pav79]   J. Pavelka. On fuzzy logic I, II, III. *Zeitschr. Math. Logik Grundlagen Math.*, 25:45–52, 119–134, 447–464, 1979.

[Pel02]   A. Pelc. Searching games with errors: fifty years of coping with liars. *Theoretical Comp. Sci.*, 270:71–109, 2002.

[Rén55]   A. Rényi. On a new axiomatic theory of probability. *Acta Math. Academiae Scientiarum Hungaricae*, 6:285–335, 1955.

[Rén76]   A. Rényi. *Napló az Információelméletről*. Gondolat, Budapest, 1976. English translation: *A Diary on Information Theory*, J.Wiley and Sons, New York, 1984.

[RR58]   A. Rose and J. B. Rosser. Fragments of many-valued statement calculi. *Trans. Amer. Math. Soc.*, 87:1–53, 1958.

[Ula76]   S. Ulam. *Adventures of a Mathematician*. Scribner's, New York, 1976.

[WD39]   M. Ward and R. P. Dilworth. Residuated lattices. *Trans. Amer. Math. Soc.*, 45(3):335–354, 1939.

Daniele Mundici
Department of Mathematics "Ulisse Dini"
University of Florence
viale Morgagni 67/A, I-50134 Florence, Italy
E-mail: mundici@math.unifi.it

On states on BL-algebras and related structures[1]

ANATOLIJ DVUREČENSKIJ AND BELOSLAV RIEČAN

Dedicated to Prof. Petr Hájek
on the occasion of his 70th birthday

ABSTRACT. In this paper, BL-algebraic states are presented as averaging processes, or coherent probabilistic assessments, of events described in various many-valued logics. In several algebraic structures arising from these logics, states have many possible definitions, like Bosbach states, or Riečan states. In BL-algebras, as well as in the "good" pseudo BL-algebras discussed in this paper, Bosbach states are the same as Riečan states. We also relate states to de Finetti coherence criterion and we present some ideas on BL-algebras with internal state.

1 Introduction

When physicists entered into the world of elementary particles, they immediately recognized that not only Newton's classical physics ceases to hold, but also quantum measurements can no longer be described by classical statistical methods. This follows from the Heisenberg uncertainty principle [Hei30] which states that the position x and the momentum p of an elementary particle cannot be measured simultaneously with an arbitrarily prescribed accuracy: If $\Delta_m p$ and $\Delta_m x$ denote the inaccuracies of the measurement of the momentum p and position x in a state m, then

$$(\Delta_m p)^2 \cdot (\Delta_m x)^2 \geq \frac{1}{4}\hbar^2 > 0, \qquad (1.1)$$

where $\hbar = h/2\pi$ and h is Planck's constant. Such an inequality does not hold for classical measurements.

In 1933, Kolmogorov presented his axiomatical system of modern probability theory, [Kol33]. While Kolmogorov axioms are still used in contemporary probability theory, they fail for quantum mechanics. The pioneering paper by

[1]The paper has been supported Center of Excellence SAS - Quantum Technologies -, ERDF OP R&D Project CE QUTE ITMS 26240120009, the grants VEGA Nos. 2/0032/09, 1/0539/08 SAV, and by the Slovak Research and Development Agency under the contract No. APVV-0071-06, Bratislava.

Birkhoff and von Neumann [BvN36] initiated a theory of the mathematical foundations of quantum mechanics that is now called the *theory of quantum structures* or *quantum logics*. In contrast to Kolmogorov's triple $(\Omega, \mathcal{S}, P)$ describing probability models, an orthodox model of quantum mechanics is the system $\mathcal{L}(H)$ of closed subspaces of a (real, complex or quaternionic) Hilbert space H with an orthogonal negation $M^{\perp}$ defined as an orthogonal complement of the subspace M, or equivalently, the system of all orthogonal projections, $\mathcal{P}(H)$. This is an orthomodular complete lattice that is not distributive.

In the last decades, physicists have been studying the more general structure $\mathcal{E}(H)$ given by the system of all Hermitian operators lying between the zero and the identity operators. This structure is not a lattice. But all existing infima and suprema in $\mathcal{P}(H)$ are preserved in $\mathcal{E}(H)$.

Nowadays quantum structures present a whole hierarchy of different algebraic structures: Boolean algebras, orthomodular lattices, orthomodular posets, orthoalgebras, D-posets, effect algebras, etc. For a reference source on quantum structures, see [DP00]. The notions of observable and *state* are fundamental for quantum structures. States are analogues of probability measures. Therefore, in order to test if a given class of algebras does provide suitable quantum structures, one may preliminarily ask if a reasonable notion of "state" can be defined on it.

In the early Twenties, a Polish logician, Łukasiewicz, presented his two-page article on a three-valued logic, [Łuk20]. Later this logic was generalized to n-valued as well as to infinitely many-valued propositional logic and then it was algebraized by Chang [Cha58] as MV-algebras in the late Fifties. While Chang's original list of axioms was lengthy, around 18, today we use a friendly list of 5-7 axioms (depending on the assumed primitive operations). Today we are witnesses of an increasing interest in MV-algebras that was initiated by a seminal paper by Mundici [Mun86] who showed that every MV-algebra is in fact an interval of a unique lattice ordered Abelian group with strong unit. For a wonderful trip through the MV-algebra reign, we recommend the monograph [CDM99].

Fuzzy sets introduced by Zadeh have also a many-valued character. This had a great influence also on quantum structures. Whereas the elements of $\mathcal{P}(H)$ have only a yes-no character because their spectrum is at most two-valued, the ones of $\mathcal{E}(H)$ have a fuzzy character and therefore, nowadays they are very popular among quantum physicists who intensively study POV-measures (POV stands for positive operator valued), an analogue of observables with fuzzy features.

In the early nineties, the fuzzy character of quantum structures was initiated by Kôpka and Chovanec [KC94] who introduced D-posets (*difference posets*) where the primary notion is a partial binary operation called *difference.* An orthodox example of D-posets is $\mathcal{E}(H)$. It was noted that MV-algebras are a particular case of D-posets and this fact was very important also for the present authors to study MV-algebras from quantum structure point of view. D-posets were immediately transformed into *effect algebras* by Foulis and Bennett [FB94] with addition as a fundamental partial binary operation. For fuzzy ideas in quantum theories, see [CGG04].

Quantum structures provide an adequate framework to model the fundamental notion of commensurability of two observables (see our discussion in 1.1). Thus for instance, in an orthomodular poset L the commensurability of any two elements amounts to saying that L is a Boolean algebra. In general quantum structures L, commensurability has a "local" character, in the sense that sets of pairwise commensurable elements in L form Boolean blocks different of L, and the orthomodular poset is a union of its blocks, Boolean algebras.

We can also observe similar features for lattice ordered effect algebras because the commensurability principle entails that such an effect algebra can be covered by commensurability blocks that are now MV-subalgebras. Thus MV-algebras play a similar role in many-valued quantum structures as Boolean algebras do for orthomodular lattices or posets.

BL-algebras were introduced in the Nineties by P. Hájek as the equivalent algebraic semantics for his basic fuzzy logic (for a wonderful trip through fuzzy logic realm, use the monograph [Háj98]). They generalize the theory of MV-algebras. Forty years after the appearance of MV-algebras, Mundici [Mun95] presented an analogue of probability, called a state, as averaging process for formulas in Łukasiewicz logic. To define states, Mundici used the same principle as that was independently used in [KC94]. In the last decade, the theory of states on MV-algebras and relative structures is intensively studied by many authors, e.g. [KM07, Kro06, DR06a, DR06b, Geo04, Pan08, Rie00, Rie02, Rie04, Rie05, RM02] and others. Today we have also non-commutative versions of MV-algebras and BL-algebras, called pseudo MV-algebras, GMV-algebras and pseudo BL-algebras.

The definition of states for orthomodular structures is straightforward because it uses a partial relation, orthogonality of two events. However, for BL-algebras and related structures, it is not clear how to find a proper definition. From the literature, we know that there are at least two notions of a state: Bosbach states [Geo04] and Riečan states [Rie00]. The present survey addresses the question of the nature of states on a given algebraic structure and also the relations between different notions of states and related problems.

This paper is organized as follows. Section 2 presents a basic idea how a state can be defined on algebraic structures, in particular for MV-algebras. Section 3 gives two definitions of states, Bosbach and Riečan states on

BL-algebras and we show a relationship between them. States on non-commutative BL-algebras, pseudo BL-algebras, are studied in Section 4. Finally, we present new trends to the theory of states, de Finetti's coherence principle introduced in [Mun06] (see also [KM07]), and state BL-algebras, that are BL-algebras with internal state.

2   States on Algebraic Structures and MV-algebras

Boole in his paper [Boo54] realized that to define a probability measure P on some algebraic structure L (Boolean algebra) it is not necessary to know all the values of $P(A \cup B)$ for all couples A and B – it is enough to know only $P(A \cup B)$ for mutually excluding events A, B, i.e. for those that $A \cap B = \emptyset$, and then to use the additivity condition

$$P(A \cup B) = P(A) + P(B). \tag{2.1}$$

Hence, we have a partial operation $+ := \cup$ on L defined only for mutually excluding events A and B. The property $A \cap B = \emptyset$ is equivalent to

$$A \subseteq B^c, \tag{2.2}$$

where B^c denotes complement.

This principle can be used in many algebraic structures, like orthomodular lattices, posets and effect algebras and also in their non-commutative versions (pseudo MV-algebras, [Dvu01], pseudo BL-algebras, [Geo04], or pseudo effect algebras, [DV01a, DV01b]). Therefore, to define a state on an algebraic structure L with smallest and greatest element, it is necessary to define a partial operation, *addition* $+$, and then a *state* on L is a mapping $s \colon L \to [0,1]$ such that $s(1) = 1$ and $s(a + b) = s(a) + s(b)$ whenever $a + b$ is defined in L.

Suppose L is an effect algebra, i.e. a partial algebraic structure $(L; +, 0, 1)$ such that (i) $+$ is commutative and associative, (ii) for any $a \in L$ there is a unique element a' called a *complement* such that $a + a' = 1$, and (iii) if $a + 1$ is defined in L, then $a = 0$. Then the latter paragraph says how we can define a state on it. One can proceed in a similar way also for pseudo effect-algebras, see [DV01a, DV01b].

Each orthomodular poset is an effect algebra if we define $a + b = a \vee b$ whenever $a \leq b^\perp$. In contrast to Boolean algebras, both orthomodular posets and effect algebras can be stateless.

If an effect algebra arises as an interval $[0, u]$, where u is a strong unit of an Abelian partially ordered group G, and with the restriction of the group addition, then $[0, u]$ has always at least one state, see [Goo86, Cor 4.4].

Suppose that $(L; \oplus, \odot, ^*, 0, 1)$ is an MV-algebra, i.e. an algebra of type $\langle 2, 2, 1, 0, 0 \rangle$ such that (i) $\oplus$ is commutative and associative, (ii) $0^* = 1$,

(iii) $x \oplus 0 = x$, (iv) $x \oplus 1 = 1$, (v) $x^{**} = x$, (vi) $y \oplus (y \oplus x^*)^* = x \oplus (x \oplus y^*)^*$, and (vi) $x \odot y = (x^* \oplus y^*)^*$.

We can define a partial operation $+$ on an MV-algebra L as follows: We say that $a + b$ is defined iff $a \leq b^*$ and then we set $a + b = a \oplus b$. Due to known Mundici's result, every MV-algebra is an interval in a unique unital Abelian ℓ-group (G, u), where u is a strong unit, so that $+$ is the restriction of the group addition on G onto $[0, u]$. Then $(L; +, 0, 1)$ is an effect algebra and in view of the above, we can define a state in the same way as for effect algebras, in addition, every MV-algebra has a state.

In this way, states for effect algebras, and in particular for MV-algebras, were defined in [KC94]. Because in MV-algebras, $a \leq b^*$ holds iff $a \odot b = 0$, Mundici in [Mun95] defined a state as a function $s : L \to [0, 1]$ such that $s(1) = 1$ and $s(a \oplus b) = s(a) + s(b)$ whenever $a \odot b = 0$. Consequently, also Mundici's definition of a state fits precisely our ideas on the partial addition.

This way of deriving a partial addition was used also for a noncommutative generalization of MV-algebras, called *pseudo MV-algebras*, [GI01], or *generalized MV-algebras*, [Rac02]. For more details, see [Dvu01] and Section 4. However, in contrast to MV-algebras, there exist stateless noncommutative MV-algebras

We denote by $\mathcal{S}(L)$ the set of all states on L. If $s_1, s_2 \in \mathcal{S}(L)$ and $\lambda \in [0, 1]$, then $\lambda s_1 + (1 - \lambda) s_2 \in \mathcal{L}(S)$, i.e. $\mathcal{S}(L)$ is a convex set. A state $s \in \mathcal{S}(L)$ is *extremal* if from $s = \lambda s_1 + (1 - \lambda) s_2$ for $\lambda \in (0, 1)$ it follows $s = s_1 = s_2$. We denote by $\partial_e \mathcal{S}(L)$ the set of extremal states on L.

If $\mathcal{S}(L) \neq \emptyset$, we endow $\mathcal{S}(L)$ with the *weak topology* of states: We say that a net of states, $\{s_\alpha\}$, converges weakly to a state, s, if $s(a) = \lim_\alpha s_\alpha(a)$ for any $a \in L$. Then $\mathcal{S}(L)$ is a convex compact Hausdorff topological space, and due to the Krein–Mil'man theorem, see e.g. [Goo86, Thm 5.17], every state s on M is a weak limit of a net of convex combinations of extremal states: i.e.

$$\mathcal{S}(M) = \mathrm{cl}(\mathrm{conv}(\partial_e \mathcal{S}(M))), \tag{2.3}$$

where cl and con denote the closure and the convex hull, respectively. Thus, extremal states are important because they contain information on the whole state space of L.

We notify that also in the case of MV-algebras or BL-algebras, $\partial_e \mathcal{S}(L)$ is a compact nonvoid set, as we will see below. However, there are examples of effect algebras such that $\partial_e \mathcal{S}(L)$ is not compact.

3   Bosbach and Riečan States on BL-algebras

When Hájek presented in the Nineties his axiomatic system of basic logic, BL-algebras, the problem of defining a state became more interesting because in this context it is not immediately clear how to define a state.

According to [Háj98], a *BL-algebra* is an algebra $(L; \wedge, \vee, \odot, \rightarrow, 0, 1)$ of the type $\langle 2, 2, 2, 2, 0, 0 \rangle$ such that $(L; \wedge, \vee, 0, 1)$ is a bounded lattice, $(L; \odot, 1)$ is a commutative monoid, and for all $a, b, c \in L$,

(1) $c \leq a \rightarrow b$ iff $a \odot c \leq b$;

(2) $a \wedge b = a \odot (a \rightarrow b)$;

(3) $(a \rightarrow b) \vee (b \rightarrow a) = 1$.

For any $a \in L$, we denote $a^- = a \rightarrow 0$, complement of a. We note that BL-algebras generalize MV-algebras, and a BL-algebra L is an MV-algebra iff $a^{--} = a$ for any $a \in L$.

If L is a BL-algebra, we can define binary operations

$$x \oplus y := (x^- \odot y^-)^-, \tag{3.1}$$

$x \ominus y := x \odot y^-$, and a function $d(x, y) := (x \rightarrow y) \odot (y \rightarrow x)$. Then $\oplus$ is commutative and

$$x \oplus y = x^- \rightarrow y^{--} = y^- \rightarrow x^{--},$$

$$x \oplus y = (x \oplus y)^{--} = x^{--} \oplus y^{--} = x^{--} \oplus y = x \oplus y^{--},$$

$$x \oplus x^- = 1,$$

$$a \leq x, b \leq y \quad \Rightarrow \quad a \oplus b \leq x \oplus y.$$

$$x \oplus 0 = x^{--}.$$

Moreover, $\oplus$ is associative. Let $\mathrm{MV}(L) := \{x \in L : x = x^{--}\}$. Then $\mathrm{MV}(L) = L^- = \{x^- : x \in L\}$, and according to [TS01], $\mathrm{MV}(L)$ is a subalgebra of L and $(\mathrm{MV}(L); \oplus, \odot, ^-, 0, 1)$ is an MV-algebra, that is the largest MV-subalgebra of L, called an *MV-skeleton* of L.

A subset F of L is a *filter* if (i) $1 \in F$, (ii) $a \leq b$ and $a \in F$ entails $b \in F$, and (iii) if $a, b \in F$, then $a \odot b \in F$. A filter F is said to be (i) proper if $F \neq L$, and (ii) *maximal* if it is proper and not properly contained in another proper filter of L.

If F is a filter of a BL-algebra L, we define the equivalence relation $x \sim_F y$ iff $(x \rightarrow y) \odot (y \rightarrow x) \in F$. Then $\sim_F$ is a congruence and the quotient algebra L/F becomes a BL-algebra. Denoting by x/F the equivalence class of x, then $x/F = 1/F$ iff $x \in F$. Conversely, if $\sim$ is a congruence, then $F_\sim := \{x \in L : x \sim 1\}$ is a filter, and $\sim_{F_\sim} = \sim$, and $F = F_{\sim_F}$.

According to [Geo04], a mapping $s : L \rightarrow [0, 1]$ is a *Bosbach state* if

(BS1) $s(x) + s(x \rightarrow y) = s(y) + s(y \rightarrow x)$, $x, y \in L$;

(BS2) $s(0) = 0$ and $s(1) = 1$.

Let s be a Bosbach state, we define a *kernel* of s as the set

$$\mathrm{Ker}(s) := \{a \in L : s(a) = 1\}.$$

Then $\mathrm{Ker}(s)$ is a filter of L.

Let s be a Bosbach state, then (i) $s(a^-) = 1 - s(a)$, (ii) $s(a^{--}) = s(a)$. Moreover, the quotient $L/\mathrm{Ker}(s)$ is an MV-algebra and the mapping $\tilde{s} \colon L/\mathrm{Ker}(s)$, defined by $\tilde{s}(a/\mathrm{Ker}(s)) = s(a)$ $(a \in L)$, is a state on $L/\mathrm{Ker}(s)$.

Another notion of state on BL-algebras was introduced in [Rie00]. It follows our original ideas. We introduce a partial relation $\perp$ as follows: Two elements $x, y \in L$ are said to be *orthogonal* and we write $x \perp y$ if $x^{--} \le y^-$. Then $x \perp y$ iff $x \le y^-$ iff $x \odot y = 0$. It is clear that $x \perp y$ iff $y \perp x$, and $x \perp 0$ for each $x \in L$. For two orthogonal elements x, y we define a partial binary operation, $+$, on L via $x + y := x \oplus y = y^- \to x^{--} (= x^- \to y^{--})$.

A function $s \colon L \to [0,1]$ is a *Riečan state* if the following conditions hold:

(RS1) if $x \perp y$, then $s(x + y) = s(x) + s(y)$;

(RS2) $s(1) = 1$.

As shown in [Geo04], every Bosbach state is a Riečan state. There was an open question whether there exists a Riečan state that is not Bosbach. The definite answer was given in [DR06a]:

THEOREM 1 *If L is a BL-algebra, then every Bosbach state is a Riečan state and vice-versa.*

Accordingly, in what follows, the term "state" will be used indifferently for Bosbach and for Riečan BL-algebraic states.

A *state-morphism* on a BL-algebra L is a function $m : L \to [0,1]$ satisfying:

(SM1) $m(0) = 0$;

(SM2) $m(x \to y) = \min\{1 - m(x) + m(y), 1\}$,

for any $x, y \in L$. (SM2) entails $s(1) = 1$.

We see that a state-morphism is in fact a BL-homomorphism from L into the BL-algebra of the real interval $[0,1]$ that is also an MV-algebra. It is easy to see that every state-morphism is a state.

The extremal states of a BL-algebra can be characterized as follows:

THEOREM 2 *Let s be state on a BL-algebra L. The following statements are equivalent:*

(i) *s is an extremal state;*

(ii) *s is a state-morphism;*

(iii) *$\mathrm{Ker}(s)$ is a maximal filter;*

(iv) *$s(x \vee y) = \max\{s(x), s(y)\}$, $x, y \in L$.*

Moreover, if F is a maximal filter, then the quotient BL-algebra L/F is isomorphic to an MV-subalgebra of the real interval $[0, 1]$, and the mapping s_F defined on L by $s_F(a) := a/F$, $a \in F$, is an extremal state on L. Therefore, the state space of any BL-algebra is nonempty. In addition, for two extremal states s_1 and s_2 on L we have $s_1 = s_2$ iff $\mathrm{Ker}(s_1) = \mathrm{Ker}(s_2)$. Hence, there is a one-to-one correspondence between the set of extremal states, $\partial_e S(L)$, and the set of extremal filters, $\mathcal{MF}(L)$, given by $s \leftrightarrow \mathrm{Ker}(s)$.

We endow $\mathcal{MF}(L)$ with the hull-kernel topology, i.e. its subbase is given by the system of open sets $\{O(a) : a \in L\}$, where $O(a) := \{F \in \mathcal{MF}(L) : a \in F\}$. Then the mapping $s \leftrightarrow \mathrm{Ker}(s)$ is a homeomorphism of $\partial_e S(L)$ onto $\mathcal{MF}(L)$.

We now present a characterization of states on BL-algebras following ideas of Kroupa [Kro06] and Panti [Pan08] for MV-algebras.

As we have already mentioned, $S(L)$ is always a nonempty compact Hausdorff space, whence we can define a Borel σ-algebra $\mathcal{B}(S(L))$ generated by all open subsets of $S(L)$. Any σ-additive measure on it is said to be a *Borel measure*. We recall that a Borel measure μ is called *regular* if

$$\inf\{\mu(O) :\ Y \subseteq O,\ O \text{ open}\} = \mu(Y) = \sup\{\mu(C) :\ C \subseteq Y,\ C \text{ closed}\}$$

for any $Y \in \mathcal{B}(S(L))$.

We note that given $a \in L$, the mapping $\tilde{a} \colon S(L) \to [0, 1]$ defined by $\tilde{a}(s) := s(a)$, $s \in S(L)$, is a continuous function on $S(L)$.

THEOREM 3 *Let s be a state on a BL-algebra L. Then there is a unique regular Borel probability measure μ on the Borel σ-algebra generated by $S(L)$ such that*

$$s(a) = \int_{\partial_e S(L)} \tilde{a}(x)\, \mathrm{d}\mu(x),\ a \in L, \tag{3.2}$$

where $\tilde{a}(x) = x(a)$, $x \in S(L)$ $(a \in L)$.

Proof Let $\mathrm{Rad}(L) = \bigcap\{\mathrm{Ker}(s) : s \in \partial_e S(L)\}$. We recall that $\mathrm{Rad}(L) = \{x \in L : (x^n)^- \le x, \forall\, n \in \mathbb{N}\}$. Then $L_0 = L/\mathrm{Rad}(L)$ is in fact an MV-algebra that is semisimple. Due to (2.3), $F = \bigcap\{\mathrm{Ker}(s) : s \in S(L)\}$. For any $s \in S(L)$, we define $\tilde{s}$ on L/L_0 via $\tilde{s}(a/\mathrm{Rad}(L)) = s(a)$ $(a \in L)$. Then $\tilde{s}$ is a state on L_0. Moreover, every (extremal) state ν on L_0 defines an (extremal) state s on L via $s(a) := \nu(a/\mathrm{Rad}(L))$, $a \in L$. By [Pan08, Prop 1.1] or [Kro06], for every state $\tilde{s}$ on L_0 there is a unique regular Borel measure μ_0 on $\mathcal{B}(S(L_0))$ such that

$$\tilde{s}(a/\mathrm{Rad}(L)) = \int_{\partial_e S(L_0)} \widehat{a/\mathrm{Rad}(L)}(x_0)\, \mathrm{d}\mu_0(x_0),\ a \in L_0,$$

where $\widehat{a/\mathrm{Rad}(L)}(x_0) = x_0(a/\mathrm{Rad}(L)) = x_0(a)$.

Because the convex compact Hausdorf state spaces $\mathcal{S}(L)$ and $\mathcal{S}(L_0)$ are affinely homeomorphic, so are also $\partial_e \mathcal{S}(L)$ and $\partial_e \mathcal{S}(L_0)$ under the homeomorphic mapping $T_0 s = \tilde{s} =: s_0$. From the last equation we conclude that the last equation gives (3.2) for $\mu = \mu_0 \circ T_0$.   □

4   States on Pseudo BL-algebras

The last decade initiated the study of noncommutative generalization of MV-algebras, known as pseudo MV-algebras [GI01] or noncommutative MV-algebras [Rac02], and BL-algebras called pseudo BL-algebras, [DGI02a, DGI02b].

According to [GI01], a *GMV-algebra* is an algebra $(L; \oplus, ^-, ^\sim, 0, 1)$ of type $\langle 2, 1, 1, 0, 0 \rangle$ such that the following axioms hold for all $x, y, z \in L$, where the operation $\odot$ appearing in the axioms (A6) and (A7) is given by

$$y \odot x = (x^- \oplus y^-)^\sim.$$

(A1) $x \oplus (y \oplus z) = (x \oplus y) \oplus z$;

(A2) $x \oplus 0 = 0 \oplus x = x$;

(A3) $x \oplus 1 = 1 \oplus x = 1$;

(A4) $1^\sim = 0$; $1^- = 0$;

(A5) $(x^- \oplus y^-)^\sim = (x^\sim \oplus y^\sim)^-$;

(A6) $x \oplus (x^\sim \odot y) = y \oplus (y^\sim \odot x) = (x \odot y^-) \oplus y = (y \odot x^-) \oplus x$;

(A7) $x \odot (x^- \oplus y) = (x \oplus y^\sim) \odot y$;

(A8) $(x^-)^\sim = x$.

Let u be an arbitrary positive element of a not necessarily Abelian ℓ-group G (= lattice ordered group). If we define

$$\Gamma(G, u) := [0, u]$$

and

$$
\begin{aligned}
x \oplus y &:= (x + y) \wedge u, \\
x^- &:= u - x, \\
x^\sim &:= -x + u, \\
x \odot y &:= (x - u + y) \vee 0,
\end{aligned}
$$

then $(\Gamma(G, u); \oplus, ^-, ^\sim, 0, u)$ is a pseudo MV-algebra

We recall that for GMV-algebras, a generalization of the Mundici theorem was proved in [Dvu02]. We note that this result states that, for every GMV-algebra L, there is a unique unital ℓ-group (G, u) such that $L \cong \Gamma(G, u)$.

Therefore, for states on GMV-algebras, we can apply the same principle mentioned in Section 2. However, as already mentioned, in contrast to MV-algebras, stateless GMV-algebras exist, [Dvu01]. This is due to the fact that a state on GMV-algebra exists iff there is a maximal filter that is normal. There are unital ℓ-groups that have no maximal ℓ-ideal that is also normal.

According to [DGI02a, DGI02b], a *pseudo BL-algebra* is an algebra $L = (L; \odot, \vee, \wedge, \rightarrow, \rightsquigarrow, 0, 1)$ of type $\langle 2, 2, 2, 2, 2, 0, 0 \rangle$ satisfying the conditions:

(i) $(L; \odot, 1)$ is a monoid (need not be commutative), i.e., $\odot$ is associative with neutral element 1;

(ii) $(L; \vee, \wedge, 0, 1)$ is a bounded lattice;

(iii) $x \odot y \leq z$ iff $x \leq y \rightarrow z$ iff $y \leq x \rightsquigarrow z$ $x, y \in L$;

(iv) $(x \rightarrow y) \odot x = x \wedge y = y \odot (y \rightsquigarrow x)$, $x, y \in L$;

(v) $(x \rightarrow y) \vee (y \rightarrow x) = 1 = (x \rightsquigarrow y) \vee (y \rightsquigarrow y)$, $x, y \in L$.

Every GMV-algebra is a pseudo BL-algebra.

Let L be a pseudo BL-algebra. Let us define two unary operations (negations) $^-$ and $^\sim$ on L such that $x^- := x \rightarrow 0$ and $x^\sim := x \rightsquigarrow 0$ for any $x \in L$. It is easy to show that

$$x \odot y = 0 \Leftrightarrow y \leq x^\sim,$$

or, equivalently, iff $x \leq y^-$.

We say that a pseudo BL-algebra L is *good* if L satisfies the identity

$$x^{-\sim} = x^{\sim-}, \quad x \in L.$$

For example, every BL-algebra is good, every GMV-algebra is good, and due to [Dvu07, Thm 4.1], every linearly ordered pseudo BL-algebra is good, it has a unique maximal filter and this is normal, (see next paragraph for all definitions). We note that it was an open problem whether every pseudo BL-algebra is good. This was answered in negative in [DGK] showing that there are uncountably many subvarieties of pseudo BL-algebras containing pseudo BL-algebras that are not good.

We recall that a *filter* in a pseudo BL-algebra L is a non-void set F of L such that (i) $x, y \in F$ implies $x \odot y \in F$ and (ii) if $x \in F$, $y \in L$ and $x \leq y$, then $y \in F$. A filter F is normal if $x \odot F = F \odot x$ for any $x \in L$. There is a one-to-one correspondence between normal filters and congruences.

Let L be a pseudo BL-algebra. A mapping $s \colon L \longrightarrow [0, 1]$ is said to be a *Bosbach state* (according to [Geo04]) if for all $x, y \in L$:

(S1) $s(x) + s(x \rightarrow y) = s(y) + s(y \rightarrow x)$;

(S2) $s(x) + s(x \rightsquigarrow y) = s(y) + s(y \rightsquigarrow x)$;

(S3) $s(0) = 0$ and $s(1) = 1$.

If L is a GMV-algebra, then this notion coincides with the notion of a state defined for GMV-algebras.

For a Bosbach state s we define $\mathrm{Ker}(s) = \{a \in L : s(a) = 1\}$; it is always a normal filter.

A mapping s from a pseudo BL-algebra L into the standard MV-algebra $[0,1]$ is called a *state-morphism* if for any $x, y \in L$:

(i) $s(x \to y) = s(x \rightsquigarrow y) = s(x) \to s(y)$;

(ii) $s(x \wedge y) = \min\{s(x), s(y)\}$;

(iii) $s(1) = 1$ and $s(0) = 0$.

According to [Geo04], we have the following characterization of Bosbach states on pseudo BL-algebras that is analogous to Theorem 2.

THEOREM 4 *Let s be a Bosbach state on a pseudo BL-algebra L. The following statements are equivalent:*

(i) *s is an extremal Bosbach state;*

(ii) *s is a state-morphism;*

(iii) *$\mathrm{Ker}(s)$ is a maximal filter;*

(iv) *$s(x \vee y) = \max\{s(x), s(y)\}$, $x, y \in L$.*

In addition, L has at least one Bosbach state if and only if L has a maximal filter that is normal.

Therefore, the set of extremal Bosbach states is compact in the weak topology of (Bosbach) states. The mappings $s \leftrightarrow \mathrm{Ker}(s) \leftrightarrow F$, where s is an extremal state and F is a maximal filter that is normal, are one-to-one. Moreover, if s is a Bosbach state, then $L/\mathrm{Ker}(s)$ is an MV-algebra and $\tilde{s}(a/\mathrm{Ker}(s)) := s(a)$ $(a \in L)$ is a state on $L/\mathrm{Ker}(s)$. Hence, also for every Bosbach state s on a pseudo BL-algebra L we can derive a formula analogous to (3.2).

Let L be a good pseudo BL-algebra. We define a total binary operation, $\oplus$, on L in analogous way as in (3.1) by

$$x \oplus y := (y^\sim \odot x^\sim)^-, \quad x, y \in L. \tag{4.1}$$

Then $(y^\sim \odot x^\sim)^- = (y^- \odot x^-)^\sim$.

We set

$$\mathrm{MV}(L) = \{x \in L : x^{-\sim} = x^{\sim-} = x\}.$$

Then $0, 1 \in \mathrm{MV}(L)$ and $\oplus$ is commutative and associative.

According to [Geo04, Prop. 1.19], if L is a good pseudo BL-algebra, then the subset $MV(L)$ endowed with $\oplus,^-,^\sim,0,1$ is a GMV-algebra called the GMV-*skeleton* of L.

Using (4.1), for a good pseudo BL-algebra, we can define a partial addition, $+$. Then $+$ follows the principle described in Section 2. We write

$$x \perp y \quad \text{iff} \quad y^{-\sim} \leq x^-. \tag{4.2}$$

We recall that (i) $x \perp y$ iff $x \odot y = 0$, (ii) $x \perp y$ iff $x^{-\sim} \leq y^\sim$, (iii) $x^\sim \perp x$, $x \perp x^-$, (iv) if $x \leq y$, then $x \perp y^-$ and $y^\sim \perp x$, and (v) if L is commutative, then $x \perp y$ iff $y \perp x$.

We say that $x + y$ is defined in L if $x \perp y$ and then we set $x + y = x \oplus y$, where $\oplus$ is defined by (4.1).

We can also apply principle (4.2) to define a state for pseudo BL-algebras. Thus, we say that a mapping $s\colon L \to [0,1]$ is a *Riečan state* if (i) $s(x + y) = s(x) + s(y)$ whenever $x + y$ is defined in L, and (ii) $s(1) = 1$. We note that every Bosbach state on a good pseudo BL-algebra is a Riečan state, [Geo04, Prop. 2.14]. The converse statement was established in [DR06a, Thm 3.8]:

THEOREM 5 *If L is a good pseudo BL-algebra, then every Bosbach state is a Riečan state and vice-versa.*

5   States on BL-algebras and de Finetti Coherence Maps, and State BL-algebras

In this section, we will follow ideas from [Mun06, KM07] and apply them to BL-algebras. In addition, we generalize some ideas on MV-algebras with an internal state introduced in [FM09] to BL-algebras with an internal state [CDH].

Let L be an arbitrary non-void set and let $\mathcal{W}$ be a fixed system of maps from $[0,1]^L$. We endow $\mathcal{W}$ with the weak topology induced from the product topology $[0,1]^L$. We say that a *coherent map* over a finite set $L' = \{a_1, \ldots, a_n\}$ is a map $\beta\colon L' \to [0,1]$ such that

$$\forall\, \sigma_1, \ldots, \sigma_n \in \mathbb{R},\ \exists\, V \in \mathcal{W}\ \text{ s.t. } \ \sum_{i=1}^{n} \sigma_i(\beta(a_i) - V(a_i)) \geq 0. \tag{5.1}$$

According to [Mun06, KM07], an interpretation of (5.1) is inspired by ideas on Dutch booking.

A mapping $\beta\colon L \to [0,1]$ is a *de Finetti map* if β is coherent over every finite subset of L and we denote by $\mathcal{F}_\mathcal{W}$ the set of all de Finetti maps on L.

THEOREM 6 *Let L be a non-void set. If $\mathcal{W}$ is closed, then*

$$\mathcal{F}_\mathcal{W} = \mathrm{cl}(\mathrm{conv}(\mathcal{W})). \tag{5.2}$$

We recall that (5.2) is in a close relation with (2.3).

Assume that L is a BL-algebra and set $W = \partial_e \mathcal{S}(L)$. Due to Theorem 2, $\partial_e \mathcal{S}(L)$ is a compact set therefore, applying Theorem 6, all the de Finetti maps are exactly states on the BL-algebra L, that is, $\mathcal{F}_{\partial_e \mathcal{S}(L)} = \mathcal{S}(L)$.

Similar statement holds also for the case of Bosbach states and extremal Bosbach states for pseudo BL-algebras.

We finish the paper with introducing state BL-algebras, see [CDH]. We extend the language of BL-algebras introducing a new unary operation, σ, called an internal state or a state-operator. This idea is inspired by state MV-algebras introduced recently in [FM09] and studied also in [DD09b, DD09a].

It presents a unified approach to states and probabilistic many-valued logic in logical and algebraic settings. For example, Hájek's approach, [Háj98], to fuzzy logic with modality Pr (interpreted as *probably*) has the following semantic interpretation: The probability of an event a is presented as the truth value of $\Pr(a)$. If s is a state, then $s(a)$ is interpreted as averaging of appearing the many-valued event a.

According to [CDH], we say that a mapping σ from a BL-algebra L into itself is a *state-operator* on L, if, for all $x, y \in L$, we have

$(1)_{BL}$ $\sigma(0) = 0$;

$(2)_{BL}$ $\sigma(x \to y) = \sigma(x) \to \sigma(x \wedge y)$;

$(3)_{BL}$ $\sigma(x \odot y) = \sigma(x) \odot \sigma(x \to x \odot y)$;

$(4)_{BL}$ $\sigma(\sigma(x) \odot \sigma(y)) = \sigma(x) \odot \sigma(y)$;

$(5)_{BL}$ $\sigma(\sigma(x) \to \sigma(y)) = \sigma(x) \to \sigma(y)$,

and the pair (L, σ) is said to be a *state BL-algebra*, or more precisely, a *BL-algebra with internal state*.

The system of all state BL-algebras forms a variety.

EXAMPLE 7 *Let L be a BL-algebra. On $L \times L$ we define two operators, σ_1 and σ_2, as follows*

$$\sigma_1(a, b) = (a, a), \quad \sigma_2(a, b) = (b, b), \quad (a, b) \in L \times L. \tag{5.3}$$

Then σ_1 and σ_2 are two state-operators on $L \times L$ that are also endomorphisms such that $\sigma_i^2 = \sigma_i$, $i = 1, 2$. Moreover, $(L \times L, \sigma_1)$ and $(L \times L, \sigma_2)$ are isomorphic state BL-algebras under the isomorphism $(a, b) \mapsto (b, a)$.

Let (L, σ) be a state BL-algebra. A filter F of a BL-algebra L is said to be a *state-filter* if $\sigma(F) \subseteq F$. If we set $\mathrm{Ker}(s) = \{a \in L : \sigma(a) = 1\}$, then $\mathrm{Ker}(\sigma)$ is a state-filter. A state-operator σ is *faithful* if $\mathrm{Ker}(\sigma) = \{1\}$.

The basic properties of state BL-algebras are as follows, [CDII]:

LEMMA 8 *In a state BL-algebra (L, σ) the following hold:*
(a) $\sigma(1) = 1$;
(b) $\sigma(x^-) = \sigma(x)^-$;
(c) *if $x \leq y$, then $\sigma(x) \leq \sigma(y)$;*
(d) $\sigma(x \odot y) \geq \sigma(x) \odot \sigma(y)$ *and if $x \odot y = 0$, then $\sigma(x \odot y) = \sigma(x) \odot \sigma(y)$;*
(e) $\sigma(x \ominus y) \geq \sigma(x) \ominus \sigma(y)$ *and if $x \leq y$, then $\sigma(x \ominus y) = \sigma(x) \ominus \sigma(y)$;*
(f) $\sigma(x \wedge y) = \sigma(x) \odot \sigma(x \rightarrow y)$;
(g) $\sigma(x \rightarrow y) \leq \sigma(x) \rightarrow \sigma(y)$ *and if x, y are comparable, then $\sigma(x \rightarrow y) = \sigma(x) \rightarrow \sigma(y)$;*
(h) $\sigma(x \rightarrow y) \odot \sigma(y \rightarrow x) \leq d(\sigma(x), \sigma(y))$;
(i) $\sigma(x) \oplus \sigma(y) \geq \sigma(x \oplus y)$ *and if $x \oplus y = 1$, then $\sigma(x) \oplus \sigma(y) = \sigma(x \oplus y) = 1$;*
(j) $\sigma(\sigma(x)) = \sigma(x)$;
(k) $\sigma(L)$ *is a BL-subalgebra of L;*
(l) $\sigma(L) = \{x \in L : x = \sigma(x)\}$;
(o) $\sigma(x \rightarrow y) = \sigma(x) \rightarrow \sigma(y)$ *iff $\sigma(y \rightarrow x) = \sigma(y) \rightarrow \sigma(x)$;*
(p) *if $\sigma(L) = L$, then σ is the identity on L;*
(q) *if σ is faithful, then $x < y$ implies $\sigma(x) < \sigma(y)$;*
(r) *if σ is faithful then either $\sigma(x) = x$ or $\sigma(x)$ and x are not comparable;*
(s) *if L is linear and σ faithful, then $\sigma(x) = x$ for any $x \in L$.*

We recall that there is a one-to-one correspondence between congruences and state-filters on a state BL-algebra (L, σ) as follows. If F is a state-filter, then the relation $\sim_F$ given by $x \sim_F y$ iff $x \rightarrow y, y \rightarrow x \in F$ is a congruence of the BL-algebra L and due to Lemma 8(h), $\sim_F$ is also a congruence of the state BL-algebra (L, σ).

In contrast to BL-algebras, where each subdirectly irreducible BL-algebra is linearly ordered, this is not a case for state BL-algebras, in general. Indeed, if in Example 7, we have $L = [0, 1]$, then $(L \times L, \sigma_i)$ gives a subdirectly irreducible state BL-algebra such that $L \times L$ is not linearly ordered. At any rate, if (L, σ) is subdirectly irreducible, then $\sigma(L)$ is linearly ordered.

We can define more types of state-operators: (i) *strong state-operator* if σ satisfies $(1)_{BL}, (2)_{BL}, (4)_{BL}, (5)_{BL}$, and $\sigma(x \odot y) = \sigma(x) \odot \sigma(x^- \vee y)$, (ii) *state-morphism-operator* if σ satisfies satisfying $(1)_{BL}, (2)_{BL}, (4)_{BL}, (5)_{BL}$, and $\sigma(x \odot y) = \sigma(x) \odot \sigma(y)$ for any $x, y \in A$.

We recall: (i) If a state-operator σ preserves $\rightarrow$, then σ is a state-morphism-operator, (ii) every state-morphism-operator is always a strong state-operator, (iii) every strong state-operator is a state-operator and if L is an MV-algebra, then these notions coincide, and (iv) every state-operator σ on a linearly ordered BL-algebra is an endomorphism such that $\sigma^2 = \sigma$.

On the other hand we do not know whether (1) there exists a state-operator that is not strong, or if (2) any state-morphism-operator preserve $\rightarrow$. A great challenge for this theory seems to be the problem of describing subdirectly irreducible state BL-algebras or state MV-algebras, see also [FM09].

State BL-algebras together with state MV-algebras seem to be very interesting and they give a new way how to axiomatize the notion of state.

Acknowledgement. The authors are strongly indebted to the anonymous referee for his very careful reading and his suggestions.

BIBLIOGRAPHY

[Boo54]   G. Boole. *An Investigation of the Laws of Thought, on Which are Founded the Mathematical Theories of Logic and Probabilities.* Macmillan, 1854. Reprinted by Dover Press, New York, 1967.

[BvN36]   G. Birkhoff and J. von Neumann. The logic of quantum mechanics. *Ann. Math. Second Series*, 37(4):823–843, 1936.

[CDH]   L. C. Ciungu, A. Dvurečenskij, and M. Hyčko. State BL-algebras. Submitted.

[CDM99]   R. Cignoli, I. M. L. D'Ottaviano, and D. Mundici. *Algebraic Foundations of Many-Valued Reasoning*, volume 7 of *Trends in Logic*. Kluwer, Dordrecht, 1999.

[CGG04]   M.-L. D. Chiara, R. Giuntini, and R. Greechie. *Reasoning in Quantum Theory. Sharp and Unsharp Quantum Logics*, volume 22 of *Trends in Logic*. Kluwer, Dordrecht, 2004.

[Cha58]   C. C. Chang. Algebraic analysis of many-valued logics. *Trans. Amer. Math. Soc.*, 88:456–490, 1958.

[DD09a]   A. Di Nola and A. Dvurečenskij. On some classes of state-morphism MV-algebras. *Math. Slovaca*, 59:517–534, 2009.

[DD09b]   A. Di Nola and A. Dvurečenskij. State-morphism MV-algebras. *Ann. Pure Appl. Logic*, 161:161–173, 2009.

[DGI02a] A. Di Nola, G. Georgescu, and A. Iorgulescu. Pseudo-BL-algebras, part I. *Multiple-Valued Logic*, 8(5–6):673–714, 2002.

[DGI02b] A. Di Nola, G. Georgescu, and A. Iorgulescu. Pseudo-BL-algebras, part II. *Multiple-Valued Logic*, 8(5–6):717–750, 2002.

[DGK]   A. Dvurečenskij, R. Giuntini, and T. Kowalski. On the structure of pseudo BL-algebras and pseudo hoops in quantum logics. To appear in *Found. Phys.* DOI 10.1007/s10701-009-9342-5.

[DP00]   A. Dvurečenskij and S. Pulmannová. *New Trends in Quantum Structures.* Mathematics and Its Applications. Kluwer and Ister Science Ltd., Dordrecht, Bratislava, 2000.

[DR06a]   A. Dvurečenskij and J. Rachůnek. On Riečan and Bosbach states for bounded non-commutative Rℓ-monoids. *Math. Slovaca*, 56:487–500, 2006.

[DR06b]   A. Dvurečenskij and J. Rachůnek. Probabilistic averaging in bounded non-commutative Rℓ-monoids. *Semigroup Forum*, 72:190–206, 2006.

[DV01a]   A. Dvurečenskij and T. Vetterlein. Pseudo-effect algebras I Basic properties. *Int. J. Theor. Phys.*, 40:685–701, 2001.

[DV01b]   A. Dvurečenskij and T. Vetterlein. Pseudo-effect algebras II Group representation. *Int. J. Theor. Phys.*, 40:703–726, 2001.

[Dvu01]   A. Dvurečenskij. States on pseudo MV-algebras. *Studia Logica*, 68:301–327, 2001.

[Dvu02]   A. Dvurečenskij. Pseudo MV-algebras are intervals in l-groups. *J. Australian Math. Soc.*, 72(3):427–445, 2002.

[Dvu07]   A. Dvurečenskij. Every linear pseudo BL-algebra admits a state. *Soft Computing*, 11:495–501, 2007.

[FB94]   D. Foulis and M. Bennett. Effect algebras and unsharp quantum logics. *Found. Phys.*, 24:1325–1346, 1994.

[FM09]   T. Flaminio and F. Montagna. MV-algebras with internal states and probabilistic fuzzy logics. *Int. J. Approximate Reasoning*, 50:138–152, 2009.

302   Anatolij Dvurečenskij and Beloslav Riečan

[Geo04]   G. Georgescu. Bosbach states on fuzzy structures. *Soft Computing*, 8:217–230, 2004.

[GI01]   G. Georgescu and A. Iorgulescu. Pseudo-MV algebras. *Multiple-Valued Logic*, 6:95–135, 2001.

[Goo86]   K. R. Goodearl. *Partially Ordered Abelian Groups with Interpolation*, volume 20 of *Mathematical Surveys and Monographs*. AMS, Providence, 1986.

[Háj98]   P. Hájek. *Metamathematics of Fuzzy Logic*, volume 4 of *Trends in Logic*. Kluwer, Dordrecht, 1998.

[Hei30]   W. Heisenberg. *The Physical Principles of Quantum Theory*. Dover Publications, New York, 1930.

[KC94]   F. Kôpka and F. Chovanec. D-posets. *Math. Slovaca*, 44:21–34, 1994.

[KM07]   J. Kühr and D. Mundici. De Finetti theorem and Borel states in [0,1]-valued algebraic logic. *Int. J. Approximate Reasoning*, 46:605–616, 2007.

[Kol33]   A. N. Kolmogorov. *Grundbegriffe der Wahrscheinlichkeitsrechnung*. Julius Springer, Berlin, 1933.

[Kro06]   T. Kroupa. Every state on semisimple MV-algebra is integral. *Fuzzy Sets Syst.*, 157(20):2771–2782, 2006.

[Łuk20]   J. Łukasiewicz. O logice trójwartościowej (On three-valued logic). *Ruch filozoficzny*, 5:170–171, 1920.

[Mun86]   D. Mundici. Interpretation of AF C^*-algebras in Łukasiewicz sentential calculus. *J. Funct. Anal.*, 65(1):15–63, 1986.

[Mun95]   D. Mundici. Averaging the truth-value in Łukasiewicz logic. *Studia Logica*, 55(1):113–127, 1995.

[Mun06]   D. Mundici. Bookmaking over infinite-valued semantics. *Int. J. Approximate Reasoning*, 43:223–240, 2006.

[Pan08]   G. Panti. Invariant measures in free MV-algebras. *Communications in Algebra*, 36:2849–2861, 2008.

[Rac02]   J. Rachůnek. A non-commutative generalization of MV-algebras. *Czechoslovak Math. J.*, 52:255–273, 2002.

[Rie00]   B. Riečan. On the probability on BL-algebras. *Acta Math. Nitra*, 4:3–13, 2000.

[Rie02]   B. Riečan. Free products of probability MV-algebras. *Atti Semin. Mat. Fis. Univ. Modena*, 50:173–186, 2002.

[Rie04]   B. Riečan. The conjugacy of probability MV-σ-algebras with the unit interval. *Atti Semin. Mat. Fis. Univ. Modena Reggio Emilia*, 52:241–248, 2004.

[Rie05]   B. Riečan. Kolmogorov-Sinaj entropy on MV-algebras. *Int. J. Theor. Phys.*, 44:1041–1052, 2005.

[RM02]   B. Riečan and D. Mundici. Probability on MV-algebras. In E. Pap, editor, *Handbook of Measure Theory, Vol. II*, pages 869–909. Elsevier, Amsterdam, 2002.

[TS01]   E. Turunen and S. Sessa. Local BL-algebras. *Multiple-Valued Logic*, 6:229–249, 2001.

Anatolij Dvurečenskij
Mathematical Institute, Slovak Academy of Sciences,
Štefánikova 49, SK-814 73 Bratislava, Slovakia
Email: dvurecen@mat.savba.sk

Beloslav Riečan
Institute of Mathematics and Informatics,
Slovak Academy of Sciences and UMB,
Ďumbierska 1, SK-974 11 Banská Bystrica, Slovakia
Email: riecan@fpv.umb.sk

A survey of generalized Basic Logic algebras

Nikolaos Galatos and Peter Jipsen

Dedicated to Petr Hájek,
on the occasion of his 70th birthday

ABSTRACT. Petr Hájek identified the logic **BL**, that was later shown to be the logic of continuous t-norms on the unit interval, and defined the corresponding algebraic models, BL-algebras, in the context of residuated lattices. The defining characteristics of BL-algebras are representability and divisibility. In this short note we survey recent developments in the study of divisible residuated lattices and attribute the inspiration for this investigation to Petr Hájek.

1 Introduction

Petr Hájek's book helped place fuzzy logic on firm mathematical ground. In particular, in view of the truth-functionality of the logic, he developed the algebraic theory of the corresponding models, which he situated within the study of residuated lattices. Moreover, he identified precisely the logic of continuous t-norms, by introducing Basic Logic **BL**.

Basic Logic generalizes three important and natural fuzzy logics, namely Łukasiewicz logic, Product logic and Gödel logic, and allows for the study of their common features. It turns out that BL-algebras, the algebraic models of **BL**, are in a sense made up of these three types of models. Arbitrary BL-algebras are subdirect products of BL-chains, which in turn are ordinal sums of MV-chains, Product-chains and Gödel-chains (though the latter are implicit within the ordinal sum construction).

Within commutative integral bounded residuated lattices, BL-algebras are exactly the representable and divisible ones. Representability as a subdirect product of chains (a.k.a. semilinearity) is often considered synonymous to fuzziness of a logic. It corresponds to the total order on the unit interval of the standard t-norms, and presents a natural view of fuzzy logic as pertaining to (linearly-ordered) degrees of truth.

On the other hand, divisibility renders the meet operation definable in terms of multiplication and its residual(s). It corresponds to the property of having a natural ordering in semigroups. Divisibility has also appeared in the study of complementary semigroups, in the work of Bosbach [Bos82], and in the study of hoops, as introduced by Büchi and Owens [BO]. This shows that it is a natural condition, appearing not only in logic, but also in algebra. While residuation corresponds to the left-continuity of a t-norm on the unit interval, basic logic captures exactly the semantics of continuous t-norms on the unit interval, as was shown later.

On a personal note, we would like to mention that Petr Hájek's book has been influential in developing some of the theory of residuated lattices and also connecting it with the study of logical systems. The book had just appeared shortly before a seminar on residuated lattices started at Vanderbilt University, organized by Constantine Tsinakis and attended by both authors of this article. The naturality of the definition of divisibility and its encompassing nature became immediately clear and *generalized BL-algebras* (or GBL-algebras) were born. These are residuated lattices (not necessarily commutative, integral, contractive or bounded) that satisfy divisibility: if $x \leq y$, there exist z, w with $x = zy = yw$. It turns out that the representable commutative bounded GBL-algebras are exactly the BL-algebras. In other words, GBL-algebras are a generalization of BL-algebras that focuses on and retains the divisibility property. Examples include lattice-ordered groups, their negative cones, (generalized) MV-algebras and Heyting algebras.

Despite their generality, GBL-algebras decompose into direct products of lattice-ordered groups and integral GBL-algebras, so it is these two subvarieties that are of main interest. From [Mun86] it follows that MV-algebras are certain intervals in abelian ℓ-groups and in [GT05] it is shown that GMV-algebras are certain convex sublattices in ℓ-groups. Also, BL-chains (the building blocks of BL-algebras) are essentially ordinal sums of parts of ℓ-groups (MV-chains and product chains). It is interesting that recent work has shown that algebras in many natural classes of GBL-algebras (including commutative, as well as k-potent) are made from parts of ℓ-groups put (in the form of a poset product) into a Heyting algebra grid. The poset product decomposition does not work for all GBL-algebras, but it is an open problem to find GBL-algebras that are not locally parts of ℓ-groups, if such algebras exist.

2   BL-algebras as residuated lattices

2.1   Residuated lattices

A *residuated lattice* is a structure $(L, \wedge, \vee, \cdot, \backslash, /, 1)$, where $(L, \wedge, \vee)$ is a lattice, $(L, \cdot, 1)$ is a monoid and the law of residuation holds; i.e., for all $a, b, c \in L$,

$$a \cdot b \leq c \quad \Leftrightarrow \quad b \leq a \backslash c \quad \Leftrightarrow \quad a \leq c/b.$$

Sometimes the expression $x \to y$ is used for $x \backslash y$, while $y \leftarrow x$ (or $x \rightsquigarrow y$) is used for y/x. The corresponding operations are called the residuals of multiplication. *FL-algebras* are expansions $(L, \wedge, \vee, \cdot, \backslash, /, 1, 0)$ of residuated lattices with an additional constant operation 0.

Residuated lattices and FL-algebras are called *commutative* if multiplication is commutative, *integral* if 1 is the greatest element, *representable* (or *semilinear*) if they are subdirect products of chains, and *divisible* if they satisfy:

If $x \leq y$, there exist z, w such that $x = zy = yw$.

In commutative residuated lattices we have $x\backslash y = y/x$, and we denote the common value by $x \to y$.

An *FL$_o$-algebra*, also known as a *bounded residuated lattice*, is an FL-algebra in which $0 \leq x$ for all x. In this case 0 is also denoted by $\perp$. Moreover, it turns out that $\top = 0/0 = 0\backslash 0$ is the top element. Integral FL$_o$-algebras are also known as FL$_w$-algebras, and in the presence of commutativity as FL$_{ew}$-algebras.[1] For more on residuated lattices see [GJKO07].

2.2   BL-algebras

It turns out that *BL-algebras*, as defined by P. Hájek, are exactly the representable and divisible algebras within the class of integral commutative bounded residuated lattices (i.e., within FL$_{ew}$-algebras).

THEOREM 1 ([BT03, JT02]) *A residuated lattice is representable iff it satisfies* $[z\backslash((x \vee y)/x)z \wedge 1] \vee [w((x \vee y)/y)/w \wedge 1] = 1$.

The next result then follows easily.

COROLLARY 2 *An integal, commutative residuated lattice is representable iff it satisfies* $(x \to y) \vee (y \to x) = 1$ *(prelinearity).*

P. Hájek defined BL-algebras using the prelinearity condition, which captures representability in the integral, commutative case. He also used the simplified form of divisibility $x \wedge y = x(x \to y)$, which we will see is equivalent to the general form in the integral commutative case; actually integrality follows from this particular form of divisibility, by setting $x = 1$. The variety (i.e. equational class) of BL-algebras is denoted by BL.

Representable Heyting algebras, known as *Gödel algebras*, are exactly the idempotent $(x^2 = x)$ BL-algebras. Moreover, Chang's MV-algebras are exactly the involutive $((x \to 0) \to 0 = x)$ BL-algebras. For additional subvarieties of FL$_w$ see Table 1 and 3, as well as Figure 1.

---

[1]The subscripts e, w are from the names of the rules *exchange* and *weakening* in proof theory that correspond to commutativity and integrality.

divisible	$x(x\backslash(x \wedge y)) = x \wedge y = ((x \wedge y)/x)x$
prelinear	$x\backslash y \vee y\backslash x = 1 = x/y \vee y/x$
linear or chain	$x \leq y$ or $y \leq x$
integral	$x \leq 1$
bottom	$0 \leq x$
commutative	$xy = yx$
idempotent	$xx = x$

Table 1. Some residuated lattice axioms

2.3   Algebraization

P. Hájek defined basic logic **BL** via a Hilbert-style system, whose sole inference rule is modus ponens; see Table 2.

(sf)	$(\varphi \to \psi) \to ((\psi \to \chi) \to (\varphi \to \chi))$	(suffixing)
(int)	$(\varphi \cdot \psi) \to \varphi$	(integrality)
(com)	$(\varphi \cdot \psi) \to (\psi \cdot \varphi)$	(commutativity)
(conj)	$(\varphi \cdot (\varphi \to \psi)) \to (\psi \cdot (\psi \to \varphi))$	(conjunction)
$(\cdot\to)$	$((\varphi \cdot \psi) \to \chi) \to (\varphi \to (\psi \to \chi))$	
$(\to\cdot)$	$(\varphi \to (\psi \to \chi)) \to ((\varphi \cdot \psi) \to \chi)$	
$(\to\text{pl})$	$((\varphi \to \psi) \to \chi) \to (((\psi \to \varphi) \to \chi) \to \chi)$	(arrow prelinearity)
(bot)	$0 \to \varphi$	
	$\varphi \wedge \psi := \varphi \cdot (\varphi \to \psi)$	(conjunction definition)
	$\varphi \vee \psi := [(\varphi \to \psi) \to \psi] \wedge [(\psi \to \varphi) \to \varphi]$	(disjunction definition)
	$\neg\varphi := \varphi \to 0$	(negation definition)
	$1 := 0 \to 0$	(unit definition)

$$\frac{\varphi \quad \varphi \to \psi}{\psi} \ (\text{mp})$$

Table 2. Hajek's basic logic.

Using this system, one defines as usual the notion of proof from assumptions. If ψ is a propositional formula in the language of **BL**, and Φ is a set of such formulas, then $\Phi \vdash_{\textbf{BL}} \psi$ denotes that ψ is provable in **BL** from (non-logical, i.e., no substitution instances are allowed) assumptions Φ.

The following results states that $\vdash_{\textbf{BL}}$ is algebraizable (in the sense of Blok and Piggozzi [BP89]) with respect to the variety BL of BL-algebras. It follows directly from the more general algebraization of substructural logics by residuated lattices given in [GO06].

As usual, propositional formulas in **BL** are identified with terms of BL. Recall that for a set $E \cup \{s = t\}$ of equations, $E \models_{\mathsf{BL}} s = t$ ($s = t$ is a *semantial consequence* of E with respect to BL) denotes that for every BL-algebra **B** and for every assignment f into **B** (i.e., for every homomorphism from the formula/term algebra into **B**), if $f(u) = f(v)$ for all $(u = v) \in E$, then $f(s) = f(t)$.

THEOREM 3 *For every set* $\Phi \cup \{\psi\}$ *of propositional formulas and every set* $E \cup \{s = t\}$ *of equations,*

(i) $\Phi \vdash_{\mathbf{BL}} \psi$ *iff* $\{\varphi = 1 : \varphi \in \Phi\} \models_{\mathsf{BL}} \psi = 1$

(ii) $E \models_{\mathsf{BL}} s = t$ *iff* $\{(u \to v) \wedge (v \to u) : (u = v) \in E\} \vdash_{\mathbf{BL}} (s \to t) \wedge (t \to s)$

(iii) $\varphi \dashv\vdash (\varphi \to 1) \wedge (1 \to \varphi)$

(iv) $s = t \dashv\models_{\mathsf{BL}} (s \to t) \wedge (t \to s) = 1$

As a corollary we obtain that the lattice of axiomatic extensions of **BL** is dually isomorphic to the subvariety lattice of BL.

3 Fuzzy logics and triangular norms: retaining representability

Truth in a fuzzy logic comes in degrees. In a so-called standard model, truth values are taken to be real numbers from the interval $[0, 1]$, while logical connectives of arity n are interpreted as functions from $[0, 1]^n$ to $[0, 1]$. In particular, multiplication in a standard model is assumed to be continuous and a *triangular norm* (*t-norm*), namely a binary operation on the interval $[0, 1]$ that is associative, commutative, monotone and has 1 as its unit element. It is easy to see that any such continuous t-norm defines a BL-algebra structure on $[0, 1]$. In fact, **BL** is complete with respect to continuous t-norms, as proved in [CEGT00], where it is shown algebraically that the variety of BL-algebras is generated by all continuous t-norms.

Lukasiewicz t-norm ($L(x, y) = \max\{x + y - 1, 0\}$), Gödel t-norm ($G(x, y) = \min\{x, y\}$), and product t-norm ($\Pi(x, y) = xy$), define three special standard models of **BL**. The corresponding logics that extend **BL** are denoted by **Ł**, **GL** and **Π**.

Chang [Cha59] proved that the standard model given by $L(x, y)$ generates the variety of MV-algebras. Thus, **Ł** is precisely the infinite-valued Lukasiewicz logic and it is axiomatized relatively to **BL** by $\neg\neg\varphi \to \varphi$. Also, **GL** is Gödel logic, namely the superintuitionistic logic defined by adding to **BL** the axiom $\varphi \to \varphi^2$, and it is the smallest superintuitionistic logic that is also a fuzzy logic. Finally, **Π** is *product logic* and is defined relatively to **BL** by $\neg\neg\varphi \to (((\varphi \to (\varphi \cdot \psi)) \to (\psi \cdot \neg\neg\psi))$; see [Cin01].

FL_w	**FL-algebras with weakening** = integral residuated lattices with bottom
FL_{ew}	**FL_w-algebras with exchange** = commutative integral residuated lattices with bottom
GBL_w	**GBL-algebras with weakening** = divisible integral residuated lattices with bottom
GBL_{ew}	**GBL_w-algebras with exchange** = commutative GBL_w-algebras
psMTL	**pseudo monoidal t-norm algebras** = integral residuated lattices with bottom and prelinearity
MTL	**monoidal t-norm algebras** = psMTL with commutativity
psBL	**pseudo BL-algebras** = psMTL with divisibility
BL	**basic logic algebras** = MTL with divisibility
HA	**Heyting algebras** = residuated lattices with bottom and $x \wedge y = xy$
psMV	**pseudo MV-algebras** = psBL with $x \vee y = x/(y\backslash x) = (x/y)\backslash x$
MV	**MV-algebras or Łukasiewicz algebras** = BL-algebras that satisfy $\neg\neg x = x$ = commutative pseudo MV-algebras
MV_n	**MV-algebras generated by $n+1$-chains** = subdirect products of the $n+1$-element MV-chain
GA	**Gödel algebras or linear Heyting algebras** = BL-algebras that are idempotent = Heyting algebras with prelinearity
GA_n	**Gödel algebras generated by $n+1$-chains** = subdirect products of the $n+1$-element Heyting chain
Π	**product algebras** = BL-algebras that satisfy $\neg\neg x \leq (x \rightarrow xy) \rightarrow y(\neg\neg y)$
BA	**Boolean algebras** = Heyting algebras that satisfy $\neg\neg x = x$ = MV-algebras that are idempotent

Table 3. Some subvarieties of FL_w

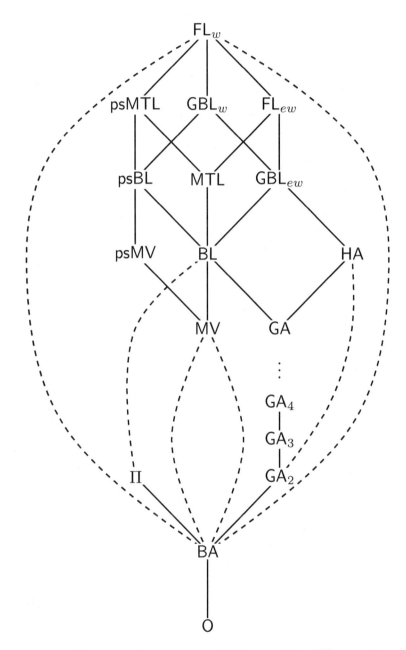

Figure 1: Some subvarieties of FL_w ordered by inclusion

It is easy to see that, due to the completeness of $[0,1]$, a t-norm is residuated iff it is left-continuous; divisibility provides right-continuity, in this context. It turns out that the variety of representable FL_{ew}-algebras is generated by left-continuous t-norms; see [JM02]. The corresponding logic is called *monoidal t-norm logic* or **MTL**. Uninorm logic, on the other hand, corresponds to representable FL_{eo}-algebras, i.e., integrality is not assumed. The introduction of **BL** lead to the study of such general fuzzy logics, where representability is the main defining property, while divisibility is dropped alltogether. In the rest of the survey, we focus on generalizations that retain the divisibility property.

4   Generalized BL-algebras: retaining divisibility

4.1   GBL-algebras

A *generalized BL-algebra* (or *GBL-algebra* for short) is defined to be a divisible residuated lattice. We begin with presenting some equivalent reformulations of divisibility:

$$\text{If } x \leq y, \text{ there exist } z, w \text{ such that } x = zy = yw.$$

LEMMA 4 ([GT05]) *The following are equivalent for a residuated lattice* **L**.

(i) **L** *is a GBL-algebra.*

(ii) **L** *satisfies* $((x \wedge y)/y)y = x \wedge y = y(y\backslash(x \wedge y))$.

(iii) **L** *satisfies the identities* $(x/y \wedge 1)y = x \wedge y = y(y\backslash x \wedge 1)$.

Proof (i) $\Leftrightarrow$ (ii): Assume that **L** is a GBL-algebra and $x, y \in L$. Since $x \wedge y \leq y$, there exist z, w such that $x \wedge y = zy = yw$. Since $zy \leq x \wedge y$, we have $z \leq (x \wedge y)/y$, so $x \wedge y = zy \leq ((x \wedge y)/y)y \leq x \wedge y$, by residuation, i.e. $((x \wedge y)/y)y = x \wedge y$. Likewise, we obtain the second equation. The other direction is obvious.

(ii) $\Leftrightarrow$ (iii): By basic properties of residuation, (ii) implies

$$x \wedge y = y(y\backslash(x \wedge y)) = y(y\backslash x \wedge y\backslash y) = y(y\backslash x \wedge 1).$$

Likewise, we get the opposite identity.

Conversely assume (iii). Note that for every element $a \geq 1$, we have $1 = a(a\backslash 1 \wedge 1) \leq a(a\backslash 1) \leq 1$; so, $a(a\backslash 1) = 1$. For $a = y\backslash y$, we have $a^2 = a$, by properties of residuation and $1 \leq a$. Thus, $a = a1 = a^2(a\backslash 1) = a(a\backslash 1) = 1$. Consequently, $y\backslash y = 1$, for every $y \in L$. Using properties of residuation, we get

$$y(y\backslash(x \wedge y)) = y(y\backslash x \wedge y\backslash y) = y(y\backslash x \wedge 1) = y \wedge x.$$

Likewise, we obtain the other equation.   □

LEMMA 5 ([GT05]) *Every GBL-algebra has a distributive lattice reduct.*

Proof Let **L** be a GBL-algebra and $x, y, z \in L$. By Lemma 4 and the fact that in residuated lattices multiplication distribures over joins, we have

$$
\begin{aligned}
x \wedge (y \vee z) \; &= [x/(y \vee z) \wedge 1](y \vee z) \\
&= [x/(y \vee z) \wedge 1]y \vee [x/(y \vee z) \wedge 1]z \\
&\leq (x/y \wedge 1)y \vee (x/z \wedge 1)z \\
&= (x \wedge y) \vee (x \wedge z),
\end{aligned}
$$

for all x, y, z. Thus, the lattice reduct of **L** is distributive. $\square$

Integral GBL-algebras, or *IGBL-algebras*, have a simpler axiomatization.

LEMMA 6 ([GT05]) *IGBL-algebras are axiomatized, relative to residuated lattices, by the equations $(x/y)y = x \wedge y = y(y \backslash x)$.*

Proof One direction holds by Lemma 4. For the converse, note that we show that the above identity implies integrality for $y = 1$. $\square$

4.2 Lattice-ordered groups and their negative cones

Lattice ordered groups, or *ℓ-groups* are defined as algebras with a lattice and a group reduct such that the group multiplication is compatible with the order, see e.g. [AF88], [Gla99]. Equivalently, they can be viewed as residuated lattices that satisfy $x(x \backslash 1) = 1$. It is easy to see that ℓ-groups are examples of (non-integral) GBL-algebras.

Given a residuated lattice **L**, its *negative cone* $\mathbf{L}^-$ is defined to be an algebra of the same type, with universe $L^- = \{x \in L : x \leq 1\}$, $x \backslash^{\mathbf{L}^-} y = x \backslash y \wedge 1$, $y /^{\mathbf{L}^-} x = y/x \wedge 1$ and where the other operations are the restrictions to L^- of the operations in **L**. With this definition $\mathbf{L}^-$ is also a residuated lattice. The map $\mathbf{L} \mapsto \mathbf{L}^-$ preserves divisibility, hence negative cones of ℓ-groups are also examples of integral GBL-algebras.

Both ℓ-groups and their negative cones are cancellative residuated lattices, namely their multiplication is cancellative. Equivalently, they satisfy the identities $xy/y = x = y \backslash yx$.

THEOREM 7 ([BCG$^+$03]) *The cancellative integral GBL-algebras are exactly the negative cones of ℓ-groups.*

Consequently, negative cones of ℓ-groups are equationally defined. As we mentioned, cancellative GBL-algebras in general (without the assumption of integrality) include ℓ-groups as well. However, we will see that every cancellative GBL-algebra is the direct product of an ℓ-group and a negative cone of an ℓ-group. More generally, we will see that a similar decomposition exists for arbitrary GBL-algebras.

4.3   GMV-algebras

Recall that an *MV-algebra* is a commutative bounded residuated lattice that satisfies the identity $x \vee y = (x \rightarrow y) \rightarrow y$. This identity implies that the residuated lattice is also integral, divisible and representable, hence MV-algebras are examples of BL-algebras.

It turns out that MV-algebras are intervals in abelian ℓ-groups. If $\mathbf{G} = (G, \wedge, \vee, \cdot, \backslash, /, 1)$ is an abelian ℓ-group and $a \leq 1$, then $\Gamma(\mathbf{G}, a) = ([a, 1], \wedge, \vee, \circ, \rightarrow, 1, a)$ is an MV-algebra, where $x \circ y = xy \vee a$, and $x \rightarrow y = x \backslash y \wedge 1$. (A version where $a \geq 1$ also yields an MV-algebra.) If $\mathbf{M}$ is an MV-algebra then there is an abelian ℓ-group $\mathbf{G}$ and an element $a \leq 1$ such that $\mathbf{M} \cong \Gamma(\mathbf{G}, a)$; see [Cha59], [Mun86].

Generalized MV-algebras, or *GMV-algebras*, are generalizations of MV-algebras in a similar way as GBL-algebras generalize BL-algebras, and are defined as residuated lattices that satisfy $x/((x \vee y) \backslash x) = x \vee y = (x/(x \vee y)) \backslash x$. As with GBL-algebras, GMV-algebras have alternative characterizations.

LEMMA 8 ([BCG$^+$03])   *A residuated lattice is a GMV algebra iff it satisfies* $x \leq y \Rightarrow y = x/(y \backslash x) = (x/y) \backslash x$.

THEOREM 9 ([BCG$^+$03])   *Every GMV-algebra is a GBL-algebra.*

Proof   We make use of the quasi-equational formulation from the preceding lemma. Assume $x \leq y$ and let $z = y(y \backslash x)$. Note that $z \leq x$ and $y \backslash z \leq x \backslash z$, hence

$$
\begin{aligned}
x \backslash z &= ((y \backslash z)/(x \backslash z)) \backslash (y \backslash z) \\
&= (y \backslash (z/(x \backslash z))) \backslash (y \backslash z) & \text{since } (u \backslash v)/w = u \backslash (v/w) \\
&= (y \backslash x) \backslash (y \backslash z) & \text{since } z \leq x \Rightarrow x = z/(x \backslash z) \\
&= y(y \backslash x) \backslash z & \text{since } u \backslash (v \backslash w) = vu \backslash w \\
&= z \backslash z.
\end{aligned}
$$

Therefore $x = z/(x \backslash z) = z/(z \backslash z) = z$, as required. The proof of $x = (x/y)y$ is similar.   □

LEMMA 10 ([BCG$^+$03])   (i) *Every integral GBL-algebra satisfies the identity* $(y/x) \backslash (x/y) = x/y$ *and its opposite.*

   (ii) *Every integral GMV-algebra satisfies the identity* $x/y \vee y/x = 1$ *and its opposite.*

   (iii) *Every integral GMV-algebra satisfies the identities* $x/(y \wedge z) = x/y \vee x/z$, $(x \vee y)/z = x/z \vee y/z$ *and the opposite ones.*

   (iv) *Every commutative integral GMV-algebra is representable.*

It will be shown in Section 5 that the assumption of integrality in condition (iv) is not needed. They do not have to be intervals, but they are convex sublattices in ℓ-groups. More details can be found in [GT05], while the standard reference for MV-algebras is [CDM99].

Ordinal sum constructions and decompositions have been used extensively for ordered algebraic structures, and we recall here the definition for integral residuated lattices. Usually the ordinal sum of two posets $\mathbf{A}_0, \mathbf{A}_1$ is defined as the disjoint union with all elements of A_0 less than all elements of A_1 (and if $\mathbf{A}_0$ has a top and $\mathbf{A}_1$ has a bottom, these two elements are often identified). However, for most decomposition results on integral residuated lattices a slightly different point of view is to *replace* the element 1 of $\mathbf{A}_0$ by the algebra $\mathbf{A}_1$. The precise definition for an arbitrary number of summands is as follows.

Let I be a linearly ordered set, and for $i \in I$ let $\{\mathbf{A}_i : i \in I\}$ be a family of integral residuated lattices such that for all $i \neq j$, $\mathbf{A}_i \cap \mathbf{A}_j = \{1\}$ and 1 is join irreducible in $\mathbf{A}_i$. Then the *ordinal sum* $\bigoplus_{i \in I} \mathbf{A}_i$ is defined on the set $\bigcup_{i \in I} A_i$ by

$$x \cdot y = \begin{cases} x \cdot_i y & \text{if } x, y \in A_i \text{ for some } i \in I \\ x & \text{if } x \in A_i \setminus \{1\} \text{ and } y \in A_j \text{ where } i < j \\ y & \text{if } y \in A_i \setminus \{1\} \text{ and } x \in A_j \text{ where } i < j. \end{cases}$$

The partial order on $\bigoplus_{i \in I} \mathbf{A}_i$ is the unique partial order $\leq$ such that 1 is the top element, the partial order $\leq_i$ on $\mathbf{A}_i$ is the restriction of $\leq$ to $\mathbf{A}_i$, and if $i < j$ then every element of $\mathbf{A}_i \setminus \{1\}$ precedes every element of $\mathbf{A}_j$. Finally, the lattice operations and the residuals are uniquely determined by $\leq$ and the monoid operation.

It is not difficult to check that this construction again yields an integral residuated lattice, and that it preserves divisibility and prelinearity. If $I = \{0, 1\}$ with $0 < 1$ then the ordinal sum is simply denoted by $\mathbf{A}_0 \oplus \mathbf{A}_1$. The assumption that 1 is join-irreducible can be omitted if $\mathbf{A}_1$ has a least element m, since if $x \vee_0 y = 1$ in $\mathbf{A}_0$ then $x \vee y$ still exists in $\mathbf{A}_0 \oplus \mathbf{A}_1$ and has value m. If 1 is join-reducible in $\mathbf{A}_0$ and if $\mathbf{A}_1$ has no minimum then the ordinal sum cannot be defined as above. However an "extended" ordinal sum may be obtained by taking the ordinal sum of $(\mathbf{A}_0 \oplus \mathbf{2}) \oplus \mathbf{A}_1$, where $\mathbf{2}$ is the 2-element MV-algebra.

The following representation theorem was proved by Agliano and Montagna [AM03].

THEOREM 11 *Every linearly ordered commutative integral GBL-algebra $\mathbf{A}$ can be represented as an ordinal sum $\bigoplus_{i \in I} \mathbf{A}_i$ of linearly ordered commutative integral GMV-algebras. Moreover $\mathbf{A}$ is a BL-algebra iff I has a minimum i_0 and $\mathbf{A}_{i_0}$ is bounded.*

Thus ordinal sums are a fundamental construction for BL-algebras, and commutative integral GMV-algebras are the building blocks.

For an overview of some subvarieties of RL see Table 4 and Figure 2.

4.4   Pseudo BL-algebras and pseudo MV-algebras

Pseudo BL-algebras [DGI02a, DGI02b] are divisible *prelinear* integral bounded residuated lattices, i.e., prelinear integral bounded GBL-algebras. Similarly, pseudo MV-algebras are integral bounded GMV-algebras (in this case prelinearity holds automatically). Hence BL-algebras and MV-algebras are exactly the commutative pseudo BL-algebras and pseudo MV-algebras respectively, but the latter do not have to be commutative. They do not have to be representable either, as prelinearity is equivalent to representability only under the assumption of commutativity and integrality. By a fundamental result of [Dvu02], all pseudo MV-algebras are obtained from intervals of ℓ-groups, as with MV-algebras in the commutative case. Hence any nonrepresentable ℓ-group provides examples of nonrepresentable pseudo MV-algebras. The tight categorical connections between GMV-algebras and ℓ-groups (with a strong order unit) have produced new results and interesting research directions in both areas [Hol05, DH07, DH09].

The relationship between noncommutative t-norms and pseudo BL-algebras is investigated in [FGI01]. In particular it is noted that any continuous t-norm must be commutative. Hájek [Háj03c] shows that noncommutative pseudo BL-algebras can be constructed on "non-standard" unit intervals, and in [Háj03a] it is shown that all BL-algebras embed in such pseudo BL-algebras.

As for Boolean algebras, Heyting algebras and MV-algebras, the constant 0 can be used to define unary negation operations $-x = 0/x$ and $\sim x = x\backslash 0$. For pseudo BL-algebras, these operations need not coincide. Pseudo MV-algebras are *involutive residuated lattices* which means the negations satisfy $\sim{-}x = x = {-}\sim x$. However, for pseudo BL-algebras this identity need not hold, and for quite some time it was an open problem whether the weaker identity $\sim{-}x = {-}\sim x$ might also fail. Recently it was noted in [DGK] that an example from [JM06] shows the identity can indeed fail, and a construction is given to show that there are uncountably many varieties of pseudo BL-algebras in which it does not hold.

We conclude this section with an important result on the structure of representable pseudo BL-algebras. Dvurečenskij [Dvu07] proved that the Aglianò-Montagna decomposition result extends to the non-commutative case.

THEOREM 12 *Every linearly ordered integral GBL-algebra* **A** *can be represented as an ordinal sum* $\bigoplus_{i \in I} \mathbf{A}_i$ *of linearly ordered integral GMV-algebras. Moreover* **A** *is a pseudo BL-algebra iff* I *has a minimum* i_0 *and* A_{i_0} *is bounded.*

As a consequence it is shown that representable pseudo BL-algebras satisfy the identity $\sim -x = -\sim x$, and that countably complete representable pseudo BL-algebras are commutative. Further results and references about pseudo BL-algebras can be found in [Háj03b], [Dvu07], [Ior08] and [DGK].

4.5 Hoops and pseudo-hoops

Hoops, originally introduced in [BO] by Büchi and Owens under an equivalent definition, can be defined as algebras $\mathbf{A} = (A, \cdot, \to, 1)$, where $(A, \cdot, 1)$ is a commutative monoid and the following identities hold:

$$x \to x = 1 \qquad x(x \to y) = y(y \to x) \qquad (xy) \to z = y \to (x \to z)$$

It is easy to see that the relation defined by $a \leq b$ iff $1 = a \to b$ is a partial order and that $\mathbf{A}$ is a hoop iff $(A, \cdot, \to, 1, \leq)$ is an integral residuated partially ordered monoid that satisfies $x(x \to y) = y(y \to x)$. Actually, if $\mathbf{A}$ is a hoop, then $(A, \leq)$ admits a meet operation defined by $x \wedge y = x(x \to y)$. Consequently, every hoop satisfies divisibility. It turns out that not all hoops have a lattice reduct; the ones that do are exactly the join-free reducts of commutative integral GBL-algebras. Also, among those, the ones that satisfy prelinearity are exactly the reducts of BL-algebras and are known as *basic hoops*. If a hoop satisfies

$$(x \to y) \to y = (y \to x) \to x$$

then it admits a join given by $x \vee y = (x \to y) \to y$. Such hoops are known as *Wajsberg hoops* and as *Łukasiewicz hoops* and they are term equivalent to commutative integral GMV-algebras.

More on hoops and Wajsberg hoops can be found in [AFM07], [BF00], [BP94] and their references. Pseudo-hoops are the non-commutative generalizations of hoops. Their basic properties are studied in [GLP05]. The join operation is definable in representable pseudo hoops, hence they are term-equivalent to integral representable GBL-algebras.

5 Decomposition of GBL-algebras

The structure of ℓ-groups has been studied extensively in the past decades. The structure of GMV-algebras has essentially been reduced to that of ℓ-groups (with a nucleus operation), and GMV-algebras are parts of ℓ-groups. The study of the structure of GBL-algebras has proven to be much more difficult. Nevertheless, the existing results indicate that ties to ℓ-groups exist. We first show that the study of GBL-algebras can be reduced to the integral case, by showing that every GBL-algebra is the direct product of an ℓ-group and an integral GBL-algebra. This result was proved in the dual setting of DRl-monoids in [Kov96] and independently in the setting of residuated

RRL	**representable residuated lattices**
	= residuated lattices that are subdirect products of residuated chains
	= residuated lattices with $1 \leq u\backslash((x \vee y)\backslash x)u \vee v((x \vee y)\backslash y)/v$
GBL	**generalized BL-algebras**
	= divisible residuated lattices
GMV	**generalized MV-algebras**
	= residuated lattices with $x \vee y = x/((x \vee y)\backslash x) = (x/(x \vee y))\backslash x$
Fleas	= integral residuated lattices with prelinearity
GBH	**generalized basic hoops**
	= divisible integral residuated lattices
BH	**basic hoops**
	= commutative prelinear generalized basic hoops
WH	**Wajsberg hoops**
	= commutative integral generalized MV-algebras
LG	**lattice-ordered groups or ℓ-groups**
	= residuated lattices that satisfy $1 = x(x\backslash 1)$
RLG	**representable ℓ-groups**
	= ℓ-groups with $1 \leq (1\backslash x)yx \vee 1\backslash y$
CLG	**commutative ℓ-groups**
LG$^-$	**negative cones of lattice-ordered groups**
	= cancellative integral generalized BL-algebras
RLG$^-$	**negative cones of representable ℓ-groups**
	= cancellative integral representable generalized BL-algebras
CLG$^-$	**negative cones of commutative ℓ-groups**
	= cancellative basic hoops
Br	**Brouwerian algebras**
	= residuated lattices with $x \wedge y = xy$
RBr	**representable Brouwerian algebras**
	= Brouwerian algebras that satisfy prelinearity
	= basic hoops that are idempotent
GBA	**generalized Boolean algebras**
	= Brouwerian algebras with $x \vee y = (x\backslash y)\backslash y$
	= Wajsberg hoops that are idempotent

Table 4. Some subvarieties of RL

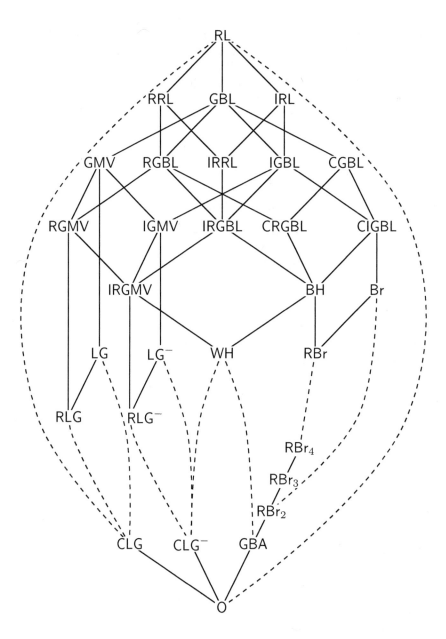

Figure 2: Some subvarieties of **RL** ordered by inclusion

lattices in [GT05]. With hindsight, the later result can, of course, be deduced by duality from the former. The presentation below follows [GT05].

An element a in a residuated lattice $\mathbf{L}$ is called *invertible* if $a(a\backslash 1) = 1 = (1/a)a$, and it is called *integral* if $1/a = a\backslash 1 = 1$. We denote the set of invertible elements of $\mathbf{L}$ by $G(\mathbf{L})$ and the set of integral elements by $I(\mathbf{L})$. Recall that a *positive* element a is an element that satisfies $a \geq 1$.

Note that a is invertible if and only if there exists an element a^{-1} such that $aa^{-1} = 1 = a^{-1}a$. In this case $a^{-1} = 1/a = a\backslash 1$. It is easy to see that multiplication by an invertible element is an order automorphism.

LEMMA 13 ([GT05]) *Let $\mathbf{L}$ be a GBL-algebra.*

(i) *Every positive element of $\mathbf{L}$ is invertible.*

(ii) *$\mathbf{L}$ satisfies the identities $x/x = x\backslash x = 1$.*

(iii) *$\mathbf{L}$ satisfies the identity $1/x = x\backslash 1$.*

(iv) *For all $x, y \in L$, if $x \vee y = 1$ then $xy = x \wedge y$.*

(v) *$\mathbf{L}$ satisfies the identity $x = (x \vee 1)(x \wedge 1)$.*

Proof For (i), note that if a is a positive element in L, $a(a\backslash 1) = 1 = (1/a)a$, by definition; that is, a is invertible. For (ii), we argue as is the proof of Lemma 4.

By (ii) and residuation, we have $x(1/x) \leq x/x = 1$, hence $1/x \leq x\backslash 1$. Likewise, $x\backslash 1 \leq 1/x$.

For (iv), we have $x = x/1 = x/(x \vee y) = x/x \wedge x/y = 1 \wedge x/y = y/y \wedge x/y = (y \wedge x)/y$. So, $xy = ((x \wedge y)/y)y = x \wedge y$.

Finally, by Lemma 4, $(1/x \wedge 1)x = x \wedge 1$. Moreover, by (i) $x \vee 1$ is invertible and $(x \vee 1)^{-1} = 1/(x \vee 1) = 1/x \wedge 1$. Thus, $(x \vee 1)^{-1}x = x \wedge 1$, or $x = (x \vee 1)(x \wedge 1)$. □

The following theorem shows that if $\mathbf{L}$ is a GBL-algebra then the sets $G(\mathbf{L})$ and $I(\mathbf{L})$ are subuniverses of $\mathbf{L}$. We denote the corresponding subalgebras by $\mathbf{G}(\mathbf{L})$ and $\mathbf{I}(\mathbf{L})$.

THEOREM 14 ([GT05]) *Every GBL-algebra $\mathbf{L}$ is isomorphic to $\mathbf{G}(\mathbf{L}) \times \mathbf{I}(\mathbf{L})$.*

Proof We begin with a series of claims.
Claim 1: $G(\mathbf{L})$ is a subuniverse of $\mathbf{L}$.

Let x, y be invertible elements. It is clear that xy is invertible. Additionally, for all $x, y \in G(\mathbf{L})$ and $z \in L$, $z \leq x^{-1}y \Leftrightarrow xz \leq y \Leftrightarrow z \leq x\backslash y$. It follows that $x\backslash y = x^{-1}y$, hence $x\backslash y$ is invertible. Likewise, $y/x = yx^{-1}$ is invertible.

Moreover, $x \vee y = (xy^{-1} \vee 1)y$. So, $x \vee y$ is invertible, since every positive element is invertible, by Lemma 13(i), and the product of two invertible elements is invertible. By properties of residuation, $x \wedge y = 1/(x^{-1} \vee y^{-1})$, which is invertible, since we have already shown that $G(\mathbf{L})$ is closed under joins and the division operations.

Claim 2: $I(\mathbf{L})$ is a subuniverse of $\mathbf{L}$.

Note that every integral element a is negative, since $1 = 1/a$ implies $1 \leq 1/a$ and $a \leq 1$. For $x, y \in I(\mathbf{L})$, we get:

$$1/xy = (1/y)/x = 1/x = 1, \text{ so } xy \in I(\mathbf{L}).$$

$$1/(x \vee y) = 1/x \wedge 1/y = 1, \text{ so } x \vee y \in I(\mathbf{L}).$$

$$1 \leq 1/x \leq 1/(x \wedge y) \leq 1/xy = 1, \text{ so } x \wedge y \in I(\mathbf{L}).$$

$$1 = 1/(1/y) \leq 1/(x/y) \leq 1/(x/1) = 1/x = 1, \text{ so } x/y \in I(\mathbf{L}).$$

Claim 3: For every $g \in (G(\mathbf{L}))^-$ and every $h \in I(\mathbf{L})$, $g \vee h = 1$.

Let $g \in (G(\mathbf{L}))^-$ and $h \in I(\mathbf{L})$. We have $1/(g \vee h) = 1/g \wedge 1/h = 1/g \wedge 1 = 1$, since $1 \leq 1/g$. Moreover, $g \leq g \vee h$, so $1 \leq g^{-1}(g \vee h)$. Thus, by the GBL-algebra identities and properties of residuation

$$\begin{aligned}
1 &= (1/[g^{-1}(g \vee h)])[g^{-1}(g \vee h)] \\
&= ([1/(g \vee h)]/g^{-1})g^{-1}(g \vee h) \\
&= (1/g^{-1})g^{-1}(g \vee h) \\
&= gg^{-1}(g \vee h) \\
&= g \vee h.
\end{aligned}$$

Claim 4: For every $g \in (G(\mathbf{L}))^-$ and every $h \in I(\mathbf{L})$, $gh = g \wedge h$.

In light of Lemma 13(iv), $g^{-1}h = (g^{-1}h \vee 1)(g^{-1}h \wedge 1)$. Multiplication by g yields $h = (h \vee g)(g^{-1}h \wedge 1)$. Using Claim 3, we have $gh = g(g^{-1}h \wedge 1) = h \wedge g$, since multiplication by an invertible element is an order automorphism.

Claim 5: For every $g \in G(\mathbf{L})$ and every $h \in I(\mathbf{L})$, $gh = hg$.

The statement is true if $g \leq 1$, by Claim 4. If $g \geq 1$ then $g^{-1} \leq 1$, thus $g^{-1}h = hg^{-1}$, hence $hg = gh$. For arbitrary g, note that both $g \vee 1$ and $g \wedge 1$ commute with h. Using Lemma 13(iv), we get $gh = (g \vee 1)(g \wedge 1)h = (g \vee 1)h(g \wedge 1) = h(g \vee 1)(g \wedge 1) = hg$.

Claim 6: For every $x \in L$, there exist $g_x \in G(\mathbf{L})$ and $h_x \in I(\mathbf{L})$, such that $x = g_x h_x$.

By Lemma 13(iv), $x = (x \vee 1)(x \wedge 1)$. Since $1 \leq x \vee 1$ and $1 \leq 1/(x \wedge 1)$, by Lemma 13(i), these elements are invertible. Set $g_x = (x \vee 1)(1/(x \wedge 1))^{-1}$ and $h_x = (1/(x \wedge 1))(x \wedge 1)$. It is clear that $x = g_x h_x$, g_x is invertible and h_x is integral.

Claim 7: For every $g_1, g_2 \in G(\mathbf{L})$ and $h_1, h_2 \in I(\mathbf{L})$, $g_1 h_1 \leq g_2 h_2$ if and only if $g_1 \leq g_2$ and $h_1 \leq h_2$.

For the non-trivial direction we have

$$g_1 h_1 \leq g_2 h_2 \Rightarrow g_2^{-1} g_1 h_1 \leq h_2 \Rightarrow g_2^{-1} g_1 \leq h_2/h_1 \leq e \Rightarrow g_1 \leq g_2.$$

Moreover,

$$
\begin{aligned}
g_2^{-1} g_1 \leq h_2/h_1 \quad &\Rightarrow e \leq g_1^{-1} g_2(h_2/h_1) \\
&\Rightarrow 1 = [1/g_1^{-1} g_2(h_2/h_1)] g_1^{-1} g_2(h_2/h_1) \\
&\Rightarrow 1 = [(1/(h_2/h_1))/g_1^{-1} g_2] g_1^{-1} g_2(h_2/h_1) \\
&\Rightarrow 1 = g_2^{-1} g_1 g_1^{-1} g_2(h_2/h_1) \\
&\Rightarrow 1 = h_2/h_1 \\
&\Rightarrow h_1 \leq h_2.
\end{aligned}
$$

By Claims 1 and 2, $\mathbf{G(L)}$ and $\mathbf{I(L)}$ are subalgebras of $\mathbf{L}$. Define $f : \mathbf{G(L)} \times \mathbf{I(L)} \to \mathbf{L}$ by $f(g, h) = gh$. We will show that f is an isomorphism. It is onto by Claim 6 and an order isomorphism by Claim 7. So, it is a lattice isomorphism, as well. To verify that f preserves the other operations note that $gg'hh' = ghg'h'$, for all $g, g' \in G(\mathbf{L})$ and $h, h' \in I(\mathbf{L})$, by Claim 5. Moreover, for all $g, g', \bar{g} \in G(\mathbf{L})$ and $h, h', \bar{h} \in I(\mathbf{L})$, $\bar{g}\bar{h} \leq gh/g'h'$ if and only if $\bar{g}\bar{h}g'h' \leq gh$. By Claim 5, this is equivalent to $\bar{g}g'\bar{h}h' \leq gh$, and, by Claim 7, to $\bar{g}g' \leq g$ and $\bar{h}h' \leq h$. This is in turn equivalent to $\bar{g} \leq g/g'$ and $\bar{h} \leq h/h'$, which is equivalent to $\bar{g}\bar{h} \leq (g/g')(h/h')$ by Claim 7. Thus, for all $g, g' \in G(\mathbf{L})$ and $h, h' \in I(\mathbf{L})$, $gh/g'h' = (g/g')(h/h')$ and, likewise, $g'h'\backslash gh = (g'\backslash g)(h'\backslash h)$.   □

COROLLARY 15 *Every GBL-algebra is the direct product of an ℓ-group and an integral GBL-algebra.*

Combining this with Theorem 7 immediately gives the following result.

COROLLARY 16 *Every cancellative GBL-algebra is the direct product of an ℓ-group and the negative cone of an ℓ-group.*

6   Further results on the structure of GBL-algebras

In this section we briefly summarize results from a series of papers [JM06] [JM09] [JM]. The collaboration that lead to these results started when the second author and Franco Montagna met at a wonderful ERCIM workshop on Soft Computing, organized by Petr Hájek in Brno, Czech Republic, in 2003. This is yet another example how Petr Hájek's dedication to the field of fuzzy logic has had impact far beyond his long list of influential research publications.

6.1   Finite GBL-algebras are commutative

LEMMA 17 *If a is an idempotent in an integral GBL-algebra A, then $ax = a \wedge x$ for all $x \in A$. Hence every idempotent is central, i.e. commutes with every element.*

Proof   Suppose $aa = a$. Then $ax \leq a \wedge x = a(a \backslash x) = aa(a \backslash x) = a(a \wedge x) \leq ax$. $\square$

In an ℓ-group only the identity is an idempotent, hence it follows from the decomposition result mentioned above that idempotents are central in all GBL-algebras.

In fact, using this lemma, it is easy to see that the set of idempotents in a GBL-algebra is a sublattice that is closed under multiplication. In [JM06] it is proved that this set is also closed under the residuals.

THEOREM 18 *The idempotents in a GBL-algebra $\mathbf{A}$ form a subalgebra, which is the largest Brouwerian subalgebra of $\mathbf{A}$.*

By the results of the preceding section, the structure of any GBL-algebra is determined by the structure of its ℓ-group factor and its integral GBL-algebra factor. Since only the trivial ℓ-group is finite, it follows that any finite GBL-algebra is integral. A careful analysis of one-generated subalgebras in a GBL-algebra gives the following result.

THEOREM 19 ([JM06]) *Every finite GBL-algebra and every finite pseudo-BL-algebra is commutative.*

Since there exist noncommutative GBL-algebras, such as any noncommutative ℓ-group, the next result is immediate.

COROLLARY 20 *The varieties* GBL *and* psBL *are not generated by their finite members, and hence do not have the finite model property.*

As mentioned earlier, BL-algebras are subdirect products of ordinal sums of commutative integral GMV-chains, and a similar result holds without commutativity for representable GBL-algebras. So it is natural to ask to what extent GBL-algebras are determined by ordinal sums of GMV-algebras and Heyting algebras. We briefly recall a construction of a GBL-algebra that does not arise from these building blocks.

For a residuated lattice $\mathbf{B}$ with top element $\top$, let $\mathbf{B}^{\partial}$ denote the dual poset of the lattice reduct of $\mathbf{B}$. Now we define $\mathbf{B}^{\dagger}$ to be the ordinal sum of $\mathbf{B}^{\partial}$ and $\mathbf{B} \times \mathbf{B}$, i.e., every element of $\mathbf{B}^{\partial}$ is below every element of $\mathbf{B} \times \mathbf{B}$ (see

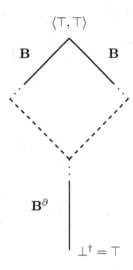

Figure 3.

Figure 3). Note that $\mathbf{B}^\dagger$ has $\top$ as bottom element, so to avoid confusion, we denote this element by $\perp^\dagger$. A binary operation $\cdot$ on $\mathbf{B}^\dagger$ is defined as follows:

$$\begin{aligned}
\langle a, b\rangle \cdot \langle c, d\rangle &= \langle ac, bd\rangle \\
\langle a, b\rangle \cdot u &= u/a \\
u \cdot \langle a, b\rangle &= b\backslash u \\
u \cdot v &= \top = \perp^\dagger
\end{aligned}$$

Note that even if $\mathbf{B}$ is a commutative residuated lattice, $\cdot$ is in general non-commutative.

LEMMA 21   (i) *For any residuated lattice $\mathbf{B}$ with top element, the algebra $\mathbf{B}^\dagger$ defined above is a bounded residuated lattice.*

   (ii) *If $\mathbf{B}$ is nontrivial, then $\mathbf{B}^\dagger$ is not a GMV-algebra, and if $\mathbf{B}$ is subdirectly irreducible, so is $\mathbf{B}^\dagger$.*

   (iii) *$\mathbf{B}^\dagger$ is a GBL-algebra if and only if $\mathbf{B}$ is a cancellative GBL-algebra.*

6.2   The Blok-Ferreirim decomposition result for normal GBL-algebras

Building on work of Büchi and Owens [BO], Blok and Ferreirim [BF00] proved the following result.

PROPOSITION 22 *Every subdirectly irreducible hoop is the ordinal sum of a proper subhoop* **H** *and a subdirectly irreducible nontrivial Wajsberg hoop* **W**.

This result was adapted to BL-algebras in [AM03]. To discuss further extensions to GBL-algebras we first recall some definitions about filters and congruences in residuated lattices.

A *filter* of a residuated lattice **A** is an upward closed subset F of **A** which contains 1 and is closed under the meet and the monoid operation. A filter F is said to be *normal* if $aF = Fa$ for all $a \in A$, or equivalently if $a\backslash(xa) \in F$ and $(ax)/a \in F$ whenever $x \in F$ and $a \in A$. A residuated lattice is said to be *normal* if every filter of it is a normal filter. A residuated lattice is said to be *n-potent* if it satisfies $x^{n+1} = x^n$, where $x^n = x \cdot \ldots \cdot x$ (n times). Note that n-potent GBL-algebras are normal ([JM09]).

In every residuated lattice, the lattice of normal filters is isomorphic to the congruence lattice: to any congruence θ one associates the normal filter $F_\theta = \uparrow\{x : (x, 1) \in \theta\}$. Conversely, given a normal filter F, the set θ_F of all pairs (x, y) such that $x\backslash y \in F$ and $y\backslash x \in F$ is a congruence such that the upward closure of the congruence class of 1 is F. Hence a residuated lattice is subdirectly irreducible if and only if it has a smallest nontrivial normal filter.

In [JM09] the following result is proved.

THEOREM 23

(i) *Every subdirectly irreducible normal integral GBL-algebra is the ordinal sum of a proper subalgebra and a non-trivial integral subdirectly irreducible GMV-algebra.*

(ii) *Every n-potent GBL-algebra is commutative and integral.*

A variety $\mathcal{V}$ has the *finite embeddability property (FEP)* iff every finite partial subalgebra of an algebra in $\mathcal{V}$ partially embeds into a finite algebra of $\mathcal{V}$. The FEP is stronger than the finite model property: for a finitely axiomatized variety $\mathcal{V}$, the finite model property implies the decidability of the equational theory of $\mathcal{V}$, while the FEP implies the decidability of the universal theory of $\mathcal{V}$. With the help of the ordinal sum decomposition, the following result is proved in [JM09].

THEOREM 24 *The variety of commutative and integral GBL-algebras has the FEP.*

However, by interpreting the quasiequational theory of ℓ-groups into that of GBL-algebras, it is shown that without the assumption of commutativity the quasiequational theory of GBL-algebras is undecidable. The decidability of the *equational* theory of GBL-algebras is currently an open problem.

6.3   Poset products

The success of ordinal sum decompositions for subclasses of GBL-algebras has prompted the use of some generalizations to obtain representation and embedding theorems for larger subclasses. The *poset product* uses a partial order on the index set to define a subset of the direct product. Specifically, let $\mathbf{I} = (I, \leq)$ be a poset, assume $\{\mathbf{A}_i : i \in I\}$ is a family of residuated lattices, and that for nonmaximal $i \in I$ each $\mathbf{A}_i$ is integral, and for nonminimal $i \in I$ each $\mathbf{A}_i$ has a least element denoted by 0. The poset product of $\{A_i : i \in I\}$ is

$$\bigotimes_{i \in \mathbf{I}} A_i = \{f \in \prod_{i \in I} A_i : f(i) = 0 \text{ or } f(j) = 1 \text{ for all } i < j \text{ in } I\}.$$

The monoid operation and the lattice operations are defined pointwise. The residuals are defined by

$(f \backslash g)(i) = f(i) \backslash_i g(i)$ if $f(j) \leq_j g(j)$ for all $j > i$, and 0 otherwise,
$(g / f)(i) = g(i) /_i f(i)$ if $f(j) \leq_j g(j)$ for all $j > i$, and 0 otherwise.

If $\mathbf{I}$ is an antichain then the poset product reduces to the direct product, and if $\mathbf{I}$ is a finite chain, the poset product gives the ordinal sum over the reverse order of $\mathbf{I}$. In [JM09] the following is proved.

THEOREM 25

(i) *The poset product of a collection of residuated lattices is a residuated lattice, which is integral (divisible, bounded respectively) when all factors are integral (divisible, bounded respectively).*

(ii) *Every finite GBL-algebra can be represented as the poset product of a finite family of finite MV-chains.*

The next result, from [JM], extends this to larger classes of GBL-algebras, but in this case one only gets an embedding theorem.

THEOREM 26 *Every n-potent GBL-algebra embeds into the poset product of a family of finite and n-potent MV-chains.*

Every normal GBL-algebra embeds into a poset product of linearly ordered integral bounded GMV-algebras and linearly ordered ℓ-groups.

For Heyting algebras the above theorem reduces to the well-known embedding theorem into the complete Heyting algebra of all upward closed subsets of some poset.

Various properties can be imposed on poset products to obtain embedding theorems for other subclasses of GBL-algebras. The follow result from [JM] collects several of them.

THEOREM 27 *A GBL-algebra is*

(i) *a BL-algebra iff it is isomorphic to a subalgebra* **A** *of a poset product* $\bigotimes_{i \in \mathbf{I}} \mathbf{A}_i$ *such that*

(a) *each* $\mathbf{A}_i$ *is a linearly ordered MV-algebra,*

(b) $\mathbf{I} = (I, \leq)$ *is a* root system, *i.e. every principal filter of* **I** *is linearly ordered, and*

(c) *the function on I which is constantly equal to 0 is in* **A***;*

(ii) *an MV-algebra iff it is isomorphic to a subalgebra* **A** *of a poset product* $\bigotimes_{i \in \mathbf{I}} \mathbf{A}_i$ *such that conditions* (a) *and* (c) *above hold and*

(d) $\mathbf{I} = (I, \leq)$ *is a poset such that* $\leq$ *is the identity on I;*

(iii) *representable iff it is embeddable into a poset product* $\bigotimes_{i \in \mathbf{I}} \mathbf{A}_i$ *such that each* $\mathbf{A}_i$ *is a linearly ordered GMV-algebra and* (b) *holds;*

(iv) *an abelian ℓ-group iff it is embeddable into a poset product* $\bigotimes_{i \in \mathbf{I}} \mathbf{A}_i$ *such that each* $\mathbf{A}_i$ *is a linearly ordered abelian ℓ-group and condition* (d) *above holds;*

(v) *n-potent iff it is embeddable into a poset product of linearly ordered n-potent MV-algebras;*

(vi) *a Heyting algebra iff it is isomorphic to a subalgebra* **A** *of a poset product* $\bigotimes_{i \in \mathbf{I}} \mathbf{A}_i$ *where condition* (c) *holds and in addition*

(e) *every* $\mathbf{A}_i$ *is the two-element MV-algebra;*

(vii) *a Gödel algebra iff it is isomorphic to a subalgebra* **A** *of a poset product* $\bigotimes_{i \in \mathbf{I}} \mathbf{A}_i$ *where* (b), (c) *and* (e) *hold;*

(viii) *a Boolean algebra iff it is isomorphic to a subalgebra* **A** *of a poset product* $\bigotimes_{i \in \mathbf{I}} \mathbf{A}_i$ *where* (c), (d) *and* (e) *hold.*

Of course this survey covers only some of the highlights of a few research papers concerned with GBL-algebras and related classes. Volumes have been written about BL-algebras, MV-algebras, ℓ-groups and Heyting algebras, as well as about many other algebras that satisfy divisibility, and the reader is encouraged to explore the literature further, starting with the references below.

BIBLIOGRAPHY

[AF88]   M. Anderson and T. Feil. *Lattice Ordered Groups, An Introduction.* D. Reidel Publishing Company, Dordrecht, Boston, Lancaster, Tokyo, 1988.

[AFM07]   P. Aglianò, I. M. A. Ferreirim, and F. Montagna. Basic hoops: An algebraic study of continuous t-norms. *Studia Logica*, 87(1):73–98, 2007.

[AM03]   P. Aglianò and F. Montagna. Varieties of BL-algebras I: General properties. *J. Pure Appl. Algebra*, 181:105–129, 2003.

[BCG$^+$03] P. Bahls, J. Cole, N. Galatos, P. Jipsen, and C. Tsinakis. Cancellative residuated lattices. *Algebra Universalis*, 50(1):83–106, 2003.

[BF00]   W. J. Blok and I. M. A. Ferreirim. On the structure of hoops. *Algebra Universalis*, 43(2–3):233–257, 2000.

[BO]   J. R. Büchi and T. M. Owens. Complemented monoids and hoops. Unpublished manuscript.

[Bos82]   B. Bosbach. Residuation groupoids. *Result. Math.*, 5:107–122, 1982.

[BP89]   W. J. Blok and D. Pigozzi. *Algebraizable Logics*, volume 396 of *Memoirs of the American Mathematical Society*. AMS, Providence, 1989.

[BP94]   W. J. Blok and D. Pigozzi. On the structure of varieties with equationally definable principal congruences III. *Algebra Universalis*, 32:545–608, 1994.

[BT03]   K. Blount and C. Tsinakis. The structure of residuated lattices. *Int. J. Algebra and Computation*, 13(4):437–461, 2003.

[CDM99]   R. Cignoli, I. M. L. D'Ottaviano, and D. Mundici. *Algebraic Foundations of Many-Valued Reasoning*, volume 7 of *Trends in Logic*. Kluwer, Dordrecht, 1999.

[CEGT00] R. Cignoli, F. Esteva, L. Godo, and A. Torrens. Basic fuzzy logic is the logic of continuous t-norms and their residua. *Soft Computing*, 4:106–112, 2000.

[Cha59]   C. C. Chang. A new proof of the completeness of Łukasiewicz axioms. *Trans. Amer. Math. Soc.*, 93:74–80, 1959.

[Cin01]   P. Cintula. The LΠ and LΠ$\frac{1}{2}$ propositional and predicate logics. *Fuzzy Sets Syst.*, 124(3):289–302, 2001.

[DGI02a] A. Di Nola, G. Georgescu, and A. Iorgulescu. Pseudo-BL-algebras, part I. *Multiple-Valued Logic*, 8(5–6):673–714, 2002.

[DGI02b] A. Di Nola, G. Georgescu, and A. Iorgulescu. Pseudo-BL-algebras, part II. *Multiple-Valued Logic*, 8(5–6):717–750, 2002.

[DGK]   A. Dvurečenskij, R. Giuntini, and T. Kowalski. On the structure of pseudo BL-algebras and pseudo hoops in quantum logics. To appear in *Found. Phys.* DOI 10.1007/s10701-009-9342-5.

[DH07]   A. Dvurečenskij and W. C. Holland. Top varieties of generalized MV-algebras and unital lattice-ordered groups. *Communications in Algebra*, 35(11):3370–3390, 2007.

[DH09]   A. Dvurečenskij and W. C. Holland. Komori's characterization and top varieties of GMV-algebras. *Algebra Universalis*, 60(1):37–62, 2009.

[Dvu02]   A. Dvurečenskij. Pseudo MV-algebras are intervals in *l*-groups. *J. Australian Math. Soc.*, 72(3):427–445, 2002.

[Dvu07]   A. Dvurečenskij. Aglianò-Montagna type decomposition of pseudo hoops and its applications. *J. Pure Appl. Algebra*, 211:851–861, 2007.

[FGI01]   P. Flondor, G. Georgescu, and A. Iorgulescu. Pseudo t-norms and pseudo-BL-algebras. *Soft Computing*, 5:355–371, 2001.

[GJKO07] N. Galatos, P. Jipsen, T. Kowalski, and H. Ono. *Residuated Lattices: An Algebraic Glimpse at Substructural Logics*, volume 151 of *Studies in Logic and the Foundations of Mathematics*. Elsevier, Amsterdam, 2007.

[Gla99]   A. M. W. Glass. *Partially Ordered Groups*, volume 7 of *Series in Algebra*. World Scientific Publishing Co. Inc., River Edge NJ, 1999.

[GLP05]   G. Georgescu, L. Leuştean, and V. Preoteasa. Pseudo hoops. *J. Mult.-Valued Logic Soft Comput.*, 11:153–184, 2005.

[GO06]   N. Galatos and H. Ono. Algebraization, parametrized local deduction theorem and interpolation for substructural logics over **FL**. *Studia Logica*, 83(1-3):279–308, 2006.

[GT05]   N. Galatos and C. Tsinakis. Generalized MV-algebras. *J. Algebra*, 283(1):254–291, 2005.

[Háj98]   P. Hájek. Basic fuzzy logic and BL-algebras. *Soft Computing*, 2:124–128, 1998.

[Háj03a]   P. Hájek. Embedding standard BL-algebras into non-commutative pseudo-BL-algebras. *Tatra Mountains Mathematical Publications*, 27:125–130, 2003.

[Háj03b]   P. Hájek. Fuzzy logics with noncommutative conjunctions. *J. Logic Comput.*, 13(4):469–479, 2003.

[Háj03c]   P. Hájek. Observations on non-commutative fuzzy logic. *Soft Computing*, 8(1):38–43, 2003.

[Hol05]   W. C. Holland. Small varieties of lattice-ordered groups and MV-algebras. *Contributions to General Algebra*, 16:107–11, 2005.

[Ior08]   A. Iorgulescu. *Algebras of Logic as BCK Algebras*. Editura ASE, Bucharest, 2008.

[JM]   P. Jipsen and F. Montagna. Embedding theorems for classes of GBL-algebras. To appear in *J. Pure Appl. Algebra*.

[JM02]   S. Jenei and F. Montagna. A proof of standard completeness for Esteva and Godo's logic MTL. *Studia Logica*, 70(2):183–192, 2002.

[JM06]   P. Jipsen and F. Montagna. On the structure of generalized BL-algebras. *Algebra Universalis*, 55(2–3):227–238, 2006.

[JM09]   P. Jipsen and F. Montagna. The Blok-Ferreirim theorem for normal GBL-algebras and its applications. *Algebra Universalis*, 60:381–404, 2009.

[JT02]   P. Jipsen and C. Tsinakis. A survey of residuated lattices. In J. Martinez, editor, *Ordered Algebraic Structures*, pages 19–56. Kluwer, Dordrecht, 2002.

[Kov96]   T. Kovář. *A General Theory of Dually Residuated Lattice Ordered Monoids*. PhD thesis, Palacký University Olomouc, 1996.

[Mun86]   D. Mundici. Interpretation of AF C^*-algebras in Lukasiewicz sentential calculus. *J. Funct. Anal.*, 65(1):15–63, 1986.

Nikolaos Galatos
Department of Mathematics
University of Denver
2360 S. Gaylord St.
Denver, CO 80208, USA
Email: ngalatos@du.edu

Peter Jipsen
Department of Mathematics and CS
Chapman University
Orange, CA 92866, USA
Email: jipsen@chapman.edu

Discrete duality for some axiomatic extensions of MTL algebras[1]

Ewa Orłowska and Anna Maria Radzikowska

Dedicated to Petr Hájek,
on the occasion of his 70th birthday

ABSTRACT. In this paper we recall a discrete duality between MTL algebras and the corresponding class of relational systems. Next, three axiomatic extensions of MTL algebras are considered, namely SMTL algebras, CMTL algebras, and IMTL algebras, obtained from MTL algebras by adding the axioms of pseudo-complementation, contraction, and involution, respectively. For these classes discrete dualities are presented. It follows that we also get a discrete duality for Gödel algebras which are SMTL algebras with the contraction axiom. As a by–product of the dualities we get Kripke–style semantics for the corresponding logics.

1 Introduction

The Esteva–Godo–Ono hierarchy of substructural and fuzzy logics ([EG01, EGGC03, GJKO07]) starts with the logic FL_{ew}, that is the full Lambek calculus with the rules of exchange and weakening. Algebraically, the FL_{ew} algebras are bounded integral residuated lattices where the product $\otimes$ is commutative and satisfies $a \otimes b \leq a$. Logic FL_{ew} coincides with the monoidal logic considered in ([Höh95]). The hierarchy is built by extending the signature of FL_{ew} algebras with a negation operation definable in terms of the residuum of the product, that is $\neg a = a \rightarrow 0$, and by postulating additional axioms. The axioms include:

- Pseudo–complementation: $a \wedge \neg a = 0$,
- Involution: $\neg\neg a \leq a$,
- Definability of join: $a \vee b = ((a \rightarrow b) \rightarrow b) \wedge ((b \rightarrow a) \rightarrow a)$
- Mutual distributivity of join and meet,
- Prelinearity: $(a \rightarrow b) \vee (b \rightarrow a) = 1$,
- Contraction: $a \leq a \otimes a$,
- Divisibility: $a \otimes (a \rightarrow b) = a \wedge b$.

---

[1] This work has been supported by the grant N N206 399134 of Polish Ministry of Science and Higher Education.

The logic MTL is in the middle of the hierarchy. It captures the most essential feature of fuzzy logics, namely, the product being the left-continuous t–norm, see [JM02], where a proof of standard completeness of logic MTL can be found. Algebraically, MTL algebras are FL_{ew} algebras with the prelinearity axiom. It follows from prelinearity that MTL algebras are distributive and satisfy the law of definability of join.

Duality theory emerged from the work of Marshall Stone ([Sto36, Sto37]) on Boolean algebras and distributive lattices in the 1930s. Jónsson and Tarski ([JT51, JT52]) extended the Stone results to Boolean algebras with operators which are now known as modal possibility operators. Later in the early 1970s Larisa Maksimova ([Mak72, Mak75]) and Hilary Priestley ([DP90, Pri70, Pri72]) developed analogous results for Heyting algebras, topological Boolean algebras, and distributive lattices. Since then establishing duality has become an important methodological problem both in algebra and in logic. All the above mentioned duality results were developed using topological spaces as dual spaces of algebras.

The paper is a contribution to the project of developing discrete representability and, if possible, discrete duality [ORD05] for all the logics of the hierarchy. Discreteness means that topology is involved neither in the construction of the representation algebras nor in the construction of dual spaces of algebras.

Discrete duality is a duality where a class of abstract relational systems is a dual counterpart to a class of algebras. These relational systems are referred to as *frames* following the terminology of non–classical logics. A topology is not involved in the construction of these frames and hence they may be thought of as having a discrete topology. Establishing discrete duality involves the following steps. Given a class Alg of algebras (resp. a class Frm of frames) we define a class Frm of frames (resp. a class Alg of algebras). Next, for an algebra $L \in$ Alg we define its canonical frame $\mathcal{X}(L)$ and for each frame $X \in$ Frm we define its complex algebra $\mathcal{C}(X)$. Then we prove that $\mathcal{X}(L) \in$ Frm and $\mathcal{C}(X) \in$ Alg. A duality between Alg and Frm holds provided that the following facts are proved:

(D1) Every algebra $L \in$ Alg is embeddable into the complex algebra $\mathcal{C}(\mathcal{X}(L))$ of its canonical frame.

(D2) Every frame $X \in$ Frm is embeddable into the canonical frame $\mathcal{X}(\mathcal{C}(X))$ of its complex algebra.

The basis for the project are the Urquhart representation theorem for bounded lattices ([Urq78]) and the Priestley duality for distributive lattices (e.g., [DP90]). The Priestley duality can be "lifted" to a discrete duality between distributive lattices and some abstract relational systems. The requirement of discreteness is motivated with logical purposes. Having a discrete duality between a class of algebras which provides the algebraic semantics of a logic and a class of relational systems, we often obtain the Kripke–style semantics for the logic.

In [OR06] discrete representability is developed for FL_{ew} algebras and for FL_{ew} algebras with the pseudo complementarity axiom, among others. In the present paper we recall the discrete duality for MTL algebras ([OR]) and we extend it to three axiomatic extensions of MTL algebras: SMTL algebras, CMTL algebras, and IMTL algebras, obtained from MTL algebras by adding the axioms of pseudo-complementation, contraction, and involution, respectively. Further work is planned on discrete duality for the logics which are above the logic MTL in the Esteva–Godo–Ono hierarchy, in particular for BL algebras, which are MTL algebras extended with the divisibility axiom. BL algebras were introduced by Petr Hajek and extensively studied by his followers. For a comprehensive exposition of BL algebras see ([Háj98]).

Discrete representability and duality results for the classes of algebras beyond the Esteva–Godo–Ono hierarchy include: discrete representability for latice–based relation algebras ([DOR06]), for lattice-based modal algebras ([OV05, Radc, Radb]), and for lattices with various negations ([DOA06a, DOA06b]); discrete duality for some algebras related to reasoning with incomplete information ([OR08, OR05, ORD05, Rada]), discrete duality for Heyting algebras with operators ([DO08, OR07]).

Throughout this paper we often use the same symbol for denoting algebras or relational systems (frames) and their corresponding universes.

2 MTL algebras

DEFINITION 1 *An* MTL algebra *is a system* $(L, \wedge, \vee, \otimes, \rightarrow, 0, 1)$ *such that*

(MTL.1) $(L, \wedge, \vee, 0, 1)$ *is a bounded distributive lattice with the top element* 1 *and the bottom element* 0

(MTL.2) $(L, \otimes, 1)$ *is a commutative monoid.*

(MTL.3) $\rightarrow$ *is the residuum of* $\otimes$, *that is for all* $a, b \in L$,

$$a \otimes b \leq c \Longleftrightarrow a \leq b \rightarrow c$$

where $\leq$ *is the natural lattice ordering*

(MTL.4) $(a \rightarrow b) \vee (b \rightarrow a) = 1.$

The conditions **(MTL.3)** *and* **(MTL.4)** *are called the* **residuation condition** *and the* **prelinearity condition**, *respectively.*

Note that an MTL algebra is an integral, commutative residuated lattice satisfying the prelinearity condition. Given an MTL algebra L, the following negation operator $\neg$ is defined as follows: for every $a \in L$

$$(1) \qquad\qquad\qquad \neg a \;=\; a \to 0.$$

The following lemma provides basic arithmetic laws of MTL algebras (see, for example, [BT03, Háj98, HRT02, JT02]).

LEMMA 2 *For every MTL algebra L and for all $a, b, c \in L$,*

(i) *if $a \leq b$, then*

$$a \otimes c \leq b \otimes c$$
$$b \to c \leq a \to c$$
$$c \to a \leq c \to b$$
$$\neg b \leq \neg a$$

(ii) $a \otimes b \leq a$

(iii) $a \otimes (b \to c) \leq b \to (a \otimes c)$

(iv) $a \otimes (b \vee c) = (a \otimes b) \vee (a \otimes c)$

(v) $a \otimes (a \to b) \leq b$

(vi) $a \to b \leq (a \otimes c) \to (b \otimes c)$

(vii) $a \to (b \to c) = (a \otimes b) \to c$

(viii) $(a \to b) \otimes (b \to c) \leq (a \to c)$

(ix) $a \to (b \wedge c) = (a \to b) \wedge (a \to c)$

(x) $(a \vee b) \to c = (a \to c) \wedge (b \to c)$

(xi) $a \otimes \neg a = 0$

(xii) $a \to \neg b = \neg(a \otimes b).$

Let L be an MTL algebra. For any subsets $A, B \subseteq L$ define

$$A \otimes B \;=\; \{a \in L : (\exists b \in A)(\exists c \in B)\, b \otimes c \leq a\}.$$

LEMMA 3 *For every MTL algebra L and for all $A, B, C \subseteq L$,*

(i) *if $A \subseteq B$, then $A \otimes C \subseteq B \otimes C$*

(ii) $A \otimes B = B \otimes A$

(iii) $A \otimes (B \otimes C) = (A \otimes B) \otimes C.$

Proof (i) Assume that $A \subseteq B$ and take $a \in L$ such that $a \in A \otimes C$. Then for some $b \in A$ and for some $c \in C$, it holds $b \otimes c \leq a$. By assumption, $b \in B$. Hence $a \in B \otimes C$. (ii) Follows from commutativity of $\otimes$. (iii) We show that $A \otimes (B \otimes C) \subseteq (A \otimes B) \otimes C$. The proof of the reverse inclusion is analogous. Let $a \in L$ be such that $a \in A \otimes (B \otimes C)$. Then there exist $b, c \in L$ such that **(iii.1)** $b \in A$, **(iii.2)** $c \in B \otimes C$, and **(iii.3)** $b \otimes c \leq a$. From **(iii.2)**, there exist $b' \in B$ and $c' \in C$ such that $b' \otimes c' \leq c$, which by monotonicity of $\otimes$ (see Lemma 2**(i)**) and **(iii.3)** gives $b \otimes (b' \otimes c') \leq b \otimes c \leq a$. Now, by associativity of $\otimes$, we get $(b \otimes b') \otimes c' = b \otimes (b' \otimes c') \leq a$. But $b \otimes b' \in A \otimes B$, since $b \in A$ and $b' \in B$. Therefore, there exist $b'' = b \otimes b' \in (A \otimes B)$ and $c' \in C$ such that $b'' \otimes c' \leq a$, whence $a \in (A \otimes B) \otimes C$, as required.   □

By a *filter* (resp. *prime filter*) of an MTL algebra $(L, \wedge, \vee, \otimes, \rightarrow, 0, 1)$ we mean a filter (resp. *prime filter*) of its lattice reduct $(L, \wedge, \vee, 0, 1)$. The following lemma is proved in [Urq96].

LEMMA 4 *Let L be an MTL algebra, let F, G be filters of L, and let P be a prime filter of L. Then*

(i) *$F \otimes G$ is also a filter of L*

(ii) *if $F \otimes G \subseteq P$, then there exist prime filters F', G' of L such that $F \subseteq F'$, $G \subseteq G'$, and $F' \otimes G' \subseteq P$.*

DEFINITION 5 *An **MTL frame** is a system $(X, \leqslant, R)$ such that $(X, \leqslant)$ is a poset and R is a ternary relation on X satisfying the following conditions for all $x, x'y, y'z, z' \in X$*

(F1.MTL) $R(x, y, z)$ & $x' \leqslant x$ & $y' \leqslant y$ & $z \leqslant z' \implies R(x', y', z')$

(F2.MTL) $R(x, y, z) \implies R(y, x, z)$

(F3.MTL) $R(x, y, z)$ & $R(z, y', z') \implies (\exists u \in X)\, R(y, y', u)$ & $R(x, u, z')$

(F4.MTL) $R(x, y, z)$ & $R(x', z, z') \implies (\exists u \in X)\, R(x', x, u)$ & $R(u, y, z')$

(F5.MTL) $(\exists u \in X)\, R(u, x, x)$

(F6.MTL) $R(x, y, z) \implies y \leqslant z$

(F7.MTL) $R(x, y, z)$ & $R(x, y', z') \implies (y \leqslant z')$ or $(y' \leqslant z)$.

As usual in duality theory, $n+1$–ary relations in the frames are the counterparts to the n–ary operations in the algebras. Consequently, the relation R in the MTL frame corresponds to the product operation. As it will be shown in Proposition 1, axiom **(F1.MTL)** reflects monotonicity properties of the product, axiom **(F2.MTL)** reflects its commutativity, and axioms **(F3.MTL)** and **(F4.MTL)** reflect its associativity. Also, **(F5.MTL)** and **(F6.MTL)** express the fact that the top element of an MTL algebra is a neutral element of $\otimes$. Finally, axiom **(F7.MTL)** corresponds to prelinearity.

Given an MTL frame $(X, \leqslant, R)$, let us define the following binary operations in 2^X: for all $A, B \subseteq X$,

(2) $\qquad A \otimes_R B = \{x \in X : (\exists y, z \in X)\ y \in A\ \&\ z \in B\ \&\ R(y, z, x)\}$

(3) $\qquad A \rightarrow_R B = \{x \in X : (\forall y, z \in X)\ R(x, y, z)\ \&\ y \in A \Rightarrow z \in B\}.$

Given an MTL frame $(X, \leqslant, R)$, let $\mathcal{C}(X) \subseteq 2^X$ stands for the set

(4) $\qquad\qquad\qquad \mathcal{C}(X) = \{A \subseteq X : A = [\leqslant]A\},$

where $[\leqslant]A = \{x \in X : (\forall y \in X)\ x \leqslant y \Rightarrow y \in A\}$.

DEFINITION 6 *The **complex algebra** of an MTL frame $(X, \leqslant, R)$ is the structure $(\mathcal{C}(X), \cap, \cup, \otimes_R, \rightarrow_R, \emptyset, X)$, where $\mathcal{C}(X)$ is given by (4) and the operations $\otimes_R$ and $\rightarrow_R$ are defined by (2) and (3), respectively.*

PROPOSITION 7 *The complex algebra of an MTL frame is an MTL algebra.*

Proof We have to show that

 (i) $\mathcal{C}(X)$ is closed under operations $\otimes_R$ and $\rightarrow_R$

 (ii) $\mathcal{C}(X)$ is a commutative monoid

 (iii) the residuation and the prelinearity conditions hold.

(i) We have to show that $A \otimes_R B = [\leqslant](A \otimes_R B)$ and $A \rightarrow_R B = [\leqslant](A \rightarrow_R B)$. In both cases the inclusion $(\supseteq)$ follows directly from reflexivity of $\leqslant$. To show that $A \otimes_R B \subseteq [\leqslant](A \otimes_R B)$, let us take $x \in X$ such that **(i.1)** $x \notin [\leqslant](A \otimes_R B)$. Then there exists $y \in X$ such that **(i.2)** $x \leqslant y$ and **(i.3)** $y \notin A \otimes_R B$. Suppose that $x \in A \otimes_R B$. Then there are $z, u \in X$ such that **(i.4)** $R(z, u, x)$, **(i.5)** $z \in A$, and **(i.6)** $u \in B$. By **(F1.MTL)** it follows from **(i.2)** and **(i.4)** that $R(z, u, y)$, which together with **(i.6)** and **(i.5)** gives $y \in A \rightarrow_R B$, a contradiction with **(i.3)**. The proof of $A \rightarrow_R \subseteq [\leqslant](A \rightarrow_R B)$ is analogous.

 (ii) Commutativity of $\otimes_R$ easily follows from **(F2.MTL)**. We show that $\otimes_R$ is associative, i.e., for all $A, B, C \in \mathcal{C}(X)$, $A \otimes_R (B \otimes_R C) = (A \otimes_R B) \otimes_R C$. $(\subseteq)$ Let $x \in X$ be such that $x \in A \otimes_R (B \otimes_R C)$. Then there exist $y, z \in X$ such that **(ii.1)** $R(y, z, x)$, **(ii.2)** $y \in A$, and **(ii.3)** $z \in B \otimes_R C$. Next, from **(ii.3)** it follows that there exist $u, t \in X$ such that **(ii.4)** $R(u, t, z)$, **(ii.5)** $u \in B$, and **(ii.6)** $t \in C$. Using **(F4.MTL)** we obtain from **(ii.1)** and **(ii.4)** that there exists $v \in X$ such that **(ii.7)** $R(y, u, v)$ and **(ii.8)** $R(v, t, x)$. Now, **(ii.2)**, **(ii.5)**, and **(ii.7)** give $v \in A \otimes_R B$, which together with **(ii.6)** and **(ii.8)** yields $x \in (A \otimes_R B) \otimes_R C$. By similar arguments, using **(F3.MTL)**, we have the proof of ($\supseteq$).

It remains to show that $X \otimes_R A = A$ for every $A \in \mathcal{C}(X)$.

($\subseteq$) Let $x \in X \otimes_R A$. Then for some y, $x \in X$, (ii.9) $z \in A$ and (ii.10) $R(y, z, x)$. By (F6.MTL) it follows from (ii.10) that $z \leqslant x$, so in view of (ii.9), $x \in A$.

($\supseteq$) Let $x \in A$. By (F5.MTL), there exists $u \in X$ such that $R(u, z, z)$. Then we get $z \in X \otimes_R A$.

(iii) Now, we show that for all $A, B, C \subseteq X$, $A \otimes_R B \subseteq C$ iff $A \subseteq (B \to_R C)$.
($\Rightarrow$) Suppose that (iii.1) $A \otimes_R B \subseteq C$ and there is $x \in X$ such that (iii.2) $x \in A$ and (iii.3) $x \notin B \to_R C$. From (iii.3) it follows that there exist $y, z \in X$ such that (iii.4) $R(x, y, z)$, (iii.5) $y \in B$, and (iii.6) $z \notin C$. Next, (iii.2), (iii.4), and (iii.5) imply $z \in A \otimes_R B$, which by (iii.1) gives $z \in C$, which contradicts (iii.6). The proof of ($\Leftarrow$) is similar.

Finally, we show that the prelinearity condition also holds, i.e., for all $A, B \in \mathcal{C}(X)$, $(A \to_R B) \cup (B \to_R A) = X$. Suppose otherwise, that is there exists $x \in X$ such that $x \notin (A \to_R B)$ and $x \notin (B \to_R A)$. Then there exist $y, y', z, z' \in X$ such that (iii.7) $R(x, y, z)$, (iii.8) $y \in A$, (iii.9) $z \notin B$, (iii.10) $R(x, y', z')$, (iii.11) $y' \in B$, and (iii.12) $z' \notin A$. By (F7.MTL), from (iii.7) and (iii.10) it follows $y \leqslant z'$ or $y' \leqslant z$. If $y \leqslant z'$, then by (iii.8), $z' \in A$, since $A \in \mathcal{C}(X)$. But this contradicts (iii.12). Similarly, if $y' \leqslant z$, then by (iii.11), $z \in B$, which contradicts (iii.9). $\qquad\square$

Let an MTL algebra L be given. By $\mathcal{X}(L)$ we denote the set of all prime filters of L. Define the following ternary relation on $\mathcal{X}(L)$ by: for all $x, y, z \in \mathcal{X}(L)$,

$$(5) \qquad\qquad R_\otimes(x, y, z) \qquad \text{iff} \qquad x \otimes y \subseteq z.$$

DEFINITION 8 *The **canonical frame** of an MTL algebra L is a relational system $(\mathcal{X}(L), \subseteq, R_\otimes)$, where the relation $R_\otimes$ is defined by (5).*

PROPOSITION 9 *The canonical frame of an MTL algebra is an MTL frame.*

Proof Let $x, x', y, y', z, z' \in \mathcal{X}(L)$ be such that $x' \subseteq x$, $y' \subseteq y$, $z \subseteq z'$, and assume that $R_\otimes(x, y, z)$ holds, i.e., $x \otimes y \subseteq z$. Then, by Lemma 3(i), we have the following inclusions: $x' \otimes y' \subseteq x' \otimes y \subseteq x \otimes y \subseteq z \subseteq z'$. Therefore, $R_\otimes(x', y', z')$ holds as well, so (F1.MTL) is shown.

Axiom (F2.MTL) easily follows from Lemma 3(ii).

In order to show (F3.MTL), let us take $x, y, y', z, z' \in \mathcal{X}(L)$ such that $R_\otimes(x, y, z)$ and $R_\otimes(z, y', z')$ hold, that is $x \otimes y \subseteq z$ and $z \otimes y' \subseteq z'$. By Lemma 3(i) it follows that $(x \otimes y) \otimes y' \subseteq z \otimes y' \subseteq z'$. Also, by Lemma 3(iii), $(x \otimes y) \otimes y' = x \otimes (y \otimes y')$, so we have $x \otimes (y \otimes y') \subseteq z'$. Next, from Lemma 4(i),

$y \otimes y'$ is a filter of L, so by Lemma 4(ii) it follows that there exists a prime filter, say u, such that $y \otimes y' \subseteq u$ and $x \otimes u \subseteq z'$, which means that $R_\otimes(y, y', u)$ and $R_\otimes(x, u, z')$ hold. By similar arguments we can prove (F4.MTL).

For (F5.MTL), take any $x \in \mathcal{X}(L)$. Note that $\{1\}$ is a filter of L, so by Lemma 4(i), $\{1\} \otimes x$ is also a filter of L. We show that $\{1\} \otimes x \subseteq x$. Indeed, let $a \in \{1\} \otimes x$. Then there is $b \in x$ such that $1 \otimes b \leq a$. But $1 \otimes b = b$. Then $b \leq a$, and since $b \in x$, we get $a \in x$, as expected. Now, from Lemma 4(ii) it follows that there exists $u \in \mathcal{X}(L)$ such that $u \otimes x \subseteq x$, that is $R_\otimes(u, x, x)$ holds.

For (F6.MTL), take $x, y, z \in \mathcal{X}(L)$ such that $R_\otimes(x, y, z)$ holds, that is, $x \otimes y \subseteq z$. Let $a \in L$ be such that $a \in y$. Since $1 \in x$ and $1 \otimes a = a$, we get $a = 1 \otimes a \in x \otimes y \subseteq z$, thus $a \in z$, as required.

Finally, we show that (F7.MTL) holds, that is for all $x, y, y', z, z' \in \mathcal{X}(L)$, if $R_\otimes(x, y, z)$ and $R_\otimes(x, y', z')$, then $y \subseteq z'$ or $y' \subseteq z$. Assume that (i) $x \otimes y \subseteq z$ and (ii) $x \otimes y' \subseteq z'$. Suppose that neither $y \subseteq z'$ nor $y' \subseteq z$. Then there exist $a, b \in L$ such that $a \in y$, $a \notin z'$, $b \in y'$, and $b \notin z$. Since $b \notin z$, by (i) it follows that $b \notin x \otimes y$, so for every $c \in x$ and for every $d \in y$, $c \otimes d \nleq b$. Next, since $a \in y$, $c \otimes a \nleq b$ for every $c \in x$. By the residuation principle this is equivalent to $c \nleq a \rightarrow b$ for every $c \in x$. This immediately implies $a \rightarrow b \notin x$. By a symmetric argument we conclude $b \rightarrow a \notin x$. Hence $(a \rightarrow b) \vee (b \rightarrow a) \notin x$, since x is a prime filter of L. But by axiom (MTL.4) $(a \rightarrow b) \vee (b \rightarrow a) = 1$, thus $1 \notin x$, a contradiction. $\square$

Let an MTL algebra $(L, \wedge, \vee, \otimes, \rightarrow, 0, 1)$ be given, let $(\mathcal{X}(L), \subseteq, R_\otimes)$ be its canonical frame, and let $(\mathcal{C}(\mathcal{X}(L)), \cap, \cup, \otimes_{R_\otimes}, \rightarrow_{R_\otimes}, \emptyset, \mathcal{X}(L))$ be the complex algebra of the canonical frame of L. The mapping $h : L \rightarrow 2^{\mathcal{X}(L)}$, usually called a *Stone mapping*, is defined by: for every $a \in L$,

$$(6) \qquad\qquad h(a) = \{x \in \mathcal{X}(L) : a \in x\}.$$

By standard arguments of M. H. Stone ([Sto36],[Sto37]) it follows that the mapping h defined above is a lattice embedding. Now, we show that it also preserves operations of MTL lattices.

LEMMA 10 *For every MTL algebra L and for all $a, b \in L$,*

(i) $h(a \otimes b) = h(a) \otimes_{R_\otimes} h(b)$

(ii) $h(a \rightarrow b) = h(a) \rightarrow_{R_\otimes} h(b)$.

Proof (i) Let $a, b \in L$ and let $x \in \mathcal{X}(L)$.
($\subseteq$) Assume that $x \in h(a \otimes b)$. Then $a \otimes b \in x$. Note that $\uparrow a \otimes \uparrow b \subseteq x$, where $\uparrow a$ (resp. $\uparrow b$) is the principal filter generated by a (resp. by b). Applying

Lemma 4(ii), there exist prime filters of L, say y and z, such that $\uparrow a \subseteq y$, $\uparrow b \subseteq z$, and $y \otimes z \subseteq x$. Then $R_\otimes(y, z, x)$ holds. Clearly, $a \in \uparrow a$ and $b \in \uparrow b$, so $a \in y$ and $b \in z$, whence $y \in h(a)$ and $z \in h(b)$. Therefore, $x \in h(a) \otimes_{R_\otimes} h(b)$.

($\supseteq$) Assume that $x \in h(a) \otimes_{R_\otimes} h(b)$. This means that there exist $y, z \in \mathcal{X}(L)$ such that $y \in h(a)$, $z \in h(b)$, and $R_\otimes(y, z, x)$. Then $a \in y$, $b \in z$, so $a \otimes b \in y \otimes z$, and $y \otimes z \subseteq x$. This implies that $a \otimes b \in x$, thus $x \in h(a \otimes b)$.

(ii) ($\subseteq$) Assume that $x \notin h(a) \to_{R_\otimes} h(b)$. Then there exist $y, z \in \mathcal{X}(L)$ such that $R_\otimes(x, y, z)$, $y \in h(a)$, and $z \notin h(b)$. Thus we have (ii.1) $x \otimes y \subseteq z$, (ii.2) $a \in y$, and (ii.3) $b \notin z$. From (ii.1) and (ii.3) it follows that $b \notin x \otimes y$, so for all $c, d \in L$, $c \in x$ and $d \in y$ imply $c \otimes d \not\leq b$. Hence, by (ii.2), $c \otimes a \not\leq b$, which by the residuation condition means that $c \not\leq a \to b$ for every $c \in x$. This implies $a \to b \notin x$, thus $x \notin h(a \to b)$.

($\supseteq$) Assume that $x \notin h(a \to b)$, i.e., (ii.4) $a \to b \notin x$. First, note that by Lemma 4(i), $x \otimes \uparrow a$ is a filter of L. We show that $b \notin x \otimes \uparrow a$. Suppose otherwise, that is $b \in x \otimes \uparrow a$. Then there exist $c, d \in L$ such that (ii.5) $c \in x$, (ii.6) $a \leq d$, and (ii.7) $c \otimes d \leq b$. By the residuation condition (ii.7) is equivalent to $c \leq d \to b$, which by (ii.5) yields $d \to b \in x$, and furthermore, by monotonicity of $\to$ (see Lemma 2(i)) and (ii.6), $a \to b \in x$, a contradiction with (ii.4). Therefore, $b \notin x \otimes \uparrow a$, so $(x \to \uparrow a) \cap \downarrow b = \emptyset$. By Prime Filter Theorem there exists $z \in \mathcal{X}(L)$ such that (ii.8) $x \otimes \uparrow a \subseteq z$ and $z \cap \downarrow b = \emptyset$, so $b \notin z$, whence (ii.9) $z \notin h(b)$. Moreover, from (ii.8), by Lemma 4(ii) it follows that there exists $y \in \mathcal{X}(L)$ such that $x \otimes y \subseteq z$, that is (10) $R_\otimes(x, y, z)$, and $\uparrow a \subseteq y$, so $a \in y$, thus (ii.11) $y \in h(a)$. Combining (ii.9), (ii.10), and (ii.11) we obtain $x \notin h(a) \to_{R_\otimes} h(b)$. $\square$

Let $(X, \leqslant, R)$ be an MTL frame, let $(\mathcal{C}(X), \cap, \cup, \otimes_R, \to_R, \emptyset, X)$ be its complex algebra, and let $(\mathcal{X}(\mathcal{C}(X)), \subseteq, R_{\otimes_R})$ be the canonical frame of the complex algebra of X. Define the mapping $k : X \to 2^{\mathcal{C}(X)}$ as follows: for every $x \in X$,

$$(7) \qquad k(x) = \{A \in \mathcal{C}(X) : x \in A\}.$$

By a straightforward verification one can easily show that for any $x \in X$, $k(x)$ is a prime filter of $\mathcal{C}(X)$.

LEMMA 11 *Let $(X, \leqslant, R)$ be an MTL frame and let the mapping k be defined by (7). Then*

(i) *k is injective*

(ii) *$(\forall x, y \in X) \; x \leqslant y \iff k(x) \subseteq k(y)$*

(iii) *$(\forall x, y, z \in X) \; R(x, y, z) \iff R_{\otimes_R}(k(x), k(y), k(z))$.*

Proof (i) Let $x, y \in X$ be such that $x \neq y$. Since $\leqslant$ is a partial ordering, it is antisymmetric, so without loss of generality we can assume that $x \not\leqslant y$. Then $y \notin \langle \geqslant \rangle \{x\}$, where $\geqslant = \leqslant^{-1}$ and $\langle \geqslant \rangle A = \{x \in X : (\exists y \in A) \ x \geqslant y\}$. By a straightforward verification, $\langle \geqslant \rangle \{x\} \in \mathcal{C}(X)$. Then $\langle \geqslant \rangle \{x\} \notin k(y)$. But $\langle \geqslant \rangle \{x\} \in k(x)$. Whence $k(x) \neq k(y)$.

(ii) The proof of ($\Rightarrow$) is straightforward. For the reverse implication, let us take $x, y \in X$ and note that $k(x) \subseteq k(y)$ iff $(\forall A \in \mathcal{C}(X)) \ A \in k(x) \Rightarrow A \in k(y)$ iff $(\forall A \in \mathcal{C}(X)) \ x \in A \Rightarrow y \in A$. Take $A = \langle \geqslant \rangle \{x\}$. Clearly, $A \in \mathcal{C}(X)$ and $x \in A$. Then we get $y \in A$, i.e., $y \in \langle \geqslant \rangle \{x\}$, whence $x \leqslant y$.

(iii) Let $x, y, z \in X$. First, note that the following equivalences hold:

$$R_{\otimes_R}(k(x), k(y), k(z))$$
$$\text{iff } k(x) \otimes_R k(y) \subseteq k(z)$$
$$\text{iff } (\forall A \in \mathcal{C}(X)) \ A \in k(x) \otimes_R k(y) \Rightarrow A \in k(z)$$
$$\text{iff } (\forall A \in \mathcal{C}(X)) \ [(\exists B, D \in \mathcal{C}(X)) \ B \in k(x) \ \& \ D \in k(y) \ \& \ B \otimes_R D \subseteq A] \Rightarrow z \in A$$
$$\text{iff } (\forall A \in \mathcal{C}(X)) \ [(\exists B, D \in \mathcal{C}(X)) \ x \in B \ \& \ y \in D \ \& \ B \otimes_R D \subseteq A] \Rightarrow z \in A.$$

($\Rightarrow$) Assume that (iii.1) $R(x, y, z)$ and consider $A \in \mathcal{C}(X)$ such that for some $B, D \in \mathcal{C}(X)$, (iii.2) $x \in B$, (iii.3) $y \in D$, and (iii.4) $B \otimes_R D \subseteq A$. From (iii.1), (iii.2), and (iii.3) it follows that $z \in B \otimes_R D$, which in view of (iii.4) gives $z \in A$. Therefore, $R_{\otimes_R}(k(x), k(y), k(z))$ holds.

($\Leftarrow$) Assume that (iii.5) $R_{\otimes_R}(k(x), k(y), k(z))$ and consider the set $A = \{t \in X : R(x, y, t)\}$. Let $t, t' \in X$ be such that $t \in A$ and $t \leqslant t'$. Then $R(x, y, t)$ holds, and by (F1.MTL), $R(x, y, t')$ also holds. Then $t' \in A$, so $t \in [\leqslant]A$, thus $A \subseteq [\leqslant]A$. By reflexivity of $\leqslant$, $[\leqslant]A \subseteq A$. Therefore, $A \in \mathcal{C}(X)$. Put $B = \langle \geqslant \rangle \{x\}$ and $D = \langle \geqslant \rangle \{y\}$. We show that $B \otimes_R D \subseteq A$. Indeed, let $u \in X$ be such that $u \in B \otimes_R D$. Then there exist $x', y' \in X$ such that $x \leqslant x'$, $y \leqslant y'$, and $R(x', y', u)$. Hence, by applying (F1.MTL), we obtain $R(x, y, u)$, thus $u \in A$. Clearly, $x \in \langle \geqslant \rangle \{x\}$ and $y \in \langle \geqslant \rangle \{y\}$. Finally, by assumption (iii.5) and the above equivalences we get $z \in A$, so $R(x, y, z)$ holds, which completes the proof. $\square$

We complete this section by the following discrete duality result between MTL algebras and MTL frames.

THEOREM 12 (Discrete duality for MTL algebras)

(i) *Every MTL algebra is embeddable into the complex algebra of its canonical frame.*

(ii) *Every MTL frame is embeddable into the canonical frame of its complex algebra.*

3   Axiomatic extensions of MTL algebras

In this section we present some axiomatic extensions of MTL algebras.

3.1   SMTL algebras

DEFINITION 13 *An **SMTL algebra** is an MTL algebra* $(L, \wedge, \vee, \otimes, \rightarrow, 0, 1)$ *satisfying the following condition for every* $a \in L$,

(SMTL) $a \wedge \neg a = 0$,

where $\neg a$ *is defined by (1), i..e,* $\neg a = a \rightarrow 0$.

DEFINITION 14 *An **SMTL frame** is an MTL frame* $(X, \leqslant, R)$ *satisfying the following condition for every* $x \in X$,

(F.SMTL) $(\exists y, z \in X) \, R(x, y, z) \, \& \, x \leqslant y$.

Given an SMTL frame, for every $A \subseteq X$, denote

$$(8) \quad \neg_R A = A \rightarrow_R \emptyset \;=\; \{x \in X : (\forall y, z \in X) \, R(x, y, z) \Rightarrow y \notin A\}.$$

PROPOSITION 15 *The complex algebra of an SMTL frame is an SMTL algebra.*

Proof Let $x \in X$ be such that $x \in \neg_R A$. By (8) this means that for all $y, z \in X$, **(i)** $R(x, y, z) \Rightarrow y \notin A$. From **(F.SMTL)** we get **(ii)** $R(x, y', z')$ and **(iii)** $x \leqslant y'$ for some $y', z' \in X$. Now, **(i)** and **(ii)** imply $y' \notin A$, which together with **(iii)** yields $x \notin A$, since $A \in \mathcal{C}(X)$. Therefore, $\neg_R A \subseteq -A$, where $-A$ is the usual set complement with respect to $\mathcal{C}(X)$. Consequently, $A \cap \neg_R A = \emptyset$, which completes the proof. □

PROPOSITION 16 *The canonical frame of an SMTL algebra is an SMTL frame.*

Proof We need to show that for every $x \in \mathcal{X}(L)$ there exist $y, z \in \mathcal{X}(L)$ such that $R_\otimes(x, y, z)$ and $x \subseteq y$.
Let $x \in \mathcal{X}(L)$. We show that $x \otimes x$ is a proper filter. For suppose otherwise, that is $0 \in x \otimes x$. Then there exist $a, b \in x$ such that $a \otimes b = 0$. By the residuation condition we get $b \leq \neg a$, thus $a \wedge \neg a \in x$. But by axiom **(SMTL)**, $a \wedge \neg a = 0$, whence $0 \in x$, a contradiction.
By Prime Filter Theorem $x \otimes x$ can be extended to a prime filter of L, say u, that is $x \otimes x \subseteq u$. Hence we have shown two prime filters of L, namely $y = x$ and $z = u$, such that $R_\otimes(x, y, z)$ and $x \subseteq y$, as required. □

Applying Theorem 12 and Propositions 15 and 16 we get discrete duality between SMTL algebras and SMTL frames.

THEOREM 17 (Discrete duality for SMTL algebras)

(i) *Every SMTL algebra is embeddable into the complex algebra of its canonical frame.*

(ii) *Every SMTL frame is embeddable into the canonical frame of its complex algebra.*

3.2 CMTL algebras

In this section we consider MTL algebras satisfying the contraction axiom.

DEFINITION 18 *A CMTL algebra is an MTL algebra* $(L, \wedge, \vee, \otimes, \rightarrow, 0, 1)$ *satisfying the following condition for every* $a \in L$,

(**CMTL**) $a \leq a \otimes a$.

DEFINITION 19 *A **CMTL frame** is an MTL frame* $(X, \leqslant, R)$ *such that for every* $x \in X$,

(**F.CMTL**) $R(x, x, x)$.

PROPOSITION 20 *The complex algebra of a CMTL frame is a CMTL algebra.*

Proof We only need to show that $A \subseteq A \otimes A$ for every $A \in \mathcal{C}(X)$. Let $x \in X$ be such that $x \notin A \otimes A$. Then for all $y, z \in X$, $R(y, z, x)$ and $y \in A$ imply $z \notin A$. In particular, $R(x, x, x)$ implies $x \notin A$. Hence, by (**F.CMTL**), $x \notin A$.   □

PROPOSITION 21 *The canonical frame of a CMTL algebra is a CMTL frame.*

Proof We only need to prove that (**F.CMTL**) is satisfied, that is for every $x \in \mathcal{X}(L)$, $R_{\otimes}(x, x, x)$ holds, i.e., $x \otimes x \subseteq x$. Let $a \in x \otimes x$. Then there exist $b, c \in x$ such that (**i**) $b \otimes c \leq a$. Since x is a filter of L, we have (**ii**) $b \wedge c \in x$. Obviously, $b \wedge c \leq b$ and $b \wedge c \leq c$, so by monotonicity of $\otimes$, it holds $(b \wedge c) \otimes (b \wedge c) \leq b \otimes c$. By axiom (**CMTL**), $b \wedge c \leq (b \wedge c) \otimes (b \wedge c)$. Therefore, $b \wedge c \leq b \otimes c$, which in view of (**ii**) gives $b \otimes c \in x$, and furthermore, by (**i**), $a \in x$, as required.   □

From Theorem 12 and Propositions 20 and 21 we obtain the next result.

THEOREM 22 (Discrete duality for CMTL algebras)

(i) *Every CMTL algebra is embeddable into the complex algebra of its canonical frame.*

(ii) *Every CMTL frame is embeddable into the canonical frame of its complex algebra.*

The algebras SMTL with the contraction axiom (CMTL) are Gödel algebras. So the dualities presented in the subsection 3.1 and the present one lead to the discrete duality for Gödel algebras.

3.3   IMTL algebras

The negation operator $\neg$ defined by (1) in MTL algebras is not necessarily involutive. In this section we consider MTL algebras with an additional axiom which makes this operator an involution.

DEFINITION 23 *An **IMTL algebra** is an MTL algebra* $(L, \wedge, \vee, \otimes, \rightarrow, 0, 1)$ *satisfying the following condition for every* $a \in L$,

(IMTL) $\neg\neg a \leq a$.

DEFINITION 24 *An **IMTL frame** is an MTL frame* $(X, \leqslant, R)$ *satisfying the following condition for all* $x, y \in X$,

(F.IMTL) $(\forall z \in X)[(\exists t \in X)\, R(x, z, t) \Rightarrow (\exists u \in X)\, R(z, y, u)] \Rightarrow y \leqslant x$

PROPOSITION 25 *The complex algebra of an IMTL frame is an IMTL algebra.*

Proof First, recall that $\neg_R A = \{x \in X : (\forall y, z \in X)\, R(x, y, z) \Rightarrow y \notin A\}$. We need to show that $\neg_R \neg_R A \subseteq A$ for every $A \in \mathcal{C}(X)$.
Let $x \in X$ be such that $x \in \neg_R \neg_R A$. Then we have:

$(\forall y, z \in X)[R(x, y, z) \Rightarrow y \notin \neg_R A]$
$\quad$ iff $(\forall y, z \in X)[R(x, y, z) \Rightarrow (\exists y', z' \in X)\, R(y, y', z') \,\&\, y' \in A)]$
$\quad$ iff $(\forall y, z \in X)[(\exists y', z' \in X)\, R(x, y, z) \Rightarrow (R(y, y', z') \,\&\, y' \in A)]$
$\quad$ iff $(\forall y, z \in X)[(\exists y', z' \in X)\, (R(x, y, z) \Rightarrow R(y, y', z') \,\&\, (R(x, y, z) \Rightarrow y' \in A)],$

which implies

$(\forall y, z \in X)[(\exists y', z' \in X) R(x, y, z) \Rightarrow R(y, y', z')] \,\&\, [(\exists y', z' \in X) R(x, y, z \Rightarrow) y' \in A]$

or equivalently,
(i) $(\forall y, z \in X)\, (\exists y', z' \in X)\, R(x, y, z) \Rightarrow R(y, y', z')$ and
(ii) $(\forall y, z \in X)\, (\exists y' \in X)\, R(x, y, z) \Rightarrow y' \in A$.
By **(F.IMTL)**, it follows from **(i)** that **(iii)** $y' \leqslant x$. Next, by **(F5.MTL)** we get $R(u, x, x)$ for some $u \in X$, whence, by **(F2.MTL)**, $R(x, u, x)$, which together with **(ii)** gives **(iv)** $y' \in A$. Since $A = [\leqslant]A$, **(iii)** and **(iv)** imply $x \in A$, as required. $\quad\square$

PROPOSITION 26 *The canonical frame of an IMTL algebra is an IMTL frame.*

Proof Let $x,y \in \mathcal{X}(L)$ be such that **(i)** $(\forall z \in \mathcal{X}(L))$ $[(\exists t \in \mathcal{X}(L))$ $x \otimes z \subseteq t] \Rightarrow$ $[(\exists u \in \mathcal{X}(L))$ $z \otimes y \subseteq u]$. Let $a \in L$ be such that **(ii)** $a \in y$. We have to show that $a \in x$. First, we show that **(iii)** $x \otimes \uparrow \neg a = L$. Suppose otherwise, i.e., $x \otimes \uparrow \neg a \neq L$. By Lemma 4**(i)** $x \otimes \uparrow \neg a$ is a filter of L and, by assumption, it is a proper filter of L, so $0 \notin x \otimes \uparrow \neg a$. Then, by Prime Filter Theorem, there exists $t \in \mathcal{X}(L)$ such that $x \otimes \uparrow \neg a \subseteq t$. Hence, by Lemma 4**(ii)**, there exists $z' \in \mathcal{X}(L)$ such that **(iv)** $x \otimes z' \subseteq t$ and **(v)** $\uparrow \neg a \subseteq z'$. Now, putting $z = z'$ in **(i)**, by **(iv)** we get $z' \otimes y \subseteq u$ for some $u \in \mathcal{X}(L)$. From **(v)**, $\neg a \in z'$, which together with **(ii)** gives $\neg a \otimes a \in z' \otimes y \subseteq u$. But $\neg a \otimes a = 0$ by Lemma 2**(xi)**, whence $0 \in u$, a contradiction. Therefore, **(iii)** holds, so $0 \in x \otimes \uparrow \neg a$. Then there exist $b,c \in L$ such that **(vi)** $b \in x$, **(vii)** $\neg a \leq c$, and **(viii)** $b \otimes c = 0$. By the residuation condition **(viii)** is equivalent to $b \leq \neg c$, which together with **(vi)** yields **(ix)** $\neg c \in x$. Furthermore, by Lemma 2**(i)**, it follows from **(vii)** that $\neg c \leq \neg \neg a$. But $\neg \neg a \leq a$ by axiom **(IMTL)**. Then $\neg c \leq a$, so by **(ix)**, $a \in x$, as required.   □

By Theorem 12 and Propositions 25 and 26 we obtain discrete duality between IMTL algebras and IMTL frames.

THEOREM 27 (Discrete duality for IMTL algebras)

  (i) *Every IMTL algebra is embeddable into the complex algebra of its canonical frame.*

  (ii) *Every IMTL frame is embeddable into the canonical frame of its complex algebra.*

4   Conclusions

In this paper we have considered MTL algebras and have recalled a discrete duality between these algebras and the corresponding class of relational structures. Next, we have developed discrete dualities for three axiomatic extensions of MTL algebras: SMTL algebras, CMTL algebras, and IMTL algebras. These dualities are based on the representation of bounded distributive lattices as rings of sets. It is known that Gödel algebras are the SMTL algebras endowed with the contraction axiom. Hence, the developments of the paper yield the discrete duality for them. The relational structures dual to the classes of algebras considered in this paper provide Kripke–style semantics for the corresponding logics. A method of proving completeness of logics with respect to a Kripke semantics, once a discrete duality involving the class of underlying relational systems is known, is described in [ORD05]. Priestley duality for MTL algebras, SMTL algebras and IMTL algebras is developed in [CC06].

BIBLIOGRAPHY

[BT03]   K. Blount and C. Tsinakis. The structure of residuated lattices. *Int. J. Algebra and Computation*, 13(4):437–461, 2003.

[CC06]   L. M. Carber and S. A. Celani. Pristley duality for some lattice-ordered algebraic structures including MTL, IMTL, and MV-algebras. *Central Eur. J. Math.*, 4(4):600–623, 2006.

[DO08]   I. Düntsch and E. Orłowska. A discrete duality between apartness algebras and apartness frames. *J. Appl. Non-Classical Logics*, 18(2–3):209–223, 2008.

[DOR06]   I. Düntsch, E. Orłowska, and A. M. Radzikowska. Lattice-based relation algebras II. In H. de Swart, E. Orłowska, G. Schmidt, and M. Roubens, editors, *Theory and Applications of Relational Structures as Knowledge Instruments II*, volume 4342 of *LNCS*, pages 267–289, Berlin, 2006. Springer.

[DOA06a]   W. Dzik, E. Orłowska, and C. J. van Alten. Relational representation theorems for general lattices with negations. In R. Schmidt, editor, *Proceedings of 9th International Conference of Relational Methods in Computer Science*, volume 4136 of *LNCS*, pages 162–176, Berlin, 2006. Springer.

[DOA06b]   W. Dzik, E. Orłowska, and C. J. van Alten. Relational representation theorems for lattices with negations: A survey. In H. de Swart, E. Orłowska, G. Schmidt, and M. Roubens, editors, *Theory and Applications of Relational Structures as Knowledge Instruments II*, volume 4342 of *LNCS*, pages 245–266, Berlin, 2006. Springer.

[DP90]   B. A. Davey and H. A. Priestley. *Introduction to Lattices and Order*. Cambridge University Press, Cambridge, 1990.

[EG01]   F. Esteva and L. Godo. Monoidal t-norm based logic: Towards a logic for left-continuous t-norms. *Fuzzy Sets Syst.*, 124(3):271–288, 2001.

[EGGC03]   F. Esteva, L. Godo, and À. García-Cerdaña. On the hierarchy of t-norm based residuated fuzzy logics. In M. C. Fitting and E. Orłowska, editors, *Beyond Two: Theory and Applications of Multiple-Valued Logic*, volume 114 of *Stud. Fuzziness Soft Comput.*, pages 251–272. Physica Verlag, Heidelberg, 2003.

[GJKO07]   N. Galatos, P. Jipsen, T. Kowalski, and H. Ono. *Residuated Lattices: An Algebraic Glimpse at Substructural Logics*, volume 151 of *Studies in Logic and the Foundations of Mathematics*. Elsevier, Amsterdam, 2007.

[Háj98]   P. Hájek. *Metamathematics of Fuzzy Logic*, volume 4 of *Trends in Logic*. Kluwer, Dordrecht, 1998.

[Höh95]   U. Höhle. Commutative, residuated l-monoids. In U. Höhle and E. P. Klement, editors, *Non-Classical Logics and Their Applications to Fuzzy Subsets*, volume 32 of *Theory Decis. Libr., Ser. B.*, pages 53–106. Kluwer, Dordrecht, 1995.

[HRT02]   J. B. Hart, L. Rafter, and C. Tsinakis. The structure of commutative residuated lattices. *Int. J. Algebra and Computation*, 12:509–524, 2002.

[JM02]   S. Jenei and F. Montagna. A proof of standard completeness for Esteva and Godo's logic MTL. *Studia Logica*, 70(2):183–192, 2002.

[JT51]   B. Jónsson and A. Tarski. Boolean algebras with operators. I. *Amer. J. Math.*, 73:891–939, 1951.

[JT52]   B. Jónsson and A. Tarski. Boolean algebras with operators. II. *Amer. J. Math.*, 74:127–162, 1952.

[JT02]   P. Jipsen and C. Tsinakis. A survey of residuated lattices. In J. Martinez, editor, *Ordered Algebraic Structures*, pages 19–56. Kluwer, Dordrecht, 2002.

[Mak72]   L. L. Maksimova. Pretabular extensions of superintuitionistic logics. *Algebra and Logic*, 11(5):558–570, 1972.

[Mak75]   L. L. Maksimova. Pretabular extensions of the Lewis' logic S4. *Algebra and Logic*, 14(1):28–55, 1975.

[OR]   E. Orłowska and I. Rewitzky. Algebras for Galois-style connections and their discrete duality. Submitted.

[OR05]   E. Orłowska and I. Rewitzky. Duality via truth: Semantic frameworks for lattice-based logics. *Logic J. of the IGPL*, 13:467–490, 2005.

[OR06]   E. Orłowska and A. M. Radzikowska. Relational representability for algebras of substructural logics. In W. MacCaull, M. Winter, and I. Düntsch, editors, *Relational Methods in Computer Science*, volume 3929 of *LNCS*, pages 212–226. Springer, 2006.

[OR07]   E. Orłowska and I. Rewitzky. Discrete dualities for Heyting algebras with operators. *Fund. Informaticae*, 81:275–295, 2007.

[OR08]   E. Orłowska and A. M. Radzikowska. Representation theorems for some fuzzy logics based on residuated non-distributive lattices. *Fuzzy Sets Syst.*, 159:1247–1259, 2008.

[ORD05]   E. Orłowska, I. Rewitzky, and I. Düntsch. Relational semantics through duality. In W. MacCaull, M. Winter, and I. Düntsch, editors, *Relational Methods in Computer Science*, volume 3929 of *LNCS*, pages 17–32. Springer, 2005.

[OV05]   E. Orłowska and D. Vakarelov. Lattice-based modal algebras and modal logics. In P. Hájek, L. Valdes-Villanueva, and D. Westerståhl, editors, *Logic, Methodology and Philosophy of Science*, pages 147–170, London, 2005. KLC Publications.

[Pri70]   H. A. Priestley. Representation of distributive lattices by means of ordered Stone spaces. *Bull. London Math. Soc.*, 2:186–190, 1970.

[Pri72]   H. A. Priestley. Ordered topological spaces and the representation of distributive lattices. *Proc. London Math. Soc. Third Series*, 24:507–530, 1972.

[Rada]   A. M. Radzikowska. Discrete dualities for some information algebras based on De Morgan lattices. Submitted.

[Radb]   A. M. Radzikowska. Discrete representation theorems for lattice-based modal algebras with negations. Submitted.

[Radc]   A. M. Radzikowska. Relational representation theoems for some modal algebras based on non-distributive lattices. Submitted.

[Sto36]   M. H. Stone. The theory of representations for Boolean algebras. *Trans. Amer. Math. Soc.*, 39:37–111, 1936.

[Sto37]   M. H. Stone. Topological representation of distributive lattices and Brouwerian logics. *Časopis pro pěstování matematiky a fyziky*, 67:1–25, 1937.

[Urq78]   A. Urquhart. A topological representation theory for lattices. *Algebra Universalis*, 8:45–58, 1978.

[Urq96]   A. Urquhart. Duality for algebras of relevant logics. *Studia Logica*, 56:263–276, 1996.

Ewa Orłowska
National Institute of Telecommunications,
Szachowa 1, 04–894 Warsaw, Poland
Email: E.Orlowska@itl.waw.pl

Anna Maria Radzikowska
Faculty of Mathematics and Information Science,
Warsaw University of Technology,
Plac Politechniki 1, 00–661 Warsaw, Poland
Email: annrad@mini.pw.edu.pl

Fuzzy logic as Hájek's comparative notion of truth and applications in preference modeling

PETER VOJTÁŠ

Dedicated to Petr Hájek,
on the occasion of his 70th birthday

ABSTRACT.   In this essay we are giving an overview of our work
with our students and colleagues on fuzzy rules and similarities originally motivated by Petr Hájek's teaching and work. Our main starting
points are Petr's visions of fuzzy logic in narrow sense and understanding of fuzzy values as a comparative notion of truth. Fuzzy logic in
narrow sense is in our work reflected by formal models of fuzzy logic
programming and similarities. Comparative notion of truth led us to
understand fuzzy values as degree of user preference. As far as mathematical fuzzification of a domain can lead to several possible models,
requirements of computer science application can help to prefer one of
them. We overview our work (originally published with several coau-
thors) on models of fuzzy logic programming and its connection to
generalized annotated programs and similarity reasoning; on fuzzy inductive logic programming; application to user preference learning and
querying; applications to web information extraction and web semantization and conclude with some observations and lessons learned.

1   Introduction, motivation, problems

Origins of fuzzy sets are connected to seminal work of Lotfi A. Zadeh [Zad65].
Original motivation was image processing, where different grades of shade of
image pixel were first practical examples of fuzzy sets, see e.g. picture 1a
([SHB98]). Fuzzy set theory has developed rapidly with main applications
in control (see e.g. pictures 1b and 1c [Wik07], where rules for an inverted
pendulum 1b are trained using fuzzy sets, e.g. 1c). In the case of fuzzy con-
trollers, the system can be in a continuum of states which are a combination
of values from an interval. That is why fuzzy sets form a partition and rules
have to describe the control action for all possible combinations. This is a
main difference in comparison with computer science applications.

Another motivation for fuzzy sets was modeling vague concepts like tall,
young, etc. Many applications needed formal models. Here is the first contribution of Petr Hájek. In [Háj07] he further develops Zadeh's terms "fuzzy

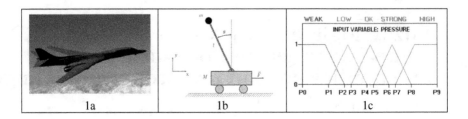

Figure 1. (a, b, c)

logic in broad or wide and narrow sense." In broad sense the term fuzzy logic
has been used as synonymous with fuzzy set theory and its applications. In
difference to Zadeh, which understood the emerging narrow sense fuzzy logic
as a theory of approximate reasoning based on many valued logic, Hájek
claims

> ... a logician will first study classical logical questions on completeness,
> decidability, complexity etc. of the symbolic calculi in question and then
> try to reduce question of Zadeh's agenda to questions of deduction as far
> as possible.

This is the approach in monograph of Petr Hájek [Háj98]. He introduced
fuzzy logic in narrow sense as a mathematical theory of many valued logic.
First contribution was done by Jan Pavelka in [Pav79] where he used graded
statements (propositions), e.g.

$$p.\,0.5 \quad p \to q.\,0.7$$

to development of an axiomatic system of Lukasiewicz logic.

Basic development by Petr Hájek [Háj98] was focused in axiomatization
of 1-tautologies of various fuzzy logics. Just to motivate our view of this
approach, let us notice that classical 0-1 tautology

$$(1) \quad (\varphi \to (\psi \to \chi)) \to ((\varphi \to \psi) \to (\psi \to \chi))$$

is no more a tautology even in a 3 valued logic. This has led Petr Hájek
to the study of metamathematics of fuzzy logic, which is detail presented in
some papers of this volume. This concerns especially introduction and study
of so called Basic Logic BL, where (1) is not provable and there are some BL
axioms replacing this form of composed implications, e.g.

$$(\varphi \to \psi) \to ((\psi \to \chi) \to (\varphi \to \chi)).$$

For tuning real world applications to data, and moreover online with response in a second, it is hard to follow this way (to look for and work with axiomatic systems).

Nevertheless our main motivation came from attending lectures of Petr Hájek on fuzzy logic [Háj94], where he introduced fuzzy values as a comparative notion of truth (see also [Háj07]). This was quite well in concordance to different representation of preferences in computer science applications. There are various ways of graphical representation of preference, importance, relevance etc. For user feedback we can use colors as in traffic lights (2a), different smiley's (2b), stars (2c) or sliders (2d, [UPr08]).

Figure 2. (a, b, c, d)

For representation of results we can use Google like graphics (3d) and mnemonics of top-down and let-right reading, enhanced by font size (3b). Another possibility is thumbs up/down mnemonic (3c) and understanding colors in geographic maps, the higher terrain the better answer (3a, [JVKH09]).

Figure 3. (a, b, c, d)

To resume our starting point, these were computer science motivated problems with a small number of truth values (usually 7+-2), data intensive, with adaptation to different users (although in formal models we often use [0,1] interval).

2   Formal models

For some approaches we have developed formal models based on our real world motivation. It is our conviction, that formal models should describe a real world problem and its formal solution should contribute to a real world solution / improvement.

2.1   Fuzzy logic programming versus fuzzy resolution

In fuzzy logic programming we first have to decide whether rules are implications or clauses; whether my deduction procedure is resolution based refutation or database querying (in the crisp case these are equivalent, [Llo93]). We have studied both approaches.

In [Voj01b] we have developed a model of fuzzy logic programming with implicative rules and database querying. Having

$$p.\,0.5 \quad p \to q.\,0.7$$

and truth function of $\to$ is an implicator $I : [0,1]^2 \to [0,1]$, then the result q can be deduced by many valued modus ponens with truth at least $C_I(0.5\ ,\ 0.7)$, where C_I is a residual conjunctor to I. For such approach we have proved correctness of fuzzy logic programming (every computed answer is correct) and approximate completeness (every correct answer can be arbitrarily precisely computed). More a fixed point theory was developed too.

In a real world situation, both for user feedback and for query results presentation, we need finite valued fuzzy logic (moreover, a very low finite number of values). This was already used by the psychologist Rensis Likert "Often five ordered response levels are used, although many psychometricians advocate using seven or nine levels; a recent empirical study found that a 5- or 7- point scale may produce slightly higher mean scores relative to the highest possible attainable score, compared to those produced from a 10-point scale, and this difference was statistically significant (see [Wik09b])". This is also a common practice when refereeing papers, usually 7 values range from strong accept to strong reject.

Small finite valued fuzzy logic programming was studied in [KLV04], where results of [Voj01b] were extended for conjunctors which are results of rounding t-norms to a finite scale. Note that rounding a conjunctor C upwards (in x axis to n values, in y axis to m values and in result z axis to k values) gives a conjunctor $C_{n,m}^k$ which need not be associative nor commutative (rounding upwards preserves these conjunctors to be left continuous). Our theory of fuzzy logic programming was extended also to this case.

In practical applications it often comes to a situation that user preferences are aggregated from particular objectives. Situation is similar as in light athletic decathlon. Here individual achievements of an athlete are first converted to points (a fuzzy degree usually between 0 and 1000) and then summed up (fuzzy aggregation without normalization). See Fig. 4 with results of a race in Götzis on 27.5.2001 where R. Šebrle established first World Record above 9000 points.

P	Athlete	Points	100m	Long	Shot	High	400m	110mh	Discus	Pole	Javelin	1500m
1	Šebrle CZE	9026	10,64	8.11	15.33	2.12	47,79	13,92	47.92	4.8	70.16	4.21,98
2	Nool EST	8604	10,73	7.8	14.37	1.97	46,89	14,46	43,32	5.3	66.94	4.39,11
3	Dvorak CZE	8527	10,84	7.69	15.83	1.97	48,76	13,99	46.74	4.7	66.66	4.33,58

Figure 4. Athletic decathlon – Götzis on 27.5.2001

Also in web shops and other user decision making situations similar phenomena appear. Namely, it can happen that even if users have similar objectives one user can have different weights for aggregation, or different users can have just opposite directions of preference.

For such situations, we have developed a model of fuzzy logic programming where fuzzy aggregations can appear in the body of rule. Such a rule can have form (in graded Prolog notation)

$$H \leftarrow @(B_1, ..., B_n).r$$

We have also showed in [KLV04] that these are, in a sense, isomorphic to rules of GAP-Generalized annotated programs (with crisp $\leftarrow$ and &, see [KS92])

$$H : @(b_1, ..., b_n) \leftarrow B_1 : b_1 \& ... \&, B_n : b_n$$

and procedural and declarative semantic are also in good connection (for more details see [KLV04]).

Concerning fuzzy resolution, we consider a rule to be a clause so instead of $H \leftarrow B$ we write $H \vee \neg B$ and deduction is a refutation process initialized by $\neg H$. We have studied this problem in [SV04] in a more general setting, namely graded many valued resolution. This means, having

$$(A \vee B).x \quad and \quad (\neg A \vee C).y$$

We are looking for a function calculating

$$(B \vee C).f_\vee(x, y)$$

correctly, that means in models of fuzzy propositional logic. Some interesting combination occurred, especially $f_\vee(x, y) = 0$ for $x + y \leq 1$, for more see [SV04]. Nevertheless a practical question remains, who will create these clauses, an untrained user probably not.

2.2   Fuzzy similarity

Fuzzy similarity is another phenomenon which is important for practical computer science applications. Somebody looking for a resource (web page, document, product, genetic entity ...) has some requirements; nevertheless it can happen that there are no objects fulfilling these requirements. Then a (most) similar resource can make user happy. Similarity is in a sense dual to distance, triangular inequality which is necessary to make a distance metric is dual to transitivity for fuzzy similarity. For a similarity s on a domain A,

$$s : A \times A \to [0,1]$$

the T-transitivity has form

$$T(s(x,y), s(y,z)) \leq s(x,z)$$

where T is a t-norm (or a conjunctor). We say that the space (A,x) is a T-similarity space if s is fulfilling T-transitivity. We say that a triple

$$s(x,y), s(y,z), s(x,z)$$

is a *nontrivial* similarity triple, if all three numbers $s(x,y)$, $s(y,z)$, and $s(x,z)$ are mutually different. We have

OBSERVATION 1 *Assume that similarity s is symmetric and (A,s) is a min-similarity space. Then there are no nontrivial similarity triples in (A,s).*

   Observation is easy to conclude. By contradiction, assume, there is a nontrivial triple. Using symmetry (and renaming if necessary) this triple can be chosen such that $s(x,z) < s(x,y) < s(y,z)$. But then $s(x,z) < min(s(x,y), s(y,z))$, a contradiction with min transitivity.
   Even notice, that min-similarity triples have to have two smaller numbers equal and only one bigger (see Fig. 5). This shows that min-similarity spaces have hierarchical structure and α-cuts form an equivalence (partition) on A. In real world situations this is very often not the case, data are more randomly distributed. A nontrivial similarity triple (Fig. 6) forces us to consider other t-norms in similarity transitivity than min. If such a triangle is colored in graph terminology we call such a triangle *colorful* if it contains all tree colors.
   This motivates us to following

PROBLEM 2 *Assume we have K_n a coloring of the complete graph on n vertices by three colors. Is there a colorful triangle?*

   Of course, in general, without any assumption it is not true (see min-transitivity generated colorings). So we can ask, under what conditions on

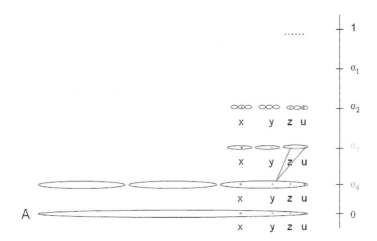

Figure 5.

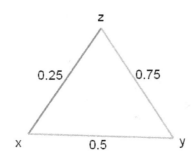

Figure 6.

colorings there is a colorful triangle? Note that in a random coloring each triangle is colorful with probability $\frac{2}{9} > 0.2$. This observation is especially useful when considering similarity spaces which are non-metric and similarity distribution is highly random. For such spaces (e.g. multimedia, genetic databases ...) we have developed in [ESV09] an indexing method based on T-similarity, where T is in a sense best similarity under which the space is still a T-similarity space. In [MOV04] we have developed the formal theory of fuzzy similarity for a general class of t-norms.

3   Tools, Data, Experiments

In this essay style paper I have to confess, that main reason why I have started to study fuzzy induction and/or data mining, was a referee, which recommended to reject my paper with an argument that it is not clear where the rules (of my fuzzy logic programming contribution) are coming from. Hence we have developed fuzzy Inductive Logic Programming (FILP) and various descendants.

3.1   Mining user preferences and top query answering

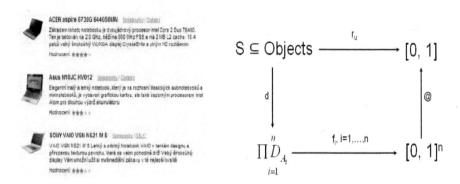

Figure 7.   Figure 8.

Main motivation for this was learning user preferences from user rating of a set sample set S of objects (see Fig. 7, stars correspond to mapping r_u), where a user interactively evaluates objects by number of stars, without making any comment on item properties. Basis for this is the object attribute representation of data (mapping d in Fig. 8), which assigns to each object its data values in the Cartesian product of domains of attributes $\prod_{i=1}^{n} D_{A_i}$. The learning task is to find user objectives on particular attributes (fuzzy sets on attribute domains f_i) and a fuzzy aggregation function combining these attribute preference degrees. In a formal model we require the whole diagram to commute, in practical setting we require it gives good advice for the user. What is a good advice can be measured in several ways, most appropriate for web search is to get best objects (with highest fuzzy degree) first. For this we use Kendal correlation coefficient

$$\tau = \frac{n_c - n_d}{\frac{1}{2}n(n-1)}$$

where n_c is the number of concordant pairs, and n_d is the number of discordant pairs in the data set (see [Wik09a]). For more on FILP and user preference mining see [HV06a, EHV07b].

There is another feature in learning user objectives on particular attributes (fuzzy sets on attribute domains f_i). This gives an ordering of domains - for user U_1 North East direction is better (red rectangle can be objects good in

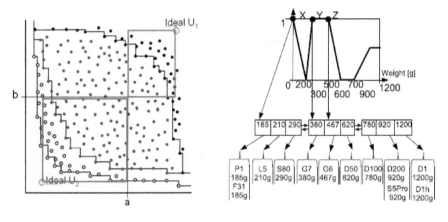

Figure 9.   Figure 10.

degree at least 0.7) and for user U_2 the South West direction is better (blue rectangle can be objects good in degree at least 0.7) in ordered domain on Fig. 9. Fuzzy aggregation @ can be glued together from GAP rules like

Good_for_U1 : 0.7 IF A1 better_for_U1 than a AND
A2 better_for_U1 than b

Most of top-k algorithm use ordered approach to data by user preference. But considering different users, this ordering can change and it would be very costly reorder data each time a new user comes. In [EPV07b] we have developed an index structure, which given a fuzzy set can search data starting from best (see Fig. 10).

3.2   Web information extraction for web semantization

Web semantization is an idea understanding process of semantization of web resources by third party annotation (as opposed to semantic web idea where it is assumed that web resource creators will annotate their pages by ontology). Of course annotating web resources by third party is a difficult task. We have tried to make a progress in this task by dividing it to smaller subtasks. First is to consider only tabular product pages and dominantly textual pages. Second idea is to split the task to a domain independent annotation and to domain dependent annotation (here user feedback is necessary, so far we do not have ontology for every domain).

A strategy for web information extraction of textual pages in depicted in Fig. 11, [DV08]. For textual domain independent annotation we use a third party linguistic annotator. We use fuzzy ILP for user feedback learning of extracted items. For tabular product pages we have a different approach

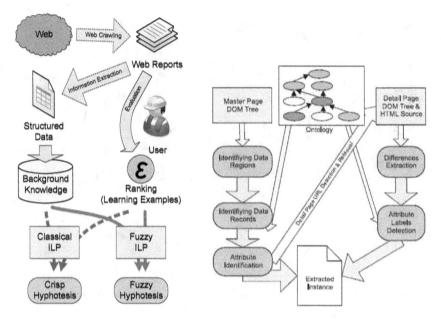

Figure 11.   Figure 12.

(Fig. 12). We parse the HTML structure by a DOM tree and looking for (fuzzy) similarities we can identify data regions and data records with a quite high accuracy. For attribute value identification we can make use of ontology and some regular expressions. If there is no ontology, we can use a heuristics which uses differences in detailed pages [NVM09]. All this projects are part of uncertainty reasoning in the web, especially of fuzzy techniques.

4   Conclusions

In this paper a tribute to Petr Hájek is given, who brought us to fuzzy logic and modeling. We have tried to give an overview of our work in this area and to Petr's influence, especially to consider fuzzy values as a comparative notion of truth which is very well suited for user preference modeling. This work was mainly done with our (former) students J. Dědek (PhD student, Charles University), A. Eckhardt (PhD student, Charles University), P. Gurský (PhD, asistent Šafárik University), T. Horváth (PhD, asistent Šafárik University), S. Krajči (PhD, associate professor Šafárik University), P. Kriško (left for industry), R. Lencses (PhD, left for industry), M. Lieskovský (left for industry), P. Marcinčák (left for industry), D. Maruščák (left for industry), M. Matula (left for industry), R. Novotný (PhD student Šafárik University),

L. Paulík (left for industry), J. Pribolová (PhD, left for industry) and V. Vaneková (PhD student Šafárik University). These can be hence considered as scientific (fuzzy) grand children of P. Hájek. We have totally ignored contributions of other authors to these fields of research, references to related work can be found in respective papers.

Concluding we can state that ideas of Petr Hájek - both on fuzzy logic in narrow sense and on fuzzy as a comparative notion of truth, have brought new insight to many problems, especially to user preference modeling and has led even to proofs of concepts of these ideas based on experimental tools and experiments on real world data.

Lessons learned show:

- In the case of applications with signal (analogue, physical) data, typically a fuzzy controller, the state of the system ranges in a continuous range of all possible data. In this case fuzzy rules are built upon fuzzy partitions of domain of controller variables. In the case of applications with human created data (e.g. web data) - state of the system contains only data of existing resources (which usually do not fill a continuum of possibilities). In this case fuzzy rules are built upon (usually one) fuzzy subset of domain of variables which indicate user preference, objective.

- Fuzzy logic in narrow sense can be developed either in the direction of axiomatization of 1-tautologies, aiming for completeness and complexity issues. Fuzzy logic in narrow sense for data intensive applications usually does not need any logical axioms. There is a difference between systems based on implicative rules (modus ponens deduction and database querying) and disjunctive rules (resolution deduction and refutation).

- Learning of implicative rules is easier, based on equivalence of FLP - fuzzy logic programming and GAP - Generalized annotated programs and understanding learning FLP rules as learning of objectives and utility function in a user preference model as in operational research (see [EV09a]).

- several application scenarios led to new requirements on indexing, top-k query answering, user preference learning and web information extraction and new measures of evaluating results / solutions.

We plan to work in this direction also in the future. In the bibliography we include additional publications from our school motivated by P. Hájek.

BIBLIOGRAPHY

[DEGV08]   J. Dědek, A. Eckhardt, L. Galambos, and P. Vojtáš. Discussion on un-
            certainty ontology for annotation and reasoning. In F. Bobillo, P. C. G.
            da Costa, C. d'Amato, N. Fanizzi, K. B. Laskey, K. J. Laskey, T. Lukasiewicz,
            T. P. Martin, M. Nickles, M. Pool, and P. S. z, editors, *URSW*, volume 423
            of *CEUR Workshop Proceedings*. CEUR-WS.org, 2008.

[DHT+08]   M. Duží, A. Heimbürger, T. Tokuda, P. Vojtáš, and N. Yoshida. Multi-agent
            knowledge modelling. In Kiyoki et al. [KTJ+09], pages 411–428.

[DV08]   J. Dědek and P. Vojtáš. Linguistic extraction for semantic annotation. In
            C. Badica, G. Mangioni, V. Carchiolo, and D. D. Burdescu, editors, *IDC*,
            volume 162 of *Studies in Computational Intelligence*, pages 85–94. Springer,
            2008.

[DV09]   J. Dědek and P. Vojtáš. Fuzzy classification of web reports with linguistic
            text mining. In WI/IAT [WI/09], pages 167–170.

[EHM+07]   A. Eckhardt, T. Horváth, D. Maruščák, R. Novotný, and P. Vojtáš. Uncer-
            tainty issues in automating process connecting web and user. In F. Bobillo,
            P. C. G. da Costa, C. d'Amato, N. Fanizzi, F. Fung, T. Lukasiewicz, T. Mar-
            tin, M. Nickles, Y. Peng, M. Pool, P. S. z, and P. Vojtáš, editors, *URSW*,
            volume 327 of *CEUR Workshop Proceedings*. CEUR-WS.org, 2007.

[EHM+08]   A. Eckhardt, T. Horváth, D. Maruščák, R. Novotný, and P. Vojtáš. Un-
            certainty issues and algorithms in automating process connecting web and
            user. In P. C. G. da Costa, C. d'Amato, N. Fanizzi, K. B. Laskey, K. J.
            Laskey, T. Lukasiewicz, M. Nickles, and M. Pool, editors, *URSW (LNCS
            Vol.)*, volume 5327 of *Lecture Notes in Computer Science*, pages 207–223.
            Springer, 2008.

[EHV07a]   A. Eckhardt, T. Horváth, and P. Vojtáš. Learning different user profile
            annotated rules for fuzzy preference top-k querying. In H. Prade and V. S.
            Subrahmanian, editors, *SUM*, volume 4772 of *Lecture Notes in Computer
            Science*, pages 116–130. Springer, 2007.

[EHV07b]   A. Eckhardt, T. Horváth, and P. Vojtáš. Phases: A user profile learning ap-
            proach for web search. In *Web Intelligence*, pages 780–783. IEEE Computer
            Society, 2007.

[EPV07a]   A. Eckhardt, J. Pokorný, and P. Vojtáš. Integrating user and group prefer-
            ences for top-k search from distributed web resources. In *DEXA Workshops*,
            pages 317–322. IEEE Computer Society, 2007.

[EPV07b]   A. Eckhardt, J. Pokorný, and P. Vojtáš. A system recommending top-k
            objects for multiple users preferences. In *FUZZ-IEEE*, pages 1–6. IEEE,
            2007.

[ESV09]   A. Eckhardt, T. Skopal, and P. Vojtáš. On fuzzy vs. metric similarity search
            in complex databases. In T. Andreasen, R. R. Yager, H. Christiansen, and
            H. Larsen, editors, *FQAS 2009 - Eighth International Conference Flexi-
            ble Query Answering Systems 2009*, volume 5822 of *LNCS*, pages 64–75.
            Springer, 2009.

[EV09a]   A. Eckhardt and P. Vojtáš. Combining various methods of automated user
            decision and preferences modelling. In *MDAI 2009 - The 6th International
            Conference on Modeling Decisions for Artificial Intelligence*, volume 5861
            of *LNCS*, pages 172–181. Springer, 2009.

[EV09b]   A. Eckhardt and P. Vojtáš. Evaluating natural user preferences for selective
            retrieval. In WI/IAT [WI/09], pages 104–107.

[GHJ+09]   P. Gurský, T. Horváth, J. Jirásek, S. Krajči, R. Novotný, J. Pribolová,
            V. Vaneková, and P. Vojtáš. User preference web search - experiments with
            a system connecting web and user. *Computing and Informatics*, 28(4):515–
            553, 2009.

[GHN+06]   P. Gurský, T. Horváth, R. Novotný, V. Vaneková, and P. Vojtáš. Upre: User preference based search system. In *Web Intelligence*, pages 841–844. IEEE Computer Society, 2006.

[GPV09]   P. Gurský, R. Pázman, and P. Vojtáš. On supporting wide range of attribute types for top-k search. *Computing and Informatics*, 4:483–513, 2009.

[GV08a]   P. Gurský and P. Vojtáš. On top-k search with no random access using small memory. In P. Atzeni, A. Caplinskas, and H. Jaakkola, editors, *ADBIS*, volume 5207 of *Lecture Notes in Computer Science*, pages 97–111. Springer, 2008.

[GV08b]   P. Gurský and P. Vojtáš. Speeding up the nra algorithm. In S. Greco and T. Lukasiewicz, editors, *SUM*, volume 5291 of *Lecture Notes in Computer Science*, pages 243–255. Springer, 2008.

[Háj94]   P. Hájek. Lectures on fuzzy logic. Technical University Vienna, 1994.

[Háj98]   P. Hájek. *Metamathematics of Fuzzy Logic*, volume 4 of *Trends in Logic*. Kluwer, Dordrecht, 1998.

[Háj07]   P. Hájek. Why fuzzy logic? In *A Companion to Philosophical Logic*, pages 595–605. Blackwell Publishing Ltd, 2007.

[HSV04]   T. Horváth, F. Sudzina, and P. Vojtáš. Mining rules from monotone classification measuring impact of information systems on business competitiveness. In L. M. Camarinha-Matos, editor, *BASYS*, volume 159 of *IFIP International Federation for Information Processing*, pages 451–458. Springer, 2004.

[HV04]   T. Horváth and P. Vojtáš. Fuzzy induction via generalized annotated programs. In B. Reusch, editor, *Fuzzy Days*, volume 33 of *Advances in Soft Computing*, pages 419–433. Springer, 2004.

[HV06a]   T. Horváth and P. Vojtáš. Induction of fuzzy and annotated logic programs. In S. Muggleton, R. P. Otero, and A. Tamaddoni-Nezhad, editors, *ILP*, volume 4455 of *Lecture Notes in Computer Science*, pages 260–274. Springer, 2006.

[HV06b]   T. Horváth and P. Vojtáš. Ordinal classification with monotonicity constraints. In P. Perner, editor, *Industrial Conference on Data Mining*, volume 4065 of *Lecture Notes in Computer Science*, pages 217–225. Springer, 2006.

[JVKH09]   P. Jenček, P. Vojtáš, M. Kopecký, and C. Höschl. Sociomapping in text retrieval systems. In T. Andreasen, R. R. Yager, H. Bulskov, H. Christiansen, and H. L. Larsen, editors, *FQAS*, volume 5822 of *Lecture Notes in Computer Science*, pages 122–133. Springer, 2009.

[KLM+02a]   S. Krajči, R. Lencses, J. Medina, M. Ojeda-Aciego, A. Valverde, and P. Vojtáš. Non-commutativity and expressive deductive logic databases. In S. Flesca, S. Greco, N. Leone, and G. Ianni, editors, *JELIA*, volume 2424 of *Lecture Notes in Computer Science*, pages 149–160. Springer, 2002.

[KLM+02b]   S. Krajči, R. Lencses, J. Medina, M. Ojeda-Aciego, and P. Vojtáš. A similarity-based unification model for flexible querying. In T. Andreasen, A. Motro, H. Christiansen, and H. L. Larsen, editors, *FQAS*, volume 2522 of *Lecture Notes in Computer Science*, pages 263–273. Springer, 2002.

[KLV02]   S. Krajči, R. Lencses, and P. Vojtáš. A data model for annotated programs. In Y. Manolopoulos and P. Návrat, editors, *ADBIS Research Communications*, pages 141–154. Slovak University of Technology, Bratislava, 2002.

[KLV04]   S. Krajči, R. Lencses, and P. Vojtáš. A comparison of fuzzy and annotated logic programming. *Fuzzy Sets and Systems*, 144(1):173–192, 2004.

[KMM+98]   P. Kriško, P. Marcinčák, P. Mihók, J. Sabol, and P. Vojtáš. Low retrieval remote querying dialogue with fuzzy conceptual, syntactical and linguistical unification. In *FQAS '98: Proceedings of the Third International Conference on Flexible Query Answering Systems*, pages 215–226, London, UK, 1998. Springer-Verlag.

358   Peter Vojtáš

[KS92]   M. Kifer and V. S. Subrahmanian. Theory of generalized annotated logic
            programming and its applications. *Journal of Logic Programming*, 12:335–
            367, 1992.
[KTJ+09]   Y. Kiyoki, T. Tokuda, H. Jaakkola, X. Chen, and N. Yoshida, editors. *In-
            formation Modelling and Knowledge Bases XX, 18th European-Japanese
            Conference on Information Modelling and Knowledge Bases (EJC 2008),
            Tsukuba, Japan, June 2-6, 2008*, volume 190 of *Frontiers in Artificial Intel-
            ligence and Applications*. IOS Press, 2009.
[Lie01]   M. Lieskovský. Reasoning strategies for best answer with tolerance. *Neural
            Network World*, 11(6):687–702, 2001.
[Llo93]   J. W. Lloyd. *Foundations of Logic Programming*. Springer, Secaucus, NJ,
            1993.
[MM00]   P. Marcinčák and M. Matula. Metadata and object-oriented approach to
            flexible querying. In *FQAS*, pages 95–102, 2000.
[MOV01a]   J. Medina, M. Ojeda-Aciego, and P. Vojtáš. A completeness theorem for
            multi-adjoint logic programming. In *FUZZ-IEEE*, pages 1031–1034, 2001.
[MOV01b]   J. Medina, M. Ojeda-Aciego, and P. Vojtáš. A multi-adjoint logic approach
            to abductive reasoning. In P. Codognet, editor, *ICLP*, volume 2237 of *Lecture
            Notes in Computer Science*, pages 269–283. Springer, 2001.
[MOV01c]   J. Medina, M. Ojeda-Aciego, and P. Vojtáš. Multi-adjoint logic programming
            with continuous semantics. In T. Eiter, W. Faber, and M. Truszczynski,
            editors, *LPNMR*, volume 2173 of *Lecture Notes in Computer Science*, pages
            351–364. Springer, 2001.
[MOV01d]   J. Medina, M. Ojeda-Aciego, and P. Vojtáš. A procedural semantics for
            multi-adjoint logic programming. In P. Brazdil and A. Jorge, editors, *EPIA*,
            volume 2258 of *Lecture Notes in Computer Science*, pages 290–297. Springer,
            2001.
[MOV01e]   J. Medina, M. Ojeda-Aciego, and P. Vojtáš. Similarity-based unification: a
            multi-adjoint approach. In J. M. Garibaldi and R. I. John, editors, *EUSFLAT
            Conf.*, pages 273–276. De Montfort University, Leicester, UK, 2001.
[MOV02]   J. Medina, M. Ojeda-Aciego, and P. Vojtáš. A multi-adjoint approach to
            similarity-based unification. *Electr. Notes Theor. Comput. Sci.*, 66(5):70–
            85, 2002.
[MOV04]   J. Medina, M. Ojeda-Aciego, and P. Vojtáš. Similarity-based unification: a
            multi-adjoint approach. *Fuzzy Sets and Systems*, 146(1):43–62, 2004.
[MOVV03]   J. Medina, M. Ojeda-Aciego, A. Valverde, and P. Vojtáš. Towards biresid-
            uated multi-adjoint logic programming. In R. Conejo, M. Urretavizcaya,
            and J.-L. P. de-la Cruz, editors, *CAEPIA*, volume 3040 of *Lecture Notes in
            Computer Science*, pages 608–617. Springer, 2003.
[NVM09]   R. Novotný, P. Vojtáš, and D. Maruščák. Information extraction from web
            pages. In WI/IAT [WI/09], pages 121–124.
[Pau96]   L. Paulík. Best possible answer is computable for fuzzy sld-resolution. In
            P. Hájek, editor, *GÖDEL'96. Logical Foundations of Mathematics, Com-
            puter Science and Physics — Kurt Gödel's Legacy*, pages 257–266, Brno,
            Czech Republic, 1996. Springer-Verlag, Lecture Notes in Logic 6.
[Pav79]   J. Pavelka. On fuzzy logic I, II, III. *Zeitschr. Math. Logik Grundlagen Math.*,
            25:45–52, 119–134, 447–464, 1979.
[PV01]   J. Pokorný and P. Vojtáš. A data model for flexible querying. In A. Caplin-
            skas and J. Eder, editors, *ADBIS*, volume 2151 of *Lecture Notes in Computer
            Science*, pages 280–293. Springer, 2001.
[SHB98]   M. Sonka, V. Hlaváč, and R. Boyle. Airplane picture. http://www.icaen.
            uiowa.edu/~dip/LECTURE/, 1998.
[SV04]   D. Smutná and P. Vojtáš. Graded many-valued resolution with aggregation.
            Fuzzy Sets and Systems, 143(1):157–168, 2004.

[UPr08]   UPreA. Method for user preferences acquisition (tool uprea). http://nazou.
            fiit.stuba.sk/home/?page=uprea, 2008.
[VE08]   P. Vojtáš and A. Eckhardt. Considering data-mining techniques in user pref-
            erence learning. In *Web Intelligence/IAT Workshops*, pages 33–36. IEEE,
            2008.
[VF00]   P. Vojtáš and Z. Fabián. Aggregating similar witnesses for flexible query
            answering. In *FQAS*, pages 220–229, 2000.
[Voj99]   P. Vojtáš. Fuzzy logic abduction. In G. Mayor and J. Suñer, editors,
            EUSFLAT-ESTYLF Joint Conf., pages 319–322. Universitat de les Illes
            Balears, Palma de Mallorca, Spain, 1999.
[Voj01a]   P. Vojtáš. Annotated and fuzzy logic programs - relationship and comparison
            of expressive power. *Neural Network World*, 11(6):661–674, 2001.
[Voj01b]   P. Vojtáš. Fuzzy logic programming. *Fuzzy Sets and Systems*, 124(3):361–
            370, 2001.
[Voj05]   P. Vojtáš. Fuzzy logic as an optimization task. In E. Montseny and P. So-
            brevilla, editors, *EUSFLAT Conf.*, pages 781–786. Universidad Polytecnica
            de Catalunya, 2005.
[Voj06a]   P. Vojtáš. EL description logic modeling querying web and learning imperfect
            user preferences. In P. C. G. da Costa, K. B. Laskey, K. J. Laskey, F. Fung,
            and M. Pool, editors, *URSW*, volume 218 of *CEUR Workshop Proceedings*.
            CEUR-WS.org, 2006.
[Voj06b]   P. Vojtáš. EL description logics with aggregation of user preference concepts.
            In M. Duzi, H. Jaakkola, Y. Kiyoki, and H. Kangassalo, editors, *EJC*, volume
            154 of *Frontiers in Artificial Intelligence and Applications*, pages 154–165.
            IOS Press, 2006.
[Voj08]   P. Vojtáš. Decathlon, conflicting objectives and user preference querying.
            In V. Snásel, K. Richta, and J. Pokorný, editors, *DATESO*, volume 330 of
            CEUR Workshop Proceedings. CEUR-WS.org, 2008.
[VP95]   P. Vojtáš and L. Paulík. Logic programming in rpl and rql. In *SOFSEM '95:
            Proceedings of the 22nd Seminar on Current Trends in Theory and Practice
            of Informatics*, pages 487–492, London, UK, 1995. Springer-Verlag.
[VP96]   P. Vojtáš and L. Paulík. Soundness and completeness of non-classical sld-
            resolution. In *ELP '96: Proceedings of the 5th International Workshop
            on Extensions of Logic Programming*, pages 289–301, London, UK, 1996.
            Springer-Verlag.
[VV08]   V. Vaneková and P. Vojtáš. A description logic with concept instance order-
            ing and top-k restriction. In Kiyoki et al. [KTJ+09], pages 139–153.
[WI/09]   *Proceedings of the 2008 IEEE/WIC/ACM International Conference on Web
            Intelligence and International Conference on Intelligent Agent Technology -
            Workshops, Milan, Italy, 15-18 September 2009*. IEEE, 2009.
[Wik07]   Wikipedia. Inverted pendulum and fuzzy control system picture from
            wikipedia, the free encyclopedia. http://en.wikipedia.org/wiki/Fuzzy_
            control_system, 2007.
[Wik09a]   Wikipedia. Kendall tau rank correlation coefficient. http://en.wikipedia.
            org/wiki/Kendall_tau_rank_correlation_coefficient, 2009.
[Wik09b]   Wikipedia. Likert scale. http://en.wikipedia.org/wiki/Likert_scale,
            2009.
[Zad65]   L. A. Zadeh. Fuzzy sets. *Information and Control*, 8:338–353, 1965.

Web pages of offspring

```
http://www.ksi.mff.cuni.cz/~dedek/
http://www.ksi.mff.cuni.cz/~eckhardt/
http://ics.upjs.sk/~gursky/
http://ics.upjs.sk/~horvath/
http://ics.upjs.sk/~krajci/
http://ics.upjs.sk/~novotnyr/
http://ics.upjs.sk/~vanekova/
```

Peter Vojtáš
Department of software engineering
Faculty of Mathematics and Physics, Charles University in Prague
Malostranské nám. 25
118 00 Prague, Czech Republic
E-mail: vojtas@ksi.mff.cuni.cz

Vagueness and ambiguity

JAN WOLEŃSKI

Dedicated to Petr Hájek
without any ambiguity or vagueness

The traditional logical account of vagueness considered it as a defect of linguistic expressions which should be improved. But how improved? Which devices are to be used? The problem is this (Sorensen [Sor06]):

> *Vagueness is standardly defined as the possession of borderline cases. For example, 'tall' is vague because a man who is 1.8 meters in height is neither clearly tall nor clearly non-tall. No amount of conceptual analysis or empirical investigation can settle whether a 1.8 meter man is tall. Borderline cases are inquiry resistant. Indeed, the inquiry resistance typically recurses. For in addition to the unclarity of the borderline case, there is normally unclarity as to where the unclarity begins. In other words 'borderline case' has borderline cases. This higher order vagueness shows that 'vague' is vague.*

If this diagnosis is correct, there is no hope to overcome vagueness, at least in ordinary language, because if we repair colloquial expressions on the first level, borderline cases re-appear on the higher level. One could say *lasciate ogni speranza* that you liquidate vagueness at all. I am not so pessimistic. Even if vagueness, due to its presence on all levels, is an universal phenomenon, various strategies of dispensing with vagueness (let us introduce the word "disvaguening") of expressions are available; almost everybody agrees that vague expressions should be locally made precise. Thus, I will disregard the higher order vagueness in what follows. On the occasion, I would like to note other simplifications in my considerations (consult the books Williamson [Wil94], Keefe [Kee00], Sorensen [Sor01], Shapiro [Sha06], Odrowąż-Sypniewska [OS00] or Smith [Smi08] as well as the anthology Keefe, Smith [KS97] for further bibliographical and substantive information concerning vagueness and its theories). I will concentrate on adjectives and words performing similar roles; otherwise speaking, I am interested in vagueness of predicates. Thus, I omit the connectives and other syncategoremata as well as the problem of vagueness of sentences. I consider vagueness as a property of linguistic items without entering into its ontological counterpart, if any.

So-called regulative definitions were typically proposed as a device of liqui-
dating vagueness. Take for example the expression "become of age", plainly
vague in the ordinary language. Clearly, due to its lack of precision (imprecision is usually considered as the most basic mark of vagueness), this term
can produce misunderstandings. Some are innocent, other not and can lead
to serious problems. Legal practice is a typical example of the area in which
vagueness cannot be tolerated. Consequently, the language of law either
avoids vague expressions or define them. In particular, the category of becoming of age is very important in law, because criminal responsibility differs
dependently of being adult (in the legal sense) or being an infant. Disregarding details, like the difference between becoming of age in penal and civil law
or exceptions due to mental sicknesses), let us define a person who becomes
of age as 18 years old. Doubtless, this definition transforms "becoming of
age" into a precise expression. On the other hand, one can maintain that the
manoeuvre of defining "become of age" as equivalent with "being 18th years
old" results with some non-intuitive consequences. In particular, a person P
who is 17 years and 364 days old is not infant, but a person P' who is 3 days
older, entirely came to age. Now remember that becoming of age is legally
(as well as ordinarily in most cases) is related to achieving a mental maturity
(ability to directing own actions by conscious choices), which is considered
as the condition sine qua non of legal responsibility. However, the difference
between P and P' as far as the matter concerns their mental maturity is
(eventual differences are not caused by the age-difference, but by different
factors) absolutely inessential. This example can be regarded as an instance
of the sorites paradox. In fact, we can say that if becoming of age by the
18th birthday suffices for achieving the mental maturity, the same concerns
everybody younger by 1 day, 2 days and so on. Yet nobody would recognize
a 10 years old boy as becoming of age, although the procedure of recurring
from n to $n-1$ seems has no limitations from the theoretical point of view.
    Consider the following statement (Ziembiński [Zie76], p. 168):

*A specific kind of ambiguity is represented by vague names. It would be
possible to speak about as many significations of a name of this kind as
there may be ways of arbitralily deciding which objects are or are not
designate of this name.*

This view regards vagueness as a special kind of ambiguity. Zygmunt Ziembiński follows the traditional account on which vagueness is a defect of languages, because the above quotation occurs in the chapter devoted to causes
of misunderstandings related to "formulation of thoughts by means of words".
He also follows Kazimierz Ajdukiewicz, who ([Ajd74], p. 54) considered vague
terms as having "a vague meaning"). Although Ziembiński does not explain
what does it mean that vague names illustrate "a specific ambiguity" (the

adjective "specific" is very unclear here), we can think about "tall" as an example (Sorensen also considers this adjective as a canonical case). This adjective is vague in the ordinary sense: there are borderline cases on which it is difficult to decide whether some objects (people, mountains, hills, towers, tree, etc.) are tall or not ([Ajd74], pp. 52–53 for a particularly clear characterization of the traditional account of vagueness). Now one might maintain that every concrete precisation of "tall" constitutes its separate meaning. Consequently, the vagueness of "tall" generates a family of meanings (or significations) with particular instances resulting from the scheme "being n centimetres (or other unit employed in measuring height)".

Speaking more technically and using the standard semantics for predicates, an adjective expresses a property as its intension and refers to a set as its extension. Let a set X be an extension of "tall". This adjective is vague because if we divide X into two subsets X_1 and X_2 such that $X_1 \cap X_2 = \emptyset$ (this means that both subsets of X are mutually complements), there is an object x such that we have no straightforward criteria to decide whether $x \in X_1$ or $x \in X_2$. Incidentally, we have here a very simple description of vagueness: an expression E is vague if and only if it extension E can divided into mutually disjoint subextesions E_1 and E_2 such than both are vague. Moreover, if E is vague, then non-E is vague too. Now, Ziembiński's view can be rephrased as the statement that vague extensions generated ambiguous intensions. If the example with "tall" fits Ziembiński's idea, we can say that regulative definitions of "tall" falling under the scheme "P is tall if and only if he or she is being of n centimetres (or his or her height is comprised by a given segment $n--m$, i order to take a more intuitive course)" discriminates particular meanings of "tall" and, thereby, disambiguates this adjective. Shortly speaking, disvaguening the extension of E disambiguates its intension, but extensional vagueness results with intensional ambiguity. I will examine this view, implicit or hidden in many theoretical treatments of the phenomenon of vagueness. For example, Joanna Odrowąż-Sypniewska (see [OS00], p. 28) says that it is difficult to imagine that precisifitation of an expression does not change its meaning. The reason is that the widening or narrowing the extension of an expression cannot be realized without changing its intension. This view assumes Frege's principle that intension functionally determines extension.

Doubtless, vagueness and ambiguity have something in common. Both are properties of expressions (modulo the assumption that the former concerns linguistic items; it is rather no doubt that words are ambiguous, not things) and they are considered as defects. Another similarity consists in being subjected to repairs by discriminating cases. Take a Polish word *zamek*, very ambiguous indeed. It means (a) castle, (b) lock, (c) zipper, (d) a manoeuvre in the ice-hockey game. Of course, we should discriminate cases of the use of *zamek*, possibly by definitions. However, there is a difference between definitional discrimination in the case of vagueness and the case of ambiguity. Whereas analytic (reportive) definitions are proper for ambiguity, regulative definitions apply to vagueness. The context-dependence is perhaps the most common for both phenomena in question. Assume that a mother repairs trousers of her son and says (in Polish) "Go to a shop and buy *zamek*!" Clearly the son will not go to the real estate agency in order to buy a castle. He exactly knows that he should walk to a place where zippers are sold. Similarly, if one hears that basket-ball or volley-ball players are tall, he or she will not think about persons being 180 cm tall, because a tall basket-ball (volley-ball) player is taller. These examples show that the ways of disvauging and disambiguating words are similar. Consequently, some authors, Ajdukiewicz, for example, speak equivalently about the sharpening of meaning and the sharpening of scope (extension). Yet, as I will argue, these similarities do not justify the view that vagueness entails ambiguity.

Guy Sorensen contrasts vagueness and ambiguity in [Sor06]:

> *"Tall" is relative. A 1.8 meter pygmy is tall for a pygmy but a 1.8 meter Masai is not tall for a Masai. Although relativization disambiguates, it does not eliminate borderline cases. There are shorter pygmies who are borderline tall for a pygmy and taller Masai who are borderline tall for a Masai. The direct bearers of vagueness are a word's full disambiguations such as 'tall for an eighteenth century French man'. Words are only vague indirectly, by virtue of having a sense that is vague. In contrast, an ambiguous word has its ambiguity directly—simply in virtue of having multiple meanings.*

> *This contrast between vagueness and ambiguity is obscured by the fact that most words are both vague and ambiguous. 'Child' is ambiguous between 'offspring' and 'immature offspring'. The latter reading of 'child' is vague because there are borderline cases of immature offspring. The contrast is further complicated by the fact that most words are also general. For instance, 'child' covers both boys and girls.*

> *Ambiguity and vagueness also contrast with respect to the speaker's discretion. If a word is ambiguous, the speaker can resolve the ambiguity without departing from literal usage. For instance, he can declare that he meant 'child' to express the concept of an immature offspring. If a word is vague, the speaker cannot resolve the borderline case. For instance, the speaker cannot make 'child' literally mean anyone under eighteen just by intending it. That concept is not, as it were, on the menu corresponding to 'child'. He would be understood as taking a special liberty with the term to suit a special purpose. This would relieve him of the obligation to defend the sharp cut-off.*

I agree that there are terms which are simultaneously ambiguous and vague, although the problem of generality seems not very essential in this context. In fact, "female child" and "male child" are sufficiently vague for own properties. Another interested thing pointed out by Sorensen concerns that disambiguating can preserve literal usage (I would prefer to say "literal meaning"). In fact, I consider this remark as having the utmost importance for the problem of vagueness and ambiguity.

On the other hand, I do not think that the contrast between vagueness and ambiguity may be successfully explained pertaining to vagueness as indirect and ambiguity as direct. The noun *zamek* sometimes appears as ambiguous directly, but its ambiguity is indirect in other cases. It seems that *zamek* in the described situation of the mother sending her son to buy an zipper (I remind that he understand her request contextually) is ambiguous indirectly (potentially) only. By contrast, when the mother simply says "buy me *zamek*" without being repairing the trousers, the son could be in troubles in understanding what is going on for direct ambiguity of *zamek* as possibly referring to a zipper or a lock. Further explanations can be required in order to disambiguate the crucial word uttered, because it is unclear without further explanations whether word and . This suggests that ambiguity is sometime direct and sometimes indirect. On the other hand, the adjective is always directly vague, if vague at all. I guess that Sorensen deceived himself by saying "Words are only vague indirectly, by virtue of having a sense that is vague.", because it is not sufficiently clear whether vagueness concerns extension, intension or both. Finally, I have reservations about Sorensen's view that relativisation does not solve the problem of borderline cases. Independently whether vagueness is only linguistic or linguistic and ontological, conventions leading to sharpening the scope or meaning of terms just decide formerly undecided cases. I will later return to this question.

Sorensen's remark about preserving literal meaning in the process of disvaguening can be developed. Consider "bald" an another canonical example of a vague adjective. If someone is asked what does it mean that a person is bald, an straightforward answer is that being bald signifies the property having no hair or having not very many hair. Now, if the questioner observes that "not very many" is a vague qualification, the asked person can observe that we can always introduce a definition or consider the problem as pointless. In fact, ordinary people, included barbers, perfectly understand what "is bald" means and do not need to making this predicate sharper. The same concerns "becoming of age". Assume now that it means "be able to leave the family house and live separately". Certainly, "becoming of age" is vague under this new meaning. Contrary to the former, legal usage, there is no need to introduce a regulative definition, because the matter is usually settled by negotiations or decisions of parents and children. Universal sharpening of vague terms is required if their uniform application has a serious social importance. Whereas differences between judges as far as the matter concerns being becoming of age as a condition of full criminal responsibility would be fatal for social equilibrium, controversies related to the age justifying separate living of young people have no special importance. Yet "becoming of age" has clear meaning in both situations. Even if we say that "becoming of age" is ambiguous, it has no influence for its vagueness. Moreover, any discrimination of various meanings does not increase or decrease vagueness associated with particular senses. The above consideration suggests that ambiguity and vagueness are fairly different phenomena, although their results for human communication can be sometimes very similar and dangerous. On the other hand, both properties are profitable to some extent.

What about the argument that sharpening of extensions results with changing meanings (intensions)? This argument must be carefully checked according to the principles of semantic interpretation, in particular, of predicates. In order to avoid possible misunderstanding, I suppose that formal semantics applies to natural languages. Assume that A is a monadic predicate and refers, due to an interpretation I, to a set X as its denotation. Thus, we write $I(A) = X$. If A is non-vague, the issue, at least as related to the present paper, reduces itself to ambiguity. If A is ambiguous, it simultaneously refers to various extensions, but this destroys the functional nature of I. In order to repair the situation, we need to discriminate various meaning, introduce new predicates, dependently on the number of meanings and, finally, ascribe new denotations to discriminate predicates. Note that two non-ambiguous predicates can refer to the same set, but the reverse connection consisting in referring of one predicate to different sets is at odds with the definition of I, if A refers to two distinct sets. Incidentally, we have here a simple explanation why ambiguity is defective and why dispensing with it requires a discrimination of meanings via introducing new predicates and new sets as their semantic interpretations. Nevertheless, nothing especially interesting occurs here. The issue is entirely pragmatic in its character and does not suggest any logical problems. Moreover, we can still say that ambiguity should be eliminated for reasons dictated by semantics, but, on the other hand, it is a valuable property of languages form the point of view their lexical economy. Instead having four words in Polish referring to castles, locks, zippers and maneuvers in ice-hockey we have one word *zamek,* which is sufficiently disambiguating by contexts. Lexical economy seems also to be a source of vagueness. Since vague words in normal colloquial circumstances do not produce major troubles in verbal communication, they can freely function in languages. Similarly as in the case of ambiguity, instead to have many particular terms related to various instances of "is tall", for instance, it is enough to operate the only item. Thus, methodological claims of the second

Vagueness and ambiguity

(later) Wittgenstein or the ordinary language philosophy insisting that ambiguity and vagueness incoherent with precisefication of words are exaggerated. These properties of linguistic expressions are not forced by "the logic of natural language", whatever it is, but stem from the obvious tendency toward linguistic economy. Yet economical aspects of language must be sacrificed in some circumstances, for example, if special terminology is required.

Yet the case of vagueness is radically different than that of ambiguity, at least from the logical point of view. If A is vague, no set X (in the standard sense) can be taken as its denotation. However, if we sharpen the extension of A, the problem of what belongs to X is entirely solved. This means that for any x, $x \in X$ or $x \notin X$ and we can retain the understanding of sets as constituted by their elements. Assume that two attempts of sharpening A, namely S and S' are available. Consequently, we obtain the sets X and X' as references of A. Thus, due to the functionality of I, we obtain two interpretations $I(A, S) = X$ and $I'(A, S') = X'$. Now, one can observe that the functionality of I is broken, because two sets are ascribed to the same predicate A. S and S' function as parameters to which I is relativized. We can rewrite our equalities as $I^S(A) = X$ and $I^{S'}(A) = X'$ and this clearly shows that we have to do with two interpretations of the same predicate having X as its denotation under I^S and X' is its reference under $I^{S'}$. Why do not say that A receives new meanings by S and S' respectively? In fact, if A is vague, one sharpening is quite enough in this respect. Suppose that "is tall" is sharpened by S (details are irrelevant here). The proponents of the view that A receives a new meaning by S should say that "is tall" changed its initial meaning and it means "is tall according to S"; introducing more sharpening choices, dependently on persons or situations, adds nothing new.

I consider the view that sharpening produces new meanings as artificial and not forced by semantics (see also Pinkal [Pin95] for contrasting vagueness and ambiguity from the point of view of linguistics); I understand semantics not as the theory of meaning, but as the theory of semantic interpretation in the indicated sense (referential or Tarskian semantics). The main reason for my opposition against Odrowąż-Sypniewska's position and similar proposals is that "is tall" preserves its original (or literal *pace* Sorensen) meaning. This obvious if we look at a possible conversation between P (working with S) and P' (working with S'). P remarks that "is tall" is vague and requires a sharpening S, but P' considers this proposal as inadequate and claims that S' is better. Yes both agree that "is tall" means exactly the same, namely expresses a property of human stature. In fact, "a is tall according to S" and "a is tall according to S" imply "a is tall". This gives a testimony that "is tall" is the core sense of "a is tall according to S" and "a is tall according to S'. We can eventually add that P and P' employ differently the predicate "is tall" to concrete cases or that P' is more precise than P (or reversely), although both use the expression "is tall" in the same way. If one observes that this view simplifies semantic issues concerning vagueness too much, because "is tall according to S" and "is tall according to S'" have to have different meanings for the difference between S and S', I would answer that sharpening always create a new language. If A is vague, we say that we have a defective jargon or a language without an interpretation in the proper sense, which should be replaced by a legitimate system with correctly described sets as references of predicates. Consequently, A is a collective label for a family $F = \{A_1, \ldots, A_n\}$, where $A_1, \ldots, A_n$ are predicates $A_1, \ldots, A_n$ interpreted in the standard way by ascribing denotations viewed as sets to them. In fact, any act of sharpening of a vague predicate results with a new language or even a collection of languages related to F. Both ways out are not exclusive. Whereas the first solution respects the receive shape of ordinary language, the second proposal fits special terminologies invented for exactly defined tasks, for instance, scientific. However, both proposals allows to solve the borderline cases. This is the main reason why I do not agree with Sorensen in this respect. He maintains, I remind by quoting once again, "If a word is vague, the speaker cannot resolve the borderline case". As I have tried to argue, conventions, motivated by ordinary usages or artificial tasks, are logically sufficient for decision where the borderline cases are to be placed, in a given set X or in its complement. Of course, there always arises a problem of how logical sufficiency is related to substantial adequacy, but this is a different matter.

It is very important to see how meanings function in referential semantics. We say that interpretations are assignments objects or their collections to terms and predicates; we assume that semantic roles of logical constants are fixed in advance. More precisely, we have a model M and its universe of discourse U and consider denotations of constants (individual names) as objects taken from U, but references of (monadic) predicates as subsets of U. No appeal to meanings of expressions appears on any stage of defining interpretations. Although referential semantics is purely extensional, I do not say that the use of words is absolutely inessential for I. In fact, our decisions are strongly related to intuitions, expectations, evaluations and similar factors governing our interpretative decisions. Theoretically speaking, we are free in inventing interpretations. On the other hand, we observe tendency to keeping the ordinary meanings of expressions so far as it is possible. Perhaps this circumstance suggests that sharpening of extensions really concerns intensions. Yet the whole reasoning looks as confusing. Vagueness is typically attributed to extension. In fact, the concept of borderline cases has no direct application to intensions, because we are speaking about objects for which it is difficult or even impossible to decide whether they are elements of X or its complement. This means that extension is involved directly, but intension indirectly at most. Thus, reasoning goes from extension to intension, not reversely. Consequently, we cannot say that intension determines extension and the Frege principle is not applicable here. In order to improve the argument, one should say that vagueness, like ambiguity, is a property of intension (meaning, content, sense, etc.). However, this manoeuvre is arbitrary and ad hoc. Anyway, since the Frege principle acts on disambiguated intensions, it is useless in the case of vagueness too. This conclusion can be further strengthened by observing that Frege's intensions are complete thoughts (in his sense) and exist as Platonic-like objects. The thesis that intensions fully determine extensions is reasonable only under this proviso. If we reject Frege's ontological assumptions concerning senses and their existence, we should rather say that extensional interpretations are motivated by various circumstances, subjective or objective.

My basic contention considers vagueness and ambiguity as different concepts. In particular, I see no reason to consider the former as a special kind of the latter. In order to give an additional argument for the difference between vagueness and ambiguity, I would like develop my former statement that ambiguity suggests no logical problem. Otherwise speaking, there is no logic of ambiguity, if we understand seriously the term "logic". The situation of vagueness differs completely. We have a lot of logical theories of vagueness (see books and papers formerly quoted). In some sense, any sharpening of vague terms embed them into standard predicate calculus. Clearly, this way out can rise doubts concerning its naturalness, because it disregards essential features of vague terms. Hence, other proposals appeared, for example, introducing supervaluations, subvaluations or truth-value gaps. All are intended to give justice to some features of vague terms. Generally speaking, all these strategies offer semantic devices appropriate for sentences (statements, propositions) in which vague terms occur. The common idea is that classical binary valuation is supplemented by additional one, which ascribes new outcomes dependently of various possible precifications of vague parts of sentences. The resulting logic is still bivalent, although some sentences are gappy (from the beginning) or lost their initial logical values. For example, some kinds of sharpening are considered as admissible other not. Let A contains a vague term. A is supertrue, if it is true under all admissible sharpening-choices; otherwise false. This is the main idea of the strategy of supervaluations.

One can complain that logics of supervaluation, subvaluation or those admitting truth-value gaps eliminate the very nature of vagueness. Hence, fuzzy logic is nowadays the most popular theory of vagueness with important applications in various branches of science and technology. Let me quote the master text in this field:

Fuzziness is imprecision (vagueness); a fuzzy proposition may be true to some degree. [...]. I hope that the book shows as following:

- *Fuzzy logic is neither a poor man's logic nor poor man's probability. Fuzzy logic (in the narrow sense) is a reasonably deep theory.*

- *Fuzzy logic is a logic. It has its syntax and semantics and notion of consequence. It is a study of consequence.*

- *There are various systems of fuzzy logic, not just one. We have one basic logic (BL) and three of its important extensions: Łukasiewicz logic, Gödel logic and the product logic.*

- *Fuzzy logic in the narrow sense is a beautiful logic, but is also important for applications; it offers foundations.*

This passage is taken from Hájek [Háj98], p. 2, p. 5. I consider Petr's diagnosis of fuzzy logic and its merits as an evidence that it constitutes the best formal theory of vague expressions. Briefly, fuzziness in the logical sense is the formal exposition of vagueness. This suggests my final word concerning the relation between vagueness and ambiguity: since vagueness has a good logical theory, but ambiguity is sterile in this respect, both must be different. *Quod erat demonstrandum.*

Appendix: Two Quotations from Tarski

Tarski 1933, p. 166/167 (page references to [Tar83], 1984):

It remains perhaps to add that we are not interested here in 'formal' languages and sciences in one special sense of the word 'formal', namely sciences to the signs and expressions of which no meaning is attached. For such sciences the problem [of truth, J. W.] here discussed has no relevance, it is not discussed. We shall always ascribe quite concrete and, for us intelligible meaning to the signs which occur in languages we shall consider. The expressions which we call sentences still remain sentences after the signs which occur in them have been translated into colloquial language. The sentences which are distinguished as axioms seem to us to be materially true, and in choosing rules of inference we are always guided by the principle that when such rules are applied to true sentences the sentences obtained by their use also should be true.

A comment: This passage suggests the pragmatic priority of ordinary (colloquial in Tarski) language even in the domain of semantics of formalized systems. Since properties of ordinary languages excludes that its elements are bearers of stable intensions, Frege's principle is very problematic.

Tarski 1933, p. 267 (page-references to [Tar83], 1984):

> *Whoever [...] wishes to pursue the semantics of colloquial language with the help of exact methods will be driven first to undertake the thankless task of a reform of this language. He will find it necessary to define its structure, to overcome the ambiguity of the terms which occur in it [...].*

It is interesting that, according to Tarski, vagueness did not occur in the catalogue of defects of ordinary language.

BIBLIOGRAPHY

[Ajd74]   K. Ajdukiewicz. *Pragmatic Logic*. D. Reidel Publishing Company, Dordrecht, 1974.

[Háj98]   P. Hájek. *Metamathematics of Fuzzy Logic*, volume 4 of *Trends in Logic*. Kluwer, Dordrecht, 1998.

[Kee00]   R. Keefe. *Theories of Vagueness*. Cambridge University Press, Cambridge, 2000.

[KS97]   R. Keefe and N. J. J. Smith, editors. *Vagueness: A Reader*. The MIT Press, Cambridge MA, 1997.

[OS00]   J. Odrowąż-Sypniewska. Zagadnienie nieostrości (The Problem of Vagueness). Technical report, Wydział Filozofii i Socjologii Uniwersytetu Warszawskiego, Warszawa, 2000.

[Pin95]   M. Pinkal. *Logic and Lexicon: The Semantics of the Indefinite*. Kluwer, 1995.

[Sha06]   S. Shapiro. *Vagueness in Context*. Oxford University Press, 2006.

[Smi08]   N. J. J. Smith. *Vagueness and Degrees of Truth*. Oxford University Press, Oxford, 2008.

[Sor01]   R. Sorensen. *Vagueness and Contradiction*. Oxford University Press, Oxford, 2001.

[Sor06]   R. Sorensen. Vagueness. In E. N. Zalta, editor, *The Stanford Encyclopedia of Philosophy*. Stanford University, Fall 2006 edition, 2006. http://plato.stanford.edu/archives/fall2006/entries/vagueness/.

[Tar83]   A. Tarski. *Logic, Semantics, Metamathematics*. Hackett, Indianapolis IN, 1983.

[Wil94]   T. Williamson. *Vagueness*. Routledge, London, 1994.

[Zie76]   Z. Ziembinski. *Practical Logic*. D. Reidel Publishing Company, Dordrecht, 1976.

Jan Woleński
Institute of Philosophy, Jagiellonian University
Grodzka st. 52
31-044 Kraków, Poland
Email: j.wolenski@iphils.uj.edu.pl

Fuzzy logic and vagueness: can philosophers learn from Petr Hájek?

CHRIS FERMÜLLER[1]

*Dedicated to Petr Hájek,
on the occasion of his 70th birthday*

Who is learning from whom?

While the topic 'fuzzy logic and vagueness' certainly needs no further comment in the given context, the peculiar subtitle of this contribution calls for some explanation. In 2003 I had submitted a small paper provocatively, as I then thought, entitled *Fuzzy Logic and Theories of Vagueness: Can Logicians Learn from Philosophers?* to a workshop on Soft Computing in Brno, organized by Petr Hájek and colleagues. It was prompted by the following observation: the very impressive work by Hájek and some of his colleagues on what is nowadays called mathematical fuzzy logic is frequently motivated by usually quite casual remarks claiming that fuzzy logic is the logic of vague propositions and notions. In particular, Hájek's deservedly very influential monograph *Metamathematics of Fuzzy Logic* [Háj98], which I had the privilege to watch growing and maturing quickly during a European COST Action 15 *Many-valued Logics in Computer Science* (and also in lectures presented at Vienna University of Technology), opens with a paragraph stating

> [...] The aim is to show that fuzzy logic as a logic of imprecise (vague) propositions does have well developed formal foundations and that most things usually named "fuzzy inference" can be naturally understood as logical deduction. [Háj98, p. vii]

While I, like so many of my colleagues, was impressed by the mathematical sophistication and clarity with which Hájek demonstrated that deductive fuzzy logic indeed can be given 'well developed formal foundations' (note that these foundations where not just already lying around, but had rather been largely provided by Hájek himself!) I also found the implicit claim that fuzzy logic is a proper account of reasoning with vague propositions problematic. In other words: while I was learning a lot from Petr Hájek's many

---

[1]Supported by Eurocores-ESF/FWF grant I143-G15 (LogICCC-LoMoReVI)

deep contributions to this type of mathematical logic, I was not sure about their immediate relevance for the challenges posed by vagueness to formal accounts of correct reasoning. The reasons for my worries had to do with the fact that I have always been a serious student of philosophy, and as such I was well aware of the fact that the principles of correct reasoning in face of vagueness has been the topic of a lively and very prolific debate on 'theories of vagueness' in analytic philosophy since decades. However most of these theories of vagueness involve an explicit dismissal of the idea that the semantics of logical connectives should be modelled by truth functions over sets of linearly ordered 'truth values' beyond just *true* and *false*. As already indicated, I thought that my hints at the wide discrepancies between fuzzy logic and philosophical accounts of vagueness might be considered a provocation to mathematical logicians and computer scientists who hardly feel the need to justify their formal models by prior conceptual analysis, and who moreover often like to make disparaging remarks on philosophical investigations. But if I felt uneasy about the reception of my contribution, I had entirely underestimated the open-mindedness and, indeed, strong intellectual curiosity of Petr Hájek. Just a few days after I had submitted my little paper he asked me to serve as an invited speaker of the meeting, entailing more time for presentation. Moreover he arranged that the essay [Fer03] corresponding to my talk was published in a special issue of *Neural Network World* journal, dedicated to contributions to the workshop. Ever since I am grateful for this turn of events, which, with hindsight, has proved to be a starting point for various joint activities in that area, bringing together viewpoints from experts with different background, all seriously interested in *Logical Models of Reasoning with Vague Information* (LoMoReVI) – to mention the title of a EUROCORES-LogICCC project that Petr Hájek, Lluís Godo, and myself are heading currently.

Of course, the mentioned subtitle of my contribution to the Soft Computing workshop and the corresponding paper [Fer03] was not meant to just imply that logicians should pay attention to the vagueness discourse in analytic philosophy. I am deeply convinced that philosophers of vagueness could profit a lot as well from a better understanding of the aims, methods, and results in mathematically fuzzy logic. But in addressing logicians, mathematicians, and computer scientists I felt it more adequate to focus only at one direction of the – up till now – somewhat unsatisfying scientific communication about reasoning under vagueness. Obviously, for the present happy occasion, it is once more not really adequate to address philosophers engaging in theories of vagueness. Nevertheless, let me right away and without hedging, uncertainty, or reference to an intermediary degree of truth answer the question chosen as subtitle of this contribution emphatically by exclaiming: of course philosophers can and should learn from Petr Hájek's work!

With this short answer I don't want to imply, as would certainly be inappropriate, that mathematical fuzzy logic by itself is a theory of vagueness that should be adopted by philosophers. Rather what I mean is that any fully developed account of vagueness, independently of whether it embraces truth functional logics based on degrees of truth or rather argues against the adequateness of using such logics as models of reasoning, should better be informed about the non-trivial generality and flexibility of Hájek's t-norm based approach to deductive fuzzy logics. More generally, there is hardly a better place to learn how and why mathematical logic progresses, not only always in a strictly technical manner, but sometimes also conceptually, than the still continually evolving work of Petr Hájek.

Until rather recently one might have been forgiven the claim that all eminent philosophers of vagueness are rather ignorant and/or very sceptical about the role that mathematical methods in non-classical logics, including t-norm based fuzzy logics, can play in theories of vagueness. (For this assessment I take the many prominent supporters of supervaluationism as best account of vagueness as explicitly defending classical logic against non-classical competitors.) However, at least two important books on vagueness have appeared since, that are all but hostile to contemporary formal approaches to non-classical logics: Stewart Shapiro's *Vagueness in Context* [Sha06] and Nicholas J.J. Smith's *Vagueness and Degrees of Truth* [Smi08]. I know that Petr Hájek is in direct contact with both authors. Very recently (in September 2009) Smith was an invited speaker at a conference in Čejkovice associated with – and bearing the same title – as the above mentioned project on *Logical Models of Reasoning with Vague Information*. I will not say much about Smith's very interesting and well written book on vagueness here. (A rather detailed review of it is forthcoming in *The Australasian Journal of Logic*.) Let me just emphasize that Smith is the only philosopher of vagueness that I am aware of that cites Petr Hájek's monograph [Háj98] as an important source and reviews the basic concepts of algebraic semantics as offered by contemporary fuzzy logics. Thus, in a sense, we now have a prominent example of a philosopher who would certainly confirm that philosophers can learn from the logician Petr Hájek. This is also the right occasion to mention that Hájek's chose the subtitle *Can Logicians Learn from Philosophers AND Can Philosophers Learn from Logicians?* for his presentation in Čejkovice.

One should be aware of the fact that the two mentioned books, while both deal with reasoning under vagueness in a manner that seriously includes non-trivial contemporary formal logic, amount to two very different accounts of vagueness and its implications for logic. While Smith explicitly defends a degree based approach, Shapiro presents a contextualist theory of vagueness that explicitly rejects many-valued logic as a basis for reasoning with vague predicates. Petr Hájek has recently presented a paper in *Studia Logica* that

seemingly has been prompted by reading Shapiro's book. In the remainder
of this paper I want to briefly discuss some aspects of Shapiro's account
of vagueness that seem to me to offer further opportunities for interaction
between theories of vagueness and mathematical fuzzy logic.

Contextualism and degrees of truth

The first page of Stewart Shapiro's *Vagueness in Context* [Sha06] contains
the following innocent looking passage:

> [...] one might begin with a person for whom it is not quite correct to
> say that she is tall, and not quite correct to say that she is short, or even
> not quite correct to say that she fails to be tall. [Sha06, p. 1]

I find this very interesting, because defenders of a degree based approach
to vagueness might argue that saying of a sentence S that neither S nor its
negation $\neg S$ are 'quite correct' amounts to assigning an intermediary degree
of truth to S. At the very least, I think, it is a defendable move to claim
that in evaluating a sentence as 'not quite true' one implicitly accepts a
concept of *partial truth*, and to point out that fuzzy logic (in the narrow
sense of a deductive formal system with truth functional semantics) is often
motivated as logic of partial truth: whereas classical logical consequence
refers to preservation of (definite) truth, fuzzy logic models preservation of
partial truth. However Shapiro, of course, has no such reading of his example
in mind. The cited passage continues as follows:

> Controversy arises when we try to say what it is to be a borderline, or
> what it is to be "not quite true". [Sha06, p. 1]

Indeed, Shapiro goes on to offer a very different analysis of reasoning in
presence of borderline cases, that focuses on a specific form of context dependence of all utterances and apparently has no role left to play for degrees
of truth. A very succinct summary of the essence of Shapiro's contextualism
can be found in the opening sentence of his paper *Reasoning with Slippery
Predicates* [Sha08]:

> On my view, vague predicates exhibit what I call 'open texture': in some
> circumstances, competent speakers can go either way in the borderline
> region. [Sha08, p. 313]

(I will continue to quote from the journal article [Sha08], rather than from
the book [Sha06]; simply since the former is a more compact, more recent,
and easily accessible account of those features of Shapiro's theory that are of
concern here.)

Petr Hájek would not be the eminent logician and thinker that we know
him to be, if he would simply ignore Shapiro's account of reasoning under

vagueness that seems to be so different from his own approach. Indeed, he not only read Shapiro's book and discussed it also directly with Shapiro, but also wrote the already mentioned paper *On Vagueness, Truth Values and Fuzzy Logics* [Háj09], in which, to paraphrase it's abstract, some aspects of vagueness as presented in [Sha06] are analyzed from a fuzzy logic perspective and corresponding generalizations of Shapiro's formal apparatus are presented. In essence, Hájek shows that Shapiro's model theory that is based on Kleene's three valued logic can alternatively be based on *t*-norm based fuzzy logic, in particular also on his own *BL*, the logic of all continuous *t*-norms.

I agree that it is very natural from the view point of fuzzy logic to generalize a three valued setting to one referring to the full real unit interval $[0,1]$ as set of truth values. However, I am also wondering whether this move does full justice to the original philosophical motivations underlying Shapiro's model. Let me emphasize that Hájek does not claim at all to engage with philosophical aspects. He states explicitly that

> [...] here we take mathematical fuzzy logic as granted and see how it combines with Shapiro's approach (concentrating on the mathematical side; philosophical interpretation would be welcome). [Háj09, p. 319]

At this point I should confess that Petr had in fact asked me at some point during the development of the just cited paper to join him as co-author by providing philosophical comments. Of course, I was grateful for this suggestion; however I finally decided that what I could possibly contribute was too meager and also too far away from Petr's treatment of Shapiro's concepts. In any case, I felt unable to come up with a convincing 'philosophical interpretation' of these results. Characteristically, far from taking offense, Petr wrote at the end of his paper 'Chris Fermüller sent me a long extremely interesting set of comments to the manuscript of the present paper' and went on to quote from my remarks:

> "My own (tentative!) 'philosophical' view about the role of fuzzy logic is that it does not directly model linguistic behavior, but it can be used to abstract away from conversational scenarios in a systematic, simple and (at least for some purposes) sufficiently adequate way". [Háj09, p. 381]

I apologize for this indirect self-quotation. My purpose here is to try to explicate this somewhat convoluted and possibly unclear statement by building on Shapiro's contextualist theory of vagueness as an example of what I think could be done in the indicated vein. To this aim I have to review some central concepts of Shapiro's model.

The gist of Shapiro's account of vagueness is that borderline cases are evaluated with respect to conversational contexts. At any given moment in a conversation every sentence is either accepted as true or judged to be false or else left undecided. Note that there are no intermediary degrees of truth involved: while 'left undecided' can technically be associated with a third truth value, the model is intended to match the observation expressed by 'open texture', which implies that the judgment about the acceptability of a sentence S in a given conversational context is binary. The vagueness of S shows up only in the dynamics of conversations: in borderline cases a previous acceptance of S can also be retracted.

> *In particular, the extension of a vague predicate shifts every time a borderline case is called, or is explicitly or implicitly retracted.* [Sha08, p. 317]

In a forced march sorites scenario, say with respect to a series of pairwise phenomenologically indistinguishable color cards starting with a clearly red one and ending with a clearly orange one, the conversationalists are forced to either explicitly accept or deny that a given card is red. Shapiro speculates with sound intuition that the conversationalists start to assert confidently that the shown card is red and will continue for some time to judge the currently shown card as red, but will eventually switch to the judgement that the card is not red. Since they realize that they should not call a card red and at the same time judge a (in the momentary situation) indistinguishable card as not red, the switch involves a retraction from previously given judgements. If, after the switch, they are shown the cards in reverse order, the conversationalists will for some time explicitly deny that the shown cards are red thereby explicitly reversing their previous judgements, which have already been implicitly retracted by the decision to switch from 'is red' to 'is not red'. In accordance with these intuitions, Shapiro introduces formal models ('*frames*') $F = \langle W, b \rangle$, that can be viewed as Kripke-style tree structures. W is a set of partial interpretations, called *sharpenings*, referring to a fixed domain D. Every $u \in W$ defines an extension $E_u^+(P) \subseteq D^n$ and an anti-extension $E_u^-(P) \subseteq D^n$ for each n-ary predicate P; b is an element of W, called *base*. A partial ordering $\preceq$ naturally arises on W by defining $u \preceq v$ (read: 'v is a sharpening of u') iff $E_u^+(P) \subseteq E_v^+(P)$ as well as $E_u^-(P) \subseteq E_v^-(P)$ for all predicates P. In $F = \langle W, b \rangle$ all elements of W must be sharpenings of its base b. Given the extensions and anti-extensions of $u \in W$, truth and falsity at u are defined in the obvious way for atomic formulas. This is extended to arbitrary formulas with classical connectives and quantifiers using Kleene's strong three valued truth functions (see, [Sha06] or [Sha08] for details). The crucial semantic notion is that of *forcing* with respect to $F = \langle W, b \rangle$: A formula ϕ is forced at $u \in W$ under a variable assignment σ iff for each sharpening v of u there is a sharpening w of v such that ϕ is true in w under σ. The underlying idea is that ϕ is forced if we are committed to accept it eventually, i.e., if it will be evaluated to true under every sufficiently progressed series of sharpenings. A formula ϕ is *weakly forced* at $u \in W$ under a variable assignment σ iff there is no sharpening of u in which ϕ is false under σ.

Note that all classically valid formulas are weakly forced already at the base of all frames. In sharp contrast, nothing is forced in all frames since there are no tautologies in the underlying three valued logic of Kleene. Nevertheless forcing is a useful notion for modelling concrete examples. Shapiro invites us to consider a scenario in which a father promises his kids to take them to a football match on Sunday if the weather is nice and that they will go to a movie if the weather is not nice. Suppose the weather on Sunday is borderline between nice and not nice. Does the father break his promise if the family neither goes to the football game nor to a movie? Shapiro's (like my own) intuition says that he does. Let N stand for the 'the weather is nice on Sunday', F for 'the family goes to a football game' and M for 'the family goes to a movie'. The father's promise should then be represented by the forcing of $(N \rightarrow F) \& (\neg N \rightarrow M)$ at the base of a frame. But in all such frames $F \vee M$ is also forced. Committing to $(N \rightarrow F) \& (\neg N \rightarrow M)$ turns out to force $N \vee \neg N$ as well; in other words the truth or falsity of N has to be decided eventually in such frames.

As already indicated above, I cannot provide a really satisfying philosophical interpretation of Hájek's generalization in [Háj09] of this model theory to a fuzzy logic setting. Let me just observe that while, technically, this generalization largely amounts to a 'fuzzification' of three valued notions, Hájek also goes beyond the strictly truth functional setting by enriching the system with modal operators. I think that this machinery can and should be employed not just to fuzzify contextualism, but rather also to contextualize fuzzy logic. In any case, it seems rather obvious to me that the insights gained into reasoning with vague notions by considering context shifts apply also to models where one takes degrees of truth for granted at individual states of a conversation. If the extensions of vague predicates can and do change during the course of a conversation, as Shapiro plausibly suggests, then contextualist versions of fuzzy logic should be useful in modelling corresponding scenarios.

Extracting truth values from context histories

Instead of contextualizing fuzzy logic in the sense just indicated above, I want to briefly consider a quite different way of combining fuzzy logic and contextualism here. To this aim we first re-visit the forced march sorites scenario involving, say, 1000 color cards that gradually lead from red to orange. Let the atom $R(c_i)$ be interpreted as 'the i-th card c_i is accepted as red'. At the base b of a corresponding frame $F = \langle W, b \rangle$ the sentences $R(c_1)$ and $\neg R(c_{1000})$ are true. Consequently $R(c_1) \& \neg R(c_{1000})$ is also forced at every sharpening of the frame. (In my version of the scenario the conversationalists are shown c_1 as well as c_{1000} in the beginning and know that they will be asked to judge on c_i for increasing i in successive states. But this presupposed knowledge seems rather irrelevant for the exercise.) What happens with respect to a corresponding frame $F = \langle W, b \rangle$ if conversationalists are are asked to either explicitly accept or deny that successively shown cards are red? According to Shapiro we expect to see that more and more previously unjudged cards are added to the extension of R as i increases. This corresponds to moving the context from the base b to successive sharpenings u_i of it: $b = u_1 \preceq u_2 \preceq \ldots$. However, the conversationalists will realize that eventually a shown card c_n, certainly the one where $n = 1000$, should not be accepted as red, given that the final card is definitely not red, if they want to stay competent and consistent in their judgements. But they also want to respect a principle of tolerance expressing that, given the indistinguishability of two successive cards c_i and c_{i+1} in the stipulated setting, one cannot accept $R(c_i)$ and explicitly deny $R(c_{i+1})$ at the same time. Consequently, together with judging c_n to be not red they also have to revise their previous judgment on c_{n-1} and probably a few more previously shown cards. The new conversational context that results from such a 'jump', as Shapiro calls it, is of course still a sharpening of b, but one that is situated at a different branch of the F with respect to the tree form of F induced by $\preceq$.

One might object to the whole forced march setting that forcing the participants into explicit acceptance or denial of vague propositions is rather artificial. But the story about the father's promise to his kids, sketched in the last section, shows that sometimes our commitments as competent and fair partners in conversations may indeed force us to decide borderline cases in one of two possible ways. There is also an interesting analogy with legal reasoning here. We expect judges and juries to take binary decisions about the alleged guilt of defendants. Indeed, we have good reasons to separate issues about the *degree of guilt* from the question whether a particular law has been violated or not. (Remember that in many legal system juries of laypersons decide 'gulity' or 'not guilty', whereupon it is left to a professional judge to decide about degrees of guilt and corresponding punishment in case of a 'guilty' verdict.) Similarly, most linguists maintain that in investigating the semantics and pragmatics of natural language there are good reasons to separate issues of hegding, emphasis, hesitation, etc., from recording whether an uttered sentence has been accepted by a hearer or not.

Nevertheless, suppose we allow conversationalists to hedge their judgements by uttering sentences like 'We accept c_n as red only tentatively and without much confidence'? We might then try to associate degrees of acceptance with the individual judgements and correspondingly replace extensions and anti-extensions as set by extensions as fuzzy sets. Something like this must be the underlying idea in 'fuzzifying' Shapiro's model theoretic apparatus along the lines of [Háj09]. While it is certainly valuable to formally explore this option as Petr Hájek did, one should also point out that it poses serious problems of interpretation; not only 'philosophical' problems, but also problems pertaining to intended applications. E.g., is it really realistic and feasible to associate degrees of acceptance (truth) in a sufficiently systematic manner? Should such degrees bear on the validity of arguments? Should we really revise the notion of logical consequence by taking into account degrees? Note that the reply, that all this is only done for the sake of arriving at a neat mathematical model, is evading the issue. It just provokes the shift of the original questions to ones about criteria for the adequateness of some formal model; nothing much is gained in practice or in principle by such a reformulation. If a fuzzy logician accepts that nothing corresponds *directly* to real numbers in $[0, 1]$ referred to as 'degrees of truth', but also wants to maintain that fuzzy logic provides a useful model nevertheless, she is left with the task of explaining how these values in $[0, 1]$ arise from features of the intended application scenario that are directly represented in the model. This challenge, of course, is usually well understood in fuzzy logic (in a broader sense). Corresponding literature on how to derive truth values and truth functions from arguably more fundamental concepts like notions of similarity, voting behavior, degrees of conformity, (approximation to) degrees of belief, etc., abounds (see, e.g., [Law98, Par97, Rus91, KGK94] and references there). In the following I want to indicate a further way to formally and systematically extract 'degrees of truth' from semantic structures that don't mention larger sets of truth values explicitly. My aim is to show that fuzzy logic might have something interesting to add to contextualism *even if we respect the principle that all semantic decisions taken in course of a conversation are binary*, i.e., even if we follow Shapiro in maintaining that every sentence that is not explicitly left undecided at a given point during a conversation is either accepted or else denied at that point.

Let me motivate my attempt by once more quoting Shapiro:

> *When reasoning about vague predicates, in ordinary contexts, we do not normally hold the conversational context fixed. It is difficult to do so since, on my view, predicates change their extension rapidly, and on the fly, as the conversation proceeds. The extensions and anti-extensions of the predicates can change in the very act of our considering them as we go through an argument.* [Sha08, p. 320]

We have seen two concrete types of (anti-)extension affecting moves operating in the forced march sorites scenario: *sharpening*, i.e. moving from a state u of a frame to another state v, where $u \preceq v$, and *jumping* to a state that is not a sharping of the preceding one and involves also retractions of previous judgements. Indeed, as the above quoted passage makes clear, the dynamics of sharpenings and jumps plays a central, if not *the* central role in Shapiro's account of reasoning under vagueness. It is therefore somewhat surprising that the temporal aspect of conversations and the particular forms of context shift do not manifest themselves as an explicit part of the formal semantic machinery that is put in place in [Sha06] and in [Sha08]. Contemporary logic certainly offers tools that allow to formalize reasoning about context shifts and, more generally, about various actions that operate on states. This, of course, can also be combined with epistemic modalities. In particular, it would be an interesting exercise to formalize reasoning about the dynamics of dialogue triggered context shifts by adapting the framework of dynamic epistemic logic (see [DHK07]). Anyway, this will hardly bring Shapiro's logic of vagueness closer to fuzzy logic.

I actually have in mind a much simpler form of making the dynamic nature of reasoning over frames explicit: let us just augment a frame $F = \langle W, b \rangle$ by a function d from (some initial sequence of) $\mathbb{N}$ into W that specifies the state $d(i) \in W$ that a conversation has reached at time i. This enables us to fix the meaning of conversational moves succintly. A *sharpening* takes place at time i iff $d(i) \preceq d(i+1)$, whereas a *jump* takes place at time i iff $d(i) \npreceq d(i+1)$ as well as $d(i+1) \npreceq d(i)$. Note that further natural conversational moves can easily be expressed. E.g., we can speak of a *proper sharpening* at time i iff $d(i) \prec d(i+1)$, short for: $d(i) \preceq d(i+1)$ and $d(i) \neq d(i+1)$. A *retraction* takes place at time i iff $d(i+1) \prec d(i)$. Since all states are sharpenings of the base of the frame, every jump can be seen as combining a retraction with a proper sharpening. Needless to say that one could go on to refine these notions by restricting the number of atoms that can be updated in a single step, and in further potentially interesting manners.

Let a *trace* of a conversation consist of a frame $F = \langle W, b \rangle$ together with a function d as just defined. Looking back at the forced march sorites scenario involving 1000 cards starting with a definitely red one and ending with a definitely orange one, we can now say that following Shapiro's intuitions we expect a corresponding trace with just one jump and otherwise only sharpenings. We expect the jump to take place somewhere in the borderline region between (still clearly) red and (already clearly) orange cards c_i (time points i). If we want to attach semantic significance to where exactly the jump occurs, we should better repeat the experiment in reverse order; i.e., start with c_{1000} definitely judged to be not red, expecting a jump from $\neg R(c_i)$ to $R(c_{i-1})$, again somewhere in the borderline region, but probably now for a different card i. Actually, this will still hardly provide us with substantial information about the general readiness of competent speakers to accept a color card as red or as not red. However, it is not unreasonable to expect that by repeating the experiment again and again with different sets of (competent) conversationalists, we do indeed gain knowledge, codified in the set of observed traces, about general tendencies to accept the individual color cards as red. There are different ways to extract from such sets of traces plausible 'truth values' for $R(c_i)$. Arguably, the simplest reasonable method is to identify the corresponding truth value for $R(c_i)$ with the relative frequency with which the i-th card is accepted as red. But, of course, more sophisticated methods of assigning truth values are conceivable. E.g., one might want to take into account the average distance of a card from the points where jumps are observed. In any case, it should be clear that we have indicated how sets of traces, i.e. 'frames in action', can be used to extract truth values, without directly involving degrees of emphasis or hesitation, and thus respecting the linguistic principle to separate semantic from pragmatic issues. Indeed we can maintain that in reasoning with vague predicates, we only need to keep track of acceptance and denial of corresponding statements on a local, heavily context dependent level. Nevertheless relevant semantic information is contained in the dynamics of context shifts, which presumably shows some statistical regularities. In this approach fuzziness is systematically related to instable binary decisions. Intermediary truth values are obtained by abstracting away from Shapiro's rapidly – 'on the fly' – changing contexts, and by taking into account only statistical parameters of the overall picture that arises on a global level over considerable spans of time.

At this point I should emphasize that I entirely agree with Petr Hájek's assertion that it is not the logician's task to assign concrete numerical values to concrete sentences, but that the task is rather to study the notion of consequence and deduction that arises from fuzziness. In this sense, the above considerations do *not directly* contribute to logic. However, let me also point out that we can hardly claim to have a full understanding of a consequence relation that is defined in direct reference to truth values from $[0, 1]$ (or some more general structure) without understanding what exactly these values are meant to represent and how they relate to classical concepts of truth and falsity. This becomes particularly evident if one has to assess whether the suggested definition of consequence is normatively or empirically adequate for concrete applications. To refer back to the initial quotation from the preface of Petr Hájek's monograph [Háj98]: if 'fuzzy inference' is to be understood as logical deduction not only in some purely formal sense, but also in a sense that corresponds to a theory of correct reasoning then we have to be able to assign more than just a purely formal meaning to truth values. (Viz. the analogous challenge for classical or for intuitionistic logic.)

Determining truth functions

We have suggested a method to associate meaning to truth values assigned to atomic sentences taken from the real unit interval, but we have not yet dealt with the problem of assigning meaning to logical connectives. In [Sha06] and in [Sha08] Shapiro specifies the semantics for not just one, but rather two full sets of propositional connectives and quantifiers in reference to frames. These definitions relate to forcing and weak forcing in different ways. While logical consequence remains classical for a large fragment of the language, it is also clear that the suggested semantics is not truth functional, but rather resembles the semantics provided for intuitionistic connectives with respect to Kripke models. Shapiro writes

> There is much fun to be had in determining the rules for reasoning with various connectives and quantifiers in this system. [Sha08, p. 235]

However it is not always easy to keep track of the corresponding conditions that often involve quantifier alternations at the meta-level. This has motivated recent work by Christoph Roschger, that illustrates the evaluation of complex formulas in Shapiro's system by reference to a Hintikka style evaluation game [Ros]. But in any case, Petr Hájek's generalization of Shapiro's models withstanding, we are quite far away from standard fuzzy logics here.

My own favorite way to formally assign meaning to logical connectives follows the dialogue based approach by Robin Giles, who in turn follows ideas that Paul Lorenzen had propagated in the context of constructive logic already since the late 1950s [Lor60]. The model of Giles [Gil77, Gil74] involves two distinct ingredients that operate on different levels:

- dialogue rules, pertaining to propositional connectives and quantifiers, that regulate the stepwise reduction of arguments about logically complex statements to arguments about sub-statements, and

- a scheme for betting on results of dispersive, binary experiments that are associated with atomic formulas and for which fixed success probabilities are assumed.

It can be shown that the expected amount of money (say, between 0 and 1 euro) that I have to pay to you, if we both play the dialogue game in an optimal, coincides with the inverse of the truth value assigned by Łukasiewicz logic to the formula representing my initial statement, under a truth value assignment to atoms that corresponds to the success probabilities assigned to the respective experiments. Łukasiewicz logic, of course, is one of the main examples of a t-norm based fuzzy logic. Giles's dialogue game was later shown to relate nicely to hypersequent based proof theory for fuzzy logics (see [CFM05, Fer09, FM09]), even beyond Łukasiewicz logic. Indeed, dialogue games can be seen as a variant of so-called proof theoretic semantics (see, e.g., [KSH06]), where the meaning of connectives is considered as essentially specified by introduction and elimination rules in natural deduction or, equivalently, the logical rules of a cut-free sequent calculus. Following this approach on top of the evaluation scenario for atomic sentences sketched in the last section, one is led to a variant of Giles's characterization of Łukasiewicz logic, in which Giles's original story about bets on the results of dispersive experiments is replaced by our story about observing the behavior of conversationalists in forced march sorites scenarios.

Of course, there are also other ways to determine or extract truth functions from basic semantic settings, instead of simply imposing them. Jeff Paris's [Par00] provides a (rather sceptical) overview of such attempts to justify particular truth functions considered in fuzzy logics with respect to arguably more fundamental stipulations about the meaning of logically complex statements. I think it is an interesting challenge to try to connect these semantic scenarios with Shapiro's model theory for reasoning with vague predicates. But this is a topic for future research.

Conclusion

Petr Hájek's investigations in [Háj09] re-introduce fuzzy logic into the contextualist approach to reasoning about vagueness, after Stewart Shapiro [Sha06, Sha08] had thrown it out from such an account of vagueness quite explicitly. In this essay we have indicated a different way to keep fuzzy logic in the ball park: intermediary truth values may arise from systematic observations about the linguistic behaviour of conversationalists in Shapiro's forced march sorites scenario. In this manner, two different logics can be seen at play in reasoning with vague propositions: a local, non-truth functional one, as presented by Shapiro, as well as a global fuzzy logic, whose truth values arise from abstracting away from the actual dynamics of conversations, recording general tendencies 'to go one way rather than the other' in context dependent decisions about predicates that exhibit 'open texture'.

I am well aware of the fact that the above remarks only relate to a tiny portion of the impressive amount of results that Petr Hájek has contributed to mathematical fuzzy logic, not to speak of his equally impressive work on many other topics in logic and related areas. However, it is very satisfying to know that Petr and me agree on the appraisal of reasoning with vague predicates and propositions as a fascinating, still challenging, and complex topic that calls for further, also interdisciplinary work. I am also well aware that my own contributions to this field will hardly ever match the degree of mathematical sophistication, precision, and fecundity of Petr's work. But this is just to say that I, whether as a logician, a computer scientist, or a philosopher will never cease to learn from Petr Hájek.

BIBLIOGRAPHY

[CFM05]   A. Ciabattoni, C. G. Fermüller, and G. Metcalfe. Uniform rules and dialogue games for fuzzy logics. In F. Baader and A. Voronkov, editors, *Logic for Programming, Artificial Intelligence, and Reasoning*, volume 3452 of *LNCS*, pages 496–510. Springer, 2005.

[Fer03]   C. G. Fermüller. Theories of vagueness versus fuzzy logic: can logicians learn from philosophers? *Neural Network World*, 13(5):455–465, 2003.

[Fer09]   C. G. Fermüller. Revisiting Giles's game—reconciling fuzzy logic and supervaluation. In O. Majer, A.-V. Pietarinen, and T. Tulenheimo, editors, *Games: Unifying Logic, Language, and Philosophy*, volume 15 of *Logic, Epistemology, and the Unity of Science*, pages 209–227. Springer, 2009.

[FM09]   C. G. Fermüller and G. Metcalfe. Giles's game and the proof theory of Łukasiewicz logic. *Studia Logica*, 92:27–61, 2009.

[Gil74]   R. Giles. A non-classical logic for physics. *Studia Logica*, 33:397–415, 1974.

[Gil77]   R. Giles. A non-classical logic for physics. In R. Wójcicki and G. Malinowski, editors, *Selected Papers on Łukasiewicz Sentential Calculi*, pages 13–51. Polish Academy of Sciences, 1977.

[Háj98]   P. Hájek. *Metamathematics of Fuzzy Logic*, volume 4 of *Trends in Logic*. Kluwer, Dordrecht, 1998.

[Háj09]   P. Hájek. On vagueness, truth values and fuzzy logics. *Studia Logica*, 91(3):367–382, 2009.

[KGK94]   R. Kruse, J. Gebhardt, and F. Klawonn. *Foundations of Fuzzy Systems*. John Wiley & Sons, New York, 1994.

[KSH06]   R. Kahle and P. Schroeder-Heister, editors. *Proof-Theoretic Semantics*, volume 148(3) of *Synthese*, 2006.

[Law98]   J. Lawry. A voting mechanism for fuzzy logic. *Int. J. Approximate Reasoning*, 19:315–333, 1998.

[Lor60]   P. Lorenzen. Logik und Agon. In *Atti Congr. Internaz. di Filosofia*, pages 187–194. Sansoni, 1960.

[Par97]   J. B. Paris. A semantics for fuzzy logic. *Soft Computing*, 1:143–147, 1997.

[Par00]   J. B. Paris. Semantics for fuzzy logic supporting truth functionality. In *Discovering the World with Fuzzy Logic*, volume 57 of *Stud. Fuzziness Soft Comput.*, pages 82–104. Springer, Heidelberg, 2000.

[Ros]   C. Roschger. Evaluation games for Shapiro's logic of vagueness in context. To appear in The Logica Yearbook 2009 (Ed. M. Peliš).

[Rus91]   E. H. Ruspini. On the semantics of fuzzy logic. *Int. J. Approximate Reasoning*, 5(1):45–88, 1991.

[Sha06]   S. Shapiro. *Vagueness in Context*. Oxford University Press, 2006.

[Sha08]   S. Shapiro. Reasoning with slippery predicates. *Studia Logica*, 90(3):313–336, 2008.

[Smi08]   N. J. J. Smith. *Vagueness and Degrees of Truth*. Oxford University Press, Oxford, 2008.

[DHK07]   H. van Ditmarsch, W. van der Hoek, and B. Kooi. *Dynamic Epistemic Logic*. Springer, 2007.

Chris Fermüller
Vienna University of Technology
Favoritenstr. 9-11/E1852
A-1040 Wien, Austria
Email: chrisf@logic.at

A poem

JONÁŠ HÁJEK

65. kilometr

Stín vytéká úhledně zarovnanou strouhou
podél plotu ze zeleně nalakovaných drátů
za nímž se otvírá pahorkatina s lemem lesa
v závěru pohledu. Nějaký dravec, než dosedne,
zakrouží kolem sloupu vedení s jediným keřem,
potok vytéká z kanálu s vyraženým letopočtem
a vytváří příznivé podmínky pro vodní rostliny.
V nedalekém ohybu přibublává jiný pramen,
zneklidňující obal od papírových kapesníků.
Úzké a široké spojuje tu nenápadná branka.

Dočasně bezejmenné místo
ilustrují řídké šmouhy hluku.

Mezi otevřenou kapotou a trojúhelníkem,
jak by ztělesňoval vyražený dech karoserie,
pár metrů od vozidla na odstavném pruhu
můj otec vytrvale vyhlíží asistenční službu.

65th kilometer

Shadow pours out in a neatly arranged ditch
Along a fence of wires painted green
With hills behind, hemmed by a forest
On the horizon. A bird of prey, about to land,
Circles a pylon with a lonely bush,
A stream pours out of a gutter with the year embossed
Creating favourable conditions for waterplants.
In a nearby bend, another spring bubbles in,
A haunting Kleenex wrapping.
The narrow and the wide is connected by an inconspicuous gate.

A temporarily nameless place
Is illustrated by thin streaks of noise.

Between the open bonnet and the safety triangle,
An impersonation of the breath taken away from the car body,
Standing a little apart from the vehicle on the hard shoulder
My father awaits the assistance service.

Bibliography of Petr Hájek

THE LIBRARY OF THE INSTITUTE OF COMPUTER SCIENCE AND
PETR CINTULA

Dedicated to Professor Petr Hájek
on the occasion of his 70th birthday

The library of the Institute of Computer Science, Academy of Sciences of the Czech Republic, has been systematically recording Petr Hájek's bibliography since 1994, mostly on the basis of original resources. Some older works or works that are difficult to access have been verified in reliable sources of information, such as Web of Knowledge, Scopus, MathSciNet, ZentralBlatt.

MONOGRAPHS

[1] P. Vopěnka and P. Hájek. *The Theory of Semisets*. Academia, Prague, 1972. 332 pages.
[2] P. Hájek and T. Havránek. *Mechanizing Hypothesis Formation: Mathematical Foundations of a General Theory*. Springer, Berlin, 1978. 396 pages.
[3] P. Hájek, T. Havránek, and M. K. Chytil. *Metoda GUHA. Automatická tvorba hypotéz*. Academia, Prague, 1983. 316 pages.
[4] P. Hájek and T. Havránek. *Avtomatičeskoje obrazovanije gipotez*. Nauka, Moscow, 1984. 280 pages.
[5] P. Hájek, T. Havránek, and R. Jiroušek. *Uncertain Information Processing in Expert Systems*. CRC Press, Boca Raton, 1992. 285 pages.
[6] P. Hájek and P. Pudlák. *Metamathematics of First-Order Arithmetic*. Perspectives in Mathematical Logic Series. Springer, Berlin, 1993. 460 pages.
[7] P. Hájek. *Metamathematics of Fuzzy Logic*, volume 4 of *Trends in Logic*. Kluwer, Dordrecht, 1998. 297 pages.

EDITED VOLUMES

[1] P. Hájek, editor. *GUHA*, volume 10 of *International Journal of Man-Machine Studies*, 1978. Special issue.
[2] P. Hájek, editor. *GUHA II*, volume 15 of *International Journal of Man-Machine Studies*, 1981. Special issue.
[3] J. Wiedermann and P. Hájek, editors. *Mathematical Foundations of Computer Science 1995. Proceeding of 20th International Symposium, MFCS'95, Prague, Czech Republic, August 28 - September 1, 1995*, volume 969 of *Lecture Notes in Computer Science*, Berlin, 1995. Springer.
[4] P. Hájek, editor. *Gödel'96. Logical Foundations of Mathematics, Computer Science and Physics - Kurt Gödel's Legacy*, volume 6 of *Lecture Notes in Logic*, Berlin, 1996. Springer.
[5] S. R. Buss, P. Hájek, and P. Pudlák, editors. *Logic Colloquium' 98. Proceedings of the Annual European Summer Meeting of the Association for Symbolic Logic, held in Prague, Czech Republic August 9-15, 1998*, volume 13 of *Lecture Notes in Logic*, Natick, 2000. AK Peters, Ltd.

[6] P. Hájek, L. Godo, and S. Gottwald, editors. *Special issue from Joint EUSFLAT-ESTYLF Conference, Palma (Mallorca) 1999 and from the Fifth International Conference Fuzzy Sets Theory and Its Applications FSTA, Liptovský Ján, 2000*, volume 124(3) of *Fuzzy Sets and Systems*, 2001.

[7] S. Gottwald and P. Hájek, editors. *Advances in Fuzzy Logic*, volume 141(1) of *Fuzzy Sets and Systems*, 2004. Special issue.

[8] S. Gottwald, P. Hájek, U. Höhle, and E. P. Klement, editors. *Fuzzy Logics and Related Structures. Volume of abstracts of the 26th Linz Seminar on Fuzzy Set Theory*, Linz, 2005. Johannes Kepler Universität.

[9] S. Gottwald, P. Hájek, and M. Ojeda-Aciego, editors. *Proceeding of The Logic of Soft Computing & Workshop of the ERCIM working group on Soft Computing*, Málaga, 2006. Universidad de Málaga.

[10] M. Ojeda-Aciego, S. Gottwald, and P. Hájek, editors. *Mathematical and Logical Foundations of Soft Computing*, volume 159(10) of *Fuzzy Sets and Systems*, 2008. Special issue.

[11] S. Gottwald, P. Hájek, U. Höhle, and E. P. Klement, editors. *Fuzzy Logics and Related Structures*, volume 161(3) of *Fuzzy Sets and Systems*, 2010. Special Issue.

JOURNAL PAPERS

[1] P. Vopěnka and P. Hájek. Über die Gultigkeit des Fundierungsaxioms in Speziellen Systemen der Mengenlehre. *Zeitschrift für Mathematische Logik und Grundlagen der Mathematik*, 9(12–15):235–241, 1963.

[2] P. Hájek. Die durch die Schwach Inneren Relationen Gegebenen Modelle der Mengenlehre. *Zeitschrift für Mathematische Logik und Grundlagen der Mathematik*, 10(9–12):151–157, 1964.

[3] P. Hájek and A. Sochor. Ein dem Fundierungsaxiom äequivalentes Axiom. *Zeitschrift für Mathematische Logik und Grundlagen der Mathematik*, 10(13–17): 261–263, 1964.

[4] P. Hájek. Die Szaszschen Gruppoide. *Matematicko-fyzikálný časopis SAV*, 15(1): 15–42, 1965.

[5] P. Hájek. Eine Bemerkung über Standarde Nichtregulare Modelle der Mengenlehre. *Commentationes Mathematicae Universitatis Carolinae*, 6(1):1–6, 1965.

[6] P. Hájek. K pojmu primitivní třídy algeber. *Časopis pro pěstování matematiky*, 90(4):477–486, 1965.

[7] P. Hájek. Modelle der Mengenlehre, in denen Mengen gegebener Gestalt Existieren. *Zeitschrift für Mathematische Logik und Grundlagen der Mathematik*, 11(2): 103–115, 1965.

[8] P. Hájek. Syntactic models of axiomatic theories. *Bulletin de l'Academie Polonaise des Sciences, Serie des Sciences Physiques et Astronomiques*, 13(4):273–278, 1965.

[9] P. Hájek and P. Vopěnka. Permutation submodels of the model ∇. *Bulletin de l'Academie Polonaise des Sciences, Serie des Sciences Physiques et Astronomiques*, 13(9):611–614, 1965.

[10] P. Hájek. The consistency of Church's alternatives. *Bulletin de l'Academie Polonaise des Sciences, Serie des Sciences Physiques et Astronomiques*, 14(8):423–430, 1966.

[11] P. Hájek. Generalized interpretability in terms of models. *Časopis pro pěstování matematiky*, 91(3):352–357, 1966.

[12] P. Hájek and L. Bukovský. On standardness and regularity of normal syntactic models of set theory. *Bulletin de l'Academie Polonaise des Sciences, Serie des Sciences Physiques et Astronomiques*, 14(3):101–105, 1966.

[13] P. Hájek, I. Havel, and M. Chytil. GUHA — metoda automatického zjišťování hypotéz. *Kybernetika*, 2(1):31–47, 1966.

[14] P. Hájek, I. Havel, and M. Chytil. The GUHA-method of automatic hypotheses determination. *Computing*, 1(4):293–308, 1966.

[15] P. Hájek and P. Vopěnka. Some permutation submodels of the model ∇. *Bulletin de l'Academie Polonaise des Sciences, Serie des Sciences Physiques et Astronomiques*, 14(1):1–7, 1966.

[16] P. Hájek, I. Havel, and M. Chytil. GUHA — metoda automatického zjišťování hypotéz II. *Kybernetika*, 3(5):430–437, 1967.

[17] P. Hájek and P. Vopěnka. Concerning the ∇-models of set theory. *Bulletin de l'Academie Polonaise des Sciences, Serie des Sciences Physiques et Astronomiques*, 15(3):113–117, 1967.

[18] P. Hájek. Problém obecného pojetí metody GUHA. *Kybernetika*, 4(4):505–515, 1968.

[19] P. Hájek, P. Vopěnka, and B. Balcar. The notion of effective sets and a new proof of the consistency of the axiom of choice. *Journal of Symbolic Logic*, 33(3):495–496, 1968.

[20] P. Hájek. O metodě GUHA automatického zjišťování hypotéz. *Československá fyziologie*, 18:137–142, 1969.

[21] P. Hájek and B. Balcar. Popis hypotetického počítacího stroje pro automatické zpracování křivek. *Československá fyziologie*, 18:151–154, 1969.

[22] P. Hájek. Logische Kategorien. *Archive for Mathematical Logic*, 13(3–4):168–193, 1970.

[23] P. Hájek. On interpretability in set theories. *Commentationes Mathematicae Universitatis Carolinae*, 12(1):73–79, 1971.

[24] P. Hájek, K. Bendová, and Z. Renc. The GUHA method and the three valued logic. *Kybernetika*, 7(6):421–435, 1971.

[25] P. Hájek. Contributions to the theory of semisets I. *Zeitschrift für Mathematische Logik und Grundlagen der Mathematik*, 18(3):241–248, 1972.

[26] P. Hájek. On interpretability in set theories II. *Commentationes Mathematicae Universitatis Carolinae*, 13(3):445–455, 1972.

[27] M. Hájková and P. Hájek. On interpretability in theories containing arithmetic. *Fundamenta Mathematicae*, 76(2):131–137, 1972.

[28] P. Hájek. Automatic listing of important observational statements I. *Kybernetika*, 9(3):187–205, 1973.

[29] P. Hájek. Automatic listing of important observational statements II. *Kybernetika*, 9(4):251–271, 1973.

[30] P. Hájek. Why semisets. *Commentationes Mathematicae Universitatis Carolinae*, 14(3):397–420, 1973.

[31] P. Hájek and D. Harmancová. On generalized credence functions. *Kybernetika*, 9(5):343–356, 1973.

[32] P. Vopěnka and P. Hájek. Existence of a generalized semantic model of Gödel-Bernays set theory. *Bulletin de l'Academie Polonaise des Sciences, Serie des Sciences Physiques et Astronomiques*, 21(12):1079–1085, 1973.

[33] P. Hájek. Automatic listing of important observational statements III. *Kybernetika*, 10(2):95–124, 1974.

[34] P. Hájek. Degrees of dependence in the theory of semisets. *Fundamenta Mathematicae*, 82:11–24, 1974.

[35] P. Hájek. Metoda GUHA a automatizace výzkumu. *Acta Polytechnica*, pages 105–118, 1975.

[36] P. Hájek. Observationsfunktorenkalküle und die Logik der Automatisierten Forschung. *Elektronische Informationsverarbeitung und Kybernetik*, 12(4–5): 181–186, 1976.

[37] P. Hájek. Another sequence of degrees of constructibility. *Notices of the American Mathematical Society*, 24(3):A–299, 1977.

[38] P. Hájek. Experimental logics and Π_3^0 theories. *Journal of Symbolic Logic*, 42(4):515–522, 1977.

[39] P. Hájek and T. Havránek. On generation of inductive hypotheses. *International Journal of Man-Machine Studies*, 9(4):415–438, 1977.

[40] P. Hájek and B. Balcar. On sequences of degrees of constructibility (solution of Friedman's problem 75). *Zeitschrift für Mathematische Logik und Grundlagen der Mathematik*, 24(4):291–296, 1978.

[41] P. Hájek and T. Havránek. The GUHA method — its aims and techniques (twenty-four questions and answers). *International Journal of Man-Machine Studies*, 10(1):3–22, 1978.

[42] P. Hájek. Arithmetical hierarchy and complexity of computation. *Theoretical Computer Science*, 8(2):227–237, 1979.

[43] P. Hájek. A note on partially conservative extensions of arithmetic. *Journal of Symbolic Logic*, 45(2):391–391, 1980.

[44] P. Hájek, P. Kalášek, and P. Kůrka. O dynamické logice. *Kybernetika*, 16(suppl): 3–41, 1980.

[45] P. Hájek. Completion closed algebras and models of Peano arithmetic. *Commentationes Mathematicae Universitatis Carolinae*, 22(3):585–594, 1981.

[46] P. Hájek. Decision problems of some statistically motivated monadic modal calculi. *International Journal of Man-Machine Studies*, 15(3):351–358, 1981.

[47] P. Hájek. On interpretability in theories containing arithmetic II. *Commentationes Mathematicae Universitatis Carolinae*, 22(4):667–688, 1981.

[48] P. Hájek and P. Kůrka. A weak second order dynamic logic with array assignments. *Fundamenta Informaticae*, 4(4):919–934, 1981.

[49] K. Bendová and P. Hájek. A logical analysis of the truth-reaction paradox. *Commentationes Mathematicae Universitatis Carolinae*, 23(4):699–713, 1982.

[50] P. Hájek and T. Havránek. GUHA 80: An application of artificial intelligence to data analysis. *Computers and Artificial Intelligence*, 1(2):107–134, 1982.

[51] P. Hájek. Arithmetical interpretations of dynamic logic. *Journal of Symbolic Logic*, 48(3):704–713, 1983.

[52] P. Hájek. Combining functions for certainty degrees in consulting systems. *International Journal of Man-Machine Studies*, 22(1):59–76, 1985.

[53] P. Hájek. Etické problémy umělé inteligence. *Filozofický časopis*, 34:467–471, 1986.

[54] P. Hájek. A simple dynamic logic. *Theoretical Computer Science*, 46(2–3):239–259, 1986.

[55] P. Hájek and J. Paris. Combinatorial principles concerning approximations of functions. *Archiv für Mathematische Logic und Grundlagenforschung*, 26(1):13–28, 1986.

[56] P. Jirků, M. Anděl, and P. Hájek. Expertní systém v diagnostice diabetu. *Časopis lékařů českých*, 125(2):49–52, 1986.

[57] P. Hájek. Partial conservativity revisited. *Commentationes Mathematicae Universitatis Carolinae*, 28(4):679–690, 1987.

[58] P. Hájek and A. Kučera. A contribution to recursion theory in fragments of arithmetic. *Journal of Symbolic Logic*, 52(3):888–889, 1987.

[59] P. Hájek. What does logic teach us? *Statistical Software Newsletter*, 14(2):67–68, 1988.

[60] P. Hájek, M. Hájková, T. Havránek, and M. Daniel. The expert system shell EQUANT-PC: Brief information. *Kybernetika*, 25(7):4–9, 1989.

[61] P. Hájek and A. Kučera. On recursion theory in $I\Sigma_1$. *Journal Symbolic of Logic*, 54(2):576–589, 1989.

[62] P. Clote, P. Hájek, and J. Paris. On some formalized conservation results in arithmetic. *Archive for Mathematical Logic*, 30(4):201–218, 1990.

[63] P. Hájek and M. Hájková. The expert system shell EQUANT-PC: Philosophy, structure and implementation. *Computational Statistics Quarterly*, 5(4):261–267, 1990.

[64] P. Hájek and F. Montagna. The logic of Π_1 conservativity. *Archive for Mathematical Logic*, 30(2):113–123, 1990.

[65] P. Hájek and J. J. Valdés. Algebraic foundations of uncertainty processing in rule-based expert systems (group-theoretic approach). *Computers and Artificial Intelligence*, 9(4):325–344, 1990.

[66] P. Hájek and V. Švejdar. A note on the normal form of closed formulas of interpretability logic. *Studia Logica*, 50(1):25–28, 1991.

[67] P. Hájek and J. J. Valdés. A generalized algebraic approach to uncertainty processing in rule-based expert systems (dempsteroids). *Computers and Artificial Intelligence*, 10(1):29–42, 1991.

[68] P. Hájek and D. Harmanec. An exercise in Dempster-Shafer theory. *International Journal of General Systems*, 20(2):137–142, 1992.

[69] P. Hájek and F. Montagna. The logic of Π_1 conservativity continued. *Archive for Mathematical Logic*, 32(1):57–63, 1992.

[70] P. Hájek. Epistemic entrenchment and arithmetical hierarchy. *Artificial Intelligence*, 62(1):79–87, 1993. Errata 65(1):191, 1994.

[71] P. Hájek. On logics of approximate reasoning. *Neural Network World*, 3(6):733–744, 1993.

[72] P. Hájek. Systems of conditional beliefs in Dempster-Shafer theory and expert systems. *International Journal of General Systems*, 22(2):113–124, 1994.

[73] P. Hájek and J. J. Valdés. An analysis of MYCIN-like expert systems. *Mathware & Soft Computing*, 1(1):45–68, 1994.

[74] D. Harmanec and P. Hájek. A qualitative belief logic. *International Journal of Uncertainty Fuzziness and Knowledge-Based Systems*, 2(2):227–236, 1994.

[75] P. Hájek. Fuzzy logic and arithmetical hierarchy. *Fuzzy Sets and Systems*, 73(3): 359–363, 1995.

[76] P. Hájek, D. Harmancová, and R. Verbrugge. A qualitative fuzzy possibilistic logic. *International Journal of Approximate Reasoning*, 12(1):1–19, 1995.

[77] P. Hájek, A. Sochorová, and J. Zvárová. GUHA for personal computers. *Computational Statistics and Data Analysis*, 19(2):149–153, 1995.

[78] P. Hájek. Getting belief functions from Kripke models. *International Journal of General Systems*, 24(3):325–327, 1996.

[79] P. Hájek. A remark on Bandler-Kohout products of relations. *International Journal of General Systems*, 25(2):165–166, 1996.

[80] P. Hájek, L. Godo, and F. Esteva. A complete many-valued logic with product-conjunction. *Archive for Mathematical Logic*, 35(3):191–208, 1996.

[81] P. Hájek and L. Kohout. Fuzzy implications and generalized quantifiers. *International Journal of Uncertainty Fuzziness and Knowledge-Based Systems*, 4(3): 225–233, 1996.

[82] M. Daniel, P. Hájek, and H. P. Nguyen. CADIAG-2 and MYCIN-like systems. *Artifical Intelligence in Medicine*, 9(3):241–259, 1997.

[83] P. Hájek. Fuzzy logic and arithmetical hierarchy II. *Studia Logica*, 58(1):129–141, 1997.

[84] P. Hájek and L. Godo. Deductive systems of fuzzy logic. Tutorial. *Tatra Mountains Mathematical Publications*, 13:35–66, 1997.

[85] P. Hájek and J. Paris. A dialogue on fuzzy logic. *Soft Computing*, 1(1):3–5, 1997.

[86] M. Baaz, P. Hájek, J. Krajíček, and D. Švejda. Embedding logics into product logic. *Studia Logica*, 61(1):35–47, 1998.

[87] P. Hájek. Basic fuzzy logic and BL-algebras. *Soft Computing*, 2(3):124–128, 1998.

[88] P. Hájek. Ten claims about fuzzy logic. *Soft Computing*, 2(1):14–15, 1998.

[89] P. Hájek. Ten questions and one problem on fuzzy logic. *Annals of Pure and Applied Logic*, 96(1):157–165, 1999.

[90] D. Dubois, P. Hájek, and H. Prade. Knowledge-driven versus data-driven logics. *Journal of Logic, Language, and Information*, 9(1):65–89, 2000.

[91] F. Esteva, L. Godo, P. Hájek, and M. Navara. Residuated fuzzy logics with an involutive negation. *Archive for Mathematical Logic*, 39(2):103–124, 2000.

[92] L. Godo, F. Esteva, and P. Hájek. Reasoning about probability using fuzzy logic. *Neural Network World*, 10(5):811–824, 2000.

[93] P. Hájek. Logics for data mining (GUHA rediviva). *Neural Network World*, 10(3):301–311, 2000.

[94] P. Hájek and D. Harmancová. A hedge for Gödel fuzzy logic. *International Journal of Uncertainty Fuzziness and Knowledge-Based Systems*, 8(4):495–498, 2000.

[95] P. Hájek, J. Paris, and J. Shepherdson. The liar paradox and fuzzy logics. *Journal of Symbolic Logic*, 65(1):339–346, 2000.

[96] P. Hájek, J. Paris, and J. Shepherdson. Rational Pavelka predicate logic is a conservative extension of Łukasiewicz predicate logic. *Journal of Symbolic Logic*, 65(2):669–682, 2000.

[97] P. Hájek. Fuzzy logic and arithmetical hierarchy III. *Studia Logica*, 68(1):129–142, 2001.

[98] P. Hájek. On very true. *Fuzzy Sets and Systems*, 124(3):329–333, 2001.

[99] P. Hájek. Properties of monadic fuzzy predicate logics. *Bulletin of Symbolic Logic*, 7(1):116–117, 2001.

[100] P. Hájek and J. Shepherdson. A note on the notion of truth in fuzzy logic. *Annals of Pure and Applied Logic*, 109(1):65–69, 2001.

[101] P. Hájek and S. Tulipani. Complexity of fuzzy probability logic. *Fundamenta Informaticae*, 45(3):207–213, 2001.

[102] M. Baaz, P. Hájek, F. Montagna, and H. Veith. Complexity of t-tautologies. *Annals of Pure and Applied Logic*, 113(1–3):3–11, 2002.

[103] J. M. Font and P. Hájek. On Łukasiewicz's four-valued modal logic. *Studia Logica*, 70(2):157–182, 2002.

[104] P. Hájek. Gödelův ontologický důkaz. *Teologické studie*, 3(2):111–113, 2002.

[105] P. Hájek. Monadic fuzzy predicate logic. *Studia Logica*, 71(2):165–175, 2002.

[106] P. Hájek. A new small emendation of Gödel's ontological proof. *Studia Logica*, 71(2):149–164, 2002.

[107] P. Hájek. Observations on the monoidal t-norm logic. *Fuzzy Sets and Systems*, 132(1):107–112, 2002.

[108] P. Hájek. Some hedges for continuous t-norm logics. *Neural Network World*, 12(2):159–164, 2002.

[109] F. Esteva, L. Godo, P. Hájek, and F. Montagna. Hoops and fuzzy logic. *Journal of Logic and Computation*, 13(4):532–555, 2003.

[110] L. Godo, P. Hájek, and F. Esteva. A fuzzy modal logic for belief functions. *Fundamenta Informaticae*, 57(2–4):127–146, 2003.

[111] P. Hájek. Basic fuzzy logic and BL-algebras II. *Soft Computing*, 7(3):179–183, 2003.

[112] P. Hájek. Briefly on the GUHA method of data mining. *Journal of Telecommunications and Information Technology*, (3):112–114, 2003.

[113] P. Hájek. Embedding standard BL-algebras into non-commutative pseudo-BL-algebras. *Tatra Mountains Mathematical Publications*, 27(3):125–130, 2003.

[114] P. Hájek. Fuzzy logics with noncommutative conjunctions. *Journal of Logic and Computation*, 13(4):469–479, 2003.

[115] P. Hájek. Mathematical fuzzy logic — state of art 2001. *Matemática Contemporanea*, 24:71–89, 2003.

[116] P. Hájek. Observations on non-commutative fuzzy logic. *Soft Computing*, 8(1): 38–43, 2003.

[117] P. Hájek and M. Holeňa. Formal logics of discovery and hypothesis formation by machine. *Theoretical Computer Science*, 292(2):345–357, 2003.

[118] P. Hájek and V. Novák. The sorites paradox and fuzzy logic. *International Journal of General Systems*, 32(4):373–383, 2003.

[119] D. Dubois, S. Gottwald, P. Hájek, J. Kacprzyk, and H. Prade. Terminological difficulties in fuzzy set theory - the case of "intuitionistic fuzzy sets". *Fuzzy Sets and Systems*, 156(3):485–491, 2005.

[120] P. Hájek. Arithmetical complexity of fuzzy predicate logics — a survey. *Soft Computing*, 9(12):935–941, 2005.

[121] P. Hájek. Fleas and fuzzy logic. *Journal of Multiple-Valued Logic and Soft Computing*, 11(1–2):137–152, 2005.

[122] P. Hájek. Making fuzzy description logic more general. *Fuzzy Sets and Systems*, 154(1):1–15, 2005.

[123] P. Hájek. A non-arithmetical Gödel logic. *Interest Group in Pure and Applied Logics. Logic Journal*, 13(4):435–441, 2005.

[124] P. Hájek. On arithmetic in the Cantor-Łukasiewicz fuzzy set theory. *Archive for Mathematical Logic*, 44(6):763–782, 2005.

[125] P. Hájek. Computational complexity of t-norm based propositional fuzzy logics with rational truth constants. *Fuzzy Sets and Systems*, 157(5):677–682, 2006.

[126] P. Hájek. Mathematical fuzzy logic — what it can learn from Mostowski and Rasiowa. *Studia Logica*, 84(1):51–62, 2006.

[127] P. Hájek. What is mathematical fuzzy logic. *Fuzzy Sets and Systems*, 157(5): 597–603, 2006.

[128] P. Hájek and P. Cintula. On theories and models in fuzzy predicate logics. *Journal of Symbolic Logic*, 71(3):863–880, 2006.

[129] P. Cintula, P. Hájek, and R. Horčík. Formal systems of fuzzy logic and their fragments. *Annals of Pure and Applied Logic*, 150(1–3):40–65, 2007.

[130] P. Hájek. Complexity of fuzzy probability logics II. *Fuzzy Sets and Systems*, 158(23):2605–2611, 2007.

[131] P. Hájek. Mathematical fuzzy logic and natural numbers. *Fundamenta Informaticae*, 81(1–3):155–163, 2007.

[132] P. Hájek. On witnessed models in fuzzy logic. *Mathematical Logic Quarterly*, 53(1):66–77, 2007.

[133] P. Hájek. On witnessed models in fuzzy logic II. *Mathematical Logic Quarterly*, 53(6):610–615, 2007.

[134] P. Hájek. On arithmetical complexity of fragments of prominent fuzzy predicate logics. *Soft Computing*, 12(4):335–340, 2008.

[135] P. Hájek. Ontological proofs of existence and non-existence. *Studia Logica*, 90(2):257–262, 2008.

[136] P. Hájek and R. Mesiar. On copulas, quasicopulas and fuzzy logic. *Soft Computing*, 12(12):1239–1243, 2008.

[137] P. Hájek and F. Montagna. A note on the first-order logic of complete BL-chains. *Mathematical Logic Quarterly*, 54(4):435–446, 2008.

[138] P. Cintula and P. Hájek. Complexity issues in axiomatic extensions of Łukasiewicz logic. *Journal of Logic and Computation*, 19(2):245–260, 2009.

[139] P. Hájek. Arithmetical complexity of fuzzy predicate logics — a survey II. *Annals of Pure and Applied Logic*, 161(2):212–219, 2009.

[140] P. Hájek. On vagueness, truth values and fuzzy logics. *Studia Logica*, 91(2):367–382, 2009.

[141] P. Cintula and P. Hájek. Triangular norm based predicate fuzzy logics. *Fuzzy Sets and Systems*, 161(3):311–346, 2010.

[142] P. Hájek, M. Holeňa, and J. Rauch. The GUHA method and its meaning for data mining. *Journal of Computer and System Sciences*, 76(1):34–48, 2010.

CHAPTERS IN BOOKS

[1] P. Hájek, T. Havránek, and M. K. Chytil. Teorie automatizovaného výzkumu. In *Metoda GUHA*, pages 65–94. ČSVTS, České Budějovice, 1976.

[2] R. Jiroušek, P. Hájek, and T. Havránek. Influenční diagramy. Integrace znalostí v pravděpodobnostních modelech expertních systémů. Experimentální systém INES. In *Zpracování nejisté informace v expertních systémech*, pages 26–54. MÚ ČSAV ČSVTS, Prague, 1989.

[3] B. Richards, P. Hájek, J. Zvárová, R. Jiroušek, F. Esteva, and M. Lubicz. MUM: managing uncertainty in medicine. In M. F. Laires, M. J. Ladeira, and J. P. Christensen, editors, *Health in the New Communications Age. Health Care Telematics for the 21st Century*, volume 24 of *Studies in Health Technology and Informatics*, pages 247–250. IOS Press, Amsterdam, 1995.

[4] P. Hájek. Gödelův důkaz existence Boha. In J. Malina and J. Novotný, editors, *Kurt Gödel*, pages 117–128. Nadace Universitas Masarykiane, Brno, 1996.

[5] P. Hájek. Matematik a logika. In J. Malina and J. Novotný, editors, *Kurt Gödel*, pages 74–92. Nadace Universitas Masarykiane, Brno, 1996.

[6] P. Hájek. Logics of knowing and believing. In U. Ratsch, M. M. Richter, and I. O. Stamatescu, editors, *Intelligence and Artificial Intelligence. An Interdisciplinary Debate*, pages 96–108. Springer, Berlin, 1998.

[7] L. Godo and P. Hájek. Fuzzy inference as deduction. In W. Carnielli, editor, *Multi-Valued Logics*, pages 37–60. Hermes Science Publications, Paris, 1999.

[8] L. Godo and P. Hájek. A note on fuzzy inference as deduction. In D. Dubois, H. Prade, and E. P. Klement, editors, *Fuzzy Sets, Logics and Reasoning about Knowledge*, volume 15 of *Applied Logic Series*, pages 237–241. Kluwer Academic Publishers, Dordrecht, 1999.

[9] P. Hájek. Fuzzy predicate calculus and fuzzy rules. In A. Da Ruan and E. E. Kerre, editors, *Fuzzy If-Then Rules in Computational Intelligence*, volume 553 of *The Kluwer International Series in Engineering and Computer Science*, pages 27–35. Kluwer Academic Publishers, Boston, 2000.

[10] P. Hájek. Many. In V. Novák and I. Perfilieva, editors, *Discovering World with Fuzzy Logic*, volume 57 of *Studies in Fuzziness and Soft Computing*, pages 302–309. Physica Verlag, Heidelberg, 2000.

[11] P. Hájek. On the metamathematics of fuzzy logic. In V. Novák and I. Perfilieva, editors, *Discovering World with Fuzzy Logic*, volume 57 of *Studies in Fuzziness and Soft Computing*, pages 155–174. Physica Verlag, Heidelberg, 2000.

[12] P. Hájek. Der Mathematiker und die Frage der Existenz Gottes. In B. Buldt, E. Köhler, M. Stöltzner, P. Weibel, C. Klein, and W. Depauli-Schimanowich-Göttig, editors, *Kurt Gödel. Wahrheit und Beweisbarkeit*, pages 325–336. öbv & htp, Vienna, 2002.

[13] P. Hájek. Gödelův ontologický důkaz. In I. Kišš, editor, *Matematika a teológia*, pages 66–69. Evanjelická bohoslovecká fakulta Univerzity Komenského, Bratislava, 2002.

[14] P. Hájek. Why fuzzy logic? In J. Dale, editor, *A Companion to Philosophical Logic*, volume 22 of *Blackwell Companions to Philosophy*, pages 595–605. Blackwell Publishers, Massachusetts, 2002.

[15] P. Hájek. Deduktivní systémy fuzzy logiky. In V. Mařík, O. Štěpánková, and J. Lažanský, editors, *Umělá inteligence 4*, pages 71–92. Academia, Prague, 2003.

[16] P. Hájek and Z. Haniková. A development of set theory in fuzzy logic. In M. Fitting and E. Orłowska, editors, *Beyond Two: Theory and Applications of Multiple-Valued Logic*, volume 114 of *Studies in Fuzziness and Soft Computing*, pages 273–285. Physica Verlag, Heidelberg, 2003.

[17] P. Hájek, M. Holeňa, and J. Rauch. The GUHA method and foundations of (relational) data mining. In H. de Swart, E. Orłowska, G. Schmidt, and M. Roubens, editors, *Theory and Applications of Relational Structures as Knowledge Instruments. COST action 274, TARSKI. Revised Papers*, pages 17–37. Springer, Berlin, 2003.

[18] P. Hájek. Fuzzy logic and arithmetical hierarchy IV. In V. Hendricks, F. Neuhaus, S. A. Pedersen, U. Scheffler, and H. Wansing, editors, *First-Order Logic Revised*, pages 107–115. Logos Verlag, Berlin, 2004.

[19] S. Gottwald and P. Hájek. Triangular norm-based mathematical fuzzy logics. In E. P. Klement and R. Mesiar, editors, *Local, Algebraic, Analytic and Probabilistic Aspects of Triangular Norms*, pages 275–299. Elsevier, Amsterdam, 2005.

[20] P. Hájek. What does mathematical fuzzy logic offer to description logic? In E. Sanchez, editor, *Fuzzy Logic and the Semantic Web*, pages 91–100. Elsevier, Amsterdam, 2006.

[21] P. Hájek. Deductive systems of fuzzy logic. In A. Gupta, R. Parikh, and J. van Benthem, editors, *Logic at the Crossroads: An Interdisciplinary View*, pages 60–74. Allied Publishers PVT, New Delhi, 2007.

[22] P. Hájek, V. W. Marek, and P. Vopěnka. Mostowski and Czech-Polish cooperation in mathematical logic. In A. Ehrenfeucht, V. W. Marek, and M. Srebrny, editors, *Andrzej Mostowski and Foundational Studies*, pages 403–405. IOS Press, Amsterdam, 2008.

ABSTRACTS AND PROCEEDING PAPERS

[1] P. Hájek. On semisets. In *Logic Colloquium '69. Proceedings of the Annual European Summer Meeting of the Association for Symbolic Logic*, pages 67–76, Amsterdam, 1971. North-Holland.

[2] P. Hájek. Sets, semisets, models. In Dana S. Scott and Thomas J. Jech, editors, *Axiomatic Set Theory*, volume XIII — Part 1 of *Proceedings of Symposia in Pure Mathematics*, pages 67–81. AMS, 1971.

[3] P. Hájek. Some logical problems of automated research. In *Proceedings of the Symposium on Mathematical Foundations of Computer Science*, pages 85–93, Bratislava, 1973. Mathematical Institute of the Slovak Academy of Sciences.

[4] P. Hájek. On logics of discovery. In J. Bečvář, editor, *Mathematical Foundations of Computer Science 1975*, volume 32 of *Lecture Notes in Computer Science*, pages 30–45, Berlin, 1975. Springer.

[5] P. Hájek. Some remarks on observational model-theoretic languages. In *Set Theory and Hierarchy Theory. A Memorial Tribute to Andrzej Mostowski*, volume 537 of *Lecture Notes in Mathematics*, pages 335–345, Berlin, 1976. Springer.

[6] P. Hájek. Arithmetical complexity of some problems in computer science. In *Mathematical Foundations of Computer Science*, volume 53 of *Lecture Notes in Computer Science*, pages 282–297, Berlin, 1977. Springer.

[7] P. Hájek. Generalized quantifiers and finite sets. In *Set Theory and Hierarchy Theory.*, number 14 in Scientific Papers of the Institute of Mathematics of Wroclaw Technical University, pages 91–104, Wroclaw, 1977. Polytechnical Edition.

[8] P. Hájek. Some results on degrees of constructibility. In *Higher Set Theory*, volume 669 of *Lecture Notes in Mathematics*, pages 55–72, Berlin, 1978. Springer.

[9] P. Hájek. On partially conservative extensions of arithmetic. In *Logic Colloquium '78. Proceedings of the Annual European Summer Meeting of the Association for Symbolic Logic*, pages 225–234, Amsterdam, 1979. North-Holland.

[10] P. Hájek and T. Havránek. O projektu GUHA-80. In *SOFSEM'79*, pages 344–346, Bratislava, 1979. Výskumné výpočtové stredisko.

[11] P. Hájek, P. Kalášek, and P. Kůrka. Systém dynamické logiky a programování. In *SOFSEM'79*, pages 79–107, Bratislava, 1979. Výskumné výpočtové stredisko.

[12] P. Hájek and P. Pudlák. Two orderings of the class of all countable models of Peano arithmetic. In *Model Theory of Algebra and Arithmetic*, volume 834 of *Lecture Notes in Mathematics*, pages 174–185, Berlin, 1980. Springer.

[13] P. Hájek. Making dynamic logic first-order. In *Mathematical Foundations of Computer Science*, volume 118 of *Lecture Notes in Computer Science*, pages 287–295, Berlin, 1981. Springer.

[14] P. Hájek. Applying artificial intelligence to data analysis. In *ECAI '82*, pages 149–150, Vienna, 1982. Physica Verlag.

[15] P. Hájek and J. Ivánek. Artificial intelligence and data analysis. In H. Caussinus, P. Ettinger, and R. Tomassone, editors, *COMPSTAT '82*, pages 54–60, Vienna, 1982. Physica Verlag.

[16] P. Hájek, B. Louvar, D. Pokorný, J. Rauch, and E. Tschernoster. Metoda GUHA - její cíle a prostředky. In *SOFSEM '82*, pages 59–84, Brno, 1982. Ústav výpočetní techniky UJEP.

[17] I. M. Havel and P. Hájek. Filozofické aspekty strojového myšlení. Úvahy nad knihou D. R. Hofstadtera "Gödel, Escher, Bach". In *SOFSEM '82*, pages 171–211, Brno, 1982. Ústav výpočetní techniky UJEP.

[18] P. Hájek and T. Havránek. Logic, statistics and computers. In *Logic in the 20th Century*, pages 56–76, Milano, 1983. Scientia.

[19] P. Hájek. Combining functions in consulting systems and dependence of premises (a remark). In *Artificial Intelligence and Information Control Systems of Robots*, pages 163–167, Amsterdam, 1984. North-Holland.

[20] P. Hájek. A new generation of the GUHA procedure ASSOC. In *COMPSTAT '84*, pages 360–365, Vienna, 1984. Physica Verlag.

[21] P. Hájek. On a new motion of partial conservativity. In *Computation and Proof Theory*, volume 1104 of *Lecture Notes in Mathematics*, pages 217–232, Berlin, 1984. Springer.

[22] P. Hájek. Teorie šíření nejisté informace v konsultačních systémech. In *Expertní systémy - principy, realizace, užití*, pages 70–80, Prague, 1984. ČSVTS - FEL ČVUT.

[23] P. Hájek and M. Hájková. Konsultační systém EQUANT - stručný popis a manuál. In *Expertní systémy - principy, realizace, užití*, pages 144–159, Prague, 1984. ČSVTS - FEL ČVUT.

[24] P. Hájek and T. Havránek. Automatická tvorba hypotéz - mezi analýzou dat a umělou inteligencí. In *Metody umělé inteligence a expertni systémy II*, pages 84–98, Prague, 1985. ČSVTS.

[25] P. Hájek and P. Jirků. Lesk a bída expertních systémů. In *SOFSEM '85*, pages 131–163, Brno, 1985. Ústav výpočetní techniky UJEP.

[26] P. Hájek. Ke struktuře prázdných konsultačních systémů založených na pravidlech. In *Teoretické základy bází znalostí*, pages 112–125, Prague, 1986. PUDIS.

[27] P. Hájek. Some conservativeness results for non-standard dynamic logic. In J. Demetrovics, editor, *Proceedings of Conference on Algebra, Combinatorics and Logic in Computer Science*, pages 443–449, Amsterdam, 1986. North-Holland.

[28] P. Hájek. Úskalí expertních systémů. In *DATASEM '86*, pages 97–108, Prague, 1986. Dům techniky ČSVTS.

[29] P. Hájek. Dempster-Shaferova teorie evidence a expertní systémy. In *Metody umělé inteligence a expertní systémy III*, pages 54–61, Prague, 1987. ČSVTS.

[30] P. Hájek. Expertní systémy a logika. In *Expertné systémy*, pages 42–49, Kežmarok, 1987.

[31] P. Hájek. Logic and plausible inference in expert systems. In *Inductive Reasoning: Managing Empirical Information in AI-systems*, Roskilde, 1987.

[32] P. Hájek. Partial conservativity revisited (abstract). *Journal of Symbolic Logic*, 52(3):1066, 1987.

[33] P. Hájek and T. Havránek. A note on the independence assumption underlying subjective Bayesian updating in expert systems. In I. Plander, editor, *Artificial Intelligence and Information Control Systems of Robots*, pages 41–47, Amsterdam, 1987. Elsevier Science Publishers.

[34] P. Hájek and J. J. Valdés. Algebraical foundation of uncertainty processing in rule based expert systems I. In *Sistemy obrabotki znanij i izobraženij*, 1987.

[35] P. Hájek and J. J. Valdés. Algebraical foundations of uncertainty processing in rule-based expert systems II. In *Aplikace umělé inteligence AI'87*, pages 9–14, Prague, 1987. ČSVTS - Ústav pro informaci a řízení v kultuře.

[36] M. Hájková, P. Hájek, and T. Havránek. Expertní systém EQUANT. In *Expertní systémy*, pages 132–137, Bratislava, 1987. ČSVTS.

[37] M. Hájková, P. Hájek, and T. Havránek. Konsultační systém EQUANT - stav 1987. In *Metody umělé inteligence a expertní systémy III*, pages 134–140, Prague, 1987. ČSVTS - FEL ČVUT.

[38] P. Hájek. Towards a probabilistic analysis of MYCIN-like expert systems. In D. Edwards and N. E. Raun, editors, *COMPSTAT '88. Proceedings in Computational Statistics*, pages 117–121, Heidelberg, 1988. Physica Verlag.

[39] P. Hájek. Towards a probabilistic analysis of MYCIN-like expert systems. In *Workshop on Uncertainty Processing in Expert Systems*, Prague, 1988. Institute of Information Theory and Automation CAS.

[40] P. Hájek, M. Hájková, T. Havránek, and M. Daniel. The expert system shell EQUANT-PC - brief description. In *Workshop on Uncertainty Processing in Expert Systems*, Prague, 1988. Institute of Information Theory and Automation CAS.

[41] P. Hájek. Towards a probabilistic analysis of MYCIN-like expert systems II. In *Artificial Intelligence and Information-Control Systems of Robots*, pages 45–54, Amsterdam, 1989. North-Holland.

[42] P. Hájek. Úvod do logiky. In *SOFSEM '89*, pages 81–92, Brno, 1989. Ústav výpočetní techniky UJEP.

[43] P. Hájek, M. Hájková, T. Havránek, and M. Daniel. Prázdný expertní systém EQUANT-PC. In *Metody umělé inteligence a expertní systémy IV*, pages 90–95, Prague, 1989. ČSVTS - FEL ČVUT.

[44] M. Hájková, P. Hájek, T. Havránek, and M. Daniel. Prázdný expertní systém EQUANT-PC. In *Dni novej techniky elektronického výskumu*, pages 60–67, Prague, 1989. TESLA VÚST.

[45] P. Hájek. A note on belief functions in MYCIN-like systems. In *Aplikace umělé inteligence AI'90*, pages 19–26, Prague, 1990. Ústav pro informační systémy v kultuře.

[46] P. Hájek and T. Havránek. Moderní metody zpracování nejisté informace v expertních systémech. In *SOFSEM'90*, pages 137–156, Brno, 1990. Ústav výpočetní techniky UJEP.

[47] P. Hájek. Dempster-Shafer theory - what it is and how (not) to use it. In *SOFSEM'92*, pages 19–24, Brno, 1992. Ústav výpočetní techniky MU.

[48] P. Hájek. Deriving Dempster's rule. In *IPMU'92 Proceedings. International Conference on Information Processing and Management of Uncertainty in Knowledge-Based Systems*, pages 73–75, Valldemossa, 1992. Universitat de les Illes Balears.

[49] P. Hájek and D. Harmanec. On belief functions (the present state of Dempster-Shafer theory). In V. Mařík, editor, *Advanced in Artificial Intelligence*, pages 286–307, Berlin, 1992. Springer.

[50] K. Bendová and P. Hájek. Possibilistic logic as tense logic. In P. Carete and M. Singh, editors, *Qualitative Reasoning and Decision Technologies*, pages 441–450, Barcelona, 1993. CIMNE Barcelona.

[51] P. Hájek. Deriving Dempster's rule. In B. Bouchon-Meunier, L. Valverde, and R. R. Yager, editors, *Uncertainty in Intelligent Systems*, pages 75–83, Amsterdam, 1993. North-Holland.

[52] P. Hájek. Interpretability and fragments of arithmetic. In P. Clote and J. Krajíček, editors, *Arithmetic, Proof Theory and Computational Complexity*, pages 185–196, Oxford, 1993. Clarendon Press.

[53] P. Hájek and D. Harmancová. A comparative fuzzy modal logic. In E. P. Klement, editor, *Fuzzy Logic in Artificial Intelligence FLAI'93*, volume 695 of *Lecture Notes in Artificial Intelligence*, pages 27–34, Berlin, 1993. Springer.

[54] P. Hájek, F. Montagna, and P. Pudlák. Abbreviating proofs using metamathematical rules. In P. Clote and J. Krajíček, editors, *Arithmetic, Proof Theory and Computational Complexity*, pages 197–221, Oxford, 1993. Clarendon Press.

[55] P. Hájek. On logics of approximate reasoning. In M. Masuch and L. Pólos, editors, *Knowledge Representation and Reasoning Under Uncertainty. Logic at Work*, volume 808 of *Lecture Notes in Artificial Intelligence*, pages 17–29, Heidelberg, 1994. Springer.

[56] P. Hájek. On logics of approximate reasoning. In *Modelling' 94. Abstracts*, page 23, Prague, 1994. ICS AS CR.

[57] P. Hájek. Possibilistic logic as interpretability logic. In *IPMU. Information Processing and Management of Uncertainty in Knowlegde-Based Systems*, pages 815–819, Paris, 1994. Cité Internationale Universitaire.

[58] P. Hájek, D. Harmancová, F. Esteva, P. García, and L. Godo. On modal logics of qualitative possibility in a fuzzy setting. In L. de Mantaras and D. Poole, editors, *Uncertainty in Artificial Intelligence. Proceedings of the 10th Conference*, pages 278–285, San Francisco, 1994. Morgan Kaufmann.

[59] J. Zvárová and P. Hájek. Medical decision making and uncertainty management. In P. Barakova and J. P. Christenson, editors, *Knowledge and Decisions in Health Telematics*, pages 224–225, Amsterdam, 1994. IOS Press.

[60] P. Hájek. Fuzzy logic as logic. In G. Coletti, D. Dubois, and R. Scozzafava, editors, *Mathematical Models of Handling Partial Knowledge in Artificial Intelligence*, pages 21–30, New York, 1995. Plenum Press.

[61] P. Hájek. Fuzzy logic from the logical point of view. In M. Bartošek, J. Staudek, and J. Wiedermann, editors, *SOFSEM'95: Theory and Practice of Informatics*, volume 1012 of *Lecture Notes in Computer Science*, pages 31–49, Berlin, 1995. Springer.

[62] P. Hájek. O Gödelově důkazu existence Boha. In J. Bečvář, E. Fuchs, D. Hrubý, and A. Trojánek, editors, *Filosofické otázky matematiky a fyziky*, pages 5–15, Brno, 1995. JČMF.

[63] P. Hájek. On logics of approximate reasoning II. In G. della Riccia, R. Kruse, and R. Viertl, editors, *Proceedings of the ISSEK94 Workshop on Mathematical and Statistical Methods in Artificial Intelligence*, volume 363 of *Courses and Lectures*, pages 147–156, Vienna, 1995. Springer.

[64] P. Hájek. Possibilistic logic as interpretability logic. In B. Bouchon-Meunier, R. R. Yager, and L. A. Zadeh, editors, *Advances in Intelligent Computing - IPMU'94*, volume 945 of *Lecture Notes in Computer Science*, pages 243–280, Berlin, 1995. Springer.

[65] P. Hájek, L. Godo, and F. Esteva. Fuzzy logic and probability. In P. Besnard and S. Hanks, editors, *Uncertainty in Artificial Intelligence. Proceedings of the 11th Conference*, pages 237–244, San Francisco, 1995. Morgan Kaufmann.

[66] P. Hájek and D. Harmancová. Medical fuzzy expert systems and reasoning about beliefs. In P. Barahona, M. Stefanelli, and J. Wyatt, editors, *Artificial Intelligence in Medicine*, pages 403–404, Berlin, 1995. Springer.

[67] L. Godo and P. Hájek. On deduction in Zadeh's fuzzy logic. In *Proceedings IPMU'96. Information Processing and Management of Uncertainty in Knowledge-Based Systems*, pages 991–996, Granada, 1996. Universidad de Granada.

[68] P. Hájek. Magari and others on Gödel's ontological proof. In A. Ursini and P. Aglianò, editors, *Logic and Algebra*, volume 180 of *Lecture Notes in Pure and Applied Mathematics*, pages 125–136, New York, 1996. Marcel Dekker.

[69] P. Hájek and D. Harmancová. A many-valued modal logic. In *Proceedings IPMU'96. Information Processing and Management of Uncertainty in Knowledge-Based Systems*, pages 1021–1024, Granada, 1996. Universidad de Granada.

[70] P. Hájek. Logic in Central and Eastern Europe. In M.-L. Dalla Chiara, editor, *Logic and Scientific Methods*, page 449, Dordrecht, 1997. Kluwer Academic Publishers.

[71] P. Hájek and D. Švejda. A strong completeness theorem for finitely axiomatized fuzzy theories. In R. Mesiar and V. Novák, editors, *Fuzzy Sets Theory and Applications*, volume 12 of *Tatra Mountains Mathematical Publications*, pages 213–219, Bratislava, 1997. Mathematical Institute of the Slovak Academy of Sciences.

[72] M. Daniel and P. Hájek. Theoretical comparison of inference in CADIAG-2 and MYCIN-like systems. In *Fourth International Conference on Fuzzy Sets Theory and Its Applications*, pages 24–25, Liptovský Mikuláš, 1998. Military Academy.

[73] P. Hájek. Logics for data mining (GUHA rediviva). In *Workshop of Japanese Society for Artificial Intelligence*, pages 27–34, Tokyo, 1998. JSAI.

[74] P. Hájek and M. Holeňa. Formal logics of discovery and hypothesis formation by machine. In S. Arikawa and H. Motoda, editors, *Discovery Science*, volume 1532 of *Lecture Notes in Artificial Intelligence*, pages 291–302, Berlin, 1998. Springer.

[75] P. Hájek, J. Paris, and J. Shepherdson. The liar's paradox and fuzzy logics. In *Fourth International Conference on Fuzzy Sets Theory and Its Applications*, page 52, Liptovský Mikuláš, 1998. Military Academy.

[76] P. Hájek. Fuzzy predicate calculus and fuzzy rules. In G. Brewka, R. Der, S. Gottwald, and A. Schierwagen, editors, *Fuzzy-Neuro Systems 99*, pages 1–8, Leipzig, 1999. Leipziger Universitätsverlag.

[77] P. Hájek. Trakhtenbrot theorem and fuzzy logic. In G. Gottlob, E. Grandjean, and K. Seyr, editors, *Computer Science Logic*, volume 1584 of *Lecture Notes in Computer Science*, pages 1–8, Berlin, 1999. Springer.

[78] P. Hájek. Function symbols in fuzzy predicate logic. In *Proceedings of East West Fuzzy Colloquium 2000*, Wissenschaftliche Berichte, pages 2–8, Zittau, Görlitz, 2000. IPM.

[79] P. Hájek. Mathematical fuzzy logic - state of art. In S. R. Buss, P. Hájek, and P. Pudlák, editors, *Logic Colloquium '98. Proceedings of the Annual European Summer Meeting of the Association for Symbolic Logic*, volume 13 of *Lecture Notes in Logic*, pages 197–205, Natick, 2000. AK Peters, Ltd.

[80] L. Godo, F. Esteva, and P. Hájek. A fuzzy modal logic for belief functions. In *Proceedings of the 17th International Joint Conference on Artificial Intelligence*, pages 723–729, Seattle, 2001. Morgan Kaufmann.

[81] P. Hájek. The GUHA method and mining association rules. In L. Kuncheva, editor, *Computational Intelligence: Methods and Applications*, pages 533–539, Canada, 2001. ICSC Academic Press.

[82] P. Hájek. Relations in GUHA style data mining. In H. C. M. de Swart, editor, *Proceedings of the sixth international workshop on Relational Methods in Computer Science RelMiCS'6 and the first international workshop of COST Action 274 Theory and Application of Relational Structures as Knowledge Instruments TARSKI*, pages 91–96, Brabant, 2001. Katholieke Universiteit.

[83] P. Hájek. Relations in GUHA style data mining. In H. C. M. de Swart, editor, *Relational Methods in Computer Science*, volume 2561 of *Lecture Notes in Computer Science*, pages 81–87, Berlin, 2001. Springer.

[84] P. Hájek and Z. Haniková. A contribution on the set theory in fuzzy logic. In *Proceedings of the 2001 ASL European Summer Meeting*, Vienna, 2001. Technische Universität.

[85] P. Hájek and Z. Haniková. A set theory within fuzzy logic. In *Multiple-Valued Logic. Proceedings*, pages 319–323, Los Alamitos, 2001. IEEE Computer Society.

[86] P. Hájek. Metoda GUHA - současný stav. In J. Antoch, G. Dohnal, and J. Klaschka, editors, *Robust'2002. Sborník prací Dvanácté zimní školy JČMF*, pages 133–135, Prague, 2002. JČMF.

[87] P. Hájek, T. Feglar, J. Rauch, and D. Coufal. The GUHA method, data preprocessing and mining. Position paper. In *Database Technologies for Data Mining. Working Notes*, pages 29–34, Prague, 2002.

[88] P. Hájek and Z. Haniková. A contribution on the set theory in fuzzy logic. *Bulletin of Symbolic Logic*, 8(1):137, 2002.

[89] P. Hájek, M. Holeňa, J. Rauch, T. Feglar, and J. Svitok. The GUHA method and foundations of (relational) data mining. In *TARSKI Meeting of the COST 274. Book of Abstracts*, 2002.

[90] P. Hájek. Fleas and fuzzy logic: a survey. In M. Wagenknecht and R. Hampel, editors, *Fuzzy Logic and Technology. Proceeding of the 3rd EUSFLAT Conference*, pages 599–603, Zittau, Görlitz, 2003. EUSFLAT.

[91] P. Hájek. On generalized quantifiers, finite sets and data mining. In M. A. Klopotek, S. T. Wierzchoń, and K. Trojanowski, editors, *Intelligent Information Processing*

and Web Mining, Advances in Soft Computing, pages 489–496, Berlin, 2003. Physica Verlag.

[92] P. Hájek. Relations in GUHA style data mining II. In R. Berghammer and B. Möller, editors, *Relational Methods in Computer Science*, pages 242–247, Kiel, 2003. Christian-Albrechts-Universität.

[93] P. Hájek. Metoda GUHA v minulém století a dnes. In V. Snášel, editor, *Znalosti 2004*, pages 10–20, Ostrava, 2004. FEI VŠB.

[94] P. Hájek. On arithmetic in the Cantor-Łukasiewicz fuzzy set theory. In *Logic Colloquium '04. Book of Abstracts*, page 178, Torino, 2004. University of Torino.

[95] P. Hájek. Relations in GUHA style data mining II. In R. Berghammer and B. Möller, editors, *Relational and Kleene-Algebraic Methods in Computer Science*, volume 3051 of *Lecture Notes in Computer Science*, pages 163–170, Berlin, 2004. Springer.

[96] P. Hájek. A true unprovable formula of fuzzy predicate logic. In W. Lenski, editor, *Logic versus Approximation. Essays Dedicated to Michael M. Richter on the Occasion of his 65th Birthday*, volume 3075 of *Lecture Notes in Computer Science*, pages 1–5, Berlin, 2004. Springer.

[97] P. Hájek, J. Rauch, D. Coufal, and T. Feglar. The GUHA method, data preprocessing and mining. In R. Meo, P. L. Lanzi, and M. Klemettinen, editors, *Database Support for Data Mining Applications: Discovering Knowledge with Inductive Queries*, volume 2682 of *Lecture Notes in Artificial Intelligence*, pages 135–153, Berlin, 2004. Springer.

[98] P. Hájek and J. Ševčík. On fuzzy predicate calculi with non-commutative conjunction. In *Proceedings of East West Fuzzy Colloquium 2004*, pages 103–110, Zittau, Görlitz, 2004. IPM.

[99] P. Hájek. Complexity of t-norm based fuzzy logics with rational truth constants. In V. Novák and M. Štěpnička, editors, *Proceedings of The Logic of Soft Computing 4*, pages 45–46, Ostrava, 2005. University of Ostrava.

[100] P. Hájek. What is mathematical fuzzy logic. In *Forging New Frontiers*, pages 199–200, Berkeley, 2005. University of California.

[101] P. Hájek and P. Cintula. On theories and models in fuzzy predicate logics. In S. Gottwald, P. Hájek, U. Höhle, and E. P. Klement, editors, *Fuzzy Logics and Related Structures*, pages 55–58, Linz, 2005. Johannes Kepler Universität.

[102] P. Cintula, P. Hájek, and R. Horčík. Fragments of prominent fuzzy logics. In *Ordered Structures in Many Valued Logics*, page 15, Salerno, 2006. Universita di Salerno.

[103] P. Hájek. Fuzzy logika v kontextu matematické logiky. In A. Trojánek, J. Novotný, and D. Hrubý, editors, *Matematika, fyzika - minulost, současnost*, pages 41–50, Brno, 2006. VUTIUM.

[104] P. Hájek. Mathematical fuzzy logic and natural numbers. In S. Gottwald, P. Hájek, and M. Ojeda-Aciego, editors, *The Logic of Soft Computing*, pages 58–60, Malaga, 2006. Universidad de Malaga.

[105] P. Hájek. On witnessed models in fuzzy predicate logic. In *Algebraic and Logical Foundations of Many-Valued Reasoning*, page 24, Milano, 2006. Universita degli Studi di Milano.

[106] P. Hájek. Complexity of fuzzy predicate logics with witnessed semantics. In *Logic Colloquium '07. Book of Abstracts*, page 49, Wroclaw, 2007. Universitet Wroclawski.

[107] P. Hájek. Mathematical fuzzy logic - a survey and some news. In V. Novák and M. Štěpnička, editors, *New Dimensions in Fuzzy Logic and Related Technologies. Proceeding of the 5th EUSFLAT Conference*, page 21, Ostrava, 2007. University of Ostrava.

[108] P. Hájek. On fuzzy theories with crisp sentences. In S. Aguzzoli, A. Ciabattoni, B. Gerla, C. Manara, and V. Marra, editors, *Algebraic and Proof-Theoretic Aspects of Non-classical Logics. Papers in Honor of Daniele Mundici on the Occasion of His 60th Birthday*, volume 4460 of *Lecture Notes in Artificial Intelligence*, pages 194–200, Berlin, 2007. Springer.

[109] P. Hájek and P. Cintula. First-order fuzzy logics: Recent developments. In J. Y. Béziau, H. He, A. Costa-Leite, and X. Z. M. Zhong, editors, *UNILOG'07 Handbook*, pages 45–46, Xian, 2007. Northwestern Polytechnical University.
[110] P. Hájek. Complexity issues in axiomatic extensions of Łukasiewicz logic. In *Residuated Structures: Algebra and Logic*, page 10, Buenos Aires, 2008. Instituto Argentino de Matemática.
[111] P. Hájek. Complexity of fuzzy predicate logics with witnessed semantics. *Bulletin of Symbolic Logic*, 14(2):141, 2008.
[112] P. Hájek. (Fuzzy) logika v širokém a úzkém slova smyslu. In *Logika mezi filosofií a matematikou*, page 10, Prague, 2008. Katedra logiky FF UK.
[113] P. Hájek. Mathematical fuzzy predicate logic - the present state of development. In *Logic, Algebra and Truth Degrees*, page 7, Siena, 2008. University of Siena.
[114] P. Hájek. An observation on (un)decidable theories in fuzzy logics. In U. Bodenhofer, B. De Baets, E. P. Klement, and S. Saminger-Platz, editors, *The Legacy of 30 Seminars - Where Do We Stand and Where Do We Go?* pages 55–56, Linz, 2009. Johannes Kepler Universität.
[115] P. Hájek. Towards metamathematics of weak arithmetics over fuzzy logic. In *Non-Classical Mathematics 2009*, pages 27–28, Prague, 2009. ICS AS CR.
[116] P. Hájek and Z. Haniková. A ZF-like theory in fuzzy logic. In *Non-Classical Mathematics 2009*, page 29, Prague, 2009. ICS AS CR.

TECHNICAL REPORTS

[1] P. Hájek and M. Hájková. New enhancements in GUHA software. Technical Report 8, Mathematical Institute CSAS, Prague, 1984.
[2] P. Hájek and J. J. Valdés. El metodo GUHA y sus aplicaciones en geofisica y geoquimia. Technical Report 9, Mathematical Institute CSAS, Prague, 1984. In Spanish.
[3] P. Hájek and J. J. Valdés. Algebraical foundations of uncertainty processing in rule-based expert systems. Technical Report 28, Mathematical Institute CSAS, Prague, 1987.
[4] A. Sochorová, J. Rauch, P. Hájek, K. Bendová, D. Harmancová, K. Hlaváčková, and J. Zvárová. GUHA'93. Technical report, ÚIVT AV ČR, Prague, 1993. In Czech.
[5] P. Hájek and D. Harmancová. Medical fuzzy expert systems and reasoning about beliefs. Technical Report V-632, ICS AS CR, Prague, 1994.
[6] P. Hájek, J. Zvárová, and F. Esteva. Managing uncertainty in medicine - state of art report. Technical report, ICS AS CR, Prague, 1994.
[7] O. Kufudaki, P. Žák, D. Harmancová, P. Hájek, J. Štuller, J. Prokop, S. Javorský, and D. Hrycej. Zpráva o výsledcích výzkumu pro centrum klastrové medicíny za rok 1993. Technical Report V-573, ÚIVT AV ČR, Prague, 1994. In Czech.
[8] P. Hájek. Possibilistic logic as interpretability logic. Technical Report V-630, ICS AS CR, Prague, 1995.
[9] P. Hájek and H.P. Nguyen. Möbius transform for CADIAG-2. Technical Report V-650, ICS AS CR, Prague, 1995.
[10] P. Hájek. Metamathematics of fuzzy logic. Technical Report V-682, ICS AS CR, Prague, 1996.
[11] D. Dubois, P. Hájek, and H. Prade. Knowledge-driven versus data-driven logics. Technical Report IRIT/97-46-R, Institut de Recherche en Informatique de Toulouse, Toulouse, 1997.
[12] P. Hájek. Basic fuzzy logic and BL-algebras. Technical Report V-736, ISC AS CR, Prague, 1997.
[13] P. Hájek. Trakhtenbrot theorem and fuzzy logic. Technical Report V-737, ISC AS CR, Prague, 1997.
[14] P. Hájek and L. Godo. Deductive systems of fuzzy logic (a tutorial). Technical Report V-707, ISC AS CR, Prague, 1997.

[15] P. Hájek and B. Richards. Managing uncertainty in medicine. Final report. Technical Report V-703, ICS AS CR, Prague, 1997.
[16] D. Pokorný, A. Sochorová, R. W. Dahlbender, P. Hájek, H. Kächele, L. Torres, and M. Zöllner. Do we use stereotypes in our relationship to the others? Technical Report V-705, ICS AS CR, Prague, 1997.
[17] P. Hájek. Mathematical fuzzy logic - state of art 2001. Technical Report 2001-027, ITI Charles University, Prague, 2001.
[18] P. Hájek. A new small emendation of Gödel's ontological proof. Technical Report 2001-015, ITI Charles University, Prague, 2001.
[19] P. Hájek. Some hedges for continuous t-norm logics. Technical Report V-857, ICS AS CR, Prague, 2001.
[20] P. Hájek. Fuzzy logic and Lindström's theorem. Technical Report V-874, ICS AS CR, Prague, 2002.
[21] P. Hájek and T. Havránek. Mechanizing hypothesis formation. Mathematical foundations for a general theory. Originally published by Springer Verlag Berlin Heidelberg New York in 1978, ISBN 3-540-08738-9. Technical Report V-859, ICS AS CR, Prague, 2002.
[22] P. Hájek. Fleas and fuzzy logic - a survey. Technical Report V-893, ICS AS CR, Prague, 2003.
[23] P. Hájek and D. Harmancová. A note on Dempster rule. Technical Report V-891, ICS AS CR, Prague, 2003.
[24] P. Hájek. One more variety in fuzzy logic: quasihoops. Technical Report V-937, ICS AS CR, Prague, 2005.

MISCELLANEA

[1] P. Hájek. Modely teorie množin s individuy. PhD thesis. Mathmetical Institute ČSAV, Prague, 1964. In Czech.
[2] P. Hájek. Syntaktické metody matematické logiky. Pokroky matematiky, fyziky & astronomie, 11(1):22–31, 1966.
[3] P. Hájek. N. Bourbaki, Éléments de mathématiques, fasc. XXII: Théorie des ensembles, chap. 4: Structures (Hermann, Paris 1966). Časopis pro pěstování matematiky, 94:491–493, 1969. In Czech. Review.
[4] P. Hájek. Bibliography of the Prague seminar on foundations of mathematics II. Czechoslovak Mathematical Journal, 23(3):521–523, 1973.
[5] P. Hájek. K nedožitým šedesátinám Ladislava Riegra. Časopis pro pěstování matematiky, 101(4):417–418, 1976. In Czech. Obituary.
[6] P. Hájek. Matematika aritmetiky prvního řádu. DrSc. thesis. ČSAV, Prague, 1988. In Czech.
[7] P. Hájek. Musikalische Rhetorik der Choralvorspiele Bachs. European Journal for Semiotic Studies, (4):659–676, 1992.
[8] P. Hájek. Práce s nejistotou v systémech umělé inteligence. Computer Echo, 4(1): 29–31, 1993. In Czech.
[9] P. Hájek and J. Rauch. Inteligentní analýza dat a GUHA. Computer Echo, 4(3): 31–33, 1993. In Czech.
[10] P. Hájek and R. Zach. Review of Leonard Bolc and Piotr Borowik: Many-valued logics: 1. theoretical foundations. Journal of Applied Non-Classical Logics, 4(2): 215–220, 1994.
[11] P. Hájek and D. Harmancová. The Prague seminar of applied mathematical logic and its work on fuzzy logic. Fuzzy Sets and Systems, 82:128–129, 1996.
[12] P. Hájek. Ještě o elementární logice. Pokroky matematiky, fyziky & astronomie, 43(4):324–325, 1998.
[13] P. Hájek and A. Sochor. Klasická logika v kontextu svých zobecnění a boj docenta Fialy proti větrným mlýnům. Pokroky matematiky, fyziky & astronomie, 43(1): 39–46, 1998.

[14] P. Hájek. Prolhaná gratulace. (Ivanu M. Havlovi k 60. narozeninám), 1999. In Czech. Unpublished.

[15] P. Hájek. P. Grim, G. Mar, and P. St. Denis: The philosophical computer. *The Bulletin of Symbolic Logic*, 6:347–348, 2000. Review.

[16] P. Hájek. Soft computing: nové informatické paradigma nebo módní slogan? *Vesmír*, 79(12):683–685, 2000. In Czech.

[17] P. Hájek, L. Godo, and S. Gottwald. Editorial. *Fuzzy Sets and Systems*, 124(3): 269–270, 2001. Special issue from Joint EUSFLAT-ESTYLF Conference, Palma (Mallorca) 1999 and from the Fifth International Conference Fuzzy Sets Theory and Its Applications FSTA, Liptovský Ján, 2000.

[18] P. Hájek. Fuzzy logic. In E.N. Zalta, editor, *Stanford Encyclopedia of Philosophy*, Fall 2002 Edition. Stanford University, The Metaphysics Research Lab, Stanford, 2002. http://plato.stanford.edu.

[19] P. Hájek. Zemřel profesor Karel Čulík. *Pokroky matematiky, fyziky & astronomie*, 47(4):344–348, 2002. In Czech. Obituary.

[20] S. Gottwald and P. Hájek. Editorial. *Fuzzy Sets and Systems*, 141(1):ii–iv, 2004. Special Issue: Advances in Fuzzy Logic.

[21] P. Hájek and D. van Dalen. Gentzens problem. Mathematische Logik im national-sozialistischen Deutschland by Eckart Menzler-Trott. *The Mathematical Intelligencer*, 26(4):64–66, 2004. Review.

[22] P. Hájek and P. Cintula. Fuzzy logics. In J.Y. Beziau and A. Costa-Leite, editors, *UNILOG'05 - Handbook*, pages 14–15, Neuchatel, 2005. University of Neuchatel. Tutorial.

[23] P. Hájek and L. Běhounek. Fuzzy logics among substructural logics. In J.Y. Béziau, H. He, A. Costa-Leite, and X.Z.M. Zhong, editors, *UNILOG'07 Handbook*, pages 20–21, Xian, 2007. Northwestern Polytechnical University. Tutorial.

[24] M. Ojeda-Aciego, S. Gottwald, and P. Hájek. Editorial. *Fuzzy Sets and Systems*, 159(10):1129–1272, 2008. Special issue: Mathematical and Logical Foundations of Soft Computing.

[25] S. Gottwald, P. Hájek, U. Höhle, and E.P. Klement. Editorial. *Fuzzy Sets and Systems*, 161(3):299–300, 2010. Special Issue: Fuzzy Logics and Related Structures.

[26] P. Hájek. *Základy matematické logiky*. Dům techniky ČSVTS, Praha, 1974. In Czech. Lecture Notes.

[27] P. Hájek, I. Havel, T. Havránek, M. Chytil, Z. Renc, and J. Rauch. *Metoda GUHA*. Dům techniky, České Budějovice, 1977. In Czech. Lecture Notes.

Ludmila Nývltová and Nina Ramešová
The Library of the Institute of Computer Science
Academy of Sciences of the Czech Republic
Pod Vodárenskou věží 2
182 07 Prague 8, Czech Republic
Email: {lidka, nina}@cs.cas.cz

Petr Cintula
Institute of Computer Science
Academy of Sciences of the Czech Republic
Pod Vodárenskou věží 2
182 07 Prague 8, Czech Republic
Email: cintula@cs.cas.cz